AF805761

Jean-Paul Pelteret, Paul Steinmann

Magneto-Active Polymers

Also of Interest

Intelligent Materials and Structures
Abramovich, 2019
ISBN 978-3-11-033801-0, e-ISBN 978-3-11-033802-7

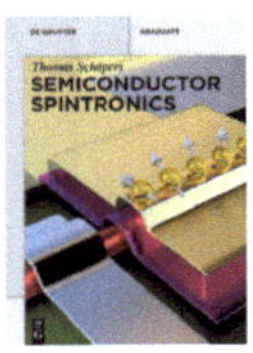

Semiconductor Spintronics
Thomas Schäpers, 2016
ISBN 978-3-11-036167-4, e-ISBN 978-3-11-042544-4

Numerical Analysis
An Introduction
Timo Heister, Leo G. Rebholz, Fei Xue, 2019
ISBN 978-3-11-057330-5, e-ISBN 978-3-11-057332-9

Advanced Materials
Theodorus van de Ven, Armand Soldera (Eds.), 2019
ISBN 978-3-11-053765-9, e-ISBN 978-3-11-053773-4

Jean-Paul Pelteret, Paul Steinmann

Magneto-Active Polymers

Fabrication, characterisation, modelling and simulation at the micro- and macro-scale

In collaboration with
George Chatzigeorgiou, Denis Davydov,
Mokarram Hossain, Ali Javili, Dmytro Pivovarov,
Prashant Saxena, Franziska Vogel, Duc Khoi Vu,
Bastian Walter, and Reza Zabihyan

DE GRUYTER

Authors
Dr. Jean-Paul Pelteret
Chair of Applied Mechanics
Friedrich-Alexander-Universität
Erlangen-Nürnberg
Paul-Gordan Straße 3
91052 Erlangen
Germany
jean-paul.pelteret@fau.de

Prof. Dr.-Ing. habil. Paul Steinmann
Chair of Applied Mechanics
Friedrich-Alexander-Universität
Erlangen-Nürnberg
Egerlandstraße 5
91058 Erlangen
Germany
paul.steinmann@fau.de

ISBN 978-3-11-041951-1
e-ISBN (PDF) 978-3-11-041857-6
e-ISBN (EPUB) 978-3-11-041862-0

Library of Congress Control Number: 2019947572

Bibliographic information published by the Deutsche Nationalbibliothek
The Deutsche Nationalbibliothek lists this publication in the Deutsche Nationalbibliografie; detailed bibliographic data are available on the Internet at http://dnb.dnb.de.

© 2020 Walter de Gruyter GmbH, Berlin/Boston
Cover image: Jean-Paul Pelteret
Typesetting: VTeX UAB, Lithuania
Printing and binding: CPI books GmbH, Leck

www.degruyter.com

To our families, for their never-ending love and support

Preface

From the mid-2000s onward, there has been an escalation in research activity on topics related to a wide variety of field-sensitive materials. Accordingly, the interest surrounding magneto-active polymer composites and their potential application in some very exciting circumstances has, in recent times, increased significantly.

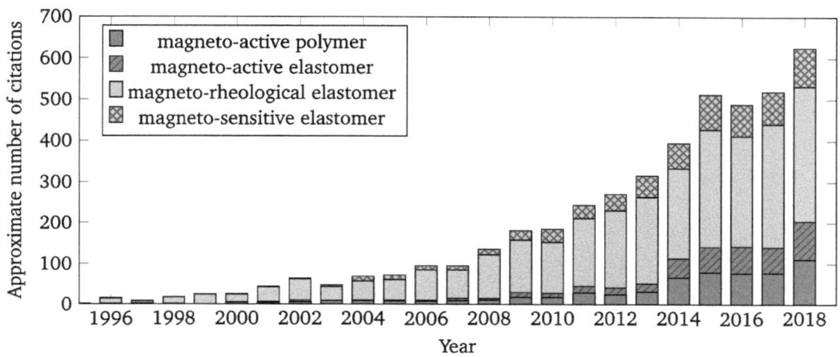

An approximation of the frequency of keywords used in academic papers according to *Google Scholar* published for the years 1995 to 2018. The data was extracted automatically using the tool located at https://github.com/Pold87/academic-keyword-occurrence.

This book was produced as part of the efforts funded by the European Research Council to better understand and characterise the nature of these composites in a laboratory setting, and to resolve some of the numerous outstanding challenges and issues as viewed from mathematical and numerical perspectives. It therefore serves as a summary of this project, which encapsulates three distinct topics in experimentation and seven facets with a basis in mathematics, numerical and computational methods. The core of the content is the peer-reviewed research articles generated by my colleagues, listed on the inside cover of the book, and ourselves. As we have primarily focussed on topics in magnetostatics, there is an added bonus that the content given herein can often be immediately (or with little effort) reformulated with electrostatics in mind.

Although Paul has been involved in the research of magneto-active polymer composites for many years, this had been a new topic for me when I initially took up this project. One of the goals that I hoped to achieve in terms of content transmission is to put forward, in as transparent a manner as possible, all of the details that I would have liked to have first known, and gotten a clear view of, when starting this line of work. As we were formulating the foundational work for some of the mathematically inclined sections, I came to realise that I had taken for granted how some of the fundamentals and key mathematical concepts surrounding magnetostatics theory had been developed and resolved. In an attempt to fill in my lack of knowledge, as well as prove to

https://doi.org/10.1515/9783110418576-201

ourselves that the content that we are sharing is indeed correct, Paul and I systematically worked through some of the more arduous theory and concepts that we discuss. Instead of simply banishing these verifications and verbose derivations to the annals of time, we decided to retain them in the small wish that they may find further use as a companion and aid to fellow scholars.

It is our sincere hope that you find the many various topics that we cover in this book interesting, clear and enlightening, and that it gives you as much joy to read as it did us to compose.

Jean-Paul Pelteret, on behalf of Paul Steinmann
Erlangen, Germany
June 2019

Acknowledgment

Nowadays most, if not all, numerical modelling leverages some collection of optimised software libraries (be they freely available or commercial in nature). In our case, we have typically utilised open-source software in order to focus the problem at hand, while relying on experts in other fields to worry about other details not directly relevant to the main focus of the current research topic. To this end, we would be remiss not to thank these projects' authors for dedicating their time and resources to provide reliable, high quality, high performance tools for us to liberally integrate into our workflow. Open-source software libraries used to generate part of the numerical content of the book include:

`deal.II`: a high-performance finite element library [29, 30, 10, 15],
Trilinos: a framework and toolbox for large scale scientific computation [193],
UMFPACK: a library for multifrontal LU factorisation of sparse matrices [107],
Metis: a graph partitioning toolbox [253],
Intel TBB: a portable library for multicore parallel processing [422], and
FEMM: a finite element solver for 2-D and axisymmetric magnetic (and other physics) problems [348];

while verification for some numerical codes was completed with the aid of:
ADOL-C: a library for automatic differentiation by overloading in `C++` [179],
Sacado: an automatic differentiation library for `C++` applications [162, 423],
SciPy: a Python-based ecosystem of software for mathematics, science and engineering [242],
SymPy: a Python library for symbolic mathematics [356], and
SymEngine: a fast symbolic manipulation library for `C++` [79];

and core infrastructures used to build, manage and run the software include:
GNU/Linux: a collection of free, open-source operating systems,
GCC and *LLVM Clang*: compilers for the `C++` programming language,
OpenMPI: a message-passing interface library facilitating distributed parallel computing, and
Spack: a package management tool for supercomputing clusters [159].

We would also like to thank the Regionales RechenZentrum Erlangen (RRZE, FAU) for providing access to their supercomputing facilities.

As a member of the `deal.II` community, I am predisposed to the idea of both collaborative research, and sharing the fruits of those labours. In this spirit, I would like to make available to the scientific community some of the code developed during the course of this project, and some specifically as a companion to this book; they may be

https://doi.org/10.1515/9783110418576-202

found within the `deal.II` code gallery at this link:

https://github.com/dealii/code-gallery/tree/master/Magneto_Active_Polymers_Book

Both Paul and I give out sincere thanks to the European Research Council (ERC) that, through the *Advanced Grant 289049 MOCOPOLY*, provided funding for the research displayed throughout the book. Paul was the primary investigator for this project conducted at the Lehrstuhl für Technische Mechanik (LTM) at the Friedrich–Alexander–Universität Erlangen–Nürnberg (FAU), and I was fortunate enough to be given the opportunity to join the research team soon after the project's inception and continue to work on it until its completion.

I would like to extend my sincerest gratitude to my coauthor and colleagues (current and former) at the *LTM*. Paul has displayed immense generosity, insights and patience to me throughout the course of the *MOCOPOLY* project, especially during the period of time in which we've been writing this book. I have learned an incredible amount from him, and will be forever grateful to him for his tutelage and guidance—dankeschön. I wish to also acknowledge the camaraderie, collaborations, support and laughter derived from my colleagues at both branches of the *LTM*. To those who gave me their time for brainstorming and problem-solving sessions—thank you, I deeply appreciate all of your input.

Lastly, to my dear wife, Kerryn, thank you for the unbelievable patience and understanding that you've shown during the many late nights and weekends lost to this endeavour. Your support for my work has always been resolute, and your confidence in me unabating. Without this foundation and you at my side, I would never have managed to achieve the successes that I have. Thank you for everything.

Jean-Paul Pelteret
Erlangen, Germany
June 2019

Contents

Acronyms

BEM boundary element method
CIP carbonyl iron particle
DoF degree-of-freedom
FE finite element
FEM finite element method
FFT fast-Fourier transform
LAOS large amplitude oscillatory shear
LBB Ladyzenskaja–Babuška–Brezzi
LVE linear viscoelastic
MAP magneto-active polymer composite
MRD magneto-rheological device
MSP magnetic scalar potential
MVP magnetic vector potential
PDMS polydimethylsiloxane
RVE representative volume element
SAOS small amplitude oscillatory shear
SEM scanning electron microscopy
SFEM stochastic finite element method

https://doi.org/10.1515/9783110418576-203

Notation

Blackboard

$\mathbb{A}$	Referential magnetic vector potential
$\mathbb{B}$	Referential magnetic induction vector
$\mathbb{D}$	Referential dielectric displacement vector
$\mathbb{E}$	Referential electric field vector
$\mathbb{H}$	Referential magnetic field vector
$\mathbb{J}$	Referential electric current density vector
$\mathbb{M}$	Referential magnetisation vector
$\mathbb{P}$	Referential polarisation vector
$\mathbb{a}$	Spatial magnetic vector potential
$\mathbb{b}$	Spatial magnetic induction vector
$\mathbb{d}$	Spatial dielectric displacement vector
$\mathbb{e}$	Spatial electric field vector
$\mathbb{h}$	Spatial magnetic field vector
$\mathbb{i}$	Spatial electric current vector
$\mathbb{j}$	Spatial electric current density vector
$\mathbb{m}$	Spatial magnetisation vector
$\mathbb{p}$	Spatial polarisation vector

Calligraphic

$\boldsymbol{\mathscr{A}}$	Fourth-order mixed elasticity tensor
$\boldsymbol{\mathscr{H}}$	Fourth-order referential elastic (material + geometric) tangent tensor
$\boldsymbol{\mathscr{I}}$	Referential fourth-order symmetric identity tensor
$\widehat{\boldsymbol{\mathscr{P}}}$	Referential isochoric projection tensor
$\mathscr{A}$	Control area
$\mathscr{B}$	Continuum body
$\mathscr{D}$	Continuum domain; Dissipative power
$\mathscr{I}$	Control surface discontinuity
$\mathscr{J}$	Control line discontinuity
$\mathscr{M}$	Magnetic power
$\mathscr{P}$	Power
$\mathscr{S}$	Free space
$\mathscr{V}$	Control volume
$\mathscr{W}$	Mechanical power
$\mathscr{C}^n$	Space of functions with n continuous derivatives
$\mathscr{H}^n$	Sobolev space with square integrable function and n derivatives
$\mathscr{L}^n$	Sobolev space with square integrable function
$\partial\mathscr{A}$	Edge of control area
$\partial\mathscr{B}$	Surface of continuum body
$\partial\mathscr{V}$	Surface of control volume
$\partial\mathscr{S}$	Far-field surface

Fraktur

$\boldsymbol{\mathfrak{L}}$	Third-order mixed magnetoelastic tensor
$\boldsymbol{\mathfrak{P}}$	Third-order referential magnetoelastic tensor

Greek

$\boldsymbol{\varepsilon}$	Infinitesimal strain tensor
$\boldsymbol{\epsilon}$	Permutation tensor (Levi–Civita symbol)
$\boldsymbol{\varphi}$	Motion
$\boldsymbol{\sigma}$	Cauchy stress tensor
$\mathbf{X}^{\xi}$	Isoparametric coordinate
$\boldsymbol{\Phi}$	Generic vector density
$\boldsymbol{\Psi}$	Referential vector flux field
$\boldsymbol{\psi}$	Spatial vector flux field
α	Degree of cure; Angular displacement (material profile)

https://doi.org/10.1515/9783110418576-204

$\Gamma_0^{\mathscr{B}\mathscr{S}}$	Boundary of a solid exposed to the free space
γ	Shear strain
δ	Kronecker delta; Phase angle between stress and strain
ε_0	Electric permittivity constant
ϑ	Angular displacement (rotor)
κ	Chain dispersion parameter
μ_0	Magnetic permeability constant
μ_s	Static friction coefficient
Π	Potential energy functional
ρ	Density
ϱ	Electric charge density
τ	Shear stress; Characteristic relaxation time
Φ	Magnetic scalar potential
$\mathit{\Phi}$	Generic scalar density
Ψ	Free energy density function
$\mathit{\Psi}$	Basis/Shape function
Ω	Stored energy density function
Ω_e	Finite element
$\Omega_\square$	Isoparametric domain
ω	Angular frequency

Italics

$\boldsymbol{K}$	Tangent stiffness matrix
$\boldsymbol{M}$	Mass matrix
$\boldsymbol{d}$	Solution vector
$\boldsymbol{f}$	Right-hand side vector
$\boldsymbol{r}$	Residual vector

Latin

$\mathbf{a}$	Acceleration vector
$\mathbf{b}$	Body force density vector
$\mathbf{C}$	Right Cauchy–Green deformation tensor
$\mathbf{D}$	Second-order magnetic tensor
$\mathbf{F}$	Deformation gradient
$\mathbf{e}_i$	Cartesian basis vector
$\mathbf{f}$	Force vector
$\mathbf{G}$	Referential structure tensor
$\mathbf{g}$	Gravity vector
$\mathbf{I}$	Referential second-order symmetric unit tensor
$\mathbf{i}$	Spatial second-order symmetric unit tensor
$\mathbf{L}$	Referential tangent vector
$\mathbf{l}$	Spatial tangent vector
$\mathbf{M}$	Referential direction of transverse isotropy; Referential manifold normal
$\mathbf{m}$	Moment vector; Spatial manifold normal
$\mathbf{N}$	Referential surface normal
$\mathbf{n}$	Spatial surface normal
$\mathbf{P}$	Piola stress tensor
$\mathbf{S}$	Piola–Kirchhoff stress tensor
$\mathbf{t}$	Traction vector
$\mathbf{u}$	Displacement vector
$\mathbf{V}$	Referential velocity vector
$\mathbf{v}$	Velocity vector
$\mathbf{X}$	Referential point
$\mathbf{x}$	Spatial point
d	Spatial dimension
e	Degree of exposure
G'	Shear storage modulus
G''	Shear loss modulus
H	Sample height/thickness
J	Volumetric Jacobian
M	Free/Stored energy density function for magnetic field
n	Number of elements in a set; Polynomial order
p	Hydrostatic pressure
q	Electric charge
R	Sample radius
s	Curing shrinkage
T	Torque
t	Time
U	Free energy function for material deformation and magnetisation
W	Stored energy density function for material deformation and magnetisation
w	Weight

Miscellaneous

FE_DGM	Discontinuous scalar-valued finite element based on monomials
FE_Q	Continuous scalar-valued Lagrange finite element

Operators

∇_0	Differential operator $\frac{\partial}{\partial \mathbf{X}}$
∇	Differential operator $\frac{\partial}{\partial \mathbf{x}}$
$\nabla \times (\bullet)$	Curl
$\nabla \cdot (\bullet)$	Divergence
$\nabla(\bullet)$	Gradient
$\overline{(\star)}$	Closure of a space
$\{\!\{\bullet\}\!\}$	Average
$[\![\bullet]\!]$	Jump
$\widehat{(\bullet)}$	Surface quantity
$\overline{\overline{(\circ)}}$	Macroscale analogue of a microscale quantity
$\langle(\circ)\rangle$	Volume average of a microscale quantity
$\lceil(\circ)\rfloor$	Surface average of a microscale quantity
$\widetilde{(\circ)}$	Fluctuation field
$\overline{(\diamond)}$	Isochoric component of a tensor
$\widetilde{(\diamond)}$	Volumetric component of a tensor
$(\diamond)\overline{\otimes}(\diamond)$	Non-standard tensor outer product
$(\diamond)\underline{\otimes}(\diamond)$	Non-standard tensor outer product
$\delta(\bullet)$	Test function; Variation
$\Delta(\bullet)$	Linearisation/Increment
$\mathrm{D}_t(\bullet)$	Material time derivative $\dot{(\bullet)}$
$\mathrm{D}_t^2(\bullet)$	Material time second derivative $\ddot{(\bullet)}$
$\mathrm{d}_t(\bullet)$	Spatial time derivative
$\partial_t(\bullet)$	Partial time derivative $\frac{\partial(\bullet)}{\partial t}$
$\mathrm{N}_t(\bullet)$	Material nominal time derivative
$\mathrm{n}_t(\bullet)$	Spatial nominal time derivative

Scripts

c	Chain
e	Element; Internal elastic (equilibrium) variable
elas	Elastic
EM	Electromagnetic
elec	Electric
eq	Equilibrium component
ext	External
h	Discrete
int	Internal
m	Matrix
mag	Magnetisation; Magnetic
max	Maxwell
mech	Mechanical
neq	Non-equilibrium component
pol	Polarisation
pon	Ponderomotive
q	(Numerical) Quadrature
s	Shrinkage
T	Transpose
t	Current (configuration)
tot	Total
v	Internal dissipative (non-equilibrium) variable
0	Reference (configuration); Amplitude
−1	Inverse
∞	Far-field
+	One-sided limit from outside
−	One-sided limit from inside
±	Periodic pairs
□	Isoparametric; Stationary
*	Prior to Legendre transformation

Tensor notation

As might be discerned from the nomenclature list, in this book we predominantly use upper case bold fonts to denote tensors, lower case bold fonts for vectors, and plain fonts for scalars (or when using indicial notation). Mechanical fields are typically styled in Roman font face, while magnetic fields in blackboard font face. When denoting material tangents, fourth-order (rank 4) tensors are styled in calligraphic fonts, third-order (rank 3) tensors in Fraktur, and second-order (rank 2) tensors in Roman font face. Linear algebra (matrices and vectors) are styled in italicised Roman font face.

Index notation

For the Einstein (index) notation, all indices represent coefficients of a two- or three-dimensional Euclidean space with a fixed Cartesian basis; that is, only a flat space is considered. We represent coefficients referencing the reference configuration by upper case characters, while those lying in the current configuration are represented by the lower case alphabet. Intermediate configurations are indicated on a case-by-case basis. We will not use the notation of raising and lowering indices to distinguish tensors and vectors with different metrics (e. g. covariant and contravariant forms).

Table of Units

Unless otherwise mentioned, in this book we will strictly adopt only SI units of measure for all quantities. Commonly used alternatives are the Lorentz–Heaviside units and Gaussian units; however, utilising these necessitates a formulation of the Maxwell equations different to that which is presented in this book.

Table 1: SI units of measure for bodies in three-dimensional space. Note that the quantities represented here have symbols presented for the spatial configuration. Alternative representations of certain units are shown in parentheses.

Quantity	Symbol	SI unit	Quantity	Symbol	SI unit
Magnetic quantities			Electric quantities		
Current	$\mathbb{i}$	A (N m V^{-1} s)	Charge	q	C (A s)
Current element	$d\mathbb{i}$	A m^{-1}			
Current density	$\mathbb{j}$	A m^{-2}	Charge density	ϱ	C m^{-3}
Magnetic scalar potential	$\boldsymbol{\Phi}$	A			
Magnetic vector potential	$\mathbb{a}$	T m			
Magnetic field	$\mathbb{h}$	A m^{-1}	Electric field	$\mathbb{e}$	V m^{-1}
Magnetic induction	$\mathbb{b}$	T (N A^{-1} m)	Dielectric displacement	$\mathbb{d}$	C m^{-2}
Magnetisation	$\mathbb{m}$	A m^{-1}	Polarisation	$\mathbb{p}$	C m^{-2}
		Mechanical quantities			
Volume element	dv	m^3	Displacement gradient	$\mathbf{F}$	–
Area element	da	m^2	Stress	$\boldsymbol{\sigma}$	Pa (N m^{-2})
Mass density	ρ	kg m^{-3}	Pressure	p	Pa
Displacement	$\mathbf{u}$	m	Body force density	$\mathbf{b}$	N m^{-3}
Velocity	$\mathbf{v}$	m s^{-1}	Traction density	$\mathbf{t}$	N m^{-2}
Acceleration	$\mathbf{a}$	m s^{-2}	Moment density	$\mathbf{m}$	N m^{-2}
Force	f	N	Moment	m	N m
		Thermodynamic quantities			
Temperature	θ	K	Entropy	η	J K^{-1}
Energy	Ψ, Ω	J (N m)	Power	$\mathscr{P}$	W (J s^{-1})

https://doi.org/10.1515/9783110418576-205

1 Introduction

Magneto-active polymer composites encompass a category of smart, field-responsive materials that typically comprise a soft polymeric substrate in which magnetically-active particulates are embedded. The fundamental behaviour of magneto-active polymer composites (MAPs) that motivates their use today was first described over three decades ago by Rigbi and Jilken [454] and Rigbi and Mark [455], with a more detailed analysis presented a decade later in the seminal papers of Ginder [172] and Jolly et al. [240]. These multi-component materials exhibit two key mechanical features that make them highly desirable in numerous applications; specifically that they demonstrate:

1. a magnetostrictive response, in that they undergo some mechanical deformation due to the external application of a magnetic field [334]; and
2. a tunable material behaviour (with large changes of various characteristics and properties, such as elastic modulus and the damping response, being attainable) that also depends on the material microstructure [528, 529].

Depending on the conditions under which the composite is manufactured, the resulting microstructure may lead the material to have different intrinsic properties and response characteristics. When the particles are homogeneously distributed throughout the material, then it exhibits isotropic behaviour; however, the magnetisable particles can be made to form imperfect chain-like microstructures thereby enduing the composite with an overall anisotropic response. Apart from the change in the stiffness properties of the MAP, the gross field-response may differ as well. This is highlighted in Figure 1.1, wherein the magnetostrictive behaviour is seen to change from being contractile (with respect to the average direction of the applied magnetic field) to expansive in nature as the material is changed from one with a statistically random particle distribution to one in which the particles appear to be arranged into chain-like structures that are aligned with the magnetic field.

The emergence of this exciting class of field-manipulable soft polymers has led to an irruption in more recent years of innovative technologies being developed in both the academic and industrial settings. Li et al. [309] provides a recent and comprehensive overview of where they are currently being applied; this expands on the early proposal of possible applications given by Ginder et al. [173] and Carlson and Jolly [76]. In the aeronautic, automotive and general mechanical engineering sectors, MAPs are seeing use as tuned or dynamic vibration absorbers working in shear [175, 114], torsion [202] and compression/elongation [246] modes, vibration isolators [174], sensors [174, 334], valves and actuators [55, 254], and adaptive sandwich-like beam structures [575, 576, 560]; while bio-inspired and bio-mechanical applications include micro- and nano-robots and swimmers [438, 561, 329, 219], and peristaltic pumps [152]; and

https://doi.org/10.1515/9783110418576-001

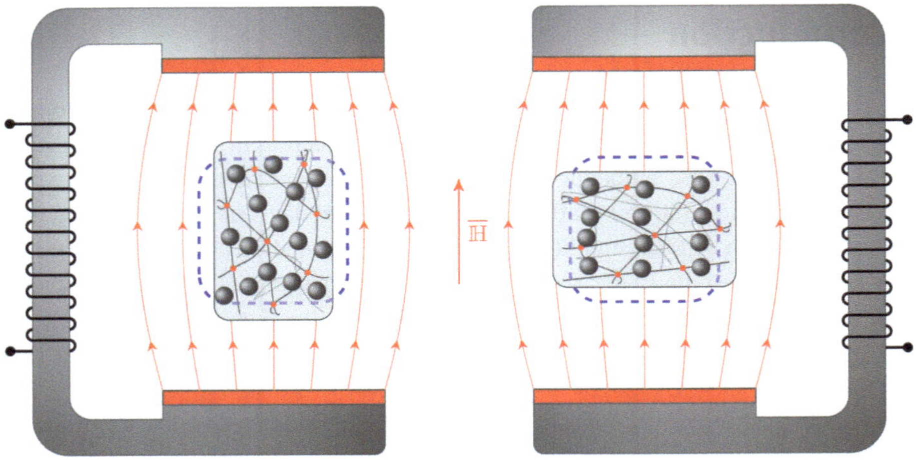

Figure 1.1: Illustration of the magnetostriction of isotropic (left) and anisotropic (right) MAPs due to the influence of an externally applied magnetic field. In this scenario, the magnetic field is generated by an electromagnet, with the composite displacing from its original configuration (outlined by the dashed blue lines) to the final depicted state. It is the arrangement of the material's microstructure that determines whether the material elongates or constricts in the direction of the magnetic field.

the civil engineering and construction industries have applied them to technologies such as seismic (base) isolators [308].

Although there is great interest and already widespread use of this class of smart materials, to date there remain numerous challenges surrounding manufacture, experimentation and their numerical simulation. Understanding the mechanical behaviour and field-responsiveness of these composites, as well as how to optimise their fabrication for a given application, is key to increasing both their effectiveness and uptake for use in novel technologies. The **M**ulti-scale, **M**ulti-physics **M**odelling and **CO**mputation of magneto-sensitive **POLY**meric materials (European Research Council Advanced Grant 289049 MOCOPOLY) project was devised with specific focus to investigate some of the difficult, interlinked facets that had been identified to be associated with magneto-active polymer composites. These included:

Fabrication: Reproducible manufacture of fully cured magneto-sensitive specimens with either isotropic or transversely isotropic distributions of the magneto-active particles while minimising the formation of particle agglomerations.

Micro-structural visualisation: Inspection, characterisation and quantification of the heterogeneous magneto-active polymer composites microstructure and particle distribution.

Testing at the macro-scale: Establishment of an accurate magneto-mechanical testing protocol that reliably measures the response of magneto-sensitive specimens, and subsequent determination of phenomenological correlations between mechanical and magnetic properties of the specimens.

Modelling and simulation: Development of continuum physics models to describe and predict the heterogeneous microstructure of magneto-active particle-filled polymer matrices after the composite material has been cured, as well as during the curing process itself.

Numerical methods: Design and implementation of finite element formulations and efficient computational tools to numerically simulate the highly non-linear, coupled composite materials in their appropriate setting while considering their field-sensitive, rate-dependent and quasi-incompressible nature.

Modelling at the micro-scale: Development of numerical models to predict the magneto-mechanical response of isotropic and anisotropic particle-filled materials that also reflect the cross-linked nature of the polymeric matrix.

Multi-physics homogenisation: Extension of two-scale computational homogenisation methods to the magneto-mechanical multi-physics problem, while considering the statistically representative microstructures appropriate for this category of materials.

Modelling at the macro-scale: Investigations of applications that make use of magneto-active polymer composites while the surrounding infinite free space has to be captured accurately in order to compute the correct Maxwell tractions acting on the surface of a magneto-sensitive body.

Parameter identification: Development of strategies for the ill-posed parameter identification problem in order to identify mechanical and magnetic properties at both the micro- and macro-scales while also accommodating the scatter and uncertainties associated with experimentally obtained data.

The overall aim of this book is to decant the greatest insights made and important knowledge gained on these topics during the course of these five years of research (2012–2017), and to summarise it in a complete and unified manner.

The content of this book has been arranged into three (more or less) distinct parts, namely (i) three chapters focussed on experimental content, followed by (ii) seven chapters dedicated to mathematical and numerical modelling, and (iii) a set of appendices to supplement the theoretical content. Starting with experimental topics, in Chapter 2 we present aspects of the fabrication of isotropic and anisotropic MAPs, and in Chapter 3 we discuss how these materials can be reliably tested by means of parallel-plate rotational rheometry. Chapter 4 is then dedicated to the characterisation of MAPs, with the purely mechanical as well as the magneto-mechanical response of both an isotropic and anisotropic MAP determined at fixed and variable frequencies. Moving on to the theoretical, continuum mechanics and numerical content, in Chapter 5 we outline the fundamental concepts of electromagnetism and magneto-mechanics, along with their associated theory as it pertains to continuum physics and thermodynamics. This is complemented by the introduction to numerical simulation using the finite element method provided in Chapter 6. Aspects on the phenomenological constitutive modelling of cured isotropic and anisotropic MAPs, as well as their rate-dependent behaviour, are given in Chapter 7, while Chapter 8 provides the counterpart description for the curing process itself. Chapter 9 is dedicated to topics in micro-to-macro-scale transition through numerical and computational homogenisation. Finally, we discuss aspects and numerical results of computational modelling conducted on the micro-scale in Chapter 10, while Chapter 11 presents numerical studies performed on the macro-scale. We conclude the main section of this book with a small discussion in Chapter 12 that outlines several areas of interest that pertain to MAPs but have not been covered in detail. To complement the theoretical content, Appendix A provides numerous tensor identities, while Appendix B lists integral theorems and identities. Building on the prior appendices is Appendix C that gives a comprehensive and detailed derivation of the core theory presented and used throughout the main chapters.

Finally, we would like to qualify and clarify a few points that may be confusing when compared with the different literature in this field. In this text, we use the term *MAP* as an acronym for *magneto-active polymer composite*, thereby emphasising the fact that it is the composite nature of these materials than endows them with their magnetic field-sensitive qualities. From the point of view of this book, the terms *magneto-active polymer* (MAP), *magneto-active elastomer* (MAE), *magneto-sensitive elastomer* (MSE), and *magneto-rheological elastomer* (MRE) are synonymous with one another. To avoid any possible confusion, in the remainder of the text we will use only the first acronym to describe rubber-like viscoelastic solids that are sensitive to magnetic fields. Similarly, we will consider the phrases *field-responsive materials* and *field-sensitive materials* to have the same meaning. In reference to a material's *viscoelastic properties*, we consider these to encapsulate those fundamental properties (e. g. the shear modulus) that are sensitive to frequency, deformation, time, and so forth, under various loading conditions. *Viscomagnetic properties* refer to those that are sensitive to the magnetic field and evolve with time. In terms of nomenclature related to continuum mechanics, we will preferentially utilise the terms *reference* (or *referential*) and *current* configuration rather than the alternate descriptors of a body's setting, specifically *material* and *spatial* configuration. This is to avoid potential ambiguities arising from the use in discussion of the state of a *material* (the physical medium), and the *spatial* (position) dependence of functions.

2 Fabrication of magneto-active polymer composites

As is the case with many aspects of materials science, the fabrication process necessary for magneto-active materials is highly dependent on their constitution, which itself is driven by its application. In this rapidly evolving sphere of research, the variety of fundamental compounds that may form a part of the final composite, each of which play a vital role in determining its overall properties, is quickly expanding. To provide a little focus to the discussion, this chapter is divided into two parts, both of which themselves will be limited to the class of rubber-like viscoelastic solids known as MAPs. In order to provide a balanced viewpoint on the topic, in the first part a relatively general presentation of the constitution, and preparation apparatus and methodology for MAPs will be made. Thereafter, in the second half of this chapter, in-depth details related to the composition and fabrication of the MAP that is characterised in Chapters 3 and 4 are provided.

2.1 A general discussion on MAP fabrication

Composition

In general, the constitution of an MAP may typically consist of three major parts: (i) a polymer base (itself comprising several components); (ii) a magnetisable filler; and (iii) additional hard fillers. Each individual component is chosen to fine-tune or accentuate a particular characteristic response from the material, and careful attention is made as to how they interact and influence each other. Due to their unique nature, there must also be consideration of the fabrication process itself to ensure that the desired end result is attainable within the restrictions of the materials' handling characteristics.

The major, and most immediately recognisable, constituent of an MAP is the polymeric matrix material (also referred to as the "continuous phase") that acts as the foundation for the other components. Typically being the component of highest volume fraction, it is more compliant than the other major components and, therefore, provides the core properties of the heterogeneous field-responsive material. Such properties include its (i) elastic stiffness; (ii) viscoelastic behaviour without a magnetic field; and (iii) range of working temperature (due to its glass transition or partial crystallisation). When in solid form, these materials are non-linear elastic in nature but, depending on their chemistry, may also exhibit significant rate-dependent (viscoelastic), yield (elastoplastic) or temperature dependent (thermoelastic) properties, or a complex combination thereof, within their working range of conditions. A typical rubber formulation consists of the base material, reinforcing filler, plasticiser, cross-linking agents and further additives [7, 9, 324, 251]. Natural rubbers, with

https://doi.org/10.1515/9783110418576-002

their inception often in the form of long sheets of material, have a relatively high elastic modulus and are viscoelastic in nature. Some polymers, such as polyurethane and liquid silicone rubber, start off as low viscosity fluid with either linear or non-linear material characteristics; they may behave like Newtonian fluids if there are no reinforcing fillers in the fluid. After a chemical reaction has taken place during a phase known as "curing," they are transformed into a relatively compliant solid material. The damping (or viscous) properties of these materials differ considerably, and is a crucial consideration with respect to their application. Both natural rubber and highly elastic synthetic elastomers display low damping characteristics; however, commercial polymers can be designed and synthesised by chemists to suit a particular application and can therefore be categorised as having chemically "tunable" damping properties. Reinforced rubbers also exhibit the Payne (amplitude-dependent dynamic stress softening that is mostly reversible) and Mullins (irreversible cyclic stress softening) effects [148, 415, 391].

The addition of a magnetisable filler to the base polymer endows the composite with its field-sensitive characteristics; for this reason these, as well as other fillers that enhance the practicality of the material, are called "functional fillers." This material often takes the form of either fibres, or spherical or ellipsoidal particles that are often several orders of magnitude stiffer than the surrounding matrix. Even when no magnetic field is present, this filler provides significant mechanical reinforcement leading to a stiffer material response. However, it is upon the application of a magnetic field that the "smart" and active nature of the new composite is revealed: due to the magnetisation of the filler, numerous interactions (both filler-matrix and filler-filler) lead to a change in the composite's mechanical properties. This enhanced material response is non-linear in nature (with respect to both the stress-strain characteristics as well as the magnetic response), coupled due to the microscopic movement of the filler upon magnetisation, and has some limiting behaviour due to the eventual magnetic saturation of the filler. To add a further degree of flexibility to its resultant properties, the material can be made anisotropic by ensuring a specific arrangement of the magnetisable filler within the matrix. This may have a significant influence on both the mechanical properties, as well as the magnetostrictive [105] and magnetisation properties [173]. Isotropic MAPs extend in the direction of the magnetic field in which it is immersed, but transversely isotropic MAPs may display a contractile response if, simultaneously, [105] (i) the material is correctly prestressed, (ii) it is under the influence of a magnetic field, and (iii) both loads are aligned with the direction of anisotropy.

Some of the readily available options for micro- and submicro-scale magnetisable fillers include magnetite (an iron-based mineral), Terfenol-D (a commercial alloy), and micro-scale carbonyl iron particles (CIPs) of which several forms exist. The factors that influence these material's magnetisation characteristics are complex, but include both their chemical composition and their microstructural arrangement. The choice of which is used in application is governed not only by its desired properties,

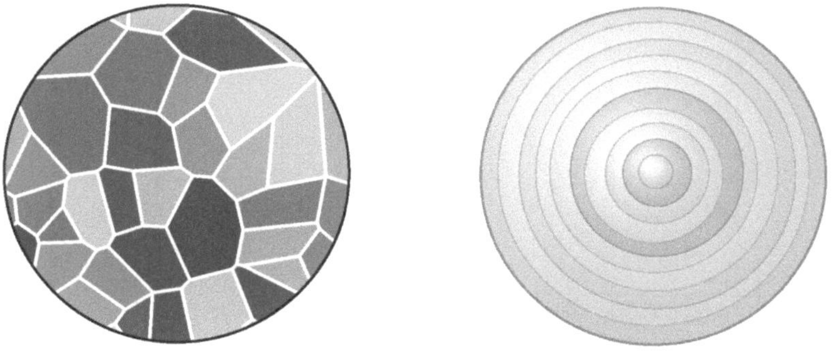

(a) Polycrystalline microstructure. (b) Onion-type microstructure.

Figure 2.1: Diagrams of microstructures that lead to varying magnetisation characteristics of fillers. The unprocessed CIP possesses the onion microstructure, but after treatment by annealing a polycrystal structure is attained.

but also by price and availability of the raw material. Shown in Figure 2.1 are two such microstructures that render very different magnetisation behaviour of CIP.

CIPs that have a polycrystalline microstructure, such as that shown in Figure 2.1a, are often characterised as having a "soft" magnetisation curve, namely being one that may display a non-linear response but has little to no magnetic hysteresis. In contrast, materials that have onion-type microstructures like that presented in Figure 2.1b also exhibit low magnetic remanence but often display weaker magnetisation characteristics. When materials that have "hard" magnetisation properties (high remanence) are removed from an applied magnetic field, they remain magnetised due to permanent changes in their microstructure; they find application in, for example, plastic-bonded permanent magnets [127]. An example of the typical magnetic response curves of these material classes is given in Figure 2.2. High magnetic remanence is often considered an undesirable property for fillers used in MAPs, as it is the fine control of the material properties under magnetisation that is specifically sought by engineers.

To further influence the effects of the filler-matrix interaction, the filler may be treated with a surface finish or surfactants. This serves to change the chemical interaction between the filler and polymer during curing as it alters the final chemical bonds between filler and matrix and may enhance the magneto-rheological effect [143, 237]. Furthermore, it can also have an effect on the cross-linking density of the polymer, which influences the viscoelastic properties of polymer matrix. Dependent on its application, enhancement of the interaction between the base matrix and embedded particles may be generated by better integrating of CIP in the polymer network, or by allowing them increased mobility [440]. The type of surface coatings that are viable for use is highly dependent on the polymer, as the two must be chemically compatible. For example, vinyl ethoxy silane can be used to treat CIP and ensure that the particles and a silicone rubber are strongly bonded after curing. Coatings that are applied to prevent oxidation of the iron-based filler [324] may also alter the magnetic properties

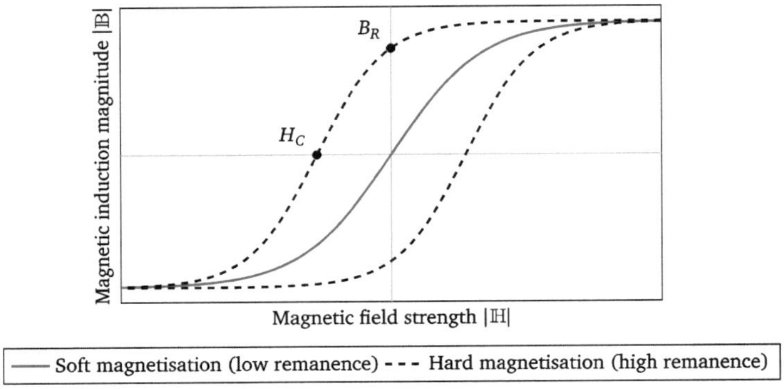

Figure 2.2: An example of magnetic remanence in a fictitious material, with the saturation and remanence curves defined in terms of hyperbolic functions [510]. While the material with soft magnetisation is fully demagnetised when the magnetic field is removed, the hard magnetising media remains partially magnetised. Both materials display characteristics of magnetic saturation at a high magnetic field. The remanence and coercive field are respectively denoted by B_R and H_C.

of the composite, for example by the promotion of oxides formation on the particle surface [4]. There have recently been some interesting developments of relevance in this area, one of which is the use of polyacrylonitril [54] to introduce an electric-field sensitivity to an already magneto-active material.

Other reinforcing fillers that deserve consideration are those traditionally used in rubber chemistry. These include, for example, fumed silica and carbon black, both of which are typically on the nano-scale. Each of these components serve to tailor not only the material's elastic modulus, but also its dynamic response, stress or elongation at break and frictional properties. Plasticisers may also be added to alter the material's compliance. Note that there exist many other options for fillers, but for the sake of brevity we mention only those listed above as they primarily pertain to the application in MAPs.

Fabrication apparatus and procedure

Key to the replication of samples with consistent chemistry is the accurate reproduction of the steps used to manufacture the material. As the raw materials are typically mixed according to a predefined volume fraction, a scale is used to ensure precise gravimetric measurement of each ingredient. When the sample size is small, the use of a high precision scale is of particular importance; due to the low volume of materials being measured, the consequent relative error of measurement imprecision is large. Thereafter, best practice is to ensure complete homogenisation of the composite before curing. In the laboratory setting, for specimens with liquid rubber formulations, this can be performed by means of a speed mixer; in contrast, within an industrial environment an internal mixer would be necessary to homogenise a large batch of raw composite material with a natural rubber base. As in all manufacturing processes, there exist a number of technical difficulties that one must be conscientious of in order to ensure that a high quality, reproducible product is created. The mixing process introduces air inclusions into the material, which would act as defects if they are not removed. If the mixture comprises a low volume of fluid, then the air inclusions can be reduced by means of a vacuum pump and desiccator. It is difficult to ensure complete homogenisation of the mixture, as the particulates may aggregate or collect in regions of the composite due to inadequate circulation. Additional energy is introduced to the system during mixing, thereby increasing its temperature. This may be problematic as it may induce premature curing in fluidic materials, thus necessitating the use of a temperature control mechanism for these compounds.

Given the nature of the homogenised material, it is common for a forming process to be employed to produce an end-product. In industrial applications, extrusion processes, and injection and compression moulds are often used in conjunction with highly viscous rubber-like materials, as they allow complex forms to be made while producing large batches of consistent products. For smaller volumes of material, a casting process may be adequate (if the material suits this process), after which identically shaped test specimens can be stamped or cut out. The latter is the approach typically used in a research environment where material characterisation is to take place. In either case, additional considerations must be made when introducing anisotropy into the material by means of aligning the filler microstructure to a preferred direction. For complex geometries, this is a challenging issue to be resolved. The preparation method may alter, hinder or completely prevent the formation of a well-defined and consistent microstructure. For simple geometries, this is typically achieved by curing the material in the presence of a uniform magnetic field. Furthermore, in both low and high viscosity systems, it is possible to introduce air voids into the system during the casting process. Inclusions may form within the material itself, or in corners where the air has not been fully displaced from the mould.

In the case of liquid rubber formulations, a temperature control device such as an oven may be utilised to initiate and/or accelerate the curing process. This may be a critical step in the fabrication process for these smart materials, as when the matrix is still of low viscosity the micro-scale particles are able to migrate easily, thereby leading to sedimentation when the curing process occurs over a long time-frame. By curing at an elevated temperature, the time taken for the low viscosity matrix to reach the gel point is reduced, thereby minimising the amount of sedimentation that may take place. This does, however, require some a priori knowledge of the system involved and likely some preliminary experimentation to optimise this procedure. Additionally, temperature control mechanisms may be of vital importance when a product with thick walls is being produced. The curing process within the material will not be homogeneous, possibly leading to the generation of defects or inhomogeneity of its properties. Additional environmental factors, such as the humidity, may need consideration if the material exhibits a sensitivity to them.

2.2 Characterisable MAP: From lab to fab

Composition

For the investigations presented in Chapter 4, a two-component silicone rubber (RTV-2) is used as the polymer base. This material has a number of properties that are beneficial within the fabrication process, as well as for the characterisation of the composite material. It is a completely unreinforced polymer, which removes confounding factors when determining the influence of other fillers, such as the magnetisable particles. By contrast, typical silicone elastomers (e. g. fumed silica reinforced liquid silicone rubber) have some reinforcement that may result in the material exhibiting the Payne effect. In its raw form, RTV-2 is a binary liquid with the first component being a vinyl-terminated polydimethylsiloxane (PDMS) with a platinum catalyst, and the second a vinyl-terminated PDMS with a cross-linker. As is shown in Figure 2.3, the formation of cross-linking bonds (curing) occurs only when the two compounds are introduced to one another. For this reason, the raw materials have a relatively long and stable shelf-life, facilitating the procurement of large batches of materials for laboratory evaluation. As both components are low viscosity fluids, an homogeneous mixture is easily obtained. Furthermore, the curing process that, for this material, occurs at room temperature requires a platinum catalyst thereby ensuring that it is a very controllable reaction with respect to the temperature and cure rate. From this base, with the addition of further components it is possible to produce an elastomer with either low or high damping properties. We have, however, noted that batch-to-batch variations in the polymer exist which leads to a discrepancy in material properties between laboratory specimens made with different batches of raw polymer. On occasion, a low molar mass PDMS plasticiser may also be added to the system in order to

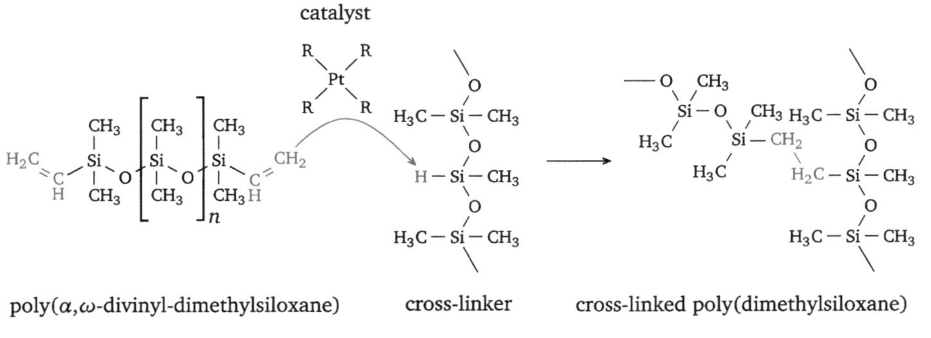

Figure 2.3: The cross-linking chemical reaction of PDMS.

tailor matrix stiffness properties. This enhances the magneto-rheological effect of the magneto-active filler component, as the matrix is made more compliant.

Micron-sized carbonyl iron particles (sourced from BASF) have been chosen as the magnetisable filler. As can be observed in the scanning electron microscopy (SEM) image in Figure 2.4a, they are spherically shaped, have soft magnetisation characteristics, and have no special surface treatment. The particles have a log-normal size distribution, as is shown in Figure 2.4b, with a mean diameter of 2.6 μm. They are therefore an ideal constituent in the construction of a minimally complex magneto-active system for which it is simple to study the influence of each individual component. Up to 40 vol.% of particles are added to the polymer base; it is not possible to add more than this as the mixture becomes too viscous to process [323]. Note that the percolation threshold is approximately 20 vol.%, but is lower when the material is cured in the presence of a magnetic field.

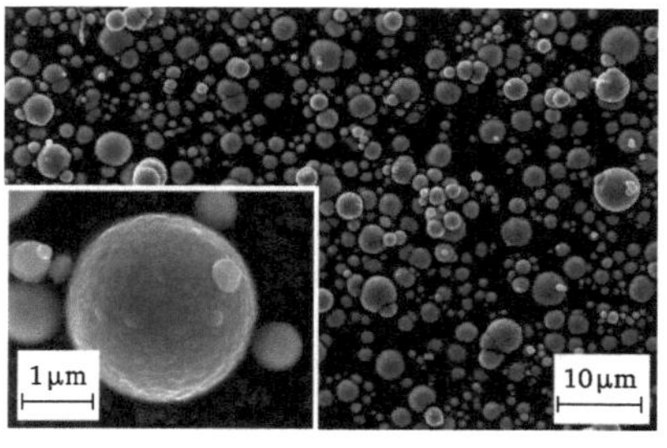

(a) Imaging of CIP by back-scattered SEM.

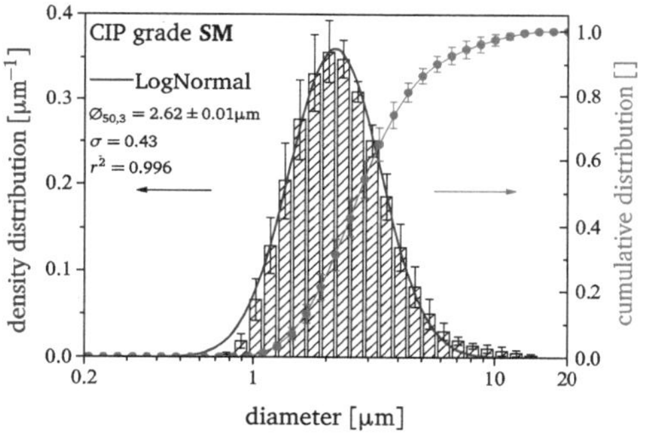

(b) The CIP has a log-normal size density distribution with a mean diameter of 2.6 μm.

Figure 2.4: Imaging is conducted by means of SEM in conjunction with the back-scattered electron detector, and the size distribution of CIPs is determined by laser diffraction particle sizing performed with a Mastersizer 2000 (Malvern, United Kingdom). Note that the particles appear white and the dark shading is air, not solid material. [542], [545, fig. 1a (reproduced with permission)]

Preparation and curing methodology

When preparing the composite mixture, all components except for the cross-linker (and all of which have a known specific density) are weighed in a beaker according to a desired volume fraction with a precision scale (XA503S, Mettler-Toledo). Each 20 ml batch is subsequently homogenised in a Speedmixer™ (DAC 150 SP, Hauschild) mixing at 2000 min^{-1} for several minutes at room conditions (25 °C at environmental humidity and pressure). The mixing time is upwards of 3 min, depending on the volume fraction of the CIP, and care is taken to ensure that the temperature remains low to prevent premature curing. Next, the mixture is deaerated using a desiccator equipped with a vacuum pump. The mixing and degassing steps are repeated at least four times in order to ensure that a completely homogeneous and deaerated mixture is obtained. Lastly, the cross-linker is introduced to the mixture and the composite is mixed and degassed one final time before it is cured.

There are a number of variables that can be manipulated when curing. For the purposes of the work presented later, we consider curing the material in a casting mould (ex situ, the approach typically adopted to manufacture these materials) or within the experimental apparatus itself (in situ), without and including the presence of an externally applied magnetic field to create isotropic and anisotropic test specimens. Highlighted in Figure 2.5 is the difference in microstructure between media cured away from and in the presence of a magnetic field. Regardless of whether the composite is cured in- or ex-situ, the resulting characteristics of the material have been found to be similar.

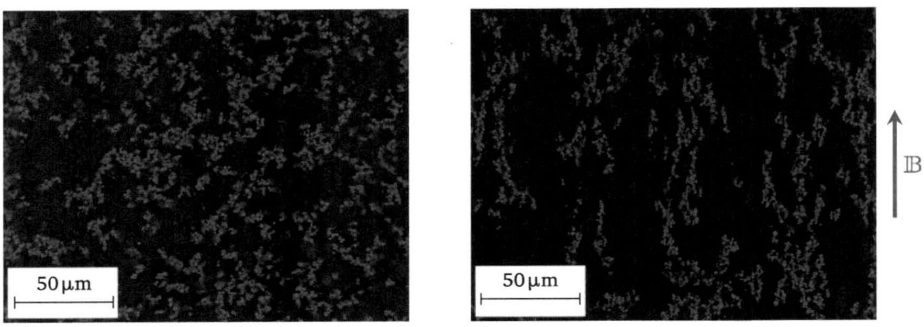

(a) Isotropic particle distribution. **(b)** Anisotropic particle distribution.

Figure 2.5: Cross-section through an ex situ pre-prepared MAP specimen with 10 % CIPs by volume. The polymeric matrix appears in black, while the micro-scale particles are white and grey. The direction of the applied magnetic field in Figure 2.5b is indicated by the arrow, and is perpendicular to the direction of applied shear in the testing apparatus. This causes the particles to migrate to form chain-like structures. [542]

When manufacturing the ex situ pre-prepared specimens, rectangular 1 mm thick sheets of material are prepared by casting using a custom made, two-part mould which is shown in Figure 2.6a. It has a Teflon™ coating in order to prevent adhesion of the polymer to its surface. Upon transfer of the viscous mixture to the mould, further deaeration is necessary in order to mitigate the formation of voids in the corners of the mould. An electromagnet, into which the mould attaches, is used to introduce a magnetic field during curing. However, the magnetic field is not provably uniform, and its intensity is imprecisely controlled. The filled mould (and, when applicable, the electromagnet) are placed in an oven (VT 6060 M, Thermo Scientific) for at least

(a) Two-part casting mould with material sheet.

(b) Cylindrical and dog-bone specimens.

Figure 2.6: Casting mould used to form ex situ pre-prepared specimens, along with examples of specimens to be tested in torsion and elongation. The mould can be placed within an electromagnet that introduces a magnetic field through the thickness of the material sheet. [542]

16 h and the material is cured at 75 °C. Lastly, cylindrical specimens of 20 mm diameter as depicted in Figure 2.6b are punched out of the sheet. With this technique, the variation in the thickness of the specimens is ±5 %.

Although this method provides a large material quantity from which to produce several test specimens, due to the thickness of the mould (and its conductivity) there exists a temperature profile in it during the initial stages of curing; consequently, the time taken for the material to reach curing temperature is relatively long. For this reason, an elevated temperature is necessary to reduce the time taken for the material to attain the temperature necessary to prevent sedimentation of the particulate. An additional disadvantage of this technique is that, when punching out the test samples, the material deforms leading to its profile having non-straight edges. This may influence the stress and magnetic field generated within the material.

An alternative technique, which is preferentially used when characterising the MAP, is to cure the material in situ within the testing apparatus (a temperature controlled parallel-plate rotational rheometer with a magneto-rheological attachment). Since this device is able to provide both thermal control and can introduce a well-defined magnetic field [293] (for the case when the parallel-plate configuration is used) to the uncured material, it serves as an ideal apparatus to cure this particular material. The magnetic field is controlled by the pre-calibrated apparatus, ensuring that a known field intensity is applied during curing.

Due to the low viscosity of the liquid rubber formulation and the geometry of the parallel plates, thin samples (with a thickness of 300 μm) are produced thereby ensuring a more homogeneous temperature profile during curing. To accelerate the curing process, a temperature of 45 °C is used. Apart from this, the preparation conditions, including mixing and curing, are identical to the protocol used to produce the ex situ pre-prepared samples. Therefore, the resulting polymer, as measured by the cross-link density, is identical in both cases. There is an additional benefit that the specimen is glued to the parallel plates, thereby ensuring that the material interface remains firmly attached to the plates during testing. In fact, after testing the base plate, rotor and specimen must be soaked in toluene to swell the MAP and debond the sample from the plates. Note that this is a non-destructive method (as the volatile toluene evaporates post-extraction), which allows further examination of the sample after testing. A more detailed discussion on the merits of using an in situ curing approach is presented in the next chapter.

3 Experimental apparatus and testing procedure

The characterisation of material behaviour in a reliable, repeatable manner is a challenging task. However, it is a critical step that is necessary to understand the material's fundamental response to external loading, and identify which parameters influence this response. Through the cycle of testing and development, the materials can be altered so that desirable attributes become more pronounced and disadvantageous ones are minimised. Furthermore, computational models are proposed and developed based on such data. Both of these applications require the measured experimental data to be completely reliable and indicative of the true material response, therefore excluding experimental artefacts or the misinterpretation of post-processed data. This task is made even more difficult when the material is inhomogeneous or anisotropic, or exhibits rate-dependent or a coupled response; MAPs demonstrate all of these behaviours.

Due to the extensive application of rubber [178] in industry, a spectrum of commercial apparatus has been developed and their use firmly established in the experimental characterisation of elastomers. A non-exhaustive list of this apparatus includes: rubber process analysers [297], servo-hydraulic actuators used in double lap simple shear experiments [548, 549], parallel-plate rotational rheometry [89], rotational rheometry in torsion-rectangular mode [268], and dynamic mechanical analysers in tension mode [349]. Collectively, these devices can measure the viscoelastic properties of rubber in tension, compression, shear and torsion modes; some are able to perform measurements during and after curing. We note specifically for rotational rheometry [89] that it is typical to use ex situ pre-prepared disc-shaped samples bonded to the measuring plates by, for example, an adhesion primer (or simply relying on friction alone).

The introduction of a well-defined, uniform magnetic field into existing commercial equipment has proved to be difficult to accomplish. For this reason, the academic and scientific communities have developed and constructed various customised apparatus to explore the properties of MAPs and other magneto-sensitive solids. Examples of magneto-mechanical loading conditions evaluated by devices specialised for magneto-sensitive media include: double lap shear tests [240, 506], a parallel-plate rotational rheometer in torsion mode [506], uniaxial compression or extension [105, 521], simple shear [105, 521], clamped uniaxial tension and compression [506, 486, 386], biaxial tests [481], and three-point bending [395, 93]. Illustrations of loading conditions examined by a selection of this custom-made equipment are provided in Figure 3.1. This graphic provides some insight into the challenge of incorporating a homogeneous magnetic load in conjunction with a mechanical load. Parts of the devices used to produce the magnetic and mechanical load may interfere with one another, and the distance over which the magnetic field must be generated can be substantial. This may

https://doi.org/10.1515/9783110418576-003

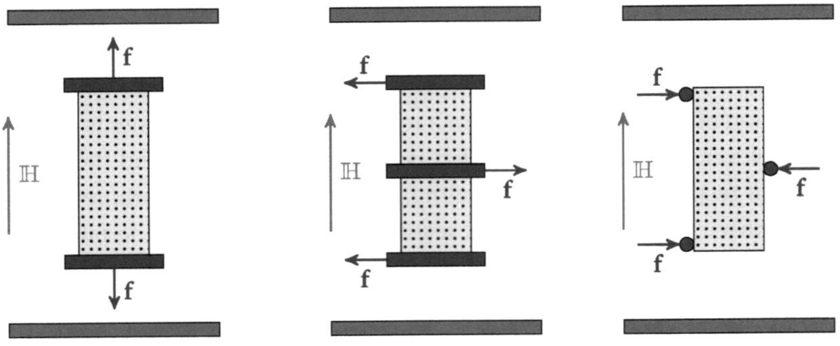

(a) Clamped tensile or compressive deformation. **(b)** Double lap shear. **(c)** Three point bending.

Figure 3.1: Examples of loading conditions produced by custom experimental apparatus. The applied magnetic field is orientated to elicit microstructural changes that most significantly influence the mechanically excited material and measured modes.

lead to both power and cooling requirements that are unobtainable and/or unsustainable should an electromagnet be used for this purpose.

To date, only two[1] sets of commercial apparatus have been engineered to evaluate and characterise MAPs. The rotational rheometer developed by Anton Paar is at present the state of the art, and can be equipped with many different measuring cells. One of these measurement units is a magneto-rheological device (MRD) [556, 290] that has been widely adopted for scientific studies of both magnetically responsive fluids and solids. Its functioning has been extensively evaluated both experimentally and computationally (under mechano-static conditions) by Laun et al. [291, 293, 295].

In Chapter 4, results derived to characterise the material described in Chapter 2 using this commercial rheometer are presented. Therefore, Section 3.1 provides a more detailed overview of the functioning of this particular device, and in Section 3.2 we describe the methodology developed and applied to conduct reliable and repeatable experimental evaluations of MAPs.

3.1 Parallel-plate rotational rheometer

Depicted in Figure 3.2 is the stress-controlled rotational rheometer (MCR 502, Anton Paar) equipped with a magneto-rheological device (MRD 170/1T, Anton Paar), a titanium parallel-plate measuring configuration, and Peltier temperature control system (H-PTD 200, Anton Paar). Primarily developed to evaluate magneto-rheological fluids

1 A rotational rheometer with a magneto-rheological accessory has very recently been introduced by TA instruments.

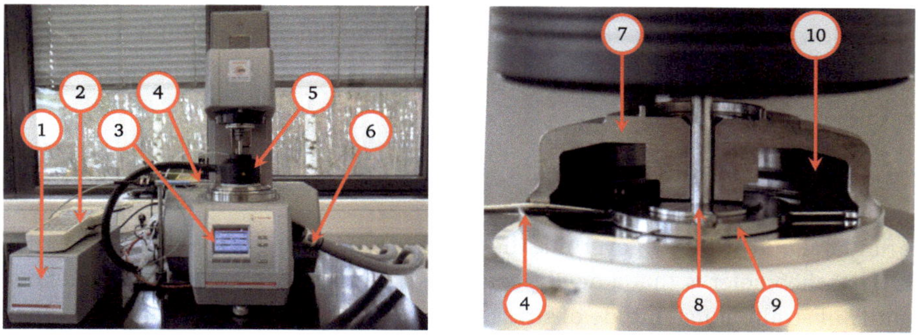

(a) A parallel-plate rotational rheometer, power and cooling units, and a teslameter.

(b) An exposed test cell showing the yoke cross-section with the rotor and hall probe.

Figure 3.2: Rotational rheometer, and detail of yoke that contains the magnetic field. Legend: (1) Power supply; (2) Teslameter; (3) Rotational rheometer; (4) Hall probe; (5) Test cell; (6) Temperature control system; (7) Upper yoke; (8) Parallel-plate rotor; (9) Base plate; (10) Free space in test cell.

[292, 154, 294] and gels [550], its use has since been extended to the characterisation of MAPs [183].

Disc-shaped samples of elastomeric materials (or, in the case of fluids, cylindrical films), located on the stator plate (lower surface), are excited through the application of a torque by the rotor (upper surface). The frictional properties of the rotor can be adjusted by tailoring the surface finish (or geometry) of the rotor and stator [295]. Figure 3.3 clearly illustrates three distinct surface finishes of the rotor. By increasing the friction coefficient between the plates and an ex situ pre-prepared elastomeric sample, it is possible to evaluate the material response at higher strain amplitudes before perfect cohesion between the sample and plates is lost (and the results of the experiment are compromised). Note that these rotor designs were primarily developed to characterise magneto-rheological fluids. A guard ring, present in all rotor types to pre-

(a) Smooth plate (PP20/MRD/Ti).

(b) Sandblasted plate (PP20/MRD/Ti/S).

(c) Serrated plate (PP20/MRD/Ti/P2).

Figure 3.3: Comparison of surface finishes of the titanium parallel plates (rotors) for use in testing of MAPs.

vent the flow of fluid around the rotor rim, is visible at the outer circumference of the plates [295].

3.1.1 Stress-controlled deformation

The fundamental principle governing the classical functioning of the stress-controlled rotational rheometer is that a device-controlled torque is applied until a predefined quantity of strain or magnitude of stress is achieved [288]. A negative-feedback loop exists that corrects the applied torque until the desired strain or stress is attained. Some rheometers also provide other control modes, such as direct strain oscillation where it applies a sinusoidal strain wave through application of real-time position control [289]. This is primarily of importance when measuring non-linear material behaviour, as the traditional controlled-stress mode does not ensure the application of sinusoidal strain to the sample within this regime.

Illustrated in Figure 3.4 is the functional geometry of parallel-plate apparatus as attached to a rotational rheometer. The disc-shaped MAP sample of radius R =10 mm and height H is compressed between the stationary stator and an oscillating rotor. Typically, the material is subjected to harmonic torsional deformation and a magnetic field is applied in the axial direction; tests can also be performed in quasi-static shear.

Through the control mechanism, a time-dependent deformation profile can be prescribed. The initial position of any point within the material is given in cylindrical coordinates by

$$R(\mathbf{X}) = \sqrt{X_1^2 + X_2^2} \quad , \quad \Theta(\mathbf{X}) = \arctan\left(\frac{X_2}{X_1}\right) \quad , \quad Z(\mathbf{X}) = X_3 \quad . \tag{3.1}$$

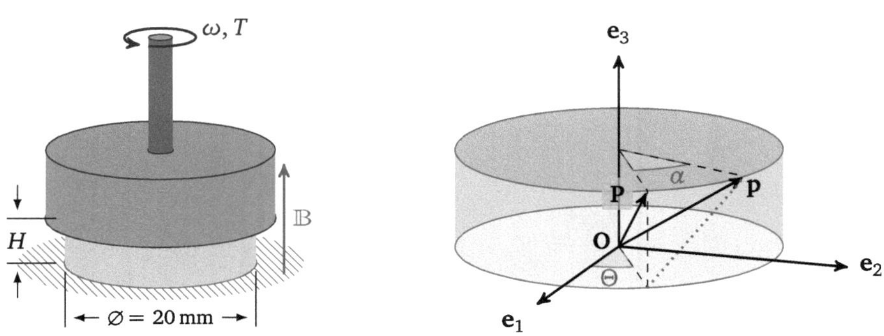

(a) Geometry. **(b)** Kinematics (no preload/compression).

Figure 3.4: Geometric description of parallel-plate configuration. The origin is assumed to be located axially and on the material's lower surface (adjacent to the stator). Due to the induced deformation, point $\mathbf{P} \in \mathscr{B}_0$ at azimuth Θ is displaced to $\mathbf{p} \in \mathscr{B}_t$ at azimuth $\theta = \Theta + \alpha$. Note that in this image it is assumed that $\lambda_3 = 1$ $(r = R)$. [418, fig. 1]

Note that the calculation of Θ specifically accommodates the quadrant in which (X_1, X_2) lies, and is aligned with the stated expression when $\mathbf{X} \cdot \mathbf{e}_1 > 0$. When assuming material incompressibility and a linear deformation profile through the sample thickness (an appropriate assumption for sufficiently small deflections), the displacement at any point $\mathbf{X}$ within the material to $\mathbf{x}$ is given in cylindrical coordinates by $\mathbf{w} = \mathbf{x}(r, \theta, z) - \mathbf{X}(R, \Theta, X_3)$ with [263, 88, 418]

$$r(\mathbf{X}) = \frac{R(X_1, X_2)}{\sqrt{\lambda_3}} \quad , \quad \theta(\mathbf{X}) = \Theta(X_1, X_2) + \underbrace{\Gamma(t)\,\overbrace{\lambda_3 X_3}^{z(\mathbf{X})}}_{\alpha(X_3, t)} \quad , \quad z(\mathbf{X}) = \lambda_3 X_3 \quad . \tag{3.2}$$

The axial stretch, denoted as λ_3, is assumed to remain constant, while the angle $\alpha(X_3, t)$ is the angular displacement at any height through the sample profile, and $\Gamma(t)$ defines the torsion angle per unit deformed height of the sample. When applying pure sinusoidal deformation,

$$\Gamma(t) = \frac{\alpha_0}{h} \sin(\omega t) \quad , \tag{3.3}$$

where ω denotes the angular frequency, $h = \lambda_3 H$ is the deformed height of the sample, and α_0 is the maximum angular displacement on the top surface $(X_3 = H)$ of the material.

The shear strain developed at a given radius (and constant throughout the material thickness) also varies sinusoidally with time, and is expressed as [263, 88]

$$\begin{aligned} \gamma(t, \omega, \gamma_0(\alpha_0)) = r \frac{\partial w_\theta}{\partial X_Z} &= r \lambda_3 \Gamma(t) \\ &\equiv \gamma_0(\alpha_0) \sin(\omega t) \quad . \end{aligned} \tag{3.4}$$

The relationship between γ_0 and α_0, valid for large deformations (under the aforementioned assumptions), is therefore

$$\gamma_0(\alpha_0) = \frac{\alpha_0 r}{H} = \frac{\alpha_0 R}{\sqrt{\lambda_3} H} \quad , \tag{3.5}$$

and the total torque generated by the rheometer and applied at the sample-rotor interface is

$$T = \underbrace{\int_0^R \tau(r)\, 2\pi r \,\mathrm{d}r}_{T_{\text{sample}}} \equiv \underbrace{\int_0^R \mu_s(r)\, p(r)\, 2\pi r \,\mathrm{d}r}_{T_{\text{rotor}}} \quad . \tag{3.6}$$

The radial profiles of the interfacial shear stress $\tau(r)$, the pressure distribution $p(r)$, and static friction coefficient $\mu_s(r)$ are governed by the geometry of the rotor and stator, and other critical assumptions. For this reason, unless taken into account, the use of rotors with complex geometries may invalidate or systematically influence measured data.

Note that, as per the German standard DIN 53018 (ISO 3210), the strain and stress is measured at $R = \frac{2}{3} R^{\max}$; this implies that the geometry factor to compute strain and stress from torque and deflection is $\frac{2}{3}$. This should be taken into consideration when interpreting the data later presented in Lissajous figures.

Small amplitude oscillatory shear (SAOS)

For case of "sufficiently small" strains, materials are assumed to exhibit a linear viscoelastic (LVE) response, namely that the material properties evolve independently of the strain amplitude. It is typically assumed that the radial shear stress profile $\tau(r) = \frac{r}{R} \tau|_{r=R}$, where $0 \leq r \leq R$, is linear [294].

For a material excited with the sinusoidal deformation given in equation (3.4), and strictly within the bounds of experimental conditions that define the LVE regime, the shear stress at the outer rim (namely that corresponding to the point **P**) is [167, 224]

$$\tau(t, \omega, \gamma_0) = \tau_0(\omega, \gamma_0) \sin(\omega t + \delta(\omega)) \tag{3.7a}$$

$$= \gamma_0 \left[G'(\omega) \sin(\omega t) + G''(\omega) \cos(\omega t) \right] \tag{3.7b}$$

where the shear storage and loss moduli are respectively denoted by G' and G'', and the phase-shift, also known as the loss angle, is

$$\delta = \arctan\left(\frac{G''}{G'}\right) \quad . \tag{3.8}$$

The magnitude of the loss factor, computed as $\tan(\delta)$, provides an indication of the ratio between the energy dissipated and stored in the viscoelastic system. Low-damping elastomers have a loss factor close to zero ($G' \gg G''$) while for high-damping elastomers $0 \ll \tan(\delta) < 1$. Note that when $\tan(\delta) \geq 1$, corresponding to $G'' \geq G'$, the material is exhibiting gel- or fluid-like characteristics.

Large amplitude oscillatory shear (LAOS)

In the event that the material is subjected to large excitation amplitudes, or displays fundamentally non-linear behaviour even at "small" strains, the stress response described in equation (3.7) is no longer valid. Conducting LAOS rheological experiments is a complex field in itself, and its intricacies are detailed in the comprehensive review paper of Hyun et al. [224], as well as [307, 142]. Analysis of the non-linear waveform data is often conducted by means of fast-Fourier transform (FFT) [552, 553, 551, 297] or basis decomposition using Chebyshev polynomials [140, 139].

In summary, corrections must be applied to compute the true shear stress, as well as the corresponding storage and loss moduli, based on this measured stress [154, 294]. These corrections derive from the rotor-stator geometry and an underlying assumption of the material constitutive law. For the cone rotor geometry, the "true" shear stress can be represented as an odd harmonic Fourier series, namely

$$\tau(t,\omega,\gamma_0) = \sum_{n=1,\text{odd}}^{\infty} \hat{\tau}_{0,n}(\omega,\gamma_0)\sin(n\omega t + \delta(\omega,\gamma_0)) \quad (3.9a)$$

$$= \gamma_0 \sum_{n=1,\text{odd}}^{\infty} \left[\widehat{G}'_n(\omega,\gamma_0)\sin(n\omega t) + \widehat{G}''_n(\omega,\gamma_0)\cos(n\omega t)\right] \quad (3.9b)$$

and the corrected shear stress amplitude, storage modulus and loss modulus corresponding to each harmonic, respectively denoted as $\hat{\tau}_{0,n}$, $\widehat{G}'_n$ and $\widehat{G}''_n$, are now dependent on strain amplitude.

The normalised stress intensity corresponding to the nth harmonic is defined as

$$I_{n/1} := \frac{\hat{\tau}_{0,n}}{\hat{\tau}_{0,1}} \quad (3.10)$$

where $\hat{\tau}_{0,1}$ represents the fundamental wave (the first harmonic) of the stress response. This approach is commonly used to quantify measured non-linearities and characterise their source.

3.1.2 Magneto-rheological device

Not visible in Figure 3.2, but illustrated by Laun et al. [293, figure 1], is the solenoid that surrounds the steel core of the bottom yoke. The strength of the magnetic induction generated by passing a current through this coil is monitored[2] by the Hall probe located 300 μm below the sample/stator interface. Tubes for the fluid circulator system run through the core of the yoke, thereby providing efficient temperature control. This is especially important when applying high magnetic loads, for which a large amplitude current (up to 5 A for the tested materials) must be passed through the solenoid.

The construction of the magneto-rheological device is such that it generates a magnetic field that is perpendicular to the axis of deformation [293]. With reference to Figure 3.4a, the magnetic induction field (within the sample) is assumed to be axially aligned and spatially uniform, and is therefore expressed as

$$\mathbb{B}(t) = \begin{bmatrix} 0 & 0 & b(t) \end{bmatrix}^T \quad . \quad (3.11)$$

The value of b is the calibrated prescribed axial magnetic induction that may change in the course of an experiment.

2 As the induction is not controlled by a feedback mechanism, the current-induction response must first be determined by the operator.

To understand aspects of its operation and to confirm the uniformity of the magnetic field within the sample, a simplified geometry of the rheometer was developed and evaluated in the finite element method (FEM) electromagnetics solver *FEMM* [348]. Figure 3.5 illustrates the magnetic field generated by a current of $|\mathbb{J}| = 5$ A passing through the solenoid, and a $H = 1000$ µm thick sample with a relative magnetic permeability of $\mu_r = 5$.

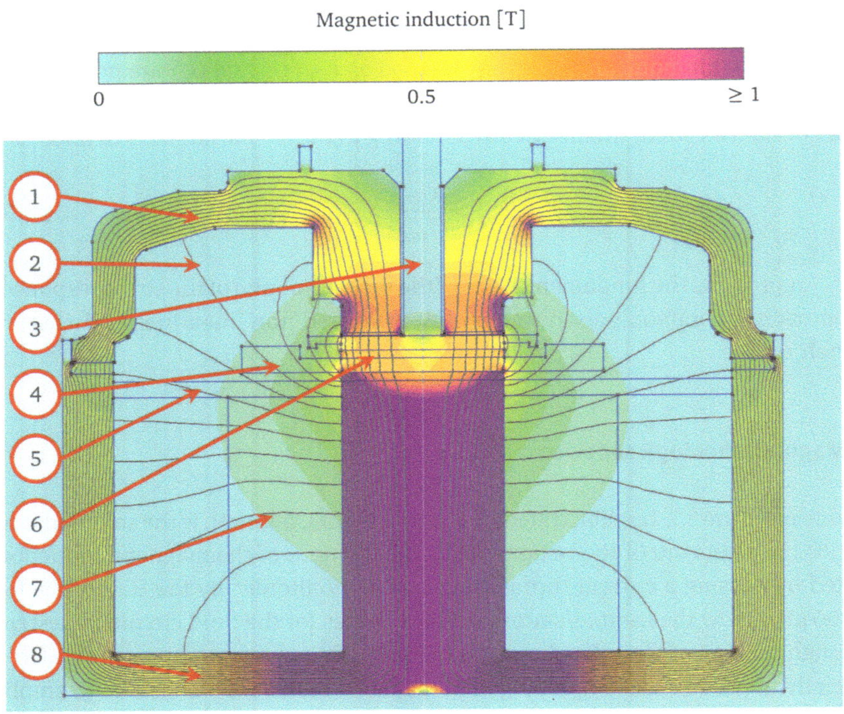

Figure 3.5: A representative static computational simulation of an axisymmetric section of the experimental domain showing isocontours of magnetic induction in the apparatus. Magnetic flux lines are drawn in black. Note that, as a simplification, fillets have not been included at many of the sharp corners of the geometry. Legend: (1) Upper yoke; (2) Free space in test cell; (3) Smooth rotor; (4) Base plate; (5) Locating plate; (6) Sample; (7) Solenoid; (8) Lower yoke.

The geometry was created from measurements of the device, specification sheets provided by the manufacturer (and, where no details could be found or easily discerned, engineering judgement[3]). The materials comprising the apparatus are considered pure iron, 316 stainless steel and titanium, and that the surrounding medium is air. It

3 We therefore remark that these results are for the purpose of illustration only.

was assumed that the number of turns of the wire in the solenoid is 635, and its gauge is 18 SWG. An axisymmetric geometry was constructed, and the 2-D magnetostatic axisymmetric Maxwell equations were solved using the vector potential formulation (see Sections 5.3.5.1 and 6.1.2). To effectively resolve magnetic singularities that arise at sharp corners, the mesh in these regions, as well as those experiencing high magnetic field gradients, was highly refined.

3.1.3 Influence of rotor geometry on magnetic field

For the parallel-plate configuration shown in Figure 3.5, the magnetic field permeating the experimental sample is relatively uniform and primarily axially aligned. This applies particularly to the outer region ($3 \leq R < 10$ mm), where the shear stress generated due to deformation is greatest. In the core of the sample ($R < 3$ mm), there is a decrease in the magnitude of the magnetic induction (a "dead-zone") due to the obstruction caused by the rotor shaft. However, the shear stress generated near the axis of oscillation is small in magnitude, and it is therefore assumed that this magnetic inhomogeneity has little effect on the measured response. A disturbance to the field is detectable at the edge of the sample ($R = 10$ mm) due to the presence of a material interface (discontinuity) that leads to the development of strong magnetic gradients.

Given the selection of available rotors (e. g. those shown in Figure 3.3), it is important to understand their influence on the magnetic field and how they might impact on the measured response of the material under magnetic loading.

Surface finish of parallel rotor

Figure 3.6 illustrates the remarkable difference in the magnetic field generated in an MAP under static displacement conditions when a smooth and grooved[4] rotor is utilised. Due to the penetration of the rotor teeth, a relatively thick (1000 µm) ex situ pre-prepared sample is required in order to prevent impact of the rotor on the stator during dynamic testing. The use of a serrated stator in conjunction with the serrated rotor was not considered because the potential for the teeth to make contact during testing of the compliant MAP.

Full penetration of the rotor teeth into the material results in the formation of large local magnetic gradients and, therefore, stress concentrations (deformation not withstanding). Within the sample itself, at the peaks of the teeth the magnetic field strength is reduced while at their valleys it becomes concentrated. Material indentation alone will result in an inhomogeneous stress response and, in the case of

4 It is important to note that a strict interpretation of the rotor geometry within the context of these simulations is one that has radially profiled grooves, as opposed to the hatch serrated pattern shown in Figure 3.3c.

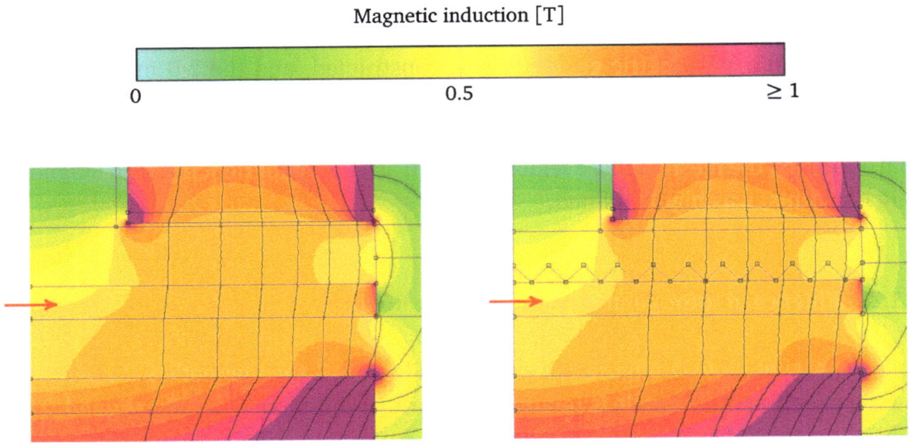

(a) Smooth parallel plate.

(b) Grooved parallel plate (no penetration).

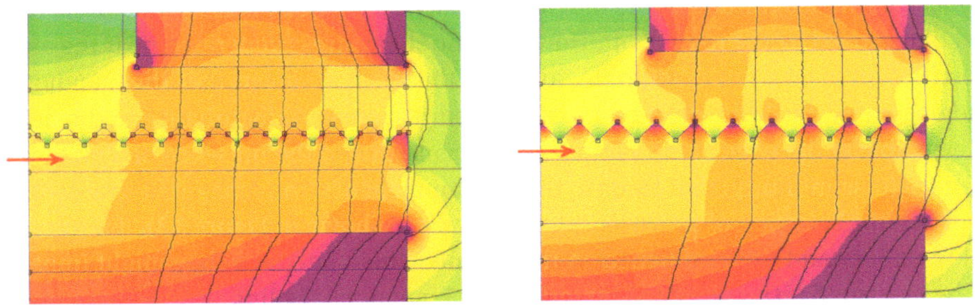

(c) Grooved parallel plate (partial penetration).

(d) Grooved parallel plate (full penetration).

Figure 3.6: Comparison of generated magnetic induction within an ex situ pre-prepared sample (indicated by the arrow; 1000 μm nominal thickness) for different smooth and radially-grooved rotor configurations. Magnetic flux lines are drawn in black. Note that the straight lines indicate the apparatus geometry; the FEM mesh is sufficiently fine to capture magnetic singularities and large magnetic gradients across material discontinuities.

anisotropic media, disturbance of the underlying chain-like particle formations. The application of a magnetic field exacerbates this phenomenon as, for anisotropic materials, the chains are no longer (locally) aligned with the magnetic field. Reducing the degree to which the teeth penetrates the material partially mitigates the issue, but a profound influence on the magnetic field is still visible at low penetration depths. Qualitatively similar computational results are shown and analysed by Laun et al. [295], confirming that the radial profile of the generated magnetic field in the sample (specifically, between the stator and fully penetrated rotor teeth) is oscillatory in nature with the peak-to-peak difference in field strength being between 20 % to 50 % of that computed for the smooth rotor.

Smooth parallel versus cone rotor

Rotors with a conical profile are designed to induce a more homogeneous stress distribution in elastomeric compounds. For a cone rotor (with a 2° draft and a given truncation length), the appropriate calibration gap size is 84 µm; results for this case are largely comparable to a sample of 300 µm thickness. In Figure 3.7, the magnetic field developed within and around the undeformed sample for these two geometries is illustrated. The difference between these two cases (especially when compared to those shown in Figure 3.6) are marginal.

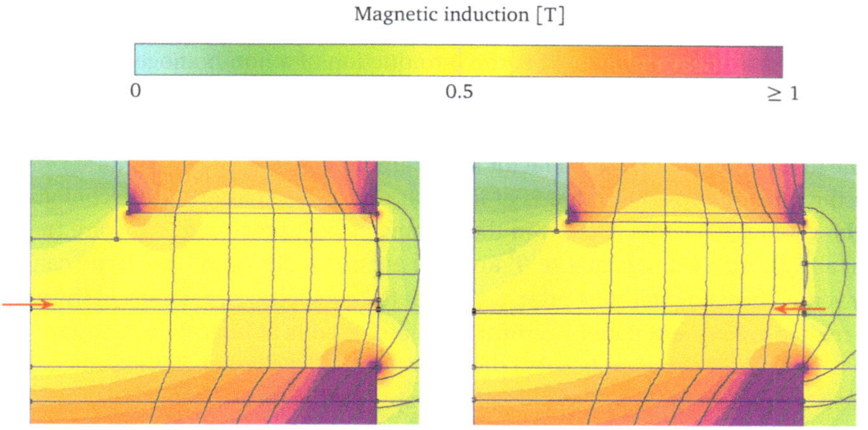

(a) Smooth parallel plate. The sample thickness is 300 µm.

(b) Cone plate. The sample thickness increase from 84 µm at the centre to 349 µm at the edge.

Figure 3.7: Comparison of generated magnetic induction within an in situ cured sample (indicated by the arrow) for the smooth and cone rotor configurations. Note that the straight lines indicate the apparatus geometry.

The more complex geometry of the conical rotor does not induce significant magnetic phenomena that were not already present for the case of the parallel rotor. Most significant is the alignment of the magnetic flux with respect to the rotor and stator surfaces. For the parallel plate, the flux lines are mostly perpendicular to the rotor surface (i. e. aligned with the direction of induced material transverse isotropy) while for the cone rotor the particle alignment direction and that of the magnetic flux are offset. It is expected that this would influence the magnetisation of the material and, therefore, its response under magnetic loading. Furthermore, considerations must be made if curing anisotropic specimens in situ with the cone rotor. In this case, the ratio of the particle diameter to the gap size becomes appreciable, and the particle formations may not be perpendicular to the rotor surface.

3.1.4 Caveats to consider when performing experiments

To obtain reliable and trustworthy experimental data, it is necessary to ensure that the constraints of the experimental apparatus are adhered to. In particular, it is vital that the experimental configuration does not invalidate some of the fundamental assumptions used to interpret and post-process data measured during the experimental procedure. Below we discuss some of these caveats, with a specific focus on those applying to the commercial rheometer used to characterise the aforedescribed MAPs.

Elastomeric polymers

As it is assumed that the stress distribution throughout the test material is homogeneous and has a linear radial profile, the introduction of a non-homogeneous stress field would lead to questionable stress data. If, however, it can be shown that a complex rotor/stator geometry leads to a systematic error in the reported data, it may be possible to introduce a correction factor to account for this.

In the non-linear regime, that is at large strains for the parallel-plate configuration, the geometric factors applied in the rheometer post-processing software (more specifically, those used to compute G' and G'') are no longer valid. This issue can be circumvented by using the cone rotor, for which the geometric corrections in the non-linear regime are present thereby returning reliable results at large deformations.

In each of the techniques listed at the beginning of the chapter, it is assumed that there exists a perfect force transfer onto the sample during the experiment at all times (and for all deformation amplitudes). Walter et al. [543, 542] demonstrate in detail that this can be a challenging requirement to ensure without resorting to the use of adhesives (which would introduce some experimental error), in particular when evaluating ex situ pre-prepared samples in conditions of low visual acuity (such as within the MRD).

In Figure 3.8, the influence of wall slip on fundamental measurements of a simple material system is illustrated. Here, the *standard protocol* refers to that in which ex situ pre-prepared samples are evaluated, while the *control protocol* is one in which the specimen was cured within the rheometer thereby firmly bonding it to the surface of the parallel plates (see Section 2.2).

Note that the material evaluated here is an unfilled, low-damping elastomer that was expected to exhibit very little non-linear behaviour throughout the range of applied strains. This is what is observed when using both the smooth and serrated rotors with the *control protocol*, with the stress-strain response being primarily linear (corresponding to a constant shear modulus), even at large shear deformations, and a low and constant loss modulus. By contrast, a spurious significantly non-linear and dissipative response is measured when using the *standard protocol*, which (in conjunction with further measurements and analysis by means of Lissajous diagrams, FFT methods and video analysis in a simplified setting) is attributable to wall slip at

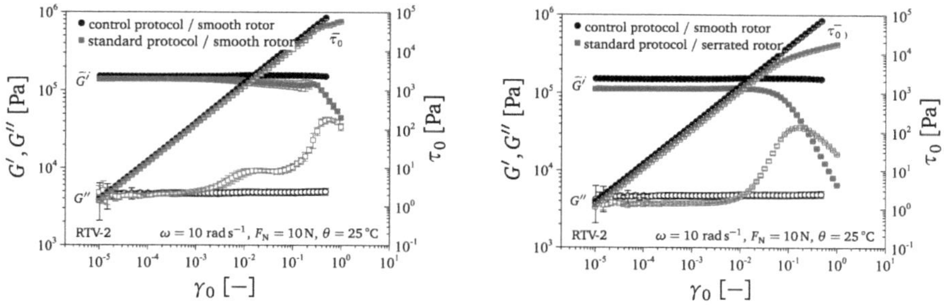

(a) Smooth rotor.

(b) Serrated rotor.

Figure 3.8: Influence of rotor geometry on wall slip of a low-damping elastomer (RTV-2) evaluated using a parallel-plate rotational rheometer in LAOS. In both cases, a smooth stator is used and a relatively high preload of $F_N = 10$ N is applied. [543, figs. 3a,II (reproduced with permission)]

the specimen-plate interface. The "critical" strain corresponding to an increase in the loss modulus was assumed to correlate to the initiation of slip. For the serrated plate slip consistently occurred at the sample-rotor interface, resulting in increased dissipation and a reduced apparent storage modulus. Slip evolved with increasing strain for this configuration as a predictable and progressive phenomenon, while for the smooth rotor a more erratic evolution of G' and G'' were observed. This was assumed to be related to a stick-slip-stick condition being attained at the material interfaces that, due to the geometry, has no preferential initiation point at the stator- or rotor-side; this assumption was subsequently verified by visual observation. Interestingly, the results shown here for both the slip and non-slip conditions were entirely repeatable, thereby masking the fact that, for the case of slip, an experimental artefact was significantly influencing the results and introducing an entirely artificial non-linearity that could easily and mistakenly be interpreted as a material effect. Observe as well that at high strains both experiments conducted with the *standard protocol* display a cross-over in the G' and G'' curves; this is an indication that the material response is no longer one of a viscoelastic solid, but rather that of a viscoelastic fluid.

The damping properties of the material can be easily identified by means of Lissajous diagrams (stress-strain plots over a single steady-state waveform of applied loading). In Figure 3.9, the normalised response of two low-damping elastomers is depicted for strains corresponding to the onset of slip, where the maximum value of G'' is recorded and a point beyond this peak value. Using the *control protocol*, the low-damping nature of the materials was confirmed as the area within the response curve was very small. By comparison, when using the *standard protocol* both of these materials appear to exhibit increased dissipation properties as the applied strain magnitude is increased.

Finally, examining the raw stress (and strain) waveforms for both conditions, and analysing them using FFT techniques can provide further insights into the degree of the (apparent) non-linearities displayed by the material under changing experimental

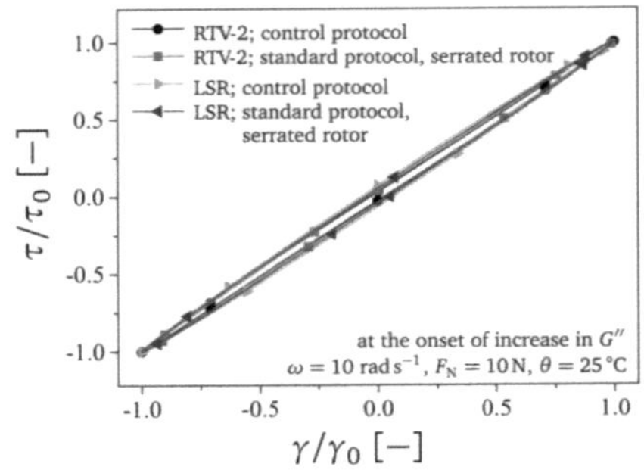

(a) Onset of increase in G''.

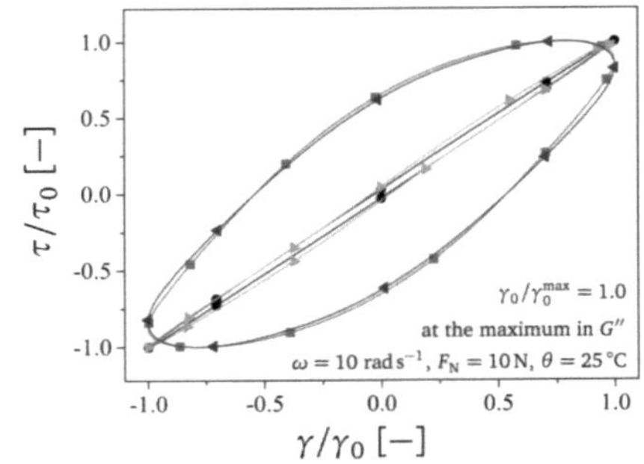

(b) At the maximum in G''.

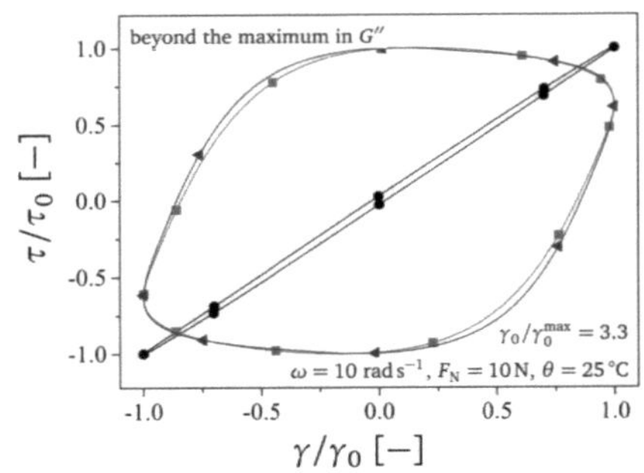

(c) Beyond the maximum in G''.

Figure 3.9: Lissajous diagrams illustrating the apparent dissipative response of highly elastic elastomers, attributable to wall slip. [543, figs. 7a,7b,7c (reproduced with permission)]

conditions. Comparison of the stress waveforms recorded under the *control* conditions and those attributable to (well-defined) slip behaviour clearly indicates that, for this material, slip leads to a significant non-linearity in the measured material response in conjunction with a pronounced phase-shift. FFT analysis confirmed that the stress waveform under slip conditions exhibited appreciable odd harmonics (but no even harmonics). In the most severe conditions (such as those reported in Figure 3.10b), the strain waveform may also exhibit non-linearities (thereby invalidating the fundamental assumption of equation (3.4)). It has been proposed [543] that this is related to the position control mechanism used by the apparatus, and could possibly be avoided through the use of a strain-controlled parallel-plate rheometer.

Collectively, the use of the three data analysis and visualisation techniques mentioned above form an insightful approach to detect wall slip in low-damping elastomers. They motivate the use of in situ cured specimens to ensure that accurate and reliable experiments for both unfilled elastomers, as well as (reinforced) MAPs that are expected to exhibit significant material non-linearities at both small and large deformations and magnetic loadings.

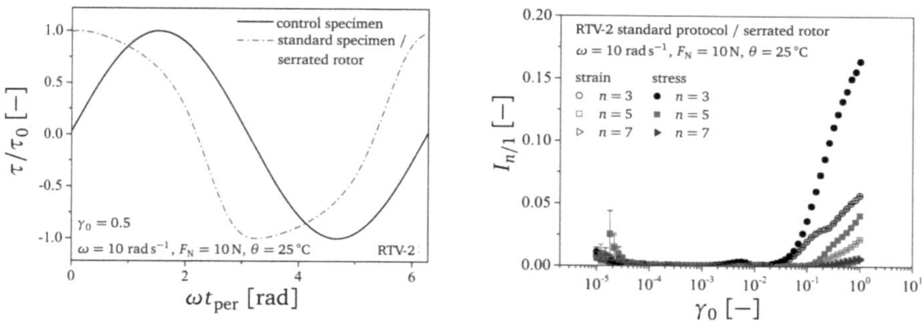

(a) Normalised stress waveform.

(b) Normalised strain and stress intensities.

Figure 3.10: Waveform and normalised harmonics as computed using FFT. [542], [543, fig. 10a (reproduced with permission)]

In the case that it is absolutely necessary to use ex situ pre-prepared specimens, they can be used to examine and inform which experimental factors can be controlled in order to mitigate potential pollution of data through the influence of wall slip. For example, in Figure 3.11 the consequence of changing the applied preload and testing frequency are examined [544, 542]. By increasing the normal force applied by the rotor, the onset of wall slip can be delayed to larger deformations. However, even when near the maximum preload (typically 50 N for commercial rotational rheometers) is applied, slip of the ex situ pre-prepared sample cannot be entirely avoided at large deformations. By comparison, when using in situ cured specimens deformations up to the torque limit of the rheometer can be applied. Contrasted to the influence of the normal force, changing the angular frequency of the conducted experiments only had a marginal effect on the onset and evolution of the slip phenomenon. Slip tended to

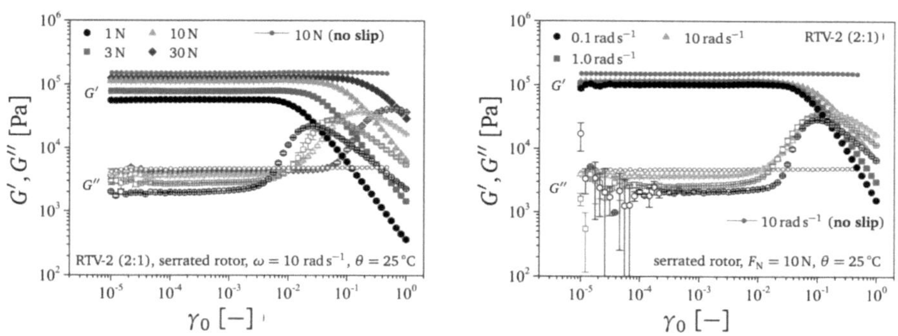

(a) Normal force F_N.

(b) Angular frequency ω.

Figure 3.11: Influence of experimental conditions on wall slip of a low-damping elastomer (RTV-2; ratio of components containing the cross-linker and catalyst is 2:1) evaluated using a parallel-plate rotational rheometer in LAOS. [544, figs. 1,6 (reproduced with permission)] occur at slightly lower deformations as the rate of deformation increased, but this was at the same order of magnitude under all tested conditions.

Magneto-active polymer composites

At large magnetic inductions, a high current must be passed through the solenoid in order to generate the magnetic field. This leads to a considerable increase in cooling requirements that must be met by the temperature control system in order to maintain a constant set of environmental conditions within the MRD.

Another thermal-related influence is that of particle sedimentation during curing [545]. As is illustrated in Figure 3.12a, the temperature at which the MAP is cured in situ has a significant impact on the time evolution of the measured properties of the material. By increasing the curing temperature from room temperature by 20 °C the time necessary to input the activation energy, as well as the total time taken to cure each sample (as measured by the complex viscosity η^*), can be reduced by over an order of magnitude.

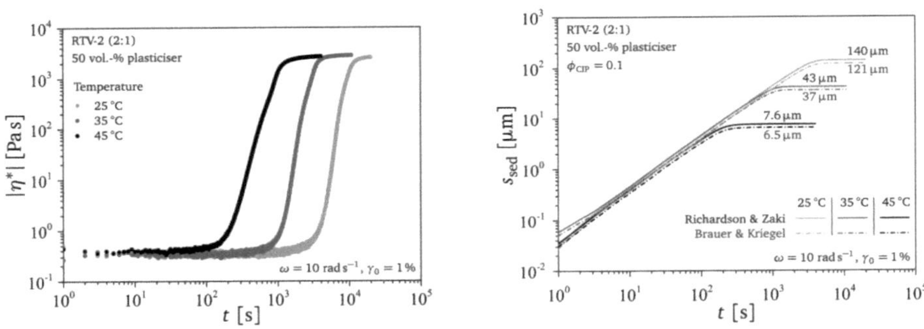

(a) Curing characteristics of the pure elastomer at preset temperatures, as given by the evolution of the complex viscosity $|\eta^*|$ over time.

(b) Approximate particle swarm sedimentation (migration) distance s_{sed} during curing ($\phi_{CIP} = 0.1$).

Figure 3.12: Curing kinetics and particle sedimentation of isotropic MAPs. [542]

From this profile, it is possible to calculate [545] the approximate distance of particle migration during curing. The velocity of a suspension of uniform, spherical particles in an infinite medium (the particle swarm velocity), valid for a large range of particle concentrations, is [452, 58]

$$v_{s,s} = v_{s,p}\,[1 - \phi_{\mathrm{CIP}}]^c \quad , \quad v_{s,p} = \frac{d^2\,[\rho_p - \rho_f]\,g}{18\eta} \tag{3.12}$$

where $v_{s,p}$ is the terminal velocity of a single particle as computed by Stoke's law, $c = 4.65$ is an factor derived experimentally for spherical particles, d is the average diameter of a CIP (given in Figure 2.4b), ρ_p and ρ_f respectively denote the particle and fluid specific densities, g is the constant of gravitational acceleration, and η is the dynamic viscosity. By integrating the suspension sedimentation velocity over the curing time, accounting for the evolution of the fluid viscosity, it is possible to compute the particle swarm sedimentation distance at different temperatures. In Figure 3.12b, results as computed from the above and an alternate approach [58, 59] for a single filler fraction are presented for different curing temperatures. It is evident that curing at room temperature may lead to a migration distance that is significant in proportion to the thickness of the testing specimen; this is confirmed in the SEM image shown Figure 3.13a. By comparison, curing at 45 °C leads to a sedimentation distance of only 2 to 3 times the average particle diameter. As is illustrated in Figure 3.13b, the resulting cured compound is considerably more uniform in terms of the particle distribution.

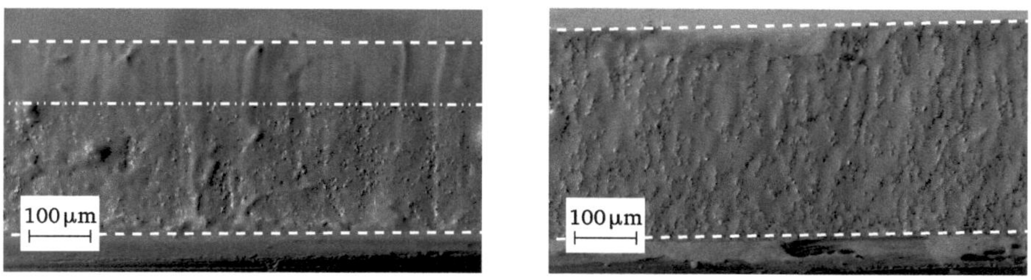

(a) Sedimentation. **(b)** No sedimentation.

Figure 3.13: Cross-section through a 300 µm thick ex situ pre-prepared MAP specimen using SEM, illustrating the result of sedimentation on the particle distribution. The dashed lines indicate the approximate location of the upper and lower edges of the specimens, while the central dash-dotted line in Figure 3.13a indicates the upper level of the particles that have migrated due to the influence of gravity during curing. [542]

3.2 Experimental methodology

Based on the knowledge derived from the body of literature, and supplemented by the fundamental studies documented in [543, 544], a conservative and rigorous approach to the characterisation of MAPs has been proposed by Walter et al. [545, 542]. The following steps are taken to obtain the data reported in Chapter 4:

1. **Material preparation:** A batch of uncured material is prepared using the procedure outlined in Section 2.2. The deaerated composition (with a fixed ratio of catalyst to cross-linker, volume fraction of CIP, plasticiser, etc.) is chosen based on the desired material properties and effects to be investigated.

2. **In situ specimen preparation:** Characterising the specimen cured in situ minimises the number of systematic errors introduced into the experimental procedure.
   (a) **Material transfer:** After 5 min has elapsed from the time that the cross-linker is added to the catalyst, the homogeneous polymer-additive-particle liquid mixture is transferred to the centre of the rheometer stator using a sterile pipette (low CIP content) or spatula (high CIP content). Although the volume of fluid is fixed by the final geometry of the rotor-stator configuration, the number of drops necessary to fill the gap varies based on the material viscosity. Excess material that flows into the guard rim and beyond cannot be "trimmed" or removed from the apparatus. Therefore, the quantity of fluid transferred is closely monitored, ensuring that experimental reproducibility is not significantly influenced by overfilling.
   (b) **Geometry formation:** The rotor is lowered onto the uncured material such that a predetermined gap distance between the rotor and stator is attained. For the smooth rotor this is 300 μm, and in the case that the cone plate is used the gap is 84 μm. Care is taken to prevent additional aeration, or the creation of voids, as the material comes into contact with the plate surfaces.
   (c) **Preconditioning:** A *time sweep* (sinusoidal oscillation at a fixed amplitude and frequency; $\gamma_0 = 20\,\%$, $\omega = 10\ \mathrm{rad\,s^{-1}}$, $b = 0\ \mathrm{mT}$) is performed on the uncured mixture for 1 min. This destroys the initial local microstructures (e. g. particle agglomerates) and, therefore, removes any local prestressing. After this step, all samples of the comparable constitution demonstrate the same stress response, indicating that their microstructures are comparable.
3. **Curing:** In situ curing takes place under predetermined, optimal environmental conditions to assure that the final specimen chemistry and microstructure is homogeneous, consistent and repeatable.
   (a) **Conditions:** The material is cured for a duration of 4000 s at a temperature of 45 °C to reduce sedimentation. During this time, a light sinusoidal deformation is applied using a time sweep configuration ($\omega = 10\ \mathrm{rad\,s^{-1}}$ and $\gamma_0 = 0.005\,\%$, for an MAP, or $\gamma_0 = 1\,\%$ for an unfilled elastomer) to measure the material characteristics during curing and detect when the curing process is complete.
   (b) **Evolution of material microstructure:** If an isotropic sample is required, no magnetic field is introduced. To create an anisotropic specimen (parallel plate only), a magnetic induction of $b = 300\ \mathrm{mT}$ is applied by the MRD over the entire curing period to promote the formation of particle chains. The applied induction is limited by the dissipation capability of the temperature control system. The current-induction relationship for the device is precali-

brated for each given material composition, namely the CIP volume fraction and required material microstructural alignment.[5]

4. **Demagnetisation:** Should a magnetic field be used during curing then a sinusoidal current is applied and its magnitude is stepwise reduced. This effectively demagnetises the apparatus, eliminating any magnetic hysteresis in the material and the rheometer itself.
5. **Delay:** A rest period of 180 min is introduced to ensure that the material is completely cured and stress-free. This also guarantees that the MRD is entirely demagnetised after the curing of anisotropic specimens.
6. **Experimental characterisation:** A collection of experiments are conducted on a single sample under carefully prescribed mechanical and magnetic loading conditions. For the smooth parallel rotor, an applied axial preload is $F_N = 1\,\mathrm{N}$, and it is desired that the temperature remains constant at $\theta = 45\,^\circ\mathrm{C}$. For the cone plate, the gap length remains fixed.[6]
   (a) **Preconditioning:** Three preliminary *amplitude sweeps* are performed without the application of any magnetic loading. The amplitude for each deformation waveform is sequentially increased from the lowest to the maximum value used during characterisation. This ensures that the material is suitably preconditioned for the increased loading condition, and minimises any possible influence of the Mullins effect [391].
   (b) **Measurement:** Three distinct steps are used in the measurement phase in order to reduce the number of confounding influences on experimental results.
      i. **Deformation at a fixed magnetic load:** Characterisation is conducted by means of amplitude sweep experiments. The range of applied shear strain magnitudes ($\gamma_0 = 1 \times 10^{-5}$ to 1, dependent on the occurrence of delamination) and a fixed frequency ($\omega \in 0.3\,\mathrm{rad\,s^{-1}}$ to $30\,\mathrm{rad\,s^{-1}}$). The test duration is determined based on the time taken for the system to reach a steady-state condition at each loading and deformation state, and is primarily influenced by the material composition. Although the timescale of a single experiment is dependent on the automatic strain control mechanism, it is typically of the order of $700 \pm 100\,\mathrm{s}$ for $\omega = 10\,\mathrm{rad\,s^{-1}}$. At high inductions (700 mT) and large deformations, a small temperature increase (at most $5\,^\circ\mathrm{C}$) may be present. However, due to the properties of silicone rubber, this has been determined to have an insignificant influence on the recorded results.

5 There is a measurably lower current required to be passed through the solenoid when conditions of higher material anisotropy and filler volume fraction are present.

6 For experiments where results obtained using the cone rotor are directly compared to those of a parallel rotor, the conditions of the latter case are altered such that the experimental conditions are qualitatively similar.

ii. **Demagnetisation and delay:** After each individual experiment, a demagnetisation process is performed thereby removing any trace of magnetic hysteresis in the unloaded material and apparatus. Thereafter, a delay time of 45 min is imposed, allowing the system to relax completely and return to its original condition.

iii. **Incrementation of magnetic load:** Each individual amplitude sweep experiment is repeated for range of magnetic inductions ($b \in \{0, 25, \dots, 100, 150, \dots, 300, 400, \dots, 700\}$mT). The applied magnetic loading is sequentially increased from no-load to the maximum value. The maximum value was chosen such that the energy dissipation requirements could be met by the temperature control system and sustained over the testing period.

7. **Specimen removal:** Due to its firm fixture to the base plate and rotor, the specimen and attached apparatus are soaked in toluene to debond it from the plates.
8. **Repetition:** Statistically significant (averaged) results are obtained by repeating each set of experiments for a single material composition. For composites that have a low filler fraction this is done at least 3 times, but up to 8 repetitions may be necessary when high filler fractions are evaluated.

4 Magneto-mechanical characterisation of magneto-active polymer composites

Following Chapters 2 and 3, which provided an outline of procedures that may be used to reliably produce and experimentally examine MAPs, in this chapter we finally present the results of the rheological characterisation of the field-responsive composite. Due to the complexity of the rheology of these materials and the vast array of techniques ultimately required to fully probe their behaviour, only a concise compendium of results (focussing on a subset of tests performed using parallel-plate rotational rheometry) will be shown. For the interested reader, significantly more detail on the characterisation of this particular material is presented by Walter [542], accompanied by an initial study documented in [545].

There are many different permutations of experimental conditions and material constitutions that can be examined, so the presentation of the material characterisation is made in discrete stages. This begins in Section 4.1 with an examination of the pure elastomer (unfilled matrix), whereafter the microstructure of the filled composite is presented in Section 4.2. This is followed in Section 4.3 by the strain dependence of both the isotropic and anisotropic media (of varying filler fractions) with and without magnetic loading, and lastly in Section 4.4 their frequency dependence with a magnetic field present and absent. To ensure that the results presented here are comparable with one another, we substitute the RTV-2 used in the preliminary studies discussed in Chapter 3 with pure PDMS. It has thereby been ensured that experiments are performed on MAP specimens fabricated from a single batch of raw materials. Overall, the system that has been examined is, purposely, a very simple one; by choosing a matrix that demonstrates the behaviour expected of cross-linked elastomers without the addition of other fillers, it is ensured that effects that are directly attributable to the CIP and the influence of an applied magnetic field are as easy to identify as can be made possible.

4.1 Unfilled matrix (PDMS)

Through the initial characterisation of the simplest material system, namely the base matrix alone, it is possible to quantify the mechanical reinforcement that the addition of CIP provides. The pure polymeric matrix is completely unreinforced (no fillers, such as fumed silica, are present) and has a low damping nature that will help highlight magneto-mechanical effects.

By performing amplitude sweeps across a range of frequencies, both the dissipative nature of the material, as well as the transition point from the linear viscoelastic to the non-linear viscoelastic response, can be ascertained. Figure 4.1a presents the amplitude dependence of the PDMS across five orders of magnitude of maximum applied

https://doi.org/10.1515/9783110418576-004

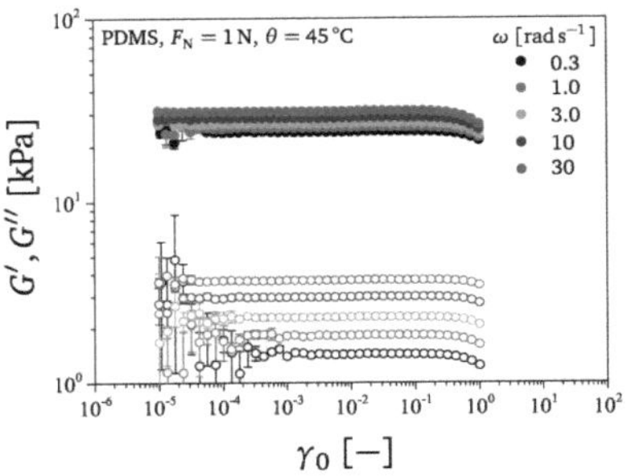

(a) Strain amplitude and frequency dependence of the storage and loss modulus.

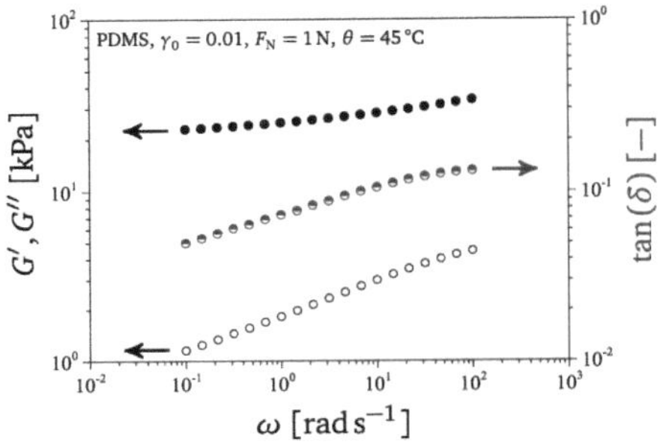

(b) Linear viscoelastic behaviour over the tested frequency range ($\gamma = 1\%$).

Figure 4.1: Characterisation of pure unreinforced PDMS using parallel-plate rotational rheometry. [542]

strain and two orders of magnitude of excitation frequencies. It is clear that this material exhibits a very broad linear regime, and that the storage modulus remains largely unaffected by the applied frequency (it is anticipated that the stress-softening that is observed at high strains will be greatly overshadowed by the magneto-mechanical stiffening exhibited by the composite system). Defining the onset of the non-linear regime to be the strain at which the storage modulus is 10 % smaller than that measured in the linear regime (i. e. as $G'\,(\gamma_0 \to 0)$), it is observed that the transition point from a linear to non-linear material response is only marginally affected by the frequency; in particular, the transition point is slightly shifted to smaller deformations at higher frequencies. By contrast, the loss modulus, which is at least an order of magnitude less than the storage modulus across the spectrum of test conditions, increases with increasing frequency. The presence of a non-linear regime indicates that the material does not behave like an ideal elastomer. This deviation from an idealised model could be influenced by factors such as the plasticiser that causes swelling of the polymeric network.

The companion result displayed in Figure 4.1b highlights the viscoelastic behaviour of the matrix (i. e. its rate-dependence) at a fixed strain amplitude. The chosen measurement point is in the linear regime under all examined conditions and sufficiently high to prevent possible resolution issues. Within these conditions, it is observed that the loss factor $\tan(\delta)$ remains on the order of 0.1. From this data, it is confirmed that the material does indeed have a low dissipative nature (as the compositional-chemistry implies) within the range of tested frequencies when compared to other viscoelastic elastomers, and that there are no artificial dissipative mechanisms in the system that negatively influence the results.

It is important to note that the spectrum of examined strain amplitudes and frequencies were specifically chosen based on a preliminary analysis of the system, after which it was determined that these conditions produced trustworthy data. At very small strains in conjunction with frequencies below $\omega = 0.3\,\mathrm{rad\,s^{-1}}$, the sensitivity of the rheometer is questionable, while at frequencies greater than $\omega = 30\,\mathrm{rad\,s^{-1}}$ resonance effects start to influence the recorded response.

4.2 Microstructure developed in the cured filled MAP

Before presenting the thorough rheological characterisation of the filled MAPs, it would be beneficial to have some exposure to the microstructure of the cured material. In this way, one gains some firm insight as to why the composite responds in the manner that it does, and will assist later in the interpretation of the results of its experimental measurement. A detailed visualisation of the particulate microstructure for a range of CIP volume fractions and two curing conditions are shown in Figure 4.2. The conditions under which the MAPs were fabricated are detailed in Section 3.2. Note as well that there was no sedimentation visible throughout the sample thickness.

Examining the isotropic samples first, it is observed that the particle distribution is quite homogeneous at and above $\phi_{\mathrm{CIP}} = 20\,\%$, while at $\phi_{\mathrm{CIP}} = 10\,\%$ a small degree of clustering (likely due to difficulties in mixing the viscous uncured material) is visible. As the CIP volume fraction is increased towards $\phi_{\mathrm{CIP}} = 40\,\%$ (above the percolation threshold), the interparticle spacing diminishes substantially. Overall, at the mesoscopic scale it is clear that the material appears isotropic, with no preferred directionality visible in terms of the particle orientation.

In contrast, the microstructure of the samples that are cured in the presence of a magnetic field clearly show preferential orientation for $\phi_{\mathrm{CIP}} = 5, 10$ and $20\,\%$. Due to their magnetisation, the particles form magnetic dipoles or multi-poles and are attracted to one another. During curing, they therefore migrate in order to attain their lowest energy state, namely one that promotes the formation of chain-like or columnar particle structures reminiscent of a transverse isotropic material. It is interesting to observe that, even at low filler fractions, the structures that form are not perfectly chain-like in nature (such as is depicted in [240, 48]), but rather more columnar (dispersed) with adjacent particles agglomerating to form clusters. Comparison to the microstructures shown in [84] indicates that the relatively low magnetic field applied during curing, along with the liquid matrix viscosity, has a large role to play in this phenomenon. At higher particle densities, one observes that the microstructure more closely resembles the isotropic case. The percolation threshold for the system magnetised during curing is significantly lower than that of the isotropic case, so adding more filler after the percolation threshold is exceeded naturally leads to more dispersion rather than increased ordering of the columnar structures.

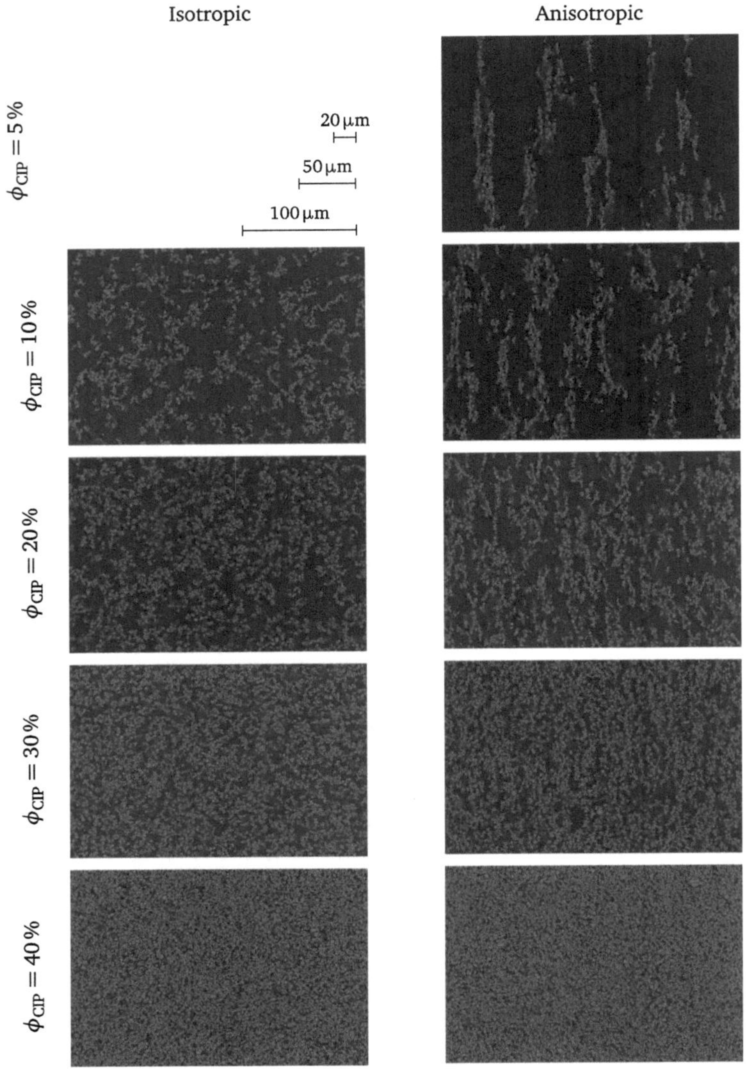

Figure 4.2: Detail of the particle distribution of isotropic and anisotropic MAP specimens for different volume fractions of CIP. The samples were cured in situ, with the anisotropic material cured under a 300 mT magnetic field (orientated vertically for the inset images). Visualisation through the cross-section was achieved using cryogenic cracking and, subsequently, back-scattered SEM imaging. [542]

4.3 Strain amplitude dependence of the composite MAP (PDMS with a CIP filling)

By investigating the shear strain amplitude dependent response of the MAP it is possible to examine the elastic and dissipative components of the composite's shear modulus in both the linear and non-linear regimes, and to determine where the transition from one to the other is. Experiments were therefore conducted by means of oscillatory strain amplitude sweeps, for which the variation of the applied shear strain amplitude γ_0 was between 0.001 % and on the order of 100 %, dependent on the material composition. For each configuration, the characterisation was performed at a fixed frequency, temperature and applied normal force. The parameters that are systematically varied within this study are the preferred orientation of the microstructure (isotropic versus anisotropic), the particle volume fraction, and, in Section 4.3.2, the applied magnetic loading.

4.3.1 Response to mechanical loading

Under conditions of purely mechanical loading ($\mathbb{B} = 0$ mT), it is possible to establish the combined influence of particle-particle and particle-polymer interactions without the complexity of any added magnetic response. Presented in Figure 4.3 is the loss and storage modulus for an isotropic and transversely isotropic MAP with a varying volume fraction of CIP. Comparing the response of the unfilled system (discussed in Section 4.1) to the highly filled systems, it is evident that the addition of the rigid filler increases the overall stiffness of the composite, as well as the dissipation (a result of friction due to interparticle interactions). Note, however, that the ratio between the storage and loss modulus indicates that the material remains highly elastic, even though the internal friction increases measurably as the particle density is raised. With reference to the discussion in Section 3.1.4, it was ensured that wall slip was not present when samples with high filler fractions ($\phi_{CIP} = 40$ %) were examined.

An increase in the filler volume fraction also leads to a shortening of the linear regime; this is linked to the "intrinsic deformation" of the material [557], namely that the deformation of the compliant matrix is significantly greater than that of the comparably stiff CIP filler. For low filler fractions ($\phi_{CIP} = 0.1$), this effect is already pronounced, while at $\phi_{CIP} = 0.4$, the transition from a linear to a non-linear material response occurs between $\gamma_0 = 1 \times 10^{-4}$ to 1×10^{-3}. This additional non-linearity also affects the loss modulus, as a similar trend is also observed for both the isotropic and anisotropic samples. However, the anisotropic material exhibits a pronounced peak in the loss modulus for high ϕ_{CIP}, while the isotropic material tends to display a near-constant loss modulus until the point of non-linearity is reached.

Comparing the response of the anisotropic to the isotropic material, it is observed that by introducing a preferred particle orientation it is possible to increase the overall

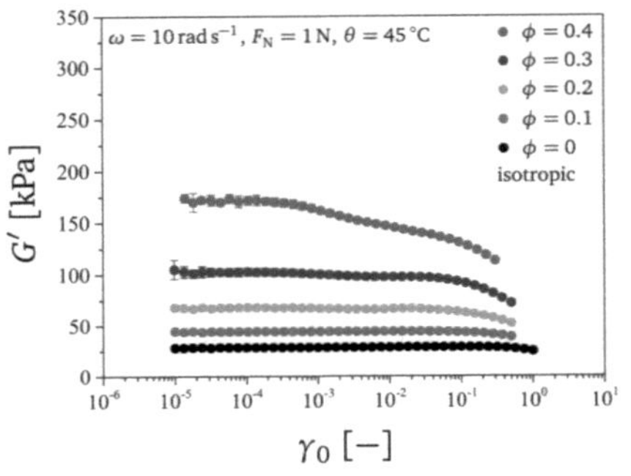

(a) Isotropic (storage modulus).

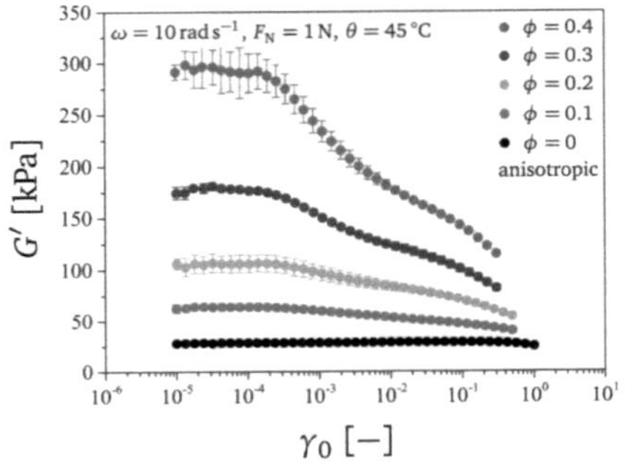

(b) Anisotropic (storage modulus).

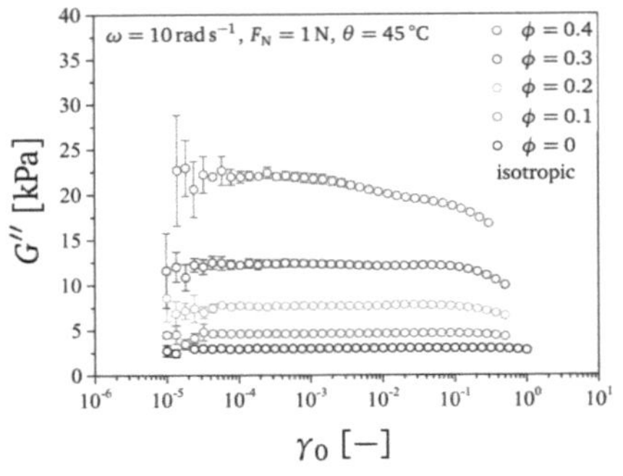

(c) Isotropic (loss modulus).

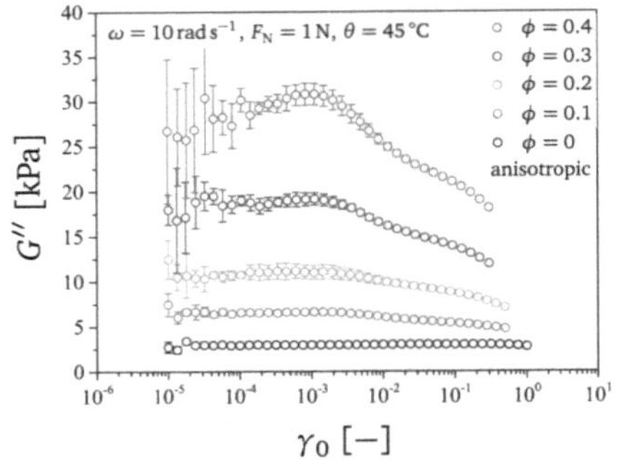

(d) Anisotropic (loss modulus).

Figure 4.3: Strain amplitude dependence of an MAP with varying composition under a purely mechanical load. The scales for the y-axis for both the isotropic and anisotropic cases are identical in order to give a more representative comparison of the mechanical reinforcement and (later) the magneto-mechanical effect that the CIP provides. [542]

stiffness of the composite by approximately 2 times within the linear regime. At high strains (i. e. within the non-linear regime), however, the contrast in the storage modulus measured for the isotropic and anisotropic samples reduces significantly due to the increased interparticle spacing and, consequently, decreased particle-particle interactions. Notably, for both material compositions the reduction in the storage modulus within the non-linear regime is monotonic.

Concerning the experimental methodology itself, we note that reproducibility of the material response for different samples was very high. Only for the anisotropic case with $\phi_{\mathrm{CIP}} = \{0.2, 0.4\}$ is there any significant scatter in the data recorded for the storage modulus. This can be most likely attributed to inconsistencies in the microstructure due to both curing and preconditioning. The data scatter in the loss modulus results, observed only at very small strains, can be ascribed to stress and strain resolutions associated with the use of a stress-controlled rheometer. In later results, the latter issue is compounded by magneto-induced effects.

Microstructural interpretation of the material response

The amplitude dependence of the composite is primarily governed by:
(i) the intrinsic properties of the matrix (more specifically, the linear and non-linear behaviour of the polymeric network),
(ii) the response (both elastic and dissipative) attributable to the microstructure (interparticle and polymer-particle interactions), and
(iii) reinforcement due to the addition of a rigid phase, namely the CIP.

The latter phenomenon is referred to in the material science literature as "hydrodynamic reinforcement" [278, 296]; in the remainder of this chapter, this stiffening due to the addition of inclusions with a greater stiffness modulus than the matrix in which they are embedded will be termed "intrinsic reinforcement". We refer the reader to Leblanc [296, figure 5.40] for a visual interpretation of the elastic contributions to material behaviour.

For this material composition, the influence of the filler dominates the MAP's response for $\phi_{CIP} \geq 0.2$ for the isotropic microstructure and for all of the tested volume fractions for the anisotropic case. Due to the presence of rigid inclusions the material stiffness increases throughout the tested strain range. However, at high strains the effect of the evolving particle network (in particular, the collapse of the network as the number of interparticle contacts is reduced) is observable. As the filler fraction increases, the transition from a linear to a non-linear response occurs at lower strains.

At low filler fractions ($\phi_{CIP} < 0.2$), the influence of item (ii) is negligible as the interparticle spacing is, on average, large and the particle surfaces are chemically inert. The additional non-linearity that appears at higher filler fractions ($\phi_{CIP} \geq 0.2$) is caused by particle-particle interactions, namely interparticle contact that occurs predominantly in the direction perpendicular to that of the applied shear. This factor becomes more significant in the region of, and exceeding, the percolation threshold; this appears to be on the order of $\phi_{CIP} = 0.2$ for the isotropic medium. The material response within the linear or small-strain regime is therefore heavily influenced by the breakdown of the percolated particle network, while that of the non-linear regime is governed by the behaviour of the elastomeric matrix. It stands to reason that the energy dissipated due to internal friction should increase with increasing ϕ_{CIP}, but it was observed that the loss factor (not illustrated) remains nearly constant in the linear regime. Conversely, in the non-linear regime the collapse of the particle network leads to a highly dissipative response [545]. By examining different materials of various constitutions, Walter et al. [545] determined that this was because the interparticle contact behaviour was elastic in nature.

4.3.2 Response to magneto-mechanical loading

After repeating the previous study while varying the strength of the applied magnetic field up to $\mathbb{B}$ = 700 mT (below the saturation threshold), Figures 4.4 and 4.5 present the magneto-mechanical counterpart data to that given in Figure 4.3 for two CIP volume fractions.

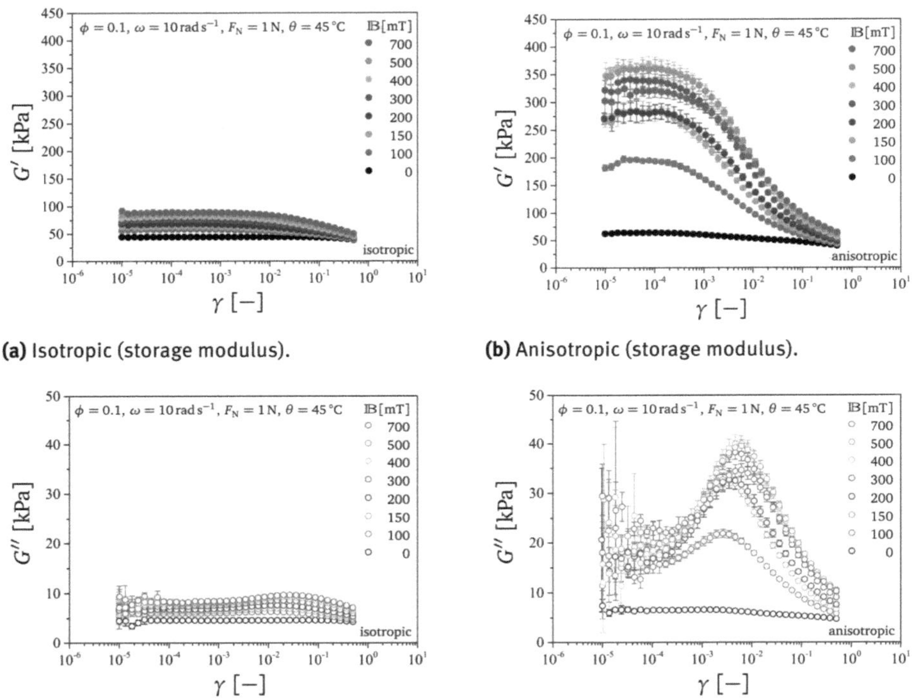

(a) Isotropic (storage modulus).

(b) Anisotropic (storage modulus).

(c) Isotropic (loss modulus).

(d) Anisotropic (loss modulus).

Figure 4.4: Strain amplitude dependence of an MAP with a particle volume fraction of ϕ_{CIP} = 0.1 under a magneto-mechanical load. [542]

Taking a broad overview of the observed magneto-mechanical effect, it is clear that the presence of the magnetic field alters the mechanical response of the composite in a non-linear manner across the entire range of applied strains. For each of the examined magnetic fields, there remains a regime in which the material exhibits a linear mechanical response; it is therein that its effect on the storage modulus is most pronounced. At higher strains, the reinforcement due to magnetic and mechanical influences decreases monotonically, similar to that of the purely mechanical case. In contrast, the dissipative qualities of the material is significantly influenced by the

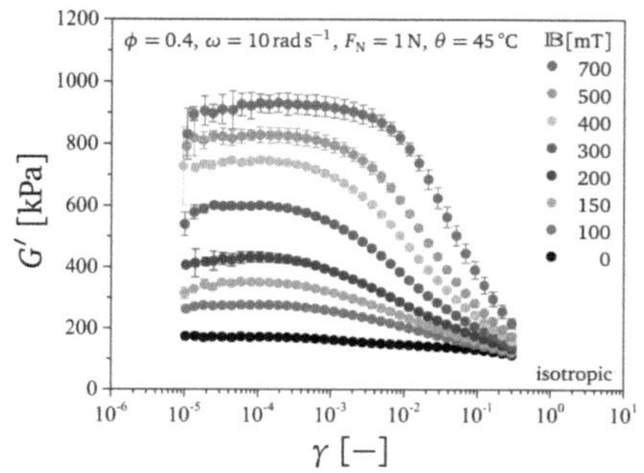

(a) Isotropic (storage modulus).

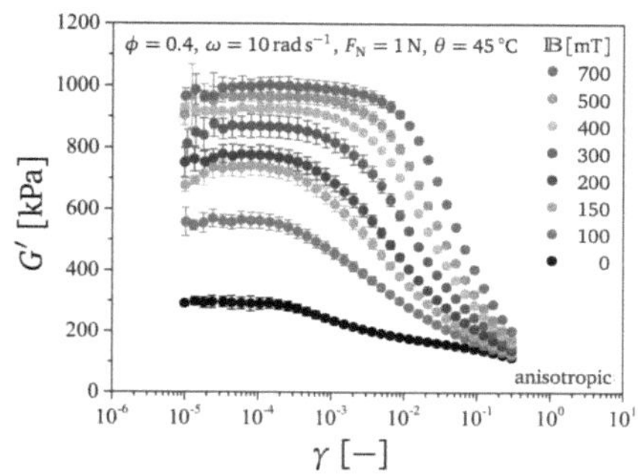

(b) Anisotropic (storage modulus).

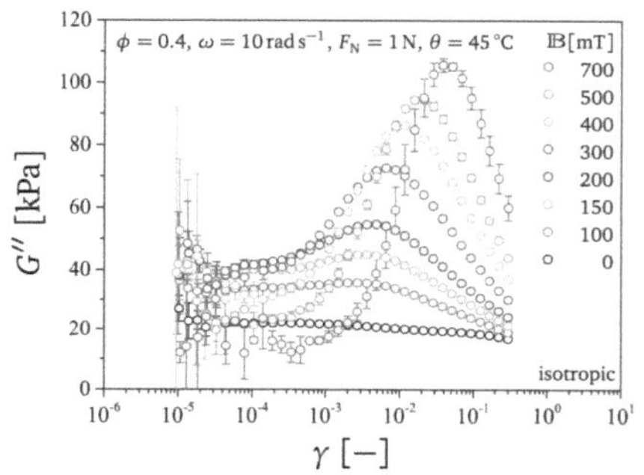

(c) Isotropic (loss modulus).

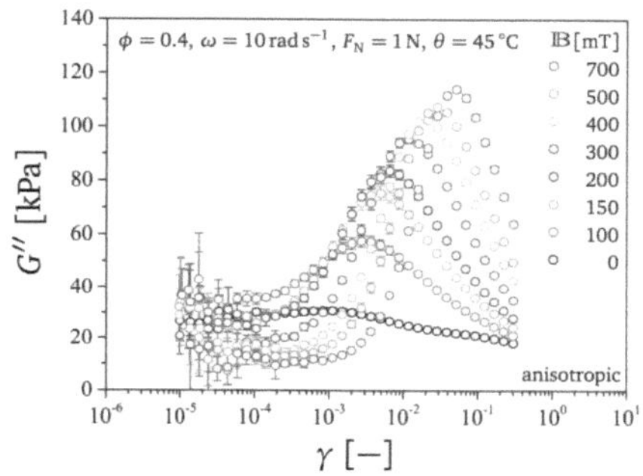

(d) Anisotropic (loss modulus).

Figure 4.5: Strain amplitude dependence of an MAP with a particle volume fraction of $\phi_{CIP} = 0.4$ under a magneto-mechanical load. [542]

magnetic field at high strains. Overall, the loss modulus increases with an increase in magnetic field, and additionally it increases to a maximum at some critical strain, after which it decreases again. This local maximum in the loss modulus typically occurs within the non-linear regime, where the storage modulus is well below its maximum value. Consequently, there is a significant difference in the loss factor measured throughout the strain range.

One observation that remains unexplained is the high magnetic field ($\mathbb{B} > 400$ mT) response of the $\phi_{CIP} = 0.1$ anisotropic sample, as shown in Figure 4.4b, in particular that the storage modulus appears to measurably decrease within the linear regime when compared to that measured at lower magnetic field strengths. Two possible, but unverified, explanations for this phenomenon are (a) a microstructural rearrangement that may occur at high fields and low particle concentrations, or (b) that this result is subject to some experimental artefact, such as may arise due to resolution issues with the rheometer at small strains or subsequent post-processing of the experimental data. To date, this phenomenon is not completely understood.

Comparing the isotropic material response to that of the anisotropic MAP, it is evident that introducing order to the microstructure increases the composite's storage and loss moduli. This effect is particularly pronounced for a low particle volume fraction, where application of even a low-magnitude external magnetic field enhances the stiffness properties considerably. As the applied strain increases, the difference between the shear stiffness properties of the material is reduced; this is noted across the range of applied magnetic fields.

An increased particle density enhances the magneto-mechanical effect significantly, especially at high magnetic field strengths for isotropic sample. Although there remains a difference in the isotropic and anisotropic sample characteristics at high filler fractions, this difference is not as pronounced for $\phi_{\mathrm{CIP}} = 0.4$ as when $\phi_{\mathrm{CIP}} = 0.1$. Given the insight into the microstructure presented in Figure 4.2, this is not entirely surprising as the increased particle density leads to decreased order in the anisotropic case. It is also observed that, for the anisotropic composite, increasing the CIP density tends to increase the strain range over which linear behaviour is exhibited. This effect becomes more noticeable at high applied magnetic loadings.

Overall, it can be concluded that the stiffest response under shear loading is attained for this MAP when it has a high volume density of CIP and is produced with a preferred orientation of its microstructure that is aligned with that of the transversely applied magnetic field. However, the largest magneto-mechanical effect (i. e. the ratio of the high-magnetic field to no-field modulus) is observed at some intermediate filler volume fraction. For the examined range of conditions, the material may exhibit up to one order of magnitude difference in the storage modulus, but this is highly influenced by both the loading conditions and material characteristics. Furthermore, the stiffness-enhancement properties attributable to the magnetisable filler decrease significantly once the mechanical loading is such that the material response is no longer linear.

Microstructural interpretation of the magneto-mechanical response

The magnetic field-induced mechanical reinforcement of the storage modulus is commonly referred to as the magneto-rheological effect, and has been extensively discussed in the literature [240, 241, 173, 76, 323, 49, 506, 147, 105]. Application of an external magnetic field leads to magnetisation of the particles, which in turn causes dipole or, dependent on the local magnetic field and particle arrangement, multi-pole interactions between nearby particles. These strong magnetic interactions are superimposed on top of any existing mechanical interactions, such as contact, that already exist between adjacent particles in the particle network (which, dependent on particle density and the existence of a preferred microstructural orientation, may be percolated).

Governed by both the intrinsic stiffness of the elastomeric network and the intensity of the magnetic field, the attractive forces between the magnetised particles displace the matrix between them in order to minimise the magnetic energy in the system. This results in an overall reduction in the average distance between the CIPs as measured along the direction of the externally applied magnetic field. The attractive forces between particles and subsequent micro-structural rearrangement both contribute to the effective mechanical reinforcement of the composite. Associated with the rearrangement of the particles is an increase in *effectively elastic* interparticle contacts, which arise in a percolated network due to the close proximity of the magnetised and highly attracting particles to one another without physical impingement. These effectively elastic contacts contribute to the decrease in the energy dissipated at small strains.

At high strains, when the interparticle spacing in the field direction is increased (and the number of particle-particle contacts decreases), the reinforcement due to magnetic and mechanical influences decreases. The former is particularly sensitive to the interparticle spacing, as the attractive force between two (idealised) magnetic dipoles scales with a factor $\frac{1}{r^4}$, where r represents the distance between the centres of the dipoles [414, 569, 180]. As the material is stretched and subsequently returned to its original configuration, there is a breakdown and reformation of the particle network that highly influences the storage and loss modulus. When the particle structure is percolated then, depending on the applied magnetic load, the structure may be seen to collapse after a critical applied mechanical deformation is achieved. This leads to the substantial dynamic stress softening that is predominately observed for the anisotropic material in particular. The loss modulus evolves with the applied mechanical strain based on the evolution of the effectively elastic microstructure to one that is dissipative in nature, primarily as a result on increased interparticle contact and polymer-particle frictional forces. As a final point, Walter et al. [545] hypothesise that the changes in the microstructure, as induced by the magnetic field in particular, leads to the field-dependence of the transition point between the linear and non-linear regimes.

4.3.3 Influence of rotor geometries on the response of isotropic MAPs

The geometry of the rotor used for in situ curing has a discernible influence on the magnetic field permeating the material, as was illustrated in Figure 3.7. The difference in the generated field as analysed using computer simulations appears marginal for an MAP with a low relative permeability. However, as is shown in Figure 4.6, its influence is measurably significant for highly filled isotropic composites exposed to high magnetic inductions. With the cone rotor, both the reported shear and loss moduli are lower for the nearly entire range of applied shear strains than their corresponding values as measured with the parallel rotor. By contrast, when no magnetic field is applied the material properties recorded with the cone and smooth parallel rotors are nearly identical (even in the non-linear regime).

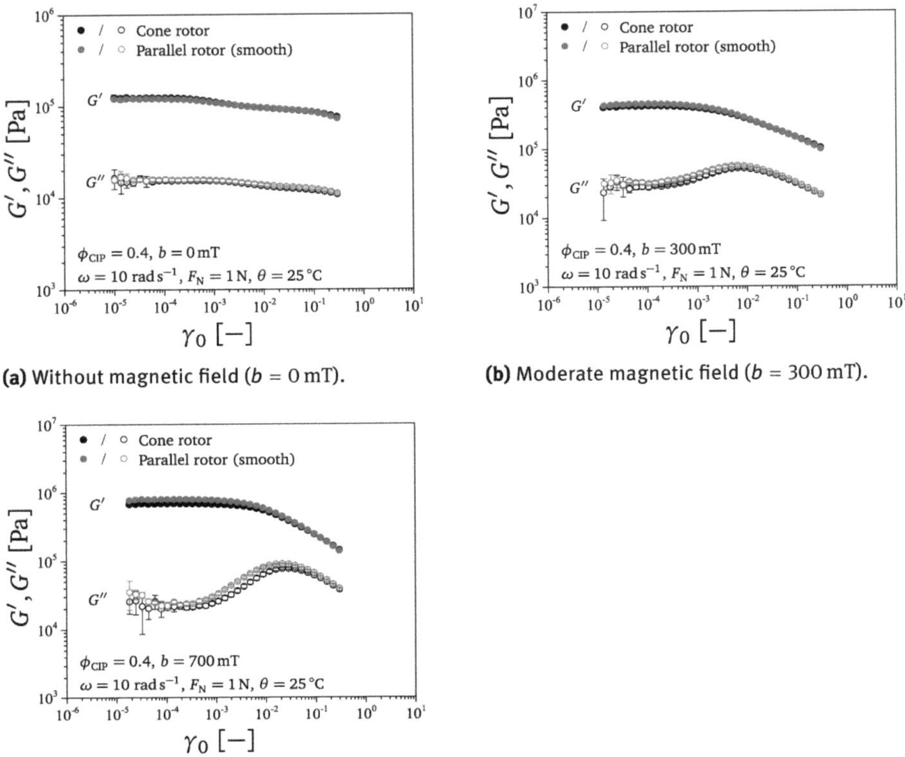

(a) Without magnetic field (b = 0 mT).

(b) Moderate magnetic field (b = 300 mT).

(c) High magnetic field (b = 700 mT).

Figure 4.6: Influence of rotor geometry on shear and loss modulus measured during magneto-rheological experiments of highly filled ($\phi_{CIP} = 0.4$) isotropic MAPs using a rotational rheometer. [542]

4.3.4 Modelling of isotropic MAPs at a fixed frequency

For the isotropic material, it has been shown that its amplitude dependence can be modelled by a combination of the Cross equation and the Ulmer modification to the Kraus model [545]. The argumentation used to derive this form of constitutive model is based on the observation that the non-linear material response mimics that of elastomers with particle reinforcement (e. g. by fumed silica or carbon black), which exhibit the dynamic stress softening behaviour also known as the "Payne" or "Fletcher–Gent" effect. Decomposed additively into its two components, the Cross component effectively models the intrinsic linear and non-linear behaviour of the silicone rubber, the intrinsic reinforcement due of the presence of the solid CIP and dissipative effects due to friction. Dynamic stress softening is modelled as a filler-network effect; thus the additional Kraus–Ulmer contribution accounts for the interparticle interactions that take place both with and without the presence of an external magnetic field. Walter et al. [545] proposed and demonstrated that the storage and loss modulus of isotropic MAPs of similar constitution are predicted to a high degree of accuracy by

$$G'(\gamma_0) = G'_\infty + \frac{G'_C}{1 + [\lambda\gamma_0]^\beta} + \frac{G'_0 - G'_\infty}{1 + [\gamma_0/\gamma_c]^{m+n}} \quad , \tag{4.1a}$$

$$G''(\gamma_0) = G''_\infty + \frac{G''_C}{1 + [\lambda\gamma_0]^\beta} + \frac{G''_0 - G''_\infty}{1 + [\gamma_0/\gamma_c]^{m+n}}$$

$$+ \left[G''_{max} - G''_\infty\right] \frac{\left[[1 + \frac{m}{n}] / [\frac{m}{n}]^{\frac{m}{m+n}}\right] [\gamma_0/\gamma_c]^m}{1 + [\gamma_0/\gamma_c]^{m+n}} \quad . \tag{4.1b}$$

The subscripted initial and infinite strain moduli are respectively denoted by G_0 and G_∞. In the second term of each equation, derived from the Cross model and with modulus contributions subscripted by C, $\lambda = \frac{G_\infty - G_0}{2}$ is the inverse of the critical strain amplitude, and β is a parameter that defines the strain-sensitivity of the non-linear material response. For the Cross terms, the infinite strain contribution was determined to be negligible and thus factored into the above. For the remaining terms, based on the Kraus–Ulmer model, γ_c represents the strain that corresponds to the equilibrium of the particle agglomeration and deagglomeration processes, $G''_{max} = G''(\gamma_c)$ is the maximum observed modulus, and m and n are model parameters that relate to the agglomeration and deagglomeration rates. Since the particle network breakdown is dominant in the anisotropic case, it is also highly plausible that one may be able to model this material behaviour using the same constitutive law; however, this was not verified in [545].

As a point of confirmation, the experimental data as well as the model given by equation (4.1) both fit the well established Guth, Gold and Simha equation [186, 44] for particle composites given by

$$\frac{G_0(\phi)}{G_0(\phi = 0)} = 1 + 2.5\phi + 14.1\phi^2 \quad , \tag{4.2}$$

which is expected since the CIP are spherically shaped and polymer-particle coupling effects are negligible.

4.4 Frequency dependence of the composite MAP (PDMS with CIP filling)

Performing and comparing multiple oscillatory strain amplitude sweeps conducted at different oscillation frequencies furnishes insight into the rate-dependence of the MAPs. More specifically, given the experimental apparatus used, the frequency sweep reliably highlights the material's rate-dependence within the linear regime or its linear viscoelastic behaviour. Although such a study can be extended into the non-linear

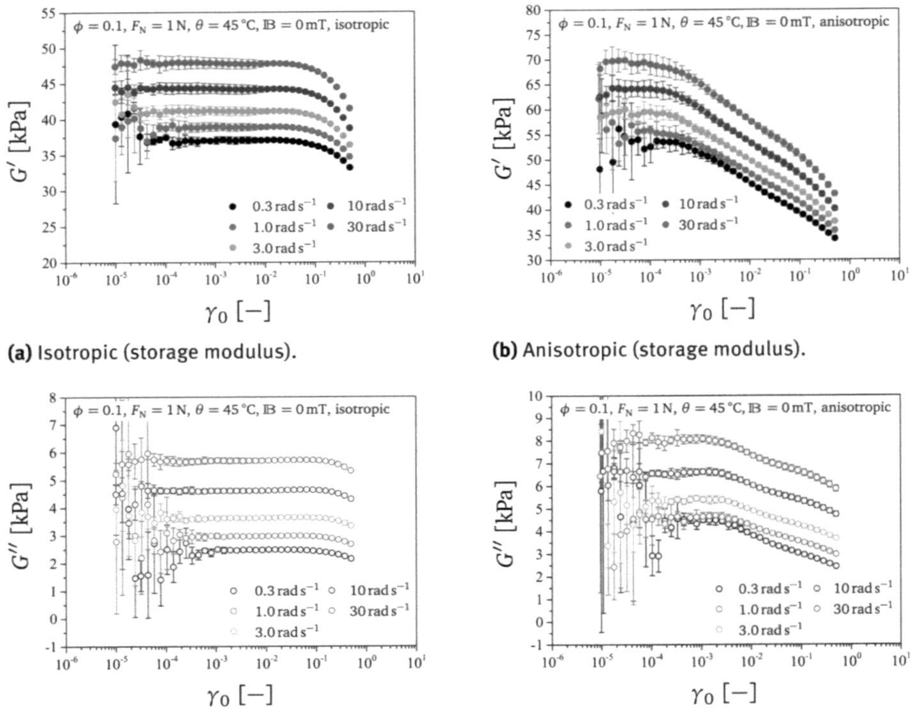

(a) Isotropic (storage modulus).

(b) Anisotropic (storage modulus).

(c) Isotropic (loss modulus).

(d) Anisotropic (loss modulus).

Figure 4.7: Frequency dependence of an MAP with a particle volume fraction of $\phi_{CIP} = 0.1$ under a purely mechanical load. [542]

regime, caution should be exercised when interpreting and comparing the results obtained for large strains; more details on the topic are provided in the section on LAOS in Section 3.1.1.

The results of amplitude sweeps[1] performed for the range $\omega = 0.3\,\mathrm{rad\,s^{-1}}$ to $30\,\mathrm{rad\,s^{-1}}$ (while keeping other experimental and material parameters fixed) will be presented. These bounds for the frequency range were chosen to ensure that the results remain reliable. At low strains and small frequencies, the sensitivity of the apparatus is insufficient to achieve a high measurement fidelity, while at higher frequencies system resonance effects become noticeable. Due to the choice of pure PDMS as the elastomeric matrix, it is to be expected that the material does not exhibit very dissipative effects in the range of tested frequencies (especially when compared to

1 The linear viscoelastic behaviour is typically examined by means of a frequency sweep within the linear regime. However, for this material the required strain amplitude is very small which leads to experiments being conducted outside the reliable resolution of the rheometer. For this reason, we conduct amplitude sweeps at different frequencies to capture some of the rate-dependent characteristics of the MAP.

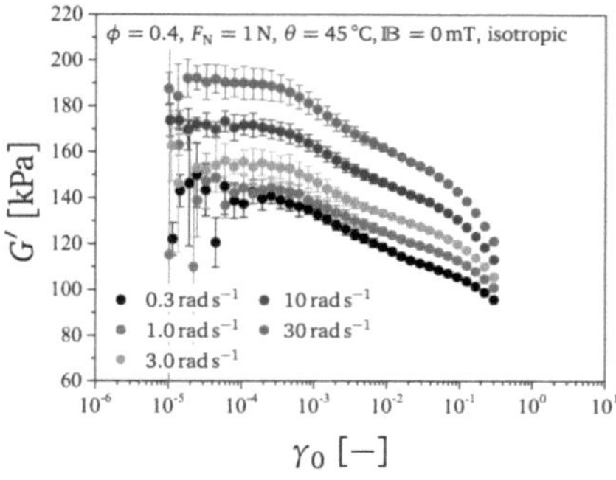

(a) Isotropic (storage modulus).

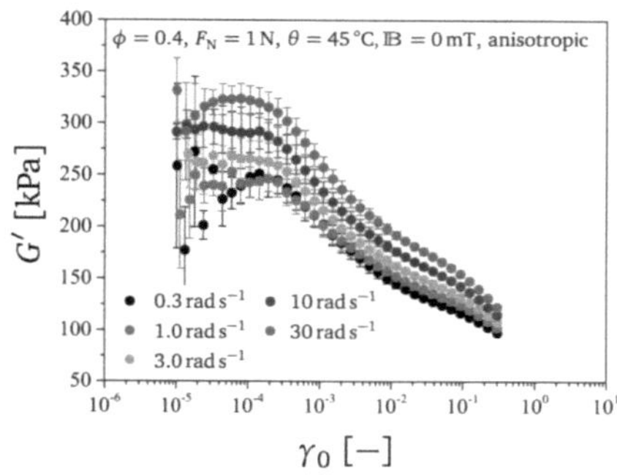

(b) Anisotropic (storage modulus).

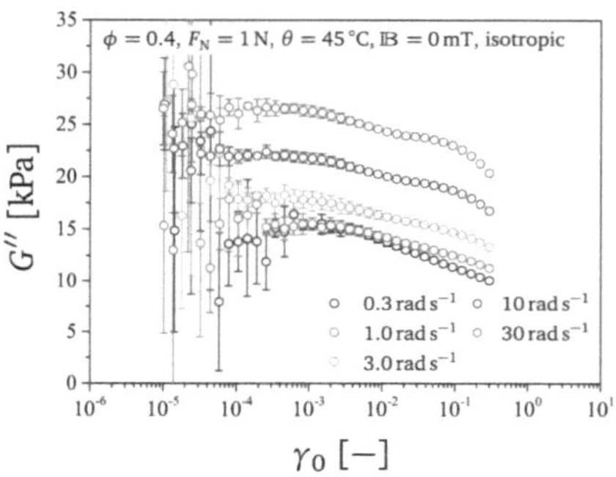

(c) Isotropic (loss modulus).

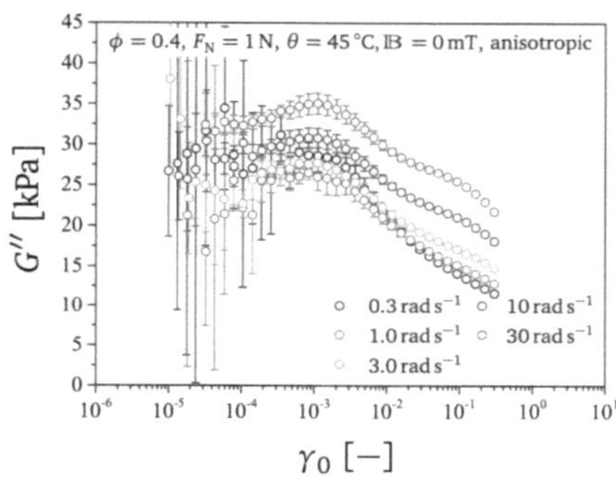

(d) Anisotropic (loss modulus).

Figure 4.8: Frequency dependence of an MAP with a particle volume fraction of $\phi_{CIP} = 0.4$ under a purely mechanical load. [542]

other viscoelastic elastomers). This was predictable from the data presented in Section 4.3.1, from which is discerned that the loss factor $\tan(\delta) < 0.1$, and it will be observed that the ratio of the storage to loss modulus remains of that order of magnitude across the tested frequency spectrum.

4.4.1 Response to mechanical loading

Comparing Figure 4.7 to Figure 4.1a, the influence of a low concentration of CIP when no magnetic field is present can be determined. Within the linear regime, the change in the storage modulus across the tested frequency range for the isotropic MAP is almost the same as that of the pure PDMS, indicating that homogeneously distributed particles play no role in the purely mechanical rate-dependent response of the composite. The strain at which the material transitions from exhibiting linear to non-linear characteristics appears, in a qualitative sense, to be largely unaffected by the strain-rate; this is again linked to the intrinsic deformation of the polymer matrix. The introduction of a chain-like microstructure has an effect even at low concentrations; however, comparing the ratio of the storage modulus in the linear regime at high and low fre-

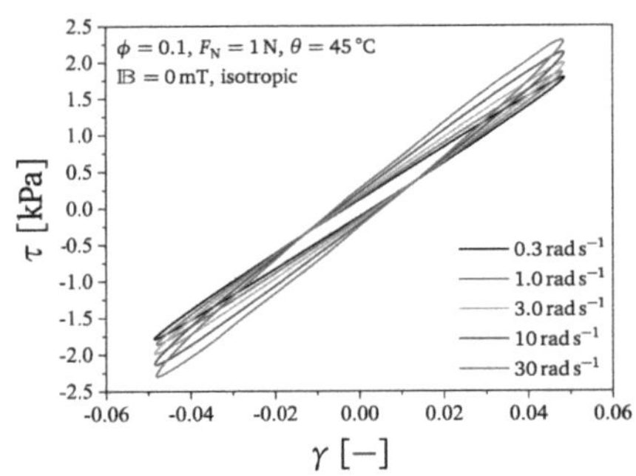

(a) Isotropic ($\phi_{CIP} = 10\,\%$).

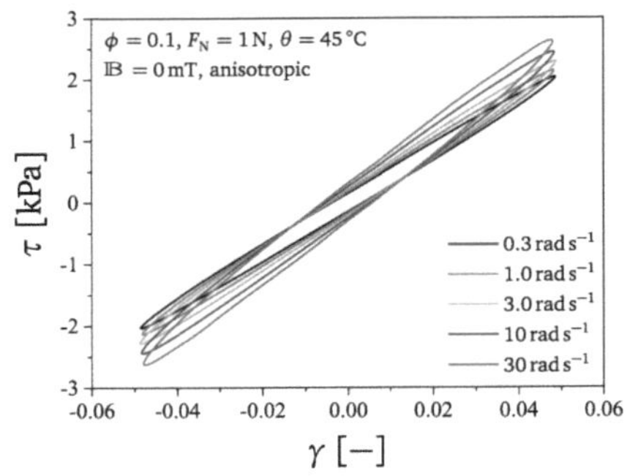

(b) Anisotropic ($\phi_{CIP} = 10\,\%$).

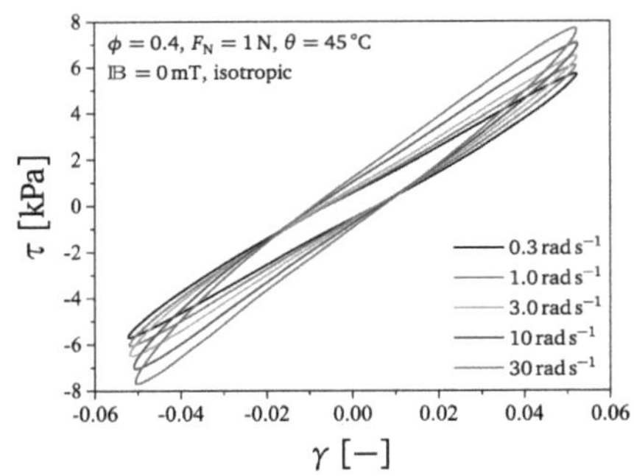

(c) Isotropic ($\phi_{CIP} = 40\,\%$).

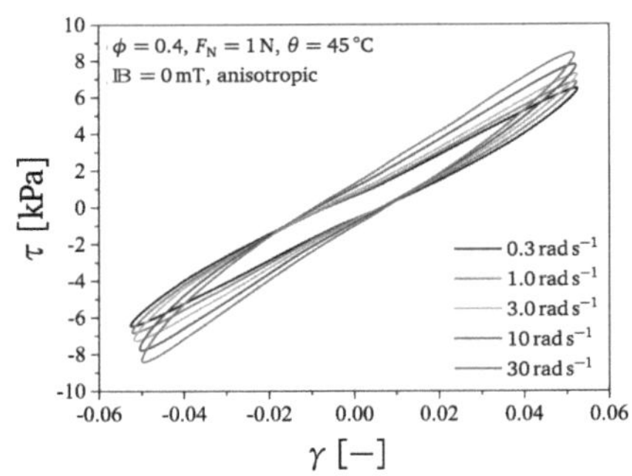

(d) Anisotropic ($\phi_{CIP} = 40\,\%$).

Figure 4.9: Lissajous figures illustrating the dissipative nature of isotropic and anisotropic MAPs under a purely mechanical load of $\gamma_0 \approx 5\%$. The data plotted here is extracted for one load cycle once a stead-state response is exhibited by the material. As the shape and bounded area of the curve are most important when comparing materials with different microstructures, the scales on each plot have been set unequally. [542]

quencies to that measured for the isotropic samples, the frequency dependence of the two materials is similar. As is evident from the SEM images shown in Figure 4.2, the percolation threshold is exceeded when curing under a magnetic field at concentrations as low as $\phi_{CIP} = 0.1$. Given the slight increase in the loss modulus exhibited by the anisotropic material, and not by the isotropic one, the slight (but measurable) change in the frequency-dependent response of the two materials can be attributed to an increase in interparticle interactions in the anisotropic MAP.

Increasing the particle density to $\phi_{CIP} = 0.4$, Figure 4.7 illustrates that the material response for both tested materials is qualitatively similar for all evaluated frequencies. The vertical shift in the response curves indicates that, in addition to the intrinsic viscoelasticity of the matrix, the filler-polymer network is also affected by the strain-rate; this is linked to the particle agglomeration and deagglomeration rate that changes with the applied angular frequency. It is also interesting to note that the increase in

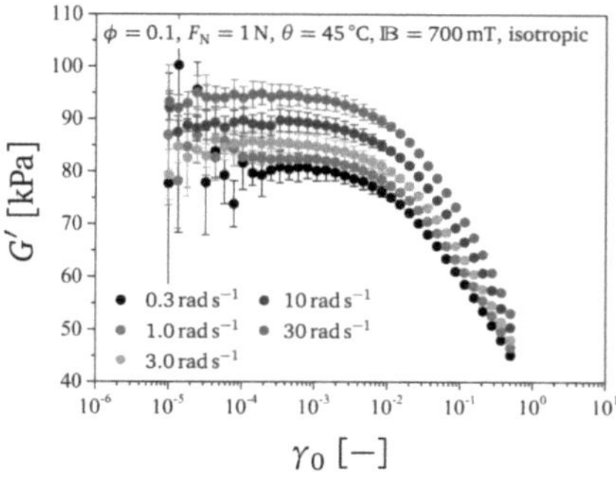

(a) Isotropic (storage modulus).

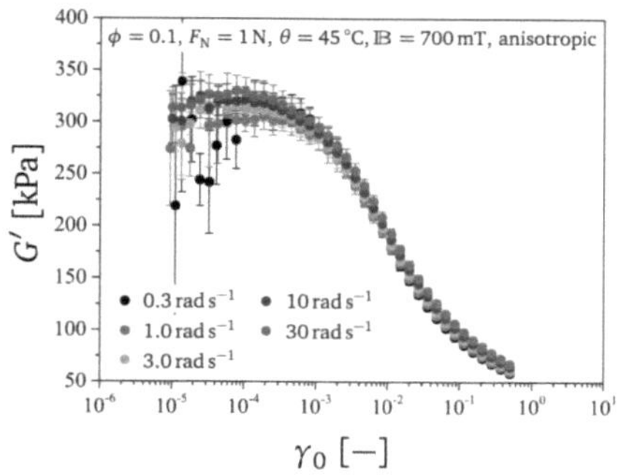

(b) Anisotropic (storage modulus).

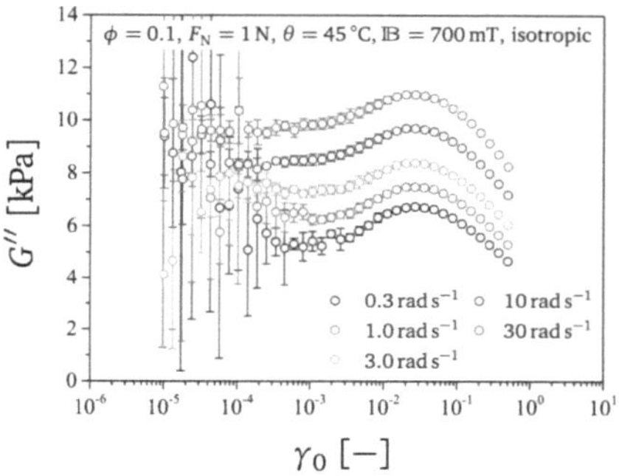

(c) Isotropic (loss modulus).

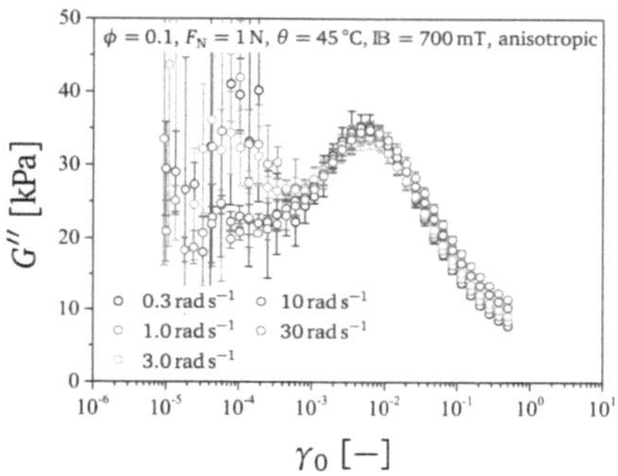

(d) Anisotropic (loss modulus).

Figure 4.10: Frequency dependence of an MAP with a particle volume fraction of $\phi_{CIP} = 0.1$ under a magneto-mechanical load. [542]

the storage modulus recorded for both compositions between low and high frequencies is proportionately similar to that measured for their $\phi_{CIP} = 0.1$ counterparts.

From these results in Figures 4.7 and 4.8, the examined materials appear to demonstrate little qualitative difference in their mechanical behaviour over the applied strain rates. Giving further scrutiny to the response measured over one loading cycle, the Lissajous curves [101] presented in Figure 4.9 convey (through the eccentricity of the loop and its deviation from the mean value through its major axis) the dissipative behaviour of the material. A number of qualities are transparently deduced from these figures. Firstly, the increase in inclination of the loop (i. e. the ratio between the linearised stress and strain) as the oscillation frequency is incremented correlates to an increase in the storage modulus. The non-elliptical nature of the curve clearly infers that the test was performed under LAOS conditions, and that the material is exhibiting a non-linear (viscoelastic) response [141, 224, 307, 142]. This is more pronounced in the anisotropic composite in comparison to the isotropic one, and at higher particle volume fractions. With this visual representation it is, however, not possible to infer whether the rate-dependent component of the response itself is linear or non-linear.

Most significantly though, the area of the bounded region in the Lissajous figure provides insight into, and some quantification of, how dissipative the material is. More specifically, the area of the Lissajous curve is directly related to the energy dissipated within one oscillation period. The small area enclosed within the curve implies that the phase angle δ is very small, and thus that the amount of energy dissipated by internal friction mechanisms is very low. Comparing the response of the different materials with one another, it is deduced that the primary mechanism of internal dissipation results from the viscoelastic matrix. Although the change in microstructure from a isotropic to anisotropic (at a fixed ϕ_{CIP}) has little effect on the dissipation, the addition of further CIP leads to increased particle-particle interactions and, therefore, energy dissipated due to contact.

4.4.2 Response to magneto-mechanical loading

Upon application of a strong magnetic field ($\mathbb{B} = 700\,\text{mT}$), the viscoelastic nature of the polymeric network is in most cases completely eclipsed by the magneto-mechanical effects. From the experimental data shown in Figure 4.10c and Figure 4.11c,

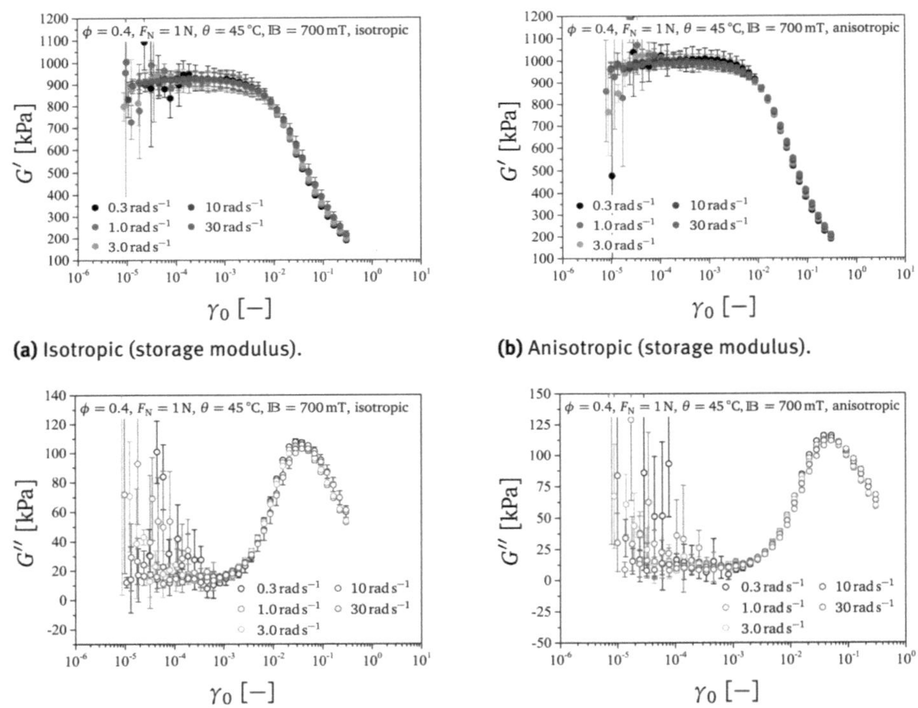

(a) Isotropic (storage modulus).

(b) Anisotropic (storage modulus).

(c) Isotropic (loss modulus).

(d) Anisotropic (loss modulus).

Figure 4.11: Frequency dependence of an MAP with a particle volume fraction of $\phi_{CIP} = 0.4$ under a magneto-mechanical load. [542]

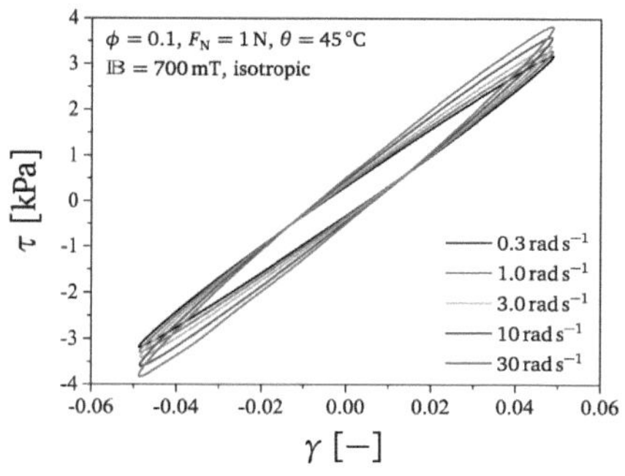

(a) Isotropic (ϕ_{CIP} = 10 %).

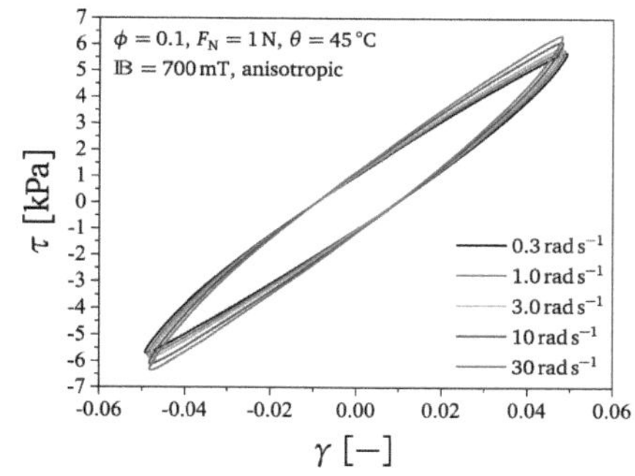

(b) Anisotropic (ϕ_{CIP} = 10 %).

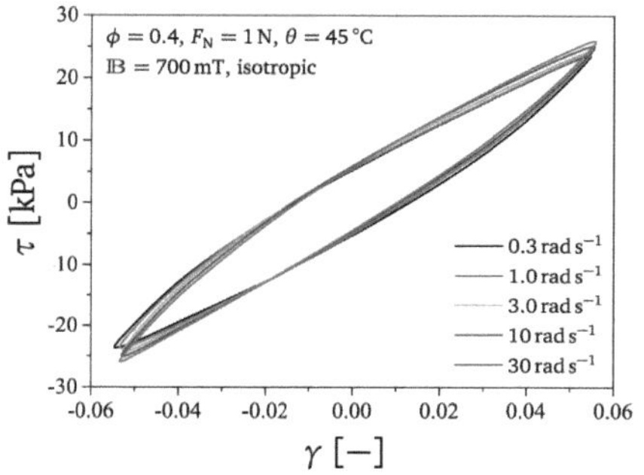

(c) Isotropic (ϕ_{CIP} = 40 %).

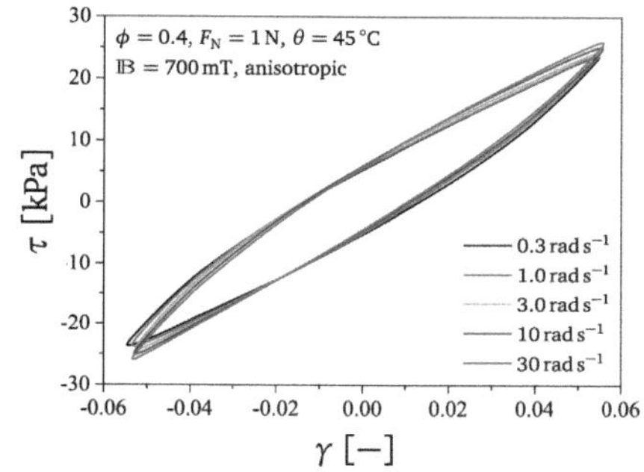

(d) Anisotropic (ϕ_{CIP} = 40 %).

Figure 4.12: Lissajous figures illustrating the dissipative nature of isotropic and anisotropic MAPs under a magneto-mechanical load with a strain amplitude of $\gamma_0 \approx 5\%$ and magnetic field strength of $\mathbb{B} = 700\,\mathrm{mT}$. As the shape and bounded area of the curve are most important when comparing materials with different microstructures, the scales on each plot have been set unequally. [542]

it is observed that only at low volume fractions (and for isotropic compositions) that the matrix viscoelasticity plays a significant role in the strain-rate dependent material response. For anisotropic MAPs, or those that have high concentrations of the particle filler, the magnetic particle-particle interactions dominate its overall behaviour.

Similar can be concluded in terms of the dissipative nature of the various compositions. In Figure 4.12, it is evident that there is little distinction in the energy dissipated over a deformation cycle within the non-linear regime for the majority of the tested materials. It is interesting to note that the Lissajous curve becomes less elliptical as the magneto-mechanical effect increases. This is due to the additional non-linearities that arise due to the strong magnetically-induced particle-particle interactions.

Comparing these results to their counterparts shown in Figure 4.9, it is also concluded that the enhancement of the magneto-mechanical effect is associated with a marginal increase in dissipation. This is likely attributable to the increased interparticle contacts that occur as the magnetisation of the ferrous microstructure increases.

5 Introduction to continuum magneto-mechanics

In this chapter, we will present the fundamental theory necessary to understand the mechanics of magneto-sensitive materials. To provide context for what follows, the reader should note that it is imperative and unavoidable that, in addition to the magnetic field, one must also consider the influence of further non-mechanical fields (such as the electric and thermal fields) in the coupled problem. We by no means aim to provide an exhaustive account of the physics involved, but rather we will showcase the tools and theories related to the understanding of general magnetoelastic media. Appendix C is provided as a companion text to this chapter, wherein we derive in full many of the equations presented here.

For a collectively comprehensive discussion of the theory on electromagnetism from various viewpoints, we refer the reader to the well-known publications of [325, 523, 64, 341, 340, 414, 133, 56, 230, 180, 277], among many other works. This is supplemented by [51, 405, 80, 204, 559] for background theory to elastic and viscoelastic deformation, and [504, 531, 125, 530, 417] when considering both coupled magnetoelastic and electroelastic materials in a static setting.

5.1 Continuum setting

5.1.1 Continuum domain

With reference to Figure 5.1, we consider the general case of a deformable solid in the reference configuration $\mathscr{B}_0$ immersed in the surrounding free space $\mathscr{S}_0$. We assume that the reference configuration is that in which the body is at rest, and is in a state of zero strain and stress. The closure [444] of the reference configuration is denoted by $\overline{\mathscr{B}_0} = \mathscr{B}_0 \cup \overline{\partial\mathscr{B}_0}$, and that of the free space by $\overline{\mathscr{S}_0} = \mathscr{S}_0 \cup \overline{\partial\mathscr{B}_0} \cup \overline{\partial\mathscr{S}_0}$. Here, $\overline{\partial\mathscr{B}_0} = \partial\mathscr{B}_0 \cup \overline{\partial\partial\mathscr{B}_0}$ is the closure of the surface that in itself has edges potentially with distinct end-points (and so with closure $\overline{\partial\partial\mathscr{B}_0} = \partial\partial\mathscr{B}_0 \cup \partial\partial\partial\mathscr{B}_0$).

Application of a load (be it mechanical, magnetic or a combination thereof) results in the body deforming to $\overline{\mathscr{B}_t}$ and the free space to $\overline{\mathscr{S}_t}$. We define a non-linear deformation function

$$\mathbf{x} = \boldsymbol{\varphi}(\mathbf{X}, t) = \mathbf{X} + \mathbf{u}(\mathbf{X}, t) \tag{5.1}$$

that maps points $\mathbf{X} \in \overline{\mathscr{B}_0 \cup \mathscr{S}_0}$ to $\mathbf{x} \in \overline{\mathscr{B}_t \cup \mathscr{S}_t}$ through the displacement vector $\mathbf{u}(\mathbf{X}, t)$ described at any time t. However, it is assumed that the far field boundary of the (truncated) domain for the free space always remains fixed ($\partial\mathscr{S}_0 = \partial\mathscr{S}_t$).

https://doi.org/10.1515/9783110418576-005

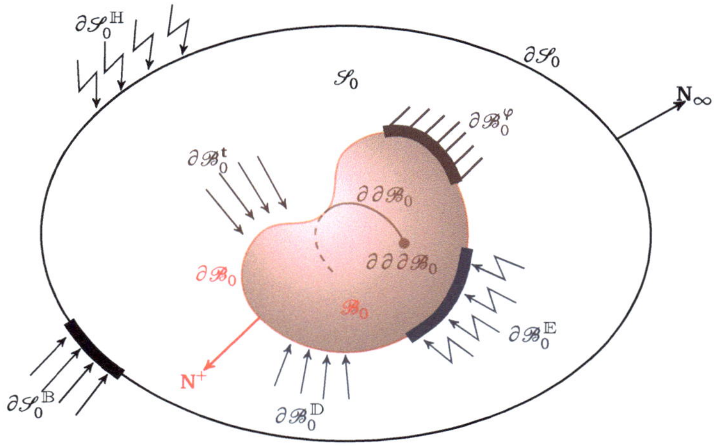

Figure 5.1: Definition of the reference domain, with illustration of the deformable body $\mathscr{B}_0$ immersed in the magnetically and electrically "permeable" free space $\mathscr{S}_0$. The subsets of the boundary to which Dirichlet constraints are applied in the primal formulation are indicated by a thick line.

Lastly, for this very general setting we assume that $\partial\mathscr{B}_0$ and $\partial\mathscr{S}_0$ admit the decomposition

$$\partial\mathscr{B}_0 = \partial\mathscr{B}_0^{\boldsymbol{\varphi}} \cup \partial\mathscr{B}_0^{\mathbf{t}} \quad \text{with} \quad \partial\mathscr{B}_0^{\boldsymbol{\varphi}} \cap \partial\mathscr{B}_0^{\mathbf{t}} = \emptyset \quad , \tag{5.2a}$$

$$= \partial\mathscr{B}_0^{\mathbb{E}} \cup \partial\mathscr{B}_0^{\mathbb{D}} \quad \text{with} \quad \partial\mathscr{B}_0^{\mathbb{E}} \cap \partial\mathscr{B}_0^{\mathbb{D}} = \emptyset \quad \text{and} \tag{5.2b}$$

$$\partial\mathscr{S}_0 = \partial\mathscr{S}_0^{\mathbb{H}} \cup \partial\mathscr{S}_0^{\mathbb{B}} \quad \text{with} \quad \partial\mathscr{S}_0^{\mathbb{H}} \cap \partial\mathscr{S}_0^{\mathbb{B}} = \emptyset \quad . \tag{5.2c}$$

Here, $\partial\mathscr{B}_0^{\boldsymbol{\varphi}}$ and $\partial\mathscr{B}_0^{\mathbf{t}}$ denote the portions of $\partial\mathscr{B}_0$ with either a prescribed deformation or traction, respectively. Likewise, the areas of the body surface that have either a prescribed electric field or electric displacement are respectively represented by $\partial\mathscr{B}_0^{\mathbb{E}}$ and $\partial\mathscr{B}_0^{\mathbb{D}}$. Lastly, the boundary of the (truncated) far field domain is decomposed into two non-overlapping regions where on $\partial\mathscr{S}_0^{\mathbb{H}}$ and $\partial\mathscr{S}_0^{\mathbb{B}}$ the magnetic field and magnetic induction are respectively prescribed. We therefore consider scenarios in which the source of the magnetic fields is either a permanent magnet or electromagnet that is encapsulated within the region of interest, as well as those in which the field is generated outside of $\mathscr{B}_0 \cup \mathscr{S}_0$.

5.1.2 Kinematics

We define two differential operators; the first with respect to $\mathbf{X}$ being represented as $\nabla_0(\bullet) = \frac{\partial(\bullet)}{\partial\mathbf{X}}$, and the second with respect to $\mathbf{x}$ being $\nabla(\bullet) = \frac{\partial(\bullet)}{\partial\mathbf{x}}$. The deformation gradient tensor is then given by

$$\mathbf{F} = \nabla_0\boldsymbol{\varphi}(\mathbf{X}, t) \tag{5.3}$$

and, following this, the volumetric Jacobian is expressed as

$$J = \det \mathbf{F} \tag{5.4}$$

with $J > 0$ necessary to ensure that $\boldsymbol{\varphi}$ remains a one-to-one map that prevents self-penetration of the material. The right Cauchy–Green deformation tensor and Green–Lagrange strain tensor are

$$\mathbf{C} = \mathbf{F}^T \cdot \mathbf{F} \quad \text{and} \tag{5.5}$$

$$\mathbf{E} = \frac{1}{2} [\mathbf{C} - \mathbf{I}] \quad , \tag{5.6}$$

respectively. Note also that the differential operators for taking the gradient or divergence of a vector field in the two configurations are related by

$$\nabla (\bullet) = \nabla_0 (\bullet) \cdot \mathbf{F}^{-1} \quad , \quad \nabla \cdot (\bullet) = \nabla (\bullet) : \mathbf{i} = \nabla_0 (\bullet) : \mathbf{F}^{-T} \quad . \tag{5.7}$$

5.1.3 Line, area and volume transformations

Infinitesimal line elements are mapped from the reference to the current configuration through the action of the deformation gradient tensor, namely by

$$\mathrm{d}\mathbf{x} = \mathbf{F} \cdot \mathrm{d}\mathbf{X} \quad . \tag{5.8}$$

Through the scalar triple product (equation (A.12)), the arbitrary referential volume element dV, a parallelepiped, can be decomposed as

$$\mathrm{dV} = \mathrm{d}\mathbf{A} \cdot \mathrm{d}\mathbf{L} \tag{5.9}$$

with the cross-section vector $\mathrm{d}\mathbf{A} = \mathrm{d}\mathbf{X}_1 \times \mathrm{d}\mathbf{X}_2$ and the length vector $\mathrm{d}\mathbf{L} = \mathrm{d}\mathbf{X}_3$. From this decomposition, one can derive Nanson's formula for the mapping of vectorial area elements,

$$\mathrm{d}\mathbf{a} = \operatorname{cof}(\mathbf{F}) \cdot \mathrm{d}\mathbf{A} \quad , \tag{5.10}$$

where the definition for the cofactor of the deformation gradient is $\operatorname{cof}(\mathbf{F}) := \det(\mathbf{F})\, \mathbf{F}^{-T}$. The corresponding transformation of volume elements reads

$$\mathrm{dv} = J\, \mathrm{dV} \quad . \tag{5.11}$$

5.1.4 Time derivatives

The material time derivative of a referential field (i. e. one that is parametrised in the reference co-ordinates) is

$$\mathrm{D}_t(\bullet)_0 = \partial_t(\bullet)_0|_{\mathbf{X}} \equiv \dot{(\bullet)}_0 \quad , \tag{5.12}$$

where we have abbreviated the partial time derivative as $\partial_t(\bullet) = \frac{\partial(\bullet)}{\partial t}$. From this, we can define the material velocity and acceleration vectors as

$$\mathrm{D}_t\boldsymbol{\varphi}(\mathbf{X},t) = \dot{\boldsymbol{\varphi}}(\mathbf{X},t) \equiv \mathrm{D}_t\mathbf{u}(\mathbf{X},t) = \mathbf{v}(\mathbf{X},t) \quad , \tag{5.13a}$$

$$\mathrm{D}_t\dot{\boldsymbol{\varphi}}(\mathbf{X},t) = \ddot{\boldsymbol{\varphi}}(\mathbf{X},t) \equiv \mathrm{D}_t\mathbf{v}(\mathbf{X},t) = \mathbf{a}(\mathbf{X},t) \quad . \tag{5.13b}$$

Similarly, the spatial time derivative of a spatial field (i. e. one that is parametrised in the spatial co-ordinates) is described in general as

$$\mathrm{d}_t(\bullet)_t = \partial_t(\bullet)_t|_{\mathbf{x}} \quad . \tag{5.14}$$

The material time derivative of a field with a spatial parametrisation $(\bullet)_t$ is therefore

$$\begin{aligned}\mathrm{D}_t(\bullet)_t &= \partial_t(\bullet)_t|_{\mathbf{x}} + \nabla(\bullet)_t|_t \cdot \partial_t\boldsymbol{\varphi}|_{\mathbf{X}} \\ &= \mathrm{d}_t(\bullet) + \nabla(\bullet)\cdot\mathbf{v} \quad .\end{aligned} \tag{5.15}$$

In order to avoid any ambiguity in their meaning, from here on we will express all time derivatives in terms of the shorthand operators $\dot{(\bullet)} \equiv \mathrm{D}_t(\bullet)$ and $\mathrm{d}_t(\bullet)$. For convenience, we define synonyms for the velocity and acceleration vectors

$$\mathbf{v}(\mathbf{X},t) := \dot{\mathbf{u}} \quad , \quad \mathbf{a}(\mathbf{X},t) := \ddot{\mathbf{u}} \tag{5.16}$$

as well as the (contravariant) pull-back of the velocity vector

$$\mathbf{V}(\mathbf{X},t) := -\mathbf{F}^{-1}\cdot\mathbf{v} \quad ; \tag{5.17}$$

this aligns with the velocity on the material manifold in the context of configurational mechanics [337].

We will see later that, in addition to the standard Lagrangian and Eulerian configurations, there are two additional intermediate settings (namely the "updated Lagrangian" and "updated Eulerian" configurations) that are meaningful in terms of deriving and describing the balance laws for electro-magneto-mechanics; we label the time derivatives taken in these two configurations *nominal* time derivatives. As motivated in Appendix B.2.1, we define the material nominal time derivatives for scalar and vector (density) functions in a body as

$$\mathrm{N}_t\Phi_0 := \mathrm{D}_t\Phi_0 + \nabla_0\cdot[\Phi_0\mathbf{V}] = J\mathrm{d}_t\left[J^{-1}\Phi_0\right] \quad \text{and} \tag{5.18a}$$

$$\mathrm{N}_t\boldsymbol{\Phi}_0 := \mathrm{D}_t\boldsymbol{\Phi}_0 + \nabla_0\cdot[\boldsymbol{\Phi}_0\otimes\mathbf{V}] = J\mathrm{d}_t\left[J^{-1}\boldsymbol{\Phi}_0\right] \quad , \tag{5.18b}$$

and the corresponding spatial nominal time derivatives are defined as

$$\mathrm{n}_t\Phi_t := \mathrm{d}_t\Phi_t + \nabla \cdot [\Phi_t \mathbf{v}] = J^{-1}\mathrm{D}_t[J\Phi_t] \quad \text{and} \tag{5.19a}$$

$$\mathrm{n}_t\boldsymbol{\Phi}_t := \mathrm{d}_t\boldsymbol{\Phi}_t + \nabla \cdot [\boldsymbol{\Phi}_t \otimes \mathbf{v}] = J^{-1}\mathrm{D}_t[J\boldsymbol{\Phi}_t] \quad . \tag{5.19b}$$

We also define the material and spatial nominal time derivatives for vector (flux) functions on the surface of a body as

$$\mathrm{N}_t\boldsymbol{\Psi} := \mathrm{D}_t\boldsymbol{\Psi} + \nabla_0 \times [\boldsymbol{\Psi} \times \mathbf{V}] + [\nabla_0 \cdot \boldsymbol{\Psi}]\,\mathbf{V} = \mathrm{d}_t\left[\boldsymbol{\Psi} \cdot \mathrm{cof}\left(\mathbf{F}^{-1}\right)\right] \cdot \mathrm{cof}\,(\mathbf{F}) \quad \text{and} \tag{5.20a}$$

$$\mathrm{n}_t\boldsymbol{\psi} := \mathrm{d}_t\boldsymbol{\psi} + \nabla \times [\boldsymbol{\psi} \times \mathbf{v}] + [\nabla \cdot \boldsymbol{\psi}]\,\mathbf{v} = \mathrm{D}_t[\boldsymbol{\psi} \cdot \mathrm{cof}\,(\mathbf{F})] \cdot \mathrm{cof}\left(\mathbf{F}^{-1}\right) \quad . \tag{5.20b}$$

When the densities, and flux vectors have the association

$$\Phi_t = J^{-1}\Phi_0 \quad , \tag{5.21a}$$

$$\boldsymbol{\Phi}_t = J^{-1}\boldsymbol{\Phi}_0 \quad \text{and} \tag{5.21b}$$

$$\boldsymbol{\psi} = \boldsymbol{\Psi} \cdot \mathrm{cof}\left(\mathbf{F}^{-1}\right) = J^{-1}\mathbf{F} \cdot \boldsymbol{\Psi} \tag{5.21c}$$

then the various time derivatives are thus related to one another in the following way:

$$\int_{\mathscr{V}_0} \mathrm{D}_t\Phi_0 \,\mathrm{dV} = \int_{\mathscr{V}_t} \mathrm{n}_t\Phi_t \,\mathrm{dv} \quad , \quad \int_{\mathscr{V}_0} \mathrm{N}_t\Phi_0 \,\mathrm{dV} = \int_{\mathscr{V}_t} \mathrm{d}_t\Phi_t \,\mathrm{dv} \quad , \tag{5.22a}$$

$$\int_{\mathscr{V}_0} \mathrm{D}_t\boldsymbol{\Phi}_0 \,\mathrm{dV} = \int_{\mathscr{V}_t} \mathrm{n}_t\boldsymbol{\Phi}_t \,\mathrm{dv} \quad , \quad \int_{\mathscr{V}_0} \mathrm{N}_t\boldsymbol{\Phi}_0 \,\mathrm{dV} = \int_{\mathscr{V}_t} \mathrm{d}_t\boldsymbol{\Phi}_t \,\mathrm{dv} \quad , \tag{5.22b}$$

$$\int_{\mathscr{A}_0} [\mathrm{D}_t\boldsymbol{\Psi}] \cdot \mathbf{N} \,\mathrm{dA} = \int_{\mathscr{A}_t} [\mathrm{n}_t\boldsymbol{\psi}] \cdot \mathbf{n} \,\mathrm{da} \quad , \quad \int_{\mathscr{A}_0} [\mathrm{N}_t\boldsymbol{\Psi}] \cdot \mathbf{N} \,\mathrm{dA} = \int_{\mathscr{A}_t} [\mathrm{d}_t\boldsymbol{\psi}] \cdot \mathbf{n} \,\mathrm{da} \quad . \tag{5.22c}$$

5.1.5 Fundamentals of electromagnetism

The basis of classical electromagnetism, and the interactions of matter with electric and magnetic fields, was provided through the experimental observations of Coulomb [97, 98] and Faraday (as documented in [344, chapter 8]), and later unified and generalised to moving bodies by Maxwell [342, 343] and advanced further by Lorentz [326]. The covariant formulation of electromagnetic theory in a relativistic setting was subsequently postulated by Einstein [131] and developed by Planck [427], Minkowski [381], and others for application in thermodynamics and electromagnetism.

In the following section, we briefly outline the derivation of the governing laws for both electric and magnetic fields in a quasi-static setting, and in addition the electromotive (Lorentz) force, as motivated from a non-relativistic microscopic perspective in a Cartesian coordinate system in flat Euclidean space. This follows the condensed description of the fundamental principles described by Truesdell and Toupin [523], Pao

[414], Eringen and Maugin [133], Jackson [230], Ogden [406], and Dorfmann and Ogden [125]. As a point of comparison to that described below, we note that Kovetz [277] (and also Truesdell and Toupin [523], Eringen and Maugin [133], and Jackson [230]) derive the electromagnetic balance laws from first principles considering a *Galilean reference frame* (one in which the laws of motion are *Galilean invariant*, that is to say that they are identical in all inertial frames). This contrasts the description of the Maxwell's equations formulated in the canonical manuscript by Lorentz [325], wherein the existence of a special stationary reference frame known as the absolute or *aether reference frame* is postulated and exploited. For further contextualisation of the aether relations for Maxwell's equations, we refer to [523, 230, 277] once again, and for a discussion on why the existence of an *aether frame* contradicts the principle of *Galilean invariance* we refer to the arguments stated by Blagojevic [47]. Furthermore, both Pao [414] and Eringen and Maugin [133] demonstrate how Maxwell's macroscopic field equations can be derived in flat Euclidean space from the atomic field equations. A collection of models for field-matter interactions have been examined in [414], each of which leads to slightly different expressions for the electromagnetic force, torque, energy, momentum density and Maxwell stress. However, [133] strictly consider the use of statistical averaging techniques to motivate the form of the macroscopic laws directly from Lorentz's electron theory.

Electrostatics: Coulomb's law for charged particles

In his experiments on electrically charged particles, Coulomb [97, 98] proposed the relationship between distinct, stationary charges and the attractive or repulsive force between them. With reference to Figure 5.2, depicting the configuration of non-overlapping, stationary point charges i and j, the Lorentz force acting on the electric charge i due to the presence of charge j was determined to be

$$\mathbf{f}_{ij} = \frac{1}{4\pi\varepsilon_0} \frac{q_i q_j}{|\mathbf{r}_{ji}|^2} \frac{\mathbf{r}_{ji}}{|\mathbf{r}_{ji}|} \quad \text{(no summation convention)} \quad , \tag{5.23}$$

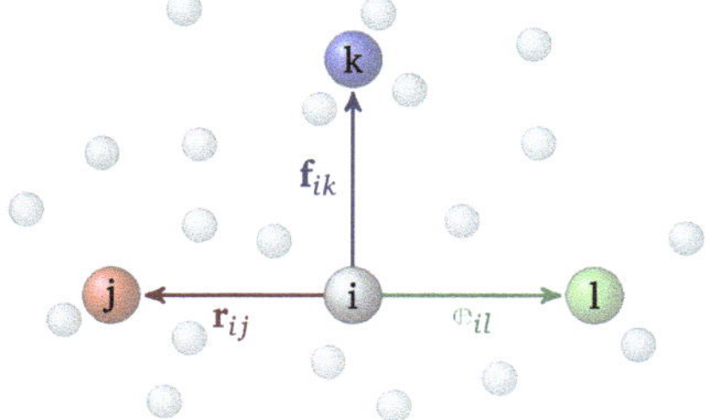

Figure 5.2: The interactions of charged particles lead to the derivation of Coulomb's law. The presence of neighbouring charged particles j, k, l results in an electric field in the vicinity of particle i and a resultant electromotive force attracting or repelling.

where $\mathbf{r}_{ij} = \mathbf{x}_j - \mathbf{x}_i$ (pointing from charge i to charge j) is the distance vector between the two charges. The charges q were found to be an integer multiple of the unit electron charge $e = -1.602 \times 10^{-19}$ C, and $\varepsilon_0 = 8.854 \times 10^{-12}$ F m^{-1} denotes the electric permittivity constant of free space. The direction vector $\frac{\mathbf{r}_{ji}}{|\mathbf{r}_{ji}|}$ was chosen such that particles of similar charge (both positive or negative) repel one another. The electric field present at the location of charge i due to the presence of charge j is defined as the force exerted on charge i normalised by its charge, that is,

$$\mathbf{f}_{ij} = q_i \mathbb{e}_{ij} \quad \text{with} \quad \mathbb{e}_{ij} := \frac{1}{4\pi\varepsilon_0} \frac{q_j}{|\mathbf{r}_{ji}|^2} \frac{\mathbf{r}_{ji}}{|\mathbf{r}_{ji}|} \quad . \tag{5.24}$$

Using the principle of superposition, the resulting force as well as the electric field present at charge i due to all surrounding charges j are

$$\mathbf{f}_i = \sum_{j=1}^{n_q} \mathbf{f}_{ij} = q_i \mathbb{e}_i \quad \text{with} \quad \mathbb{e}_i := \sum_{j=1}^{n_q} \mathbb{e}_{ij} = \frac{1}{4\pi\varepsilon_0} \sum_{j=1}^{n_q} \frac{q_j}{|\mathbf{r}_{ji}|^2} \frac{\mathbf{r}_{ji}}{|\mathbf{r}_{ji}|} \quad . \tag{5.25}$$

If it is then assumed that the collection of charges form a continuous distribution, then the electric field at position $\mathbf{x}$ within a volume of interest $\mathscr{D}_t$ may be expressed as [414, 133, 230]

$$\mathbb{e}(\mathbf{x}) = \lim_{\substack{n_q \to \infty \\ q_j \to \mathrm{d}q}} \mathbb{e}_i = \frac{1}{4\pi\varepsilon_0} \int_{\mathscr{D}_t} \frac{\varrho_t(\mathbf{x}')}{|\mathbf{x}-\mathbf{x}'|^2} \frac{\mathbf{x}-\mathbf{x}'}{|\mathbf{x}-\mathbf{x}'|} \,\mathrm{dv} \quad \text{with} \quad \mathrm{d}q(\mathbf{x}') = \varrho_t(\mathbf{x}')\,\mathrm{dv} \tag{5.26}$$

and where $\varrho_t(\mathbf{x}')$ represents the charge density per unit volume at position $\mathbf{x}'$ away from $\mathbf{x}$. The force on a stationary test point charge q due to its immersion in a continuous electric field is therefore [414, 277]

$$\mathbf{f}_{\mathbb{e}}(\mathbf{x}) = \lim_{\substack{n_q \to \infty \\ q_j \to \mathrm{d}q \\ q_i \to q}} \mathbf{f}_i = q\,\mathbb{e}(\mathbf{x}) \quad ; \tag{5.27}$$

this quantity represents the electric part of the Lorentz force acting on a test charge.

Also from equation (5.26), using the microscopic divergence identity stated in equation (B.4), Gauss' (in German, Gauß) flux theorem for electric fields

$$\nabla \cdot \mathbb{e}(\mathbf{x}) = \frac{\varrho_t(\mathbf{x})}{\varepsilon_0} \quad , \tag{5.28}$$

can be derived. With consideration of the identity given in equation (B.3), the existence of an electric potential field [325, 230]

$$\Phi(\mathbf{x}) = \Phi^{\text{ext}} + \frac{1}{4\pi\varepsilon_0} \int_{\mathscr{D}_t} \frac{\varrho_t(\mathbf{x}')}{|\mathbf{x}-\mathbf{x}'|} \,\mathrm{dv} \tag{5.29}$$

is proposed. The second term defines the potential field arising due to the charge distribution within the domain of interest, while the first onto which the second is superimposed encompasses that of a fixed field present due to external influences that exist outside of $\mathscr{D}_t$ (such as a second system of charges). Equation (5.29) is then related to the electric field by

$$\mathbb{e}(\mathbf{x}) = -\nabla\Phi(\mathbf{x}) \quad \text{with} \quad \nabla \times \mathbb{e}(\mathbf{x}) = \mathbf{0} \tag{5.30}$$

from equation (B.6).

Magnetostatics: Ampére's law for electric circuits

Faraday's experiments with electric circuits lead to the understanding of the interaction between magnetic fields and electric circuits, and the electromotive force that arises between them. Maxwell [344] used this as the basis for his derivation of the Maxwell–Ampére circuital law. In Figure 5.3, we consider a collection of neighbouring electric circuits as formed by stationary, infinitesimally thin charge carrying wires. Associated with each moving charge q in a circuit is an elementary electric current $\mathrm{d}\mathbb{i} = q\mathbb{v}$ [133, 230], where $\mathbb{v}$ defines the velocity of the charge. The force acting on the electric circuit i due to the presence of circuit j is [230]

$$\mathbf{f}_{ij} = -\frac{\mu_0}{4\pi} \oint_{\mathscr{C}_i} \oint_{\mathscr{C}_j} \frac{\mathrm{d}\mathbb{i}_i \cdot \mathrm{d}\mathbb{i}_j}{|\mathbf{r}_{ji}|^2} \frac{\mathbf{r}_{ji}}{|\mathbf{r}_{ji}|} \quad , \tag{5.31}$$

where $\mathbf{r}_{ij} = \mathbf{x}_j - \mathbf{x}_i$ is the distance vector between the two differential current elements $\mathrm{d}\mathbb{i}_i$ and $\mathrm{d}\mathbb{i}_j$, and the free space magnetic permeability is $\mu_0 = 4\pi \times 10^{-7}\,\mathrm{H\,m^{-1}}$. The sign of this equation originates from the experimental observation that two parallel wires with currents flowing in the same direction attract one another; this is the essence of Lenz's law. Note the resemblance between the structure of equation (5.31) and equation (5.23). Ampére's law expressed in equation (5.31) can, as shown in equation (C.4),

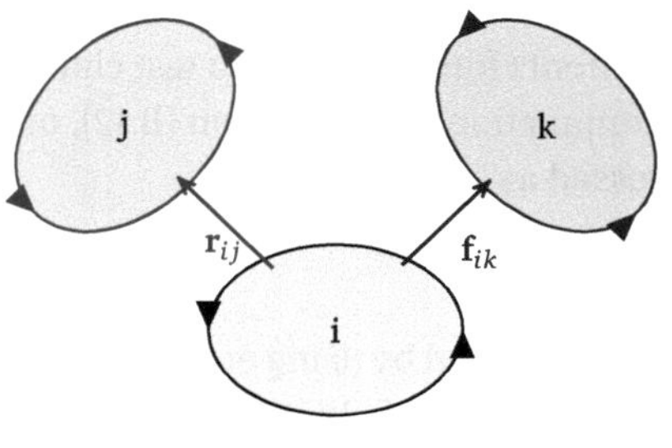

Figure 5.3: The interactions of electric circuits leads to the derivation of Ampére's law. The flow of currents in neighbouring circuits j, k leads to an induced magnetic field in circuit i and a resultant electromotive force on it.

be reformulated as

$$\mathbf{f}_{ij} = \oint_{\mathscr{C}_i} \mathrm{d}\mathbb{i}_i \times \frac{\mu_0}{4\pi} \oint_{\mathscr{C}_j} \frac{\mathrm{d}\mathbb{i}_j}{|\mathbf{r}_{ji}|^2} \times \frac{\mathbf{r}_{ji}}{|\mathbf{r}_{ji}|} \quad . \tag{5.32}$$

The magnetic induction that arises at i due to the influence of the current element at j is defined through the interaction force

$$\mathbf{f}_{ij} = \oint_{\mathscr{C}_i} \mathrm{d}\mathbb{i}_i \times \mathbb{b}_{ij} \quad \text{as} \quad \mathbb{b}_{ij} := \frac{\mu_0}{4\pi} \oint_{\mathscr{C}_j} \frac{\mathrm{d}\mathbb{i}_j}{|\mathbf{r}_{ji}|^2} \times \frac{\mathbf{r}_{ji}}{|\mathbf{r}_{ji}|} \quad . \tag{5.33}$$

Again, adopting the superposition principle we can express the magnetic induction resulting at i due to the collection of neighbouring current elements j through the resulting force

$$\mathbf{f}_i = \sum_{j=1}^{n_{\mathscr{C}}} \mathbf{f}_{ij} = \oint_{\mathscr{C}_i} \mathrm{d}\mathbb{i}_i \times \mathbb{b}_i \quad \text{with} \quad \mathbb{b}_i := \sum_{j=1}^{n_{\mathscr{C}}} \mathbb{b}_{ij} = \frac{\mu_0}{4\pi} \sum_{j=1}^{n_{\mathscr{C}}} \oint_{\mathscr{C}_i} \frac{\mathrm{d}\mathbb{i}_j}{|\mathbf{r}_{ji}|^2} \times \frac{\mathbf{r}_{ji}}{|\mathbf{r}_{ji}|} \quad . \tag{5.34}$$

The magnetic induction at point $\mathbf{x}$ due to a continuous distribution of currents is then encapsulated in the Biot–Savart law [414]

$$\mathbb{b}(\mathbf{x}) = \lim_{\substack{n_{\mathscr{C}_j} \to \infty \\ \mathrm{d}\mathbb{i}_j \to \mathrm{d}\mathbb{i} \\ |\mathscr{C}_j| \to 0}} \mathbb{b}_i = \frac{\mu_0}{4\pi} \int_{\mathscr{D}_t} \frac{\mathbb{j}(\mathbf{x}')}{|\mathbf{x}-\mathbf{x}'|^2} \times \frac{\mathbf{x}-\mathbf{x}'}{|\mathbf{x}-\mathbf{x}'|} \, \mathrm{dv} \quad \text{with} \quad \mathrm{d}\mathbb{i}(\mathbf{x}') = \mathbb{j}(\mathbf{x}') \, \mathrm{dv} \quad , \tag{5.35}$$

and where $\mathbb{j}(\mathbf{x}')$ represents the electric current density per unit volume at position $\mathbf{x}'$ away from $\mathbf{x}$. The force on a moving test charge due to the magnetic induction is therefore

$$\mathbf{f}_{\mathbb{b}}(\mathbf{x}) = \lim_{\substack{n_{\mathscr{C}_j} \to \infty \\ \mathrm{d}\mathbb{i}_j \to \mathrm{d}\mathbb{i} \\ \mathrm{d}\mathbb{i}_i \to q\mathbb{v} \\ |\mathscr{C}_{i,j}| \to 0}} \mathbf{f}_i = q\mathbb{v}(\mathbf{x}) \times \mathbb{b}(\mathbf{x}) \quad ; \tag{5.36}$$

this quantity represents the magnetic part of the Lorentz force acting on a test charge.

From the identity given in equation (B.6) in conjunction with equation (B.12), one can deduce Gauss' magnetism law which is expressed as

$$\nabla \cdot \mathbb{b}(\mathbf{x}) = 0 \quad . \tag{5.37}$$

Furthermore, due to the identity stated in equation (A.25) and by using equation (B.3), one can postulate the existence of a magnetic vector potential field [325, 414]

$$\mathbb{a}(\mathbf{x}) = \mathbb{a}^{\text{ext}} + \frac{\mu_0}{4\pi} \int_{\mathscr{D}_t} \frac{\mathbb{j}(\mathbf{x}')}{|\mathbf{x}-\mathbf{x}'|} \, \mathrm{dv} \tag{5.38}$$

which is then related to the magnetic induction by

$$\mathbb{b}(\mathbf{x}) = \nabla \times \mathbb{a}(\mathbf{x}) \quad . \tag{5.39}$$

The magnetic vector potential field may derive as a superposition of an externally generated fixed field (for instance, due to another set of current circuits) and that arising due to the action of circuits, internal to the domain of interest, on one another.

Lorentz force

From equations (5.27) and (5.36), the resultant electromotive force acting on the system, collectively due to the presence of charges and flow of currents through existing fields, is therefore

$$\mathbf{f}_t^{\text{pon}} = \int_{\mathscr{D}_t} \left[\varrho_t(\mathbf{x})\, \mathbb{e}(\mathbf{x}) + \mathbb{j}(\mathbf{x}) \times \mathbb{b}(\mathbf{x})\right] \, \mathrm{dv} \quad . \tag{5.40}$$

As it stands, equation (5.40) is strictly valid for a distribution of static charges and stationary current circuits through externally generated fixed fields in a vacuum [414]. However, it is from this basic concept that the expression for the ponderomotive force on a moving, magnetisable and polarisable body can be derived. We will return to this expression in the magneto-elastostatic context in Section 5.3.4; when what has been denoted by $\mathscr{D}_t$ is surrounded by a another medium, it requires further extension by

$$\int_{\partial\mathscr{D}_t} \left[\widehat{\varrho}_t \, \{\!\{\mathbb{e}\}\!\} + \widehat{\mathbb{j}} \times \{\!\{\mathbb{b}\}\!\}\right] \, \mathrm{da} \tag{5.41}$$

to incorporate the surface effects that arise due to the existence of surface charges $\widehat{\varrho}_t$ and currents $\widehat{\mathbb{j}}$ on $\partial\mathscr{D}_t$. To supplement this, the extensions described in [414, 133, 119, 522, 333] express this force in moving or deformable matter with particular consideration of the dipole-current circuit model.

5.2 Continuum theorems for materials with discontinuities

Polar materials are those that experience electromagnetic force and moment couples in addition to those solely attributable to mechanical actions [80]. In the context of electro-magneto-mechanical materials, it is the distribution of charges and flow of currents that lead to the additional body forces acting on the magnetisable or polarisable materials. In order to derive the local balance equations and jump conditions for polar media, it is necessary to extend the classical description of Reynolds transport theorem for the case where there exist moving surface and/or line discontinuities. Using this, one is able to link the expressions for the conservation laws to the known global form, and posit the form of various terms that easily renders the equivalent local statements within a volume or at a material discontinuity.

In this section, we provide a concise derivation that leads to expressions that render the localisation and jump conditions for the conservation laws. For the sake of simplicity, the derivations are conducted assuming the absence of external tractions; where appropriate, we will state the necessary modifications for the case where external tractions are considered. The theory and identities for the balance laws for domains with surface discontinuities, and surfaces with line discontinuities, is presented in equal or greater detail in [523, 133, 180, 277, 184, 406, 530], among other sources.

Note that, for what follows, one of the generic subdomains $\mathscr{B}_t^i$ that will be introduced could be readily replaced such that we evaluate fields at the discontinuity between the body $\mathscr{B}_t$ and the free space $\mathscr{S}_t$. The reader may also observe that in each case removal of the material discontinuity collapses the general statements (namely Gauss'-, Kelvin–Stokes'-, and Reynolds' transport theorems) to their more common representations that are given in Appendix B.2.1. To deduce the localisation of balance laws when considering materials with moving interfaces, the standard approach (detailed in [406, 133, 530]) utilises the "pillbox" control volume and" Ampérian loop" control area argumentation. We will adopt a slightly different approach that is based on extended versions of the Gauss and Kelvin–Stokes theorems leading to the identical fundamental results.

5.2.1 Control volumes with surface discontinuities

With reference to Figure 5.4, consider a body in the spatial setting composed of two material subdomains such that $\mathscr{B}_t = \mathscr{B}_t^+ \cup \mathscr{B}_t^-$. We then place a control volume such

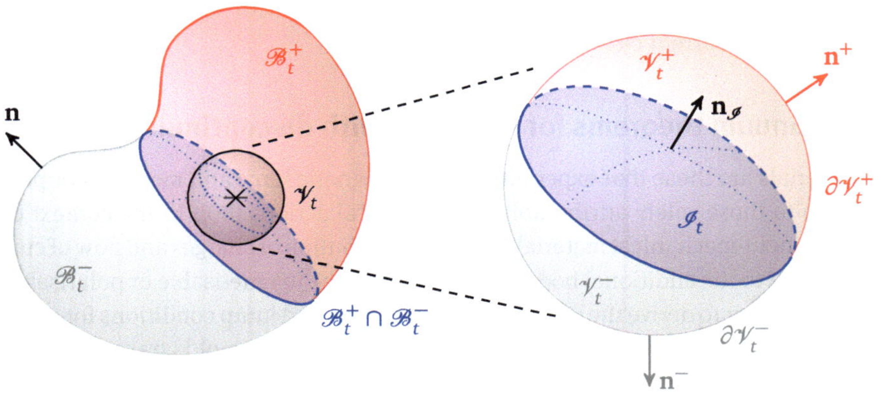

Figure 5.4: A general spatial control volume. The surface $\mathscr{I}_t$ represents the interface between two volume domains $\mathscr{V}_t^+$ and $\mathscr{V}_t^-$. The outwards facing normal to the control volume is represented by the unit vector $\mathbf{n}_\mathscr{I}$.

that it is divided into $\mathscr{V}_t = \mathscr{V}_t^+ \cup \mathscr{V}_t^-$ by the material discontinuity at $\mathscr{B}_t^+ \cap \mathscr{B}_t^-$. We denote the interface of the material discontinuity, which exists at the intersection of the two subdomain boundaries, as $\mathscr{I}_t = \partial\mathscr{V}_t^+ \cap \partial\mathscr{V}_t^-$, and the exterior surface of the control volume $\partial\mathscr{V}_t = [\partial\mathscr{V}_t^+ \cup \partial\mathscr{V}_t^-]\backslash\mathscr{I}_t$. The outward normal to the interface is at any point $\mathbf{x} \in \mathscr{I}$ denoted by $\mathbf{n}_{\mathscr{I}}$.

Kinematics of moving interfaces

Viewed from its current configuration, the (independently) migrating interface $\mathbf{x}(\mathbf{X}(t), t) = \mathbf{x} \in \overline{\mathscr{B}_t^+ \cap \mathscr{B}_t^-}$ moves with the velocity $\mathbf{w}$. We can associate each point $\mathbf{x}$ with a material position $\mathbf{X}(t) \in \overline{\mathscr{B}_0^+ \cap \mathscr{B}_0^-}$ and a material velocity $\mathbf{W}$. These material and spatial interface velocities are therefore

$$\mathbf{W} = \dot{\mathbf{X}} \tag{5.42a}$$

$$\mathbf{w} = \dot{\mathbf{x}} = \partial_t \mathbf{x}|_{\mathbf{X}} + \left.\frac{\partial \mathbf{x}}{\partial \mathbf{X}}\right|_t \cdot \partial_t \mathbf{X}$$

$$\equiv \mathbf{v} + \mathbf{F} \cdot \mathbf{W} \quad , \tag{5.42b}$$

respectively. From this, we can identify the fundamental relationship

$$\mathbf{F} \cdot \mathbf{W} = \mathbf{w} - \mathbf{v} \tag{5.43}$$

representing the difference between the velocity of the moving interface and that of continuum points along the interface.

Gauss' (divergence) theorem

Considering the divergence of a vector or tensorial quantity $[\bullet]$ over the two subvolumes separately, namely

$$\int_{\mathscr{V}_t^+} \nabla \cdot [\bullet] \; \mathrm{dv} = \int_{\partial\mathscr{V}_t^+\backslash\mathscr{I}_t} [\bullet] \cdot \mathbf{n}^+ \; \mathrm{da} - \int_{\mathscr{I}_t} [\bullet]^+ \cdot \mathbf{n}_{\mathscr{I}} \; \mathrm{da} \quad \text{and}$$

$$\int_{\mathscr{V}_t^-} \nabla \cdot [\bullet] \; \mathrm{dv} = \int_{\partial\mathscr{V}_t^-\backslash\mathscr{I}_t} [\bullet] \cdot \mathbf{n}^- \; \mathrm{da} + \int_{\mathscr{I}_t} [\bullet]^- \cdot \mathbf{n}_{\mathscr{I}} \; \mathrm{da} \quad ,$$

the general form of equation (B.16) is established to be

$$\int_{\mathscr{V}_t} \nabla \cdot [\bullet] \; \mathrm{dv} = \int_{\partial\mathscr{V}_t} [\bullet] \cdot \mathbf{n} \; \mathrm{da} - \int_{\mathscr{I}_t} [\![\bullet]\!] \cdot \mathbf{n}_{\mathscr{I}} \; \mathrm{da} \quad \forall\, \mathscr{V}_t\,,\, \mathscr{I}_t \tag{5.44}$$

when a surface discontinuity is located within the control volume. Here, we define the jump and (for later convenience) the average of a quantity as

$$[\![\bullet]\!] := [\bullet]^+ - [\bullet]^- \tag{5.45a}$$

$$\{\!\{\bullet\}\!\} := \frac{1}{2}\left[[\bullet]^+ + [\bullet]^-\right] \tag{5.45b}$$

where $[\bullet]^+$ denotes the one-sided limit of the function $(\bullet)$ from the outward normal direction, and similarly $[\bullet]^-$ the limit from the inside normal direction. As a corollary of this generalisation of the divergence theorem, through the relation given in equation (5.10) the area theorem states that

$$\mathbf{0} = \int_{\partial \mathscr{V}_t} \mathbf{n} \, \mathrm{da} = \int_{\partial \mathscr{V}_0} \mathrm{cof}(\mathbf{F}) \cdot \mathbf{N} \, \mathrm{dA} = \int_{\mathscr{V}_0} \nabla_0 \cdot [\mathrm{cof}(\mathbf{F})] \, \mathrm{dV} + \int_{\mathscr{I}_0} [\![\mathrm{cof}(\mathbf{F})]\!] \cdot \mathbf{N}_{\mathscr{I}} \, \mathrm{dA} \quad . \tag{5.46}$$

After localisation and under the consideration of the Piola identity stated in equation (B.19), we deduce that

$$\left.\begin{array}{lll} \nabla_0 \cdot [\mathrm{cof}(\mathbf{F})] = \mathbf{0} & \forall \, \mathbf{X} & \text{in} \quad \mathscr{V}_0 \\ [\![\mathrm{cof}(\mathbf{F})]\!] \cdot \mathbf{N}_{\mathscr{I}} = \mathbf{0} & \forall \, \mathbf{X} & \text{on} \quad \mathscr{I}_0 \end{array}\right\} \quad \forall \, \mathscr{V}_0 \, , \, [\mathscr{I}_0 \subset \mathscr{V}_0] \quad , \tag{5.47}$$

which also implies that

$$\left.\begin{array}{lll} \nabla \cdot \left[\mathrm{cof}(\mathbf{F}^{-1})\right] = \mathbf{0} & \forall \, \mathbf{x} & \text{in} \quad \mathscr{V}_t \\ [\![\mathrm{cof}(\mathbf{F}^{-1})]\!] \cdot \mathbf{n}_{\mathscr{I}} = \mathbf{0} & \forall \, \mathbf{x} & \text{on} \quad \mathscr{I}_t \end{array}\right\} \quad \forall \, \mathscr{V}_t \, , \, [\mathscr{I}_0 \subset \mathscr{V}_t] \quad . \tag{5.48}$$

Reynolds transport theorem

Accounting for the material discontinuity, Reynolds transport theorem for a spatial scalar density field $\Phi_t(\mathbf{x}, t)$ is

$$\mathrm{D}_t \int_{\mathscr{V}_t} \Phi_t \, \mathrm{dv} = \int_{\mathscr{V}_t} \mathrm{n}_t \Phi_t \, \mathrm{dv} - \int_{\mathscr{I}_t} [\![\Phi_t \, [\mathbf{w} - \mathbf{v}]]\!] \cdot \mathbf{n}_{\mathscr{I}} \, \mathrm{da} \tag{5.49}$$

and for a spatial vector density field $\boldsymbol{\Phi}_t(\mathbf{x}, t)$ it is stated as

$$\mathrm{D}_t \int_{\mathscr{V}_t} \boldsymbol{\Phi}_t \, \mathrm{dv} = \int_{\mathscr{V}_t} \mathrm{n}_t \boldsymbol{\Phi}_t \, \mathrm{dv} - \int_{\mathscr{I}_t} [\![\boldsymbol{\Phi}_t \otimes [\mathbf{w} - \mathbf{v}]]\!] \cdot \mathbf{n}_{\mathscr{I}} \, \mathrm{da} \quad . \tag{5.50}$$

As is demonstrated in Appendix B.2.2, these are the generalised forms for equations (B.21) and (B.22), derived by considering equation (5.44) in substitution of equation (B.16).

Global balance law and localisation

The volume source and surface flux terms contributing to a change in scalar and vector fields have the magneto-mechanical decomposition

$$s_{\Phi}(\Phi_t) = s_{\Phi}^{\mathrm{mech}} + s_{\Phi}^{\mathrm{mag}} \quad , \quad \mathbf{s}_{\boldsymbol{\Phi}}(\boldsymbol{\Phi}_t) = \mathbf{s}_{\boldsymbol{\Phi}}^{\mathrm{mech}} + \mathbf{s}_{\boldsymbol{\Phi}}^{\mathrm{mag}} \quad , \tag{5.51a}$$

$$\mathbf{s}_{\Phi}(\Phi_t) = \mathbf{s}_{\Phi}^{\mathrm{mech}} + \mathbf{s}_{\Phi}^{\mathrm{mag}} \quad , \quad \mathbf{S}_{\boldsymbol{\Phi}}(\boldsymbol{\Phi}_t) = \mathbf{S}_{\boldsymbol{\Phi}}^{\mathrm{mech}} + \mathbf{S}_{\boldsymbol{\Phi}}^{\mathrm{mag}} \quad . \tag{5.51b}$$

Importantly, we assume that the electromagnetic fluxes $\mathbf{s}_{\Phi}^{\mathrm{mag}}, \mathbf{S}_{\boldsymbol{\Phi}}^{\mathrm{mag}}$ arise only in the presence of a surface discontinuity, and $s_{\Phi}^{\mathrm{mag}}, \mathbf{s}_{\boldsymbol{\Phi}}^{\mathrm{mag}}$ are continuous throughout the volume.

Through the substitution of equation (5.49) for the left-hand side of equation (B.28), and the application of equation (5.44) to its right-hand side, the generic description of the balance law for spatial scalar density fields in a control volume with a material discontinuity is

$$\begin{aligned} \mathrm{D}_t \int_{\mathcal{V}_t} \Phi_t \,\mathrm{dv} &= \int_{\mathcal{V}_t} \mathrm{n}_t \Phi_t \,\mathrm{dv} - \int_{\mathcal{I}_t} [\![\Phi_t \,[\mathbf{w} - \mathbf{v}]]\!] \cdot \mathbf{n}_{\mathcal{I}} \,\mathrm{da} \\ &\equiv \int_{\mathcal{V}_t} [s_{\Phi} + \nabla \cdot [\mathbf{s}_{\Phi}]] \,\mathrm{dv} + \int_{\mathcal{I}_t} [\![\mathbf{s}_{\Phi}]\!] \cdot \mathbf{n}_{\mathcal{I}} \,\mathrm{da} \quad . \end{aligned} \tag{5.52}$$

After localisation, we deduce that

$$\left. \begin{aligned} \mathrm{n}_t \Phi_t &= s_{\Phi}(\Phi_t) + \nabla \cdot [\mathbf{s}_{\Phi}(\Phi_t)] && \forall\, \mathbf{x} \;\; \text{in} \;\; \mathcal{V}_t \\ [\![\Phi_t \,[\mathbf{v} - \mathbf{w}]]\!] \cdot \mathbf{n}_{\mathcal{I}} &= [\![\mathbf{s}_{\Phi}(\Phi_t)]\!] \cdot \mathbf{n}_{\mathcal{I}} && \forall\, \mathbf{x} \;\; \text{on} \;\; \mathcal{I}_t \end{aligned} \right\} \quad \forall\, \mathcal{V}_t \,,\; [\mathcal{I}_t \subset \mathcal{V}_t] \quad . \tag{5.53}$$

We can achieve a similar expression for spatial vector density fields by replacing the left-hand side of equation (B.29) by equation (5.50), and the application of equation (5.44) to its right-hand side. This leads to the result that

$$\begin{aligned} \mathrm{D}_t \int_{\mathcal{V}_t} \boldsymbol{\Phi}_t \,\mathrm{dv} &= \int_{\mathcal{V}_t} \mathrm{n}_t \boldsymbol{\Phi}_t \,\mathrm{dv} - \int_{\mathcal{I}_t} [\![\boldsymbol{\Phi}_t \otimes [\mathbf{w} - \mathbf{v}]]\!] \cdot \mathbf{n}_{\mathcal{I}} \,\mathrm{da} \\ &\equiv \int_{\mathcal{V}_t} [\mathbf{s}_{\boldsymbol{\Phi}} + \nabla \cdot [\mathbf{S}_{\boldsymbol{\Phi}}]] \,\mathrm{dv} + \int_{\mathcal{I}_t} [\![\mathbf{S}_{\boldsymbol{\Phi}}]\!] \cdot \mathbf{n}_{\mathcal{I}} \,\mathrm{da} \end{aligned} \tag{5.54}$$

which, upon localisation, renders

$$\left. \begin{aligned} \mathrm{n}_t \boldsymbol{\Phi}_t &= \mathbf{s}_{\boldsymbol{\Phi}}(\boldsymbol{\Phi}_t) + \nabla \cdot [\mathbf{S}_{\boldsymbol{\Phi}}(\boldsymbol{\Phi}_t)] && \forall\, \mathbf{x} \;\; \text{in} \;\; \mathcal{V}_t \\ [\![\boldsymbol{\Phi}_t \otimes [\mathbf{v} - \mathbf{w}]]\!] \cdot \mathbf{n}_{\mathcal{I}} &= [\![\mathbf{S}_{\boldsymbol{\Phi}}(\boldsymbol{\Phi}_t)]\!] \cdot \mathbf{n}_{\mathcal{I}} && \forall\, \mathbf{x} \;\; \text{on} \;\; \mathcal{I}_t \end{aligned} \right\} \quad \forall\, \mathcal{V}_t \,,\; [\mathcal{I}_t \subset \mathcal{V}_t] \quad . \tag{5.55}$$

If it is assumed that the interface moves coincidentally with the body, then the interface represents a convected material surface discontinuity. In such a scenario, there exists no relative difference in the interface and body velocities, then the localised balance laws on the interface can be further simplified to

$$[\![\mathbf{s}_{\Phi}(\Phi_t)]\!] \cdot \mathbf{n} = 0 \quad \forall\, \mathbf{x} \;\; \text{on} \;\; \partial\mathcal{B}_t \;\; \text{and other material discontinuities} \quad , \tag{5.56a}$$

$$[\![\mathbf{S}_{\boldsymbol{\Phi}}(\boldsymbol{\Phi}_t)]\!] \cdot \mathbf{n} = \mathbf{0} \quad \forall\, \mathbf{x} \;\; \text{on} \;\; \partial\mathcal{B}_t \;\; \text{and other material discontinuities} \quad . \tag{5.56b}$$

In the case where external traction are considered, the normal flux of the sources in equations (5.56a) and (5.56b) are balanced by the applied loads.

5.2.2 Control surfaces with line discontinuities

With reference to Figure 5.5, again consider a body in the spatial setting composed of two material subdomains such that $\mathscr{B}_t = \mathscr{B}_t^+ \cup \mathscr{B}_t^-$. We then consider an arbitrary control area $\mathscr{A}_t$ with outwards normal $\mathbf{m}$ that is divided into $\mathscr{A}_t = \mathscr{A}_t^+ \cup \mathscr{A}_t^-$ by the material discontinuity at $\partial\mathscr{B}_t^+ \cap \partial\mathscr{B}_t^-$. We denote the line interface of this surface material discontinuity as $\mathscr{J}_t = \partial\mathscr{A}_t^+ \cap \partial\mathscr{A}_t^-$, and the exterior curve of the control area $\partial\mathscr{A}_t = [\partial\mathscr{A}_t^+ \cup \partial\mathscr{A}_t^-]\backslash\mathscr{J}_t$. The tangent to the line interface, that is at any point $\mathbf{x} \in \mathscr{J}_t$, is denoted by $\mathbf{l}_{\mathscr{J}}$.

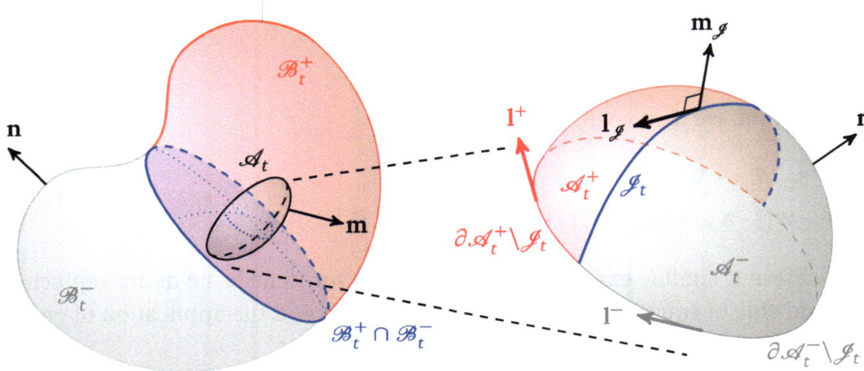

Figure 5.5: A general spatial control area. The line $\mathscr{J}_t$ represents the interface between two area surfaces $\mathscr{A}_t^+$ and $\mathscr{A}_t^-$, and the curve $\partial\mathscr{A}_t\backslash\mathscr{J}_t$ is closed and smooth. The outwards facing normal to the control area is represented by the unit vector **m**, while the tangent vector along the line discontinuity is denoted by $\mathbf{l}_{\mathscr{J}}$.

Kelvin–Stokes theorem

Considering the Kelvin–Stokes theorem as applied to the two sub-areas separately, namely

$$\int_{\mathscr{A}_t^+} [\nabla \times [\bullet]] \cdot \mathbf{m} \, \mathrm{da} = \int_{\partial\mathscr{A}_t^+\backslash\mathscr{J}_t} [\bullet] \cdot \mathbf{l}^+ \, \mathrm{dl} - \int_{\mathscr{J}_t} [\bullet]^+ \cdot \mathbf{l}_{\mathscr{J}} \, \mathrm{dl} \quad \text{and}$$

$$\int_{\mathscr{A}_t^-} [\nabla \times [\bullet]] \cdot \mathbf{m} \, \mathrm{da} = \int_{\partial\mathscr{A}_t^-\backslash\mathscr{J}_t} [\bullet] \cdot \mathbf{l}^- \, \mathrm{dl} + \int_{\mathscr{J}_t} [\bullet]^- \cdot \mathbf{l}_{\mathscr{J}} \, \mathrm{dl} \quad ,$$

the general form of equation (B.14) is rendered as

$$\int_{\mathscr{A}_t} [\nabla \times [\bullet]] \cdot \mathbf{m} \, \mathrm{da} = \oint_{\partial\mathscr{A}_t} [\bullet] \cdot \mathbf{l} \, \mathrm{dl} - \int_{\mathscr{J}_t} [\![\bullet]\!] \cdot \mathbf{l}_{\mathscr{J}} \, \mathrm{dl} \quad \forall\, \mathscr{A}_t \, , \, \mathscr{J}_t \quad . \tag{5.57}$$

As a corollary of this generalisation of the Kelvin–Stokes theorem, through the relation given in equation (5.8), the line theorem states that

$$\mathbf{0} = \oint_{\partial\mathscr{A}_t} \mathrm{d}\mathbf{x} = \oint_{\partial\mathscr{A}_0} \mathbf{F}\cdot\,\mathrm{d}\mathbf{X} = \int_{\mathscr{A}_0} [\nabla_0 \times \mathbf{F}]\cdot\mathbf{M}\,\mathrm{dA} + \int_{\mathscr{I}_0} [\![\mathbf{F}]\!]\cdot\underbrace{\mathbf{L}_{\mathscr{I}}\,\mathrm{dL}}_{\mathrm{d}\mathbf{X}} \tag{5.58}$$

from which the localised relations are given by

$$\left.\begin{array}{l} \left.\begin{array}{ll} [\nabla_0 \times \mathbf{F}]\cdot\mathbf{M} & = \mathbf{0} \\ \Longrightarrow \quad \nabla_0 \times \mathbf{F} & = \mathbf{0} \end{array}\right\} \quad \forall\,\mathbf{X} \quad \text{on} \quad \mathscr{A}_0 \\ \left.\begin{array}{ll} [\![\mathbf{F}]\!]\cdot\,\mathrm{d}\mathbf{X} & = \mathbf{0} \\ \Longrightarrow \quad [\![\mathbf{F}]\!]\times\mathbf{M} & = \mathbf{0} \end{array}\right\} \quad \forall\,\mathbf{X} \quad \text{on} \quad \mathscr{I}_0 \end{array}\right\} \quad \forall\,\mathscr{A}_0\,,\ [\mathscr{I}_0 \subset \mathscr{A}_0] \quad . \tag{5.59}$$

The inferred relationship for the jump terms may be deduced when noting that the reference normal **M**, related to the spatial normal **m** through equation (B.8), is orthogonal to the direction vector $\mathrm{d}\mathbf{X}$.

Reynolds transport theorem

Accommodating the presence of a migrating material discontinuity, Reynolds transport theorem for a spatial vector flux density field $\boldsymbol{\psi}(\mathbf{x}, t)$ is

$$\mathrm{D}_t \int_{\mathscr{A}_t} \boldsymbol{\psi}\cdot\mathbf{m}\,\mathrm{da} = \int_{\mathscr{A}_t} [\mathrm{n}_t\boldsymbol{\psi}]\cdot\mathbf{m}\,\mathrm{da} - \int_{\mathscr{I}_t} [\![\boldsymbol{\psi}\times[\mathbf{w}-\mathbf{v}]]\!]\cdot\mathbf{l}_{\mathscr{I}}\,\mathrm{dl} \quad . \tag{5.60}$$

As is demonstrated in Appendix B.2.2, this is the generalised form of equation (B.25), derived by considering equation (5.57) in substitution of equation (B.14) to transform the curl-related term.

Global balance law and localisation

The general balance law for fluxes through surfaces with material discontinuities can be established by substituting the left-hand side of equation (B.30) with equation (5.60) and applying equation (5.57) to its right-hand side. This substitution renders the balance law

$$\begin{aligned} \mathrm{D}_t \int_{\mathscr{A}_t} \boldsymbol{\psi}\cdot\mathbf{m}\,\mathrm{da} &= \int_{\mathscr{A}_t} [\mathrm{n}_t\boldsymbol{\psi}]\cdot\mathbf{m}\,\mathrm{da} - \int_{\mathscr{I}_t} [\![\boldsymbol{\psi}\times[\mathbf{w}-\mathbf{v}]]\!]\cdot\mathbf{l}_{\mathscr{I}}\,\mathrm{dl} \\ &\equiv \int_{\mathscr{A}_t} [\widehat{\mathbf{s}}_{\boldsymbol{\psi}} + \nabla\times\widetilde{\mathbf{s}}_{\boldsymbol{\psi}}]\cdot\mathbf{m}\,\mathrm{da} + \int_{\mathscr{I}_t} [\![\widetilde{\mathbf{s}}_{\boldsymbol{\psi}}]\!]\cdot\mathbf{l}_{\mathscr{I}}\,\mathrm{dl} \quad \forall\,\mathscr{A}_t\,,\ \mathscr{I}_t \quad . \end{aligned} \tag{5.61}$$

The surface and line fluxes are again decomposed into components respectively of mechanical and magnetic nature,

$$\widehat{\mathbf{s}}_{\boldsymbol{\psi}}(\boldsymbol{\psi}) = \widehat{\mathbf{s}}_{\boldsymbol{\psi}}^{\mathrm{mech}} + \widehat{\mathbf{s}}_{\boldsymbol{\psi}}^{\mathrm{mag}} \quad , \quad \widetilde{\mathbf{s}}_{\boldsymbol{\psi}}(\boldsymbol{\psi}) = \widetilde{\mathbf{s}}_{\boldsymbol{\psi}}^{\mathrm{mech}} + \widetilde{\mathbf{s}}_{\boldsymbol{\psi}}^{\mathrm{mag}} \quad . \tag{5.62}$$

Furthermore, we note that the electromagnetic flux $\widetilde{\mathbf{s}}_{\boldsymbol{\psi}}^{\mathrm{mag}}$ develops only in the presence of a line discontinuity, and $\widehat{\mathbf{s}}_{\boldsymbol{\psi}}^{\mathrm{mag}}$ is continuous on the surface. Localisation of the global balance law renders the relations

$$\left.\begin{array}{lll} & \left.\begin{array}{ll}[\mathrm{n}_t\boldsymbol{\psi}]\cdot\mathbf{m} & = [\widehat{\mathbf{s}}_{\psi}(\boldsymbol{\psi}) + \nabla\times\widetilde{\mathbf{s}}_{\psi}(\boldsymbol{\psi})]\cdot\mathbf{m} \\ \Longrightarrow \quad \mathrm{n}_t\boldsymbol{\psi} & = \widehat{\mathbf{s}}_{\psi}(\boldsymbol{\psi}) + \nabla\times\widetilde{\mathbf{s}}_{\psi}(\boldsymbol{\psi})\end{array}\right\} & \forall\,\mathbf{x} \quad \text{on} \quad \mathscr{A}_t \\ & \left.\begin{array}{ll}[\![\boldsymbol{\psi}\times[\mathbf{w}-\mathbf{v}]]\!]\cdot\mathbf{1}_{\mathscr{I}} & = [\![\widetilde{\mathbf{s}}_{\psi}(\boldsymbol{\psi})]\!]\cdot\mathbf{1}_{\mathscr{I}} \\ \Longrightarrow \quad [\![\boldsymbol{\psi}\times[\mathbf{w}-\mathbf{v}]]\!]\times\mathbf{m} & = [\![\widetilde{\mathbf{s}}_{\psi}(\boldsymbol{\psi})]\!]\times\mathbf{m}\end{array}\right\} & \forall\,\mathbf{x} \quad \text{on} \quad \mathscr{I}_t\end{array}\right\} \quad \forall\,\mathscr{A}_t\,,\ [\mathscr{I}_t\subset\mathscr{A}_t]\ . \tag{5.63}$$

If we assume that the line of discontinuity moves coincidentally with the body, then the line represents a convected material line discontinuity. In such a scenario, the localised balance laws can be further simplified to

$$[\![\,\widetilde{\mathbf{s}}_{\psi}(\boldsymbol{\psi})]\!]\times\mathbf{m} = \mathbf{0} \quad \forall\,\mathbf{x} \quad \text{on} \quad \partial\mathscr{B}_t \quad . \tag{5.64}$$

When external surface tractions are to be considered, these terms balance those stated on the left-hand side of equation (5.64).

Summary of localised balance laws

As a glance towards the scalar and vector conservation equations stated later in the chapter, we summarise the relevant localisation statements in Tables 5.1 to 5.2. Their companion results, as derived in Appendix C.5, are also conveniently tabulated for direct comparison.

5.3 Governing equations

The localisation of the balance laws as derived from fundamental principles and the equations stated in Section 5.2 leads to a spatial formulation of the governing laws. The statement of the coupled problem in the current configuration is well documented in the cited literature, and its derivation is outlined in Appendices C.5.1 and C.5.2. However, on many occasions the algorithmic implementation of finite element problems in solid mechanics is most easily performed (or most convenient) in the Lagrangian framework. We therefore preferentially focus on the less-well described Lagrangian statement of the governing equations for electromagnetics and, thereafter, magneto-mechanics. The two-point description of the balance laws expressed in the first part of this section are derived through the pull-back transformation of Tables 5.1 to 5.2; this is documented in Appendix C.5.4. Section 5.3.2 then provides a summary of continuum quantities and balance laws in the spatial setting. This is followed in Section 5.3.4 by a discussion on the magnetically-induced force and moment acting on material bodies.

Table 5.1: Identified sources for relevant scalar balance laws and continuity conditions for control volumes, derived through application of equations (5.53) and (5.56a). Equations for which only an indirect result is given are indicated by square brackets. The electromagnetic energy flux $\mathbf{s}_{\mathrm{EM}} = \mathbf{v} \cdot \boldsymbol{\sigma}^{*\mathrm{pon}} + [\mathbb{e} \times \mathbb{m}]$ and supply of electromagnetic energy $\Phi_{\mathrm{EM}} = \mathbf{v} \cdot \mathbf{b}_t^{\mathrm{pon}} + \boldsymbol{\sigma}^{*(\mathrm{mag+pol})} : \nabla \mathbf{v} + \mathbb{j}^{\mathrm{b}} \cdot \mathbb{e}$ can be inferred from equation (5.196). The symbols U_t defines the internal energy density per unit current volume and is related to its referential counterpart defined in equation $(5.193)_2$, while r_t and $\mathbf{q}$, respectively denoting the thermal flux and the thermal source, are related to quantities used in equation (5.195). Similarly, the spatial entropy flux vector $\mathbf{h}$ relates to its referential counterpart postulated in equation (5.205).

Balance law / Conservation law	Equation	Field Φ_t	Volume sources s_Φ^{mech}	Volume sources s_Φ^{mag}	Flux sources $\mathbf{s}_\Phi^{\mathrm{mech}}$	Flux sources $\mathbf{s}_\Phi^{\mathrm{mag}}$
Electric charge	5.103	ϱ_t	0	0	$\mathbf{0}$	$-\mathbb{j}$
Gauss' law	5.104_2	0	0	ϱ_t^{f}	$\mathbf{0}$	$-\mathbb{d}$
Gauss' magnetism law	5.105_2	0	0	0	$\mathbf{0}$	$\mathbb{b}$
Mass	5.119	ρ_t	0	0	$\mathbf{0}$	$\mathbf{0}$
Energy	[5.192–5.198]	$U_t + \frac{1}{2}\rho_t \mathbf{v} \cdot \mathbf{v}$	$\mathbf{v} \cdot \mathbf{b}_t^{\mathrm{mech}} + r_t$	Φ_{EM}	$\mathbf{v} \cdot \boldsymbol{\sigma}^{\mathrm{mech,ext}} + \mathbf{q}$	$\mathbf{s}_{\mathrm{EM}}$
Entropy inequality	[5.202–5.204]	η_t	K_t	0	$\mathbf{h}$	$\mathbf{0}$

Table 5.2: Identified sources for relevant vector balance laws and continuity conditions for control volumes, derived through application of equations (5.55) and (5.56b). Equations for which only an indirect result is given are indicated by square brackets.

Balance law / Conservation law	Equation	Field $\boldsymbol{\Phi}_t$	Volume sources $\mathbf{s}_\Phi^{\mathrm{mech}}$	Volume sources $\mathbf{s}_\Phi^{\mathrm{mag}}$	Flux sources $\mathbf{S}_\Phi^{\mathrm{mech}}$	Flux sources $\mathbf{S}_\Phi^{\mathrm{mag}}$
Linear momentum	5.120	$\rho_t \mathbf{v}$	$\mathbf{b}_t^{\mathrm{mech}}$	$\mathbf{0}$	$\boldsymbol{\sigma}^{\mathrm{mech}}$	$\boldsymbol{\sigma}^{*\mathrm{pon}}$
Angular momentum	[5.121]	$\mathbf{r} \times \rho_t \mathbf{v}$	$\mathbf{r} \times \mathbf{b}_t^{\mathrm{mech}}$	$\mathbf{m}_t^{\mathrm{mag}} - \boldsymbol{\epsilon} : [\boldsymbol{\sigma}^{*\mathrm{pon}}]^T$ → 0	$\mathbf{r} \times \boldsymbol{\sigma}^{\mathrm{mech}}$	$\mathbf{r} \times \boldsymbol{\sigma}^{*\mathrm{pon}}$

Table 5.3: Identified sources for relevant vector balance laws and continuity conditions for control areas, derived through application of equations (5.63) and (5.64).

Balance law / Conservation law	Equation	Field $\boldsymbol{\psi}$	Surface sources $\hat{\mathbf{s}}_\psi^{\mathrm{mech}}$	Surface sources $\hat{\mathbf{s}}_\psi^{\mathrm{mag}}$	Flux sources $\tilde{\mathbf{s}}_\psi^{\mathrm{mech}}$	Flux sources $\tilde{\mathbf{s}}_\psi^{\mathrm{mag}}$
Faraday's law	5.104_1	$\mathbb{b}$	$\mathbf{0}$	$\mathbf{0}$	$\mathbf{0}$	$-\mathbb{e}$
Ampére's law	5.105_1	$\mathbb{d}$	$\mathbf{0}$	$-\mathbb{j}^{\mathrm{f}}$	$\mathbf{0}$	$\mathbb{h}$

In Section 5.3.6, we derive the weak formulation of the governing equations directly from their strong form, which can be compared to that shown in Section 5.3.7, produced from a variational perspective for non-dissipative materials using the potential fields described in Section 5.3.5.

5.3.1 Balance laws

The balance equations of continuum physics may be expressed either in terms of (i) referential and current vector and tensor fields, and (ii) material or spatial time derivatives; this results in four different (but otherwise equivalent) formulations. For the set of Maxwell equations, these are assembled in Tables 5.5 and 5.6 in the overall summary Section 5.3.3. In the following, we will present the balance laws in the referential setting and using material time derivatives, as this is the most convenient setting for the purpose of material modelling.

Electromagnetics

The fundamental balance laws that govern magnetoelastic materials are the conservation of electric charge, Maxwell's equations and the balance of linear and angular momentum. The conservation of charge as defined in a Lagrangian framework is

$$\dot{\varrho}_0 + \nabla_0 \cdot \mathbb{J} = 0 \quad \text{in} \quad \mathscr{B}_0 \cup \mathscr{S}_0 \quad . \tag{5.65}$$

The general coupled Maxwell's equations are defined by Faraday's law and Gauss' law and are expressed in referential quantities by

$$\nabla_0 \times \mathbb{E} = -\dot{\mathbb{B}} \quad , \quad \nabla_0 \cdot \mathbb{D}_\varepsilon = \varrho_0 \quad \text{in} \quad \mathscr{B}_0 \cup \mathscr{S}_0 \quad , \tag{5.66}$$

in conjunction with Ampére's law and Gauss' magnetism law

$$\nabla_0 \times \mathbb{H}_\mu = \dot{\mathbb{D}}_\varepsilon + \mathbb{J} \quad , \quad \nabla_0 \cdot \mathbb{B} = 0 \quad \text{in} \quad \mathscr{B}_0 \cup \mathscr{S}_0 \quad , \tag{5.67}$$

the latter of which dictates that there exist no magnetic monopoles. The referential magnetic induction and electric field vectors are given by $\mathbb{B}$ and $\mathbb{E}$, respectively, and we define the quantities

$$\mathbb{D}_\varepsilon := \varepsilon_0 J \mathbf{C}^{-1} \cdot \mathbb{E} \quad \text{and} \tag{5.68}$$

$$\mathbb{H}_\mu := \frac{1}{\mu_0} J^{-1} \mathbf{C} \cdot \mathbb{B} \tag{5.69}$$

that represent the referential free space electric displacement vector and the referential free space magnetic field vector. The total referential charge density and current vectors expand as

$$\varrho_0 = \varrho_0^{\mathrm{b}} + \varrho_0^{\mathrm{f}} \quad , \quad \mathbb{J} = \mathbb{J}^{\mathrm{b}} + \mathbb{J}^{\mathrm{f}} \quad , \tag{5.70}$$

where the referential bound- and free-charge densities are respectively denoted as ϱ_0^{b} and ϱ_0^{f}, while the bound- and free-current vectors are represented by $\mathbb{J}^{\mathrm{b}}$ and $\mathbb{J}^{\mathrm{f}}$, respectively.

We introduce two additional referential quantities, namely $\mathbb{M}$ that represents the magnetisation and $\mathbb{P}$ the polarisation of the solid media (both of which, by definition, vanish in the free space). By defining the polarisation and magnetisation vectors as [406]

$$\nabla_0 \cdot \mathbb{P} := -\varrho_0^{\mathrm{b}} \quad , \quad \nabla_0 \times \mathbb{M} := \mathbb{J}^{\mathrm{b}} - \dot{\mathbb{P}} \quad , \tag{5.71}$$

Gauss' flux theorem and Ampére's law are re-expressed as

$$\nabla_0 \cdot [\mathbb{D}_\varepsilon + \mathbb{P}] = \varrho_0^{\mathrm{f}} \quad , \quad \nabla_0 \times [\mathbb{H}_\mu - \mathbb{M}] = \dot{\mathbb{D}} + \mathbb{J}^{\mathrm{f}} \quad \text{in} \quad \mathscr{B}_0 \cup \mathscr{S}_0 \quad . \tag{5.72}$$

Linking the various magnetic and electric quantities by the fundamental constitutive equations [277, 120, 122, 537]

$$\mathbb{D} = \varepsilon_0 J \mathbf{C}^{-1} \cdot \mathbb{E} + \mathbb{P} \quad \text{and} \tag{5.73}$$

$$J^{-1}\mathbf{C} \cdot \mathbb{B} = \mu_0 \, [\mathbb{H} + \mathbb{M}] \quad , \tag{5.74}$$

for which the referential magnetic field and electric displacement vectors are given by $\mathbb{H}$ and $\mathbb{D}$, respectively, we arrive at the reduced expression of Gauss' and Ampére's laws

$$\nabla_0 \cdot \mathbb{D} = \varrho_0^{\mathrm{f}} \quad , \quad \nabla_0 \times \mathbb{H} = \dot{\mathbb{D}} + \mathbb{J}^{\mathrm{f}} \quad \text{in} \quad \mathscr{B}_0 \cup \mathscr{S}_0 \quad . \tag{5.75}$$

The general continuity conditions for Maxwell's equations are [406]

$$\mathbf{N} \times [\![\mathbb{E}]\!] = \mathbf{0} \quad , \quad \mathbf{N} \cdot [\![\mathbb{D}]\!] = \widehat{\varrho}_0^{\mathrm{f}} \quad \text{on} \quad \partial\mathscr{B}_0 \quad , \tag{5.76}$$

$$\mathbf{N} \times [\![\mathbb{H}]\!] = \widehat{\mathbb{J}}^{\mathrm{f}} \quad , \quad \mathbf{N} \cdot [\![\mathbb{B}]\!] = 0 \quad \text{on} \quad \partial\mathscr{B}_0 \quad , \tag{5.77}$$

and other material discontinuities. The bound surface charge and bound surface current that arises due to different polarisations and magnetisations on a material discontinuity are

$$\mathbf{N} \cdot [\![\mathbb{P}]\!] = -\widehat{\varrho}_0^{\mathrm{b}} \quad \text{on} \quad \partial\mathscr{B}_0 \quad , \tag{5.78}$$

$$\mathbf{N} \times [\![\mathbb{M}]\!] = \widehat{\mathbb{J}}^{\mathrm{b}} \quad \text{on} \quad \partial\mathscr{B}_0 \quad , \tag{5.79}$$

and other material discontinuities. Surface quantities are denoted by $\widehat{(\bullet)}$, so $\widehat{\varrho}_0$ and $\widehat{\mathbb{J}}$ therefore respectively represent the surface charge and current; by definition, $\widehat{\mathbb{J}}$ is orthogonal to $\mathbf{N}$. The magnetic conditions at the far field boundary of the truncated domain are

$$\mathbf{N}_\infty \times \mathbb{H} = \mathbf{N}_\infty \times \mathbb{H}_\infty \quad \text{on} \quad \partial\mathscr{S}_0^{\mathbb{H}} \quad , \tag{5.80}$$

$$\mathbf{N}_\infty \cdot \mathbb{B} = \mathbf{N}_\infty \cdot \mathbb{B}_\infty \quad \text{on} \quad \partial\mathscr{S}_0^{\mathbb{B}} \quad . \tag{5.81}$$

If we assume that the electromagnetic fields change so slowly that they can be characterised as being in a steady-state ($\dot{\mathbb{D}} = \dot{\mathbb{B}} = \mathbf{0}$) then equation $(5.66)_1$ and equation $(5.75)_1$ reduce to

$$\nabla_0 \times \mathbb{E} = \mathbf{0} \quad , \quad \nabla_0 \times \mathbb{H} = \mathbb{J}^{\mathrm{f}} \quad \text{in} \quad \mathscr{B}_0 \cup \mathscr{S}_0 \tag{5.82}$$

which decouples the magnetic and electric fields. The relevant magnetic continuity conditions are

$$\mathbf{N} \times [\![\mathbb{H}]\!] = \hat{\mathbb{J}}^{\mathrm{f}} \quad \text{on} \quad \partial\mathscr{B}_0 \tag{5.83}$$

and other material discontinuities, along with equation $(5.77)_2$. If it is also assumed that there is no free electric current ($\mathbb{J}^{\mathrm{f}} = \mathbf{0}$), then equation $(5.82)_2$ is further simplified to

$$\nabla_0 \times \mathbb{H} = \mathbf{0} \quad \text{in} \quad \mathscr{B}_0 \cup \mathscr{S}_0 \quad . \tag{5.84}$$

The magnetostatic balance laws are then given by equation (5.84) in conjunction with equation $(5.67)_2$. The quasi-static magnetic continuity conditions enforce that the normal component of the magnetic induction as well as the tangential component of the magnetic field remain continuous on the boundary $\partial\mathscr{B}_0$ with outward normal $\mathbf{N}$. These continuity conditions are expressed as

$$\mathbf{N} \times [\![\mathbb{H}]\!] = \mathbf{0} \quad , \quad \mathbf{N} \cdot [\![\mathbb{B}]\!] = 0 \quad \text{on} \quad \partial\mathscr{B}_0 \tag{5.85}$$

and other material discontinuities.

Magneto-mechanics

For the purely mechanical case, the balance of linear and angular momentum are expressed in Lagrangian form as [80, 559, 204]

$$\nabla_0 \cdot \mathbf{P}^{\mathrm{mech}} + \mathbf{b}_0^{\mathrm{mech}} = \rho_0 \mathbf{a} \quad , \quad \mathbf{P}^{\mathrm{mech}} \cdot \mathbf{F}^T = \mathbf{F} \cdot \left[\mathbf{P}^{\mathrm{mech}}\right]^T \quad \text{in} \quad \mathscr{B}_0 \tag{5.86}$$

where $\mathbf{P}^{\mathrm{mech}}$ denotes the mechanical Piola stress tensor, $\mathbf{b}_0^{\mathrm{mech}} := \rho_0 \mathbf{g}$ is the referential mechanical body force vector, and ρ_0 represents the reference mass density. For magneto-mechano-static, consideration of the influence of the magnetic field both permeating and surrounding the medium is necessary. Assuming quasi-static conditions, equation $(5.86)_1$ is augmented to [64, 341, 133, 277, 502, 120]

$$\nabla_0 \cdot \mathbf{P}^{\mathrm{mech}} + \mathbf{b}_0^{\mathrm{pon}} + \mathbf{b}_0^{\mathrm{mech}} = \nabla_0 \cdot \mathbf{P}^{\mathrm{tot}} + \mathbf{b}_0^{\mathrm{mech}} = \rho_0 \mathbf{a} \to \mathbf{0} \quad \text{in} \quad \mathscr{B}_0 \cup \mathscr{S}_0 \tag{5.87}$$

and is valid for both inside and outside the solid body. Here we note the existence of an additional body force known as the ponderomotive body force, which for magnetostatic conditions is defined as

$$\mathbf{b}_0^{\mathrm{pon}} = \mathbf{F}^{-T} \cdot \left[\left[\mathbb{J}^{\mathrm{f}} \times \mathbb{B}\right] + J\mathbb{M} \cdot \left[\nabla_0 \left[J^{-1}\mathbf{F} \cdot \mathbb{B}\right] \cdot \mathbf{F}^{-1}\right]\right] \quad , \tag{5.88}$$

where the total referential current vector $\mathbb{J} = \mathbb{J}^{f} + \mathbb{J}^{b}$. As described by Pao [414], it represents the force exerted on the solid body due to the permeating magnetic field and can be shown to be expressed as the divergence of a stress tensor

$$\mathbf{b}_0^{\text{pon}} = \nabla_0 \cdot \mathbf{P}^{*\text{pon}} \quad . \tag{5.89}$$

We introduce the total stress $\mathbf{P}^{\text{tot}}$ as a decomposition of the mechanical and ponderomotive stresses

$$\mathbf{P}^{\text{tot}} = \mathbf{P}^{\text{mech}} + \mathbf{P}^{*\text{pon}} = \mathbf{P}^{\text{mech}} + \mathbf{P}^{*\text{mag}} + \mathbf{P}^{*\text{max}} \tag{5.90}$$

and further decompose the ponderomotive stress into a Maxwell stress tensor and a secondary component directly resulting from the magnetisation of the solid material. (We note here that in what follows it is strictly assumed that the stresses relate to a free energy that has not undergone any Legendre transformations.[1]) The ponderomotive, Maxwell and magnetisation stresses can be expressed as [533, 530]

$$\mathbf{P}^{*\text{pon}} = \left[\frac{J^{-1}}{2\mu_0}\mathbf{C} : [\mathbb{B} \otimes \mathbb{B}]\right]\mathbf{F}^{-T} - [\mathbb{H} \cdot \mathbb{B}]\,\mathbf{F}^{-T} + \left[\mathbf{F}^{-T} \cdot \mathbb{H}\right] \otimes \mathbb{B} \tag{5.91a}$$

$$\mathbf{P}^{*\text{mag}} = [\mathbb{M} \cdot \mathbb{B}]\,\mathbf{F}^{-T} - \left[\mathbf{F}^{-T} \cdot \mathbb{M}\right] \otimes \mathbb{B} \tag{5.91b}$$

$$\mathbf{P}^{*\text{max}} = \left[\frac{J^{-1}}{\mu_0}\mathbf{F} \cdot \mathbb{B}\right] \otimes \mathbb{B} - \left[\frac{J^{-1}}{2\mu_0}\mathbf{C} : [\mathbb{B} \otimes \mathbb{B}]\right]\mathbf{F}^{-T} \quad ; \tag{5.91c}$$

the derivation of these stress tensors is performed in Appendix C.3.1. For a more detailed discussion on the stress tensor $\mathbf{P}^{*\text{pon}}$, we refer the reader to [121, 248, 502, 338, 537, 533] and the references therein.

Incorporating the magnetic forces, the associated augmented angular momentum balance equation derived from equation $(5.86)_2$ is therefore

$$\mathbf{P}^{\text{tot}} \cdot \mathbf{F}^{T} = \mathbf{F} \cdot \left[\mathbf{P}^{\text{tot}}\right]^{T} \quad \text{in} \quad \mathscr{B}_0 \cup \mathscr{S}_0 \tag{5.92}$$

or, equivalently [414, 119, 406],

$$\mathbf{0} = \boldsymbol{\epsilon} : \left[J^{-1}\mathbf{P}^{\text{tot}} \cdot \mathbf{F}^{T}\right]^{T} = \boldsymbol{\epsilon} : \left[J^{-1}\mathbf{P}^{\text{mech}} \cdot \mathbf{F}^{T}\right]^{T} + \left[\mathbf{F}^{-T} \cdot \mathbb{M}\right] \times \left[J^{-1}\mathbf{F} \cdot \mathbb{B}\right] \tag{5.93}$$

with the permutation tensor $\epsilon_{ijk} := [\mathbf{e}_i \times \mathbf{e}_j] \cdot \mathbf{e}_k$ and the second term in equation (5.93) encapsulating the magnetic moment $\mathbf{m}_t^{\text{mag}}$. At this point, it is important to indicate that the notion of deformation of the free space is an artificial concept that will be subsequently exploited for convenience within numerical simulations.

---

1 There are numerous plausible parameterisations in magneto-mechanics, and here the statement of the ponderomotive, magnetisation and Maxwell stresses hold only for the canonical formulation, parameterised in terms of mechanical deformation and the magnetic induction. We elucidate and elaborate more on other meaningful parameterisations later in Section 5.4.3.

For the mechanical problem, the kinematic compatibility condition that dictates that **F** remains a valid map from which to deduce a continuous displacement field is from equation (5.59),

$$\nabla_0 \times \mathbf{F} = \mathbf{0} \quad \text{in} \quad \mathscr{B}_0 \quad . \tag{5.94}$$

Furthermore, assuming the decomposition of a body into two material subdomains $\mathscr{B}_0 = \mathscr{B}_0^1 \cup \mathscr{B}_0^2$, tangential continuity of the deformation gradient across the interface is ensured by satisfying

$$[\![\mathbf{F}]\!] \times \mathbf{N} = \mathbf{0} \quad \text{on} \quad \text{material discontinuities} \quad \mathscr{B}_0^1 \cap \mathscr{B}_0^2 \quad . \tag{5.95}$$

Additionally, the mechanical Dirichlet boundary condition is simply

$$\boldsymbol{\varphi} = \overline{\boldsymbol{\varphi}} \quad \text{on} \quad \partial\mathscr{B}_0^{\boldsymbol{\varphi}} \quad . \tag{5.96}$$

The Cauchy stress theorem states that any traction vector can be expressed in terms of a rank-2 stress tensor and a unit direction vector by

$$\mathbf{t}_0 := \mathbf{P} \cdot \mathbf{N} \quad . \tag{5.97}$$

It is necessary that any mechanical traction applied at $\partial\mathscr{B}_0$ be balanced by the total Piola stress jump across a material interface [504]

$$-\left[\!\left[\mathbf{P}^{\text{tot}}\right]\!\right] \cdot \mathbf{N}^{+} = \mathbf{t}_0^{\text{mech,ext}} \quad \text{on} \quad \partial\mathscr{B}_0^{\mathbf{t}} \quad . \tag{5.98}$$

Lastly, the pointwise volumetric Jacobian J is unity in the case of fully incompressible media.

5.3.2 Spatial relationships

In order to re-express the governing equations in a spatial setting, it is necessary to define the push-forward operations for the various stress and magnetic quantities. This can be achieved with the assistance of the relationships given in Section 5.1.3. A generic Cauchy stress tensor is given by

$$\boldsymbol{\sigma} = \mathbf{P} \cdot \operatorname{cof}\left(\mathbf{F}^{-1}\right) = J^{-1}\mathbf{F} \cdot \mathbf{S} \cdot \mathbf{F}^{T} \tag{5.99}$$

where $\mathbf{S} = \mathbf{F}^{-1} \cdot \mathbf{P}$ is a generic Piola–Kirchhoff stress tensor. Through equation (5.11), the spatial mass and electric charge densities are related to their material counterparts by

$$\rho_t = J^{-1}\rho_0 \quad \text{and} \tag{5.100}$$

$$\varrho_t = J^{-1}\varrho_0 \quad . \tag{5.101}$$

The spatial equivalents of the magnetic field, magnetisation and induction vectors, as well as the electric current density vector, can be respectively computed by

$$
\begin{aligned}
\mathbb{h} = \mathbf{F}^{-T} \cdot \mathbb{H} \quad &, \quad \mathbb{m} = \mathbf{F}^{-T} \cdot \mathbb{M} \quad , \\
\mathbb{b} = \mathbb{B} \cdot \operatorname{cof}\left(\mathbf{F}^{-1}\right) = J^{-1}\mathbf{F} \cdot \mathbb{B} \quad &, \quad \mathbb{j} = \mathbb{J} \cdot \operatorname{cof}\left(\mathbf{F}^{-1}\right) = J^{-1}\mathbf{F} \cdot \mathbb{J} \quad .
\end{aligned}
\tag{5.102}
$$

The material quantities $\mathbb{H}, \mathbb{M}$ therefore undergo covariant transformation to form their spatial counterparts, while $\mathbb{B}, \mathbb{J}$ are pushed forward using a Piola transformation [204].

5.3.3 Summary of balance laws

As was previously mentioned, the derivation of the balance laws in the spatial setting is presented in Appendices C.5.1 and C.5.2. Briefly summarised in Table 5.4 are the

Table 5.4: Summary of the fully-coupled electromagnetic balance laws in the fully Eulerian setting.

Dynamic conditions (fully coupled electromagnetic balance laws)

Conservation of electric charge

$$\mathrm{d}_t \varrho_t + \nabla \cdot \mathbb{j}' = 0 \quad \text{in} \quad \mathscr{B}_t \cup \mathscr{S}_t \tag{5.103}$$

Maxwell's equations

$$\nabla \times \mathbb{e}' = -\mathrm{d}_t \mathbb{b} \quad , \quad \nabla \cdot \mathbb{d} = \varrho_t^{\mathrm{f}} \quad \text{in} \quad \mathscr{B}_t \cup \mathscr{S}_t \tag{5.104}$$

$$\nabla \times \mathbb{h}' = \mathbb{j}^{\mathrm{f}\prime} + \mathrm{d}_t \mathbb{d} \quad , \quad \nabla \cdot \mathbb{b} = 0 \quad \text{in} \quad \mathscr{B}_t \cup \mathscr{S}_t \tag{5.105}$$

Polarisation and magnetisation

$$\nabla \cdot \mathbb{p} := -\varrho_t^{\mathrm{b}} \quad , \quad \nabla \times \mathbb{m}' := \mathbb{j}^{\mathrm{b}\prime} - \mathrm{d}_t \mathbb{p} \tag{5.106}$$

$$\mathbf{n} \times [\![\mathbb{p}]\!] := -\widehat{\varrho}_t^{\mathrm{b}} \quad , \quad \mathbf{n} \times [\![\mathbb{m}']\!] := \widehat{\mathbb{j}}^{\mathrm{b}\prime} + [\![\mathbb{p}]\!] \left[\mathbf{v} \cdot \mathbf{n}\right] \tag{5.107}$$

Fundamental constitutive relations

$$\mathbb{d} = \varepsilon_0 \mathbb{e} + \mathbb{p} \quad , \quad \mathbb{b} = \mu_0 \left[\mathbb{h} + \mathbb{m}\right] \tag{5.108}$$

Continuity conditions

$$\mathbf{n} \times [\![\mathbb{e}']\!] = [\![\mathbb{b}]\!] \left[\mathbf{v} \cdot \mathbf{n}\right] \quad , \quad \mathbf{n} \cdot [\![\mathbb{d}]\!] = \widehat{\varrho}_t^{\mathrm{f}} \quad \text{on} \quad \partial \mathscr{B}_t \tag{5.109}$$

and other material discontinuities

$$\mathbf{n} \times [\![\mathbb{h}']\!] = \widehat{\mathbb{j}}^{\mathrm{f}\prime} - [\![\mathbb{d}]\!] \left[\mathbf{v} \cdot \mathbf{n}\right] \quad , \quad \mathbf{n} \cdot [\![\mathbb{b}]\!] = 0 \quad \text{on} \quad \partial \mathscr{B}_t \tag{5.110}$$

and other material discontinuities

Total charge and current densities

$$\varrho_t = \varrho_t^{\mathrm{b}} + \varrho_t^{\mathrm{f}} \quad , \quad \mathbb{j} = \mathbb{j}^{\mathrm{b}} + \mathbb{j}^{\mathrm{f}} \tag{5.111}$$

general spatial counterparts of the electromagnetic balance laws presented in Section 5.3.1. These statements for the balance laws are those expressed in the Eulerian configuration, which is the natural setting in which to derive them.

Although the laws of electro-magnetism are most widely expressed in an Eulerian framework, those governing solid mechanics typically follow a Lagrangian formulation. To consolidate the similarities and differences between the two settings, and to assist in transforming between them, we provide Tables 5.5 and 5.6 as a unified summary of Maxwell's equations in the fully Lagrangian and fully Eulerian settings, along with their description in two intermediate (or "updated") configurations.

Therein, in order to preserve and highlight the symmetry of the descriptions, we define the spatial free space electric displacement vector (related to equation $(5.108)_1$) and the spatial free space magnetic field vector (related to equation $(5.108)_2$) as

$$\mathbb{d}_\varepsilon := \varepsilon_0 \mathbb{e} \quad \text{and} \tag{5.112}$$

$$\mathbb{h}_\mu := \frac{1}{\mu_0} \mathbb{b} \quad ; \tag{5.113}$$

these are analogous to their referential counterparts that are related to equations (5.73) and (5.74) and which were previously defined in equations (5.68) and (5.69).

Through inspection of the transformations listed in Tables 5.5 and 5.6, we can observe that the Lagrangian and Eulerian descriptions of the governing equations coincide when the velocity of motion is negligible (as all convective terms in both the field transformations and time derivatives vanish). As noted earlier, should the time rate of change of fields also be negligible then the coupling between the electric and magnetic fields (through Faraday's law and Ampére's law) also vanishes. Table 5.7 lists the decoupled magneto-elastodynamic balance laws, constitutive relationships and decompositions from which we can draw a few key observations. The balance of angular momentum given in equation (5.121) illustrates that the total Cauchy stress is always symmetric. Due to the magnetic couple, from equation (5.130) it is deduced that the mechanical Cauchy stress $\boldsymbol{\sigma}^{\text{mech}}$ is not necessarily symmetric. However, the Maxwell stress is symmetric ensuring that in the free space (where there are no mechanical stresses nor a magnetisable medium) the total Cauchy stress remains symmetric.

To conclude this summary, in Table 5.8 are the primary results obtained when assuming quasi-static conditions in conjunction with the existence of no free currents; these will be the conditions under which the majority of the forthcoming chapters will be considered.

Table 5.5: Unified summary of the governing equations for electromagnetism in the general material domain $\mathscr{D} = \mathscr{B} \cup \mathscr{S}$. As tabulated, the governing equations are the electric charge conservation law, Faraday's law, Ampére's law, Gauss' electric law, and Gauss' magnetic law. Red arrows indicate the direction of application of a Piola transformation (Lagrangian ⇄ Eulerian setting), and blue arrows indicate the application of Reynolds transport theorem (referential ⇄ spatial variables). [325, 523, 522, 414, 133, 230, 119, 122, 541, 73, 406, 504, 532]

Referential configuration		**Current configuration**	
Transformations	**Governing equations**	**Governing equations**	**Transformations**
Free space contributions (in domain $\mathscr{S}$)			
$\mathbb{H}_\mu = \frac{1}{\mu_0} J^{-1}\mathbf{C}\cdot\mathbb{B}$ $\mathbb{D}_\varepsilon = \varepsilon_0 J\mathbf{C}^{-1}\cdot\mathbb{E}$	$\mathsf{D}_t\varrho_0 + \nabla_0\cdot\mathbb{J} = 0$ $\nabla_0\times\mathbb{E} + \mathsf{D}_t\mathbb{B} = \mathbf{0}$ $\nabla_0\times\mathbb{H}_\mu - \mathsf{D}_t\mathbb{D}_\varepsilon = \mathbb{J}$ $\nabla_0\cdot\mathbb{D}_\varepsilon = \varrho_0$ $\nabla_0\cdot\mathbb{B} = 0$	$\mathsf{n}_t\varrho_t + \nabla\cdot\mathbb{j} = 0$ $\nabla\times\mathbb{e} + \mathsf{n}_t\mathbb{b} = \mathbf{0}$ $\nabla\times\mathbb{h}_\mu - \mathsf{n}_t\mathbb{d}_\varepsilon = \mathbb{j}$ $\nabla\cdot\mathbb{d}_\varepsilon = \varrho_t$ $\nabla\cdot\mathbb{b} = 0$	$\mathbb{h}_\mu = \frac{1}{\mu_0}\mathbb{b}$ $\mathbb{d}_\varepsilon = \varepsilon_0\mathbb{e}$
$\mathbb{J}' = \mathbb{J} - \varrho_0\mathbf{V}$ $\mathbb{E}' = \mathbb{E} - \mathbb{B}\times\mathbf{V}$ $\mathbb{H}_\mu{}' = \mathbb{H}_\mu + \mathbb{D}_\varepsilon\times\mathbf{V}$	$\mathsf{N}_t\varrho_0 + \nabla_0\cdot\mathbb{J}' = 0$ $\nabla_0\times\mathbb{E}' + \mathsf{N}_t\mathbb{B} = \mathbf{0}$ $\nabla_0\times\mathbb{H}_\mu{}' - \mathsf{N}_t\mathbb{D}_\varepsilon = \mathbb{J}'$ $\nabla_0\cdot\mathbb{D}_\varepsilon = \varrho_0$ $\nabla_0\cdot\mathbb{B} = 0$	$\mathsf{d}_t\varrho_t + \nabla\cdot\mathbb{j}' = 0$ $\nabla\times\mathbb{e}' + \mathsf{d}_t\mathbb{b} = \mathbf{0}$ $\nabla\times\mathbb{h}_\mu{}' - \mathsf{d}_t\mathbb{d}_\varepsilon = \mathbb{j}'$ $\nabla\cdot\mathbb{d}_\varepsilon = \varrho_t$ $\nabla\cdot\mathbb{b} = 0$	$\mathbb{j}' = \mathbb{j} + \varrho_t\mathbf{v}$ $\mathbb{e}' = \mathbb{e} + \mathbb{b}\times\mathbf{v}$ $\mathbb{h}_\mu{}' = \mathbb{h}_\mu - \mathbb{d}_\varepsilon\times\mathbf{v}$
Matter contributions (in domain $\mathscr{B}$)			
$\mathbb{J} = \mathbb{J}^{\mathrm{b}} + \mathbb{J}^{\mathrm{f}}$ $\varrho_0 = \varrho_0^{\mathrm{b}} + \varrho_0^{\mathrm{f}}$	$\mathsf{D}_t\varrho_0^{\mathrm{b}} + \nabla_0\cdot\mathbb{J}^{\mathrm{b}} = 0$ – $-\nabla_0\times\mathbb{M} - \mathsf{D}_t\mathbb{P} = -\mathbb{J}^{\mathrm{b}}$ $\nabla_0\cdot\mathbb{P} = -\varrho_0^{\mathrm{b}}$ –	$\mathsf{n}_t\varrho_t^{\mathrm{b}} + \nabla\cdot\mathbb{j}^{\mathrm{b}} = 0$ – $-\nabla\times\mathbb{m} - \mathsf{n}_t\mathbb{p} = -\mathbb{j}^{\mathrm{b}}$ $\nabla\cdot\mathbb{p} = -\varrho_t^{\mathrm{b}}$ –	$\mathbb{j} = \mathbb{j}^{\mathrm{b}} + \mathbb{j}^{\mathrm{f}}$ $\varrho_t = \varrho_t^{\mathrm{b}} + \varrho_t^{\mathrm{f}}$
$\mathbb{M}' = \mathbb{M} - \mathbb{P}\times\mathbf{V}$ $\mathbb{J}^{\mathrm{b}\prime} = \mathbb{J}^{\mathrm{b}} - \varrho_0^{\mathrm{b}}\mathbf{V}$	$\mathsf{N}_t\varrho_0^{\mathrm{b}} + \nabla_0\cdot\mathbb{J}^{\mathrm{b}\prime} = 0$ – $-\nabla_0\times\mathbb{M}' - \mathsf{N}_t\mathbb{P} = -\mathbb{J}^{\mathrm{b}\prime}$ $\nabla_0\cdot\mathbb{P} = -\varrho_0^{\mathrm{b}}$ –	$\mathsf{d}_t\varrho_t^{\mathrm{b}} + \nabla\cdot\mathbb{j}^{\mathrm{b}\prime} = 0$ – $-\nabla\times\mathbb{m}' - \mathsf{d}_t\mathbb{p} = -\mathbb{j}^{\mathrm{b}\prime}$ $\nabla\cdot\mathbb{p} = -\varrho_t^{\mathrm{b}}$ –	$\mathbb{m}' = \mathbb{m} + \mathbb{p}\times\mathbf{v}$ $\mathbb{j}^{\mathrm{b}\prime} = \mathbb{j}^{\mathrm{b}} + \varrho_t^{\mathrm{b}}\mathbf{v}$
Combined contributions (free space + matter in domain $\mathscr{B}\cup\mathscr{S}$)			
$\mathbb{H} = \mathbb{H}_\mu - \mathbb{M}$ $\mathbb{D} = \mathbb{D}_\varepsilon + \mathbb{P}$	$\mathsf{D}_t\varrho_0^{\mathrm{f}} + \nabla_0\cdot\mathbb{J}^{\mathrm{f}} = 0$ $\nabla_0\times\mathbb{E} + \mathsf{D}_t\mathbb{B} = \mathbf{0}$ $\nabla_0\times\mathbb{H} - \mathsf{D}_t\mathbb{D} = \mathbb{J}^{\mathrm{f}}$ $\nabla_0\cdot\mathbb{D} = \varrho_0^{\mathrm{f}}$ $\nabla_0\cdot\mathbb{B} = 0$	$\mathsf{n}_t\varrho_t^{\mathrm{f}} + \nabla\cdot\mathbb{j}^{\mathrm{f}} = 0$ $\nabla\times\mathbb{e} + \mathsf{n}_t\mathbb{b} = \mathbf{0}$ $\nabla\times\mathbb{h} - \mathsf{n}_t\mathbb{d} = \mathbb{j}^{\mathrm{f}}$ $\nabla\cdot\mathbb{d} = \varrho_t^{\mathrm{f}}$ $\nabla\cdot\mathbb{b} = 0$	$\mathbb{h} = \mathbb{h}_\mu - \mathbb{m}$ $\mathbb{d} = \mathbb{d}_\varepsilon + \mathbb{p}$
$\mathbb{J}^{\mathrm{f}\prime} = \mathbb{J}^{\mathrm{f}} - \varrho_0^{\mathrm{f}}\mathbf{V}$ $\mathbb{H}' = \mathbb{H} + \mathbb{D}\times\mathbf{V}$	$\mathsf{N}_t\varrho_0^{\mathrm{f}} + \nabla_0\cdot\mathbb{J}^{\mathrm{f}\prime} = 0$ $\nabla_0\times\mathbb{E}' + \mathsf{N}_t\mathbb{B} = \mathbf{0}$ $\nabla_0\times\mathbb{H}' - \mathsf{N}_t\mathbb{D} = \mathbb{J}^{\mathrm{f}\prime}$ $\nabla_0\cdot\mathbb{D} = \varrho_0^{\mathrm{f}}$ $\nabla_0\cdot\mathbb{B} = 0$	$\mathsf{d}_t\varrho_t^{\mathrm{f}} + \nabla\cdot\mathbb{j}^{\mathrm{f}\prime} = 0$ $\nabla\times\mathbb{e}' + \mathsf{d}_t\mathbb{b} = \mathbf{0}$ $\nabla\times\mathbb{h}' - \mathsf{d}_t\mathbb{d} = \mathbb{j}^{\mathrm{f}\prime}$ $\nabla\cdot\mathbb{d} = \varrho_t^{\mathrm{f}}$ $\nabla\cdot\mathbb{b} = 0$	$\mathbb{j}^{\mathrm{f}\prime} = \mathbb{j}^{\mathrm{f}} + \varrho_t^{\mathrm{f}}\mathbf{v}$ $\mathbb{h}' = \mathbb{h} - \mathbb{d}\times\mathbf{v}$

Table 5.6: Unified summary of the jump terms to the governing equations for electromagnetism on the boundary of the general material domain $\partial\mathscr{D} = \partial\mathscr{B} \cup \partial\mathscr{S}$. As tabulated, the jump terms related to governing equations are the electric charge conservation law, Faraday's law, Ampére's law, Gauss' electric law, and Gauss' magnetic law. We have also made use of the identities $\mathbf{N} \times \llbracket (\bullet) \times \mathbf{V} \rrbracket = \llbracket (\bullet) \rrbracket V_N - [\llbracket (\bullet) \rrbracket \cdot \mathbf{N}]\, \mathbf{V}$ and $\mathbf{n} \times \llbracket (\bullet) \times \mathbf{v} \rrbracket = \llbracket (\bullet) \rrbracket v_n - [\llbracket (\bullet) \rrbracket \cdot \mathbf{n}]\, \mathbf{v}$ as part of the transformations. Red arrows indicate the direction of application of a Piola transformation (Lagrangian ⇄ Eulerian setting), and blue arrows indicate the application of Reynolds transport theorem (referential ⇄ spatial variables). [523, 414, 133, 119, 122, 541, 73, 406, 504, 532]

Referential configuration		**Current configuration**	
Transformations	**Governing equations**	**Governing equations**	**Transformations**
Free space contributions (on boundary $\partial\mathscr{S}$)			
$\mathbb{H}_\mu = \frac{1}{\mu_0} J^{-1}\mathbf{C}\cdot\mathbb{B}$ $\mathbb{D}_\varepsilon = \varepsilon_0 J\mathbf{C}^{-1}\cdot\mathbb{E}$	$\mathbf{N}\cdot\llbracket\mathbb{J}\rrbracket = 0$ $\mathbf{N}\times\llbracket\mathbb{E}\rrbracket = \mathbf{0}$ $\mathbf{N}\times\llbracket\mathbb{H}_\mu\rrbracket = \widehat{\mathbb{J}}$ $\mathbf{N}\cdot\llbracket\mathbb{D}_\varepsilon\rrbracket = \widehat{\varrho}_0$ $\mathbf{N}\cdot\llbracket\mathbb{B}\rrbracket = 0$	$\mathbf{n}\cdot\llbracket\mathbb{j}\rrbracket = 0$ $\mathbf{n}\times\llbracket\mathbb{e}\rrbracket = \mathbf{0}$ $\mathbf{n}\times\llbracket\mathbb{h}_\mu\rrbracket = \widehat{\mathbb{j}}$ $\mathbf{n}\cdot\llbracket\mathbb{d}_\varepsilon\rrbracket = \widehat{\varrho}_t$ $\mathbf{n}\cdot\llbracket\mathbb{b}\rrbracket = 0$	$\mathbb{h}_\mu = \frac{1}{\mu_0}\mathbb{b}$ $\mathbb{d}_\varepsilon = \varepsilon_0\mathbb{e}$
$V_N = \mathbf{N}\cdot\mathbf{V}$ $\widehat{\mathbb{J}}' = \widehat{\mathbb{J}} - \widehat{\varrho}_0\mathbf{V}$ $\mathbb{J}' = \mathbb{J} - \varrho_0\mathbf{V}$ $\mathbb{E}' = \mathbb{E} - \mathbb{B}\times\mathbf{V}$ $\mathbb{H}_\mu{}' = \mathbb{H}_\mu + \mathbb{D}_\varepsilon\times\mathbf{V}$	$\mathbf{N}\cdot\llbracket\mathbb{J}'\rrbracket = -\llbracket\varrho_0\rrbracket V_N$ $\mathbf{N}\times\llbracket\mathbb{E}'\rrbracket = -\llbracket\mathbb{B}\rrbracket V_N$ $\mathbf{N}\times\llbracket\mathbb{H}_\mu{}'\rrbracket = \widehat{\mathbb{J}}' + \llbracket\mathbb{D}_\varepsilon\rrbracket V_N$ $\mathbf{N}\cdot\llbracket\mathbb{D}_\varepsilon\rrbracket = \widehat{\varrho}_0$ $\mathbf{N}\cdot\llbracket\mathbb{B}\rrbracket = 0$	$\mathbf{n}\cdot\llbracket\mathbb{j}'\rrbracket = \llbracket\varrho_t\rrbracket v_n$ $\mathbf{n}\times\llbracket\mathbb{e}'\rrbracket = \llbracket\mathbb{b}\rrbracket v_n$ $\mathbf{n}\times\llbracket\mathbb{h}_\mu{}'\rrbracket = \widehat{\mathbb{j}}' - \llbracket\mathbb{d}_\varepsilon\rrbracket v_n$ $\mathbf{n}\cdot\llbracket\mathbb{d}_\varepsilon\rrbracket = \widehat{\varrho}_t$ $\mathbf{n}\cdot\llbracket\mathbb{b}\rrbracket = 0$	$v_n = \mathbf{n}\cdot\mathbf{v}$ $\widehat{\mathbb{j}}' = \widehat{\mathbb{j}} + \widehat{\varrho}_t\mathbf{v}$ $\mathbb{j}' = \mathbb{j} + \varrho_t\mathbf{v}$ $\mathbb{e}' = \mathbb{e} + \mathbb{b}\times\mathbf{v}$ $\mathbb{h}_\mu{}' = \mathbb{h}_\mu - \mathbb{d}_\varepsilon\times\mathbf{v}$
Matter contributions (on boundary $\partial\mathscr{B}$)			
$\widehat{\mathbb{J}} = \widehat{\mathbb{J}}^{\mathrm{b}} + \widehat{\mathbb{J}}^{\mathrm{f}}$ $\widehat{\varrho}_0 = \widehat{\varrho}_0^{\mathrm{b}} + \widehat{\varrho}_0^{\mathrm{f}}$	$\mathbf{N}\cdot\llbracket\mathbb{J}^{\mathrm{b}}\rrbracket = 0$ – $-\mathbf{N}\times\llbracket\mathbb{M}\rrbracket = -\widehat{\mathbb{J}}^{\mathrm{b}}$ $\mathbf{N}\cdot\llbracket\mathbb{P}\rrbracket = -\widehat{\varrho}_0^{\mathrm{b}}$ –	$\mathbf{n}\cdot\llbracket\mathbb{j}^{\mathrm{b}}\rrbracket = 0$ – $-\mathbf{n}\times\llbracket\mathbb{m}\rrbracket = -\widehat{\mathbb{j}}^{\mathrm{b}}$ $\mathbf{n}\cdot\llbracket\mathbb{p}\rrbracket = -\widehat{\varrho}_t^{\mathrm{b}}$ –	$\widehat{\mathbb{j}} = \widehat{\mathbb{j}}^{\mathrm{b}} + \widehat{\mathbb{j}}^{\mathrm{f}}$ $\widehat{\varrho}_t = \widehat{\varrho}_t^{\mathrm{b}} + \widehat{\varrho}_t^{\mathrm{f}}$
$\widehat{\mathbb{J}}^{\mathrm{b}\prime} = \widehat{\mathbb{J}}^{\mathrm{b}} - \widehat{\varrho}_0^{\mathrm{b}}\mathbf{V}$ $\mathbb{M}' = \mathbb{M} - \mathbb{P}\times\mathbf{V}$	$\mathbf{N}\cdot\llbracket\mathbb{J}^{\mathrm{b}\prime}\rrbracket = -\llbracket\varrho_0^{\mathrm{b}}\rrbracket V_N$ – $-\mathbf{N}\times\llbracket\mathbb{M}'\rrbracket = -\widehat{\mathbb{J}}^{\mathrm{b}\prime} + \llbracket\mathbb{P}\rrbracket V_N$ $\mathbf{N}\cdot\llbracket\mathbb{P}\rrbracket = -\widehat{\varrho}_0^{\mathrm{b}}$ –	$\mathbf{n}\cdot\llbracket\mathbb{j}^{\mathrm{b}\prime}\rrbracket = \llbracket\varrho_t^{\mathrm{b}}\rrbracket v_n$ – $-\mathbf{n}\times\llbracket\mathbb{m}'\rrbracket = -\widehat{\mathbb{j}}^{\mathrm{b}\prime} - \llbracket\mathbb{p}\rrbracket v_n$ $\mathbf{n}\cdot\llbracket\mathbb{p}\rrbracket = -\widehat{\varrho}_t^{\mathrm{b}}$ –	$\widehat{\mathbb{j}}^{\mathrm{b}\prime} = \widehat{\mathbb{j}}^{\mathrm{b}} + \widehat{\varrho}_t^{\mathrm{b}}\mathbf{v}$ $\mathbb{m}' = \mathbb{m} + \mathbb{p}\times\mathbf{v}$
Combined contributions (free space + matter on boundary $\partial\mathscr{B}\cup\partial\mathscr{S}$)			
$\mathbb{H} = \mathbb{H}_\mu - \mathbb{M}$ $\mathbb{D} = \mathbb{D}_\varepsilon + \mathbb{P}$	$\mathbf{N}\cdot\llbracket\mathbb{J}^{\mathrm{f}}\rrbracket = 0$ $\mathbf{N}\times\llbracket\mathbb{E}\rrbracket = \mathbf{0}$ $\mathbf{N}\times\llbracket\mathbb{H}\rrbracket = \widehat{\mathbb{J}}^{\mathrm{f}}$ $\mathbf{N}\cdot\llbracket\mathbb{D}\rrbracket = \widehat{\varrho}_0^{\mathrm{f}}$ $\mathbf{N}\cdot\llbracket\mathbb{B}\rrbracket = 0$	$\mathbf{n}\cdot\llbracket\mathbb{j}^{\mathrm{f}}\rrbracket = 0$ $\mathbf{n}\times\llbracket\mathbb{e}\rrbracket = \mathbf{0}$ $\mathbf{n}\times\llbracket\mathbb{h}\rrbracket = \widehat{\mathbb{j}}^{\mathrm{f}}$ $\mathbf{n}\cdot\llbracket\mathbb{d}\rrbracket = \widehat{\varrho}_t^{\mathrm{f}}$ $\mathbf{n}\cdot\llbracket\mathbb{b}\rrbracket = 0$	$\mathbb{h} = \mathbb{h}_\mu - \mathbb{m}$ $\mathbb{d} = \mathbb{d}_\varepsilon + \mathbb{p}$
$\widehat{\mathbb{J}}^{\mathrm{f}\prime} = \widehat{\mathbb{J}}^{\mathrm{f}} - \widehat{\varrho}_0^{\mathrm{f}}\mathbf{V}$ $\mathbb{J}^{\mathrm{f}\prime} = \mathbb{J}^{\mathrm{f}} - \varrho_0^{\mathrm{f}}\mathbf{V}$ $\mathbb{H}' = \mathbb{H} + \mathbb{D}\times\mathbf{V}$	$\mathbf{N}\cdot\llbracket\mathbb{J}^{\mathrm{f}\prime}\rrbracket = -\llbracket\varrho_0^{\mathrm{f}}\rrbracket V_N$ $\mathbf{N}\times\llbracket\mathbb{E}'\rrbracket = -\llbracket\mathbb{B}\rrbracket V_N$ $\mathbf{N}\times\llbracket\mathbb{H}'\rrbracket = \widehat{\mathbb{J}}^{\mathrm{f}\prime} + \llbracket\mathbb{D}\rrbracket V_N$ $\mathbf{N}\cdot\llbracket\mathbb{D}\rrbracket = \widehat{\varrho}_0$ $\mathbf{N}\cdot\llbracket\mathbb{B}\rrbracket = 0$	$\mathbf{n}\cdot\llbracket\mathbb{j}^{\mathrm{f}\prime}\rrbracket = \llbracket\varrho_t^{\mathrm{f}}\rrbracket v_n$ $\mathbf{n}\times\llbracket\mathbb{e}'\rrbracket = \llbracket\mathbb{b}\rrbracket v_n$ $\mathbf{n}\times\llbracket\mathbb{h}'\rrbracket = \widehat{\mathbb{j}}^{\mathrm{f}\prime} - \llbracket\mathbb{d}\rrbracket v_n$ $\mathbf{n}\cdot\llbracket\mathbb{d}\rrbracket = \widehat{\varrho}_t$ $\mathbf{n}\cdot\llbracket\mathbb{b}\rrbracket = 0$	$\widehat{\mathbb{j}}^{\mathrm{f}\prime} = \widehat{\mathbb{j}}^{\mathrm{f}} + \widehat{\varrho}_t^{\mathrm{f}}\mathbf{v}$ $\mathbb{j}^{\mathrm{f}\prime} = \mathbb{j}^{\mathrm{f}} + \varrho_t^{\mathrm{f}}\mathbf{v}$ $\mathbb{h}' = \mathbb{h} - \mathbb{d}\times\mathbf{v}$

Table 5.7: Summary of the decoupled magneto-elastodynamic balance laws in the spatial setting.

Elastodynamic (decoupled) magnetostatics with free currents

Maxwell's equations

$$\nabla \times \mathbb{h} = \mathbb{j}^{\mathrm{f}} \quad , \quad \nabla \cdot \mathbb{b} = 0 \quad \text{in} \quad \mathscr{B}_t \cup \mathscr{S}_t \tag{5.114}$$

Far-field boundary conditions

$$\mathbf{n}_\infty \times \mathbb{h} = \mathbf{n}_\infty \times \mathbb{h}_\infty \quad \text{on} \quad \partial \mathscr{S}_0^{\mathbb{H}} \tag{5.115}$$

$$\mathbf{n}_\infty \cdot \mathbb{b} = \mathbf{n}_\infty \cdot \mathbb{b}_\infty \quad \text{on} \quad \partial \mathscr{S}_0^{\mathbb{B}} \tag{5.116}$$

Magnetisation

$$\nabla \times \mathbb{m} := \mathbb{j}^{\mathrm{b}} \quad , \quad \mathbf{n} \times [\![\mathbb{m}]\!] := \widehat{\mathbb{j}}^{\mathrm{b}} \tag{5.117}$$

Fundamental constitutive relation

$$\mathbb{b} = \mu_0 \left[\mathbb{h} + \mathbb{m}\right] \tag{5.118}$$

Balance of mass

$$\mathrm{D}_t \rho_t + \rho_t \left[\nabla \cdot \mathbf{v}\right] = 0 \quad \text{in} \quad \mathscr{B}_t \cup \mathscr{S}_t \tag{5.119}$$

Balance of linear momentum

$$\nabla \cdot \boldsymbol{\sigma}^{\mathrm{mech}} + \mathbf{b}_t^{\mathrm{pon}} + \mathbf{b}_t^{\mathrm{mech}} = \nabla \cdot \boldsymbol{\sigma}^{\mathrm{tot}} + \mathbf{b}_t^{\mathrm{mech}} = \rho_t \mathbf{a}\left(\mathbf{X}, t\right) \quad \text{in} \quad \mathscr{B}_t \cup \mathscr{S}_t \tag{5.120}$$

Balance of angular momentum

$$\boldsymbol{\epsilon} : \left[\boldsymbol{\sigma}^{\mathrm{tot}}\right]^T = \mathbf{0} \quad \Rightarrow \quad \boldsymbol{\sigma}^{\mathrm{tot}} = \left[\boldsymbol{\sigma}^{\mathrm{tot}}\right]^T \quad \text{in} \quad \mathscr{B}_t \cup \mathscr{S}_t \tag{5.121}$$

Compatibility condition

$$\nabla_0 \times \mathbf{F} = \mathbf{0} \quad \text{in} \quad \mathscr{B}_0 \tag{5.122}$$

Body force

$$\mathbf{b}_t^{\mathrm{mech}} := \rho_t \mathbf{g} \tag{5.123}$$

Cauchy stress theorem

$$\mathbf{t}_t := \boldsymbol{\sigma} \cdot \mathbf{n} \tag{5.124}$$

Continuity conditions

$$[\![\mathbf{F}]\!] \times \mathbf{N} = \mathbf{0} \quad \text{on} \quad \mathscr{B}_0^1 \cap \mathscr{B}_0^2 \tag{5.125}$$

$$-[\![\boldsymbol{\sigma}^{\mathrm{tot}}]\!] \cdot \mathbf{n}^+ = \mathbf{t}_t^{\mathrm{mech,ext}} \quad \text{on} \quad \partial \mathscr{B}_t^{\mathbf{t}} \tag{5.126}$$

Ponderomotive body force and traction force, and magnetic moment densities

$$/\mathbf{b}_0^{\mathrm{pon}} \equiv \mathbf{b}_t^{\mathrm{pon}} \quad , \quad \mathbf{b}_t^{\mathrm{pon}} := \nabla \cdot \boldsymbol{\sigma}^{*\mathrm{pon}} = \mathbb{j}^{\mathrm{f}} \times \mathbb{b} + \mathbb{m} \cdot [\nabla \mathbb{b}] \tag{5.127}$$

$$\mathbf{t}_t^{\mathrm{pon}} := [\![\boldsymbol{\sigma}^{*\mathrm{pon}}]\!] \cdot \mathbf{n}^+ = \widehat{\mathbb{j}}^{\mathrm{f}} \times \{\!\{\mathbb{b}\}\!\} + [\{\!\{\mathbb{m}\}\!\} \cdot [\![\mathbb{b}]\!]] \cdot \mathbf{n}^+ \tag{5.128}$$

$$\mathbf{m}_t^{\mathrm{mag}} = \mathbb{m} \times \mathbb{b} \tag{5.129}$$

Decomposition of the mechanical and ponderomotive stresses

$$\boldsymbol{\sigma}^{\mathrm{tot}} = \boldsymbol{\sigma}^{\mathrm{mech}} + \boldsymbol{\sigma}^{*\mathrm{pon}} = \boldsymbol{\sigma}^{\mathrm{mech}} + \boldsymbol{\sigma}^{*\mathrm{mag}} + \boldsymbol{\sigma}^{*\mathrm{max}} \tag{5.130}$$

Ponderomotive, magnetisation and Maxwell stress tensors

$$\boldsymbol{\sigma}^{*\mathrm{pon}} = -\frac{1}{2\mu_0}\left[\mathbb{b} \cdot \mathbb{b}\right] \mathbf{i} + \left[\mathbb{m} \cdot \mathbb{b}\right] \mathbf{i} + \mathbb{h} \otimes \mathbb{b} = \frac{1}{2\mu_0}\left[\mathbb{b} \cdot \mathbb{b}\right] \mathbf{i} - \left[\mathbb{h} \cdot \mathbb{b}\right] \mathbf{i} + \mathbb{h} \otimes \mathbb{b} \tag{5.131a}$$

$$\boldsymbol{\sigma}^{*\mathrm{mag}} = \left[\mathbb{m} \cdot \mathbb{b}\right] \mathbf{i} - \mathbb{m} \otimes \mathbb{b} \tag{5.131b}$$

$$\boldsymbol{\sigma}^{*\mathrm{max}} = \frac{1}{\mu_0}\left[\mathbb{b} \otimes \mathbb{b} - \frac{1}{2}\left[\mathbb{b} \cdot \mathbb{b}\right] \mathbf{i}\right] \tag{5.131c}$$

Table 5.8: Summary of the decoupled magneto-elastostatic balance laws in the spatial setting: Quasi-static conditions with no free currents.

Quasi-static conditions with no free currents	
Conservation of electric charge	
$\nabla \cdot \mathbb{j} = 0 \quad \text{in} \quad \mathcal{B}_t \cup \mathcal{S}_t$	(5.132)
Maxwell's equations	
$\nabla \times \mathbb{h} = \mathbf{0} \quad , \quad \nabla \cdot \mathbb{b} = 0 \quad \text{in} \quad \mathcal{B}_t \cup \mathcal{S}_t$	(5.133)
Continuity conditions	
$\mathbf{n} \times [\![\mathbb{h}]\!] = \mathbf{0} \quad , \quad \mathbf{n} \cdot [\![\mathbb{b}]\!] = 0 \quad \text{on} \quad \partial\mathcal{B}_t \quad \text{and other material discontinuities}$	(5.134)
Balance of linear momentum	
$\nabla \cdot \boldsymbol{\sigma}^{\text{mech}} + \mathbf{b}_t^{\text{pon}} + \mathbf{b}_t^{\text{mech}} = \nabla \cdot \boldsymbol{\sigma}^{\text{tot}} + \mathbf{b}_t^{\text{mech}} = \mathbf{0} \quad \text{in} \quad \mathcal{B}_t \cup \mathcal{S}_t$	(5.135)
Ponderomotive body and traction force densities	
$\mathbf{b}_t^{\text{pon}} = \mathbb{m} \cdot [\nabla \mathbb{b}] \quad , \quad \mathbf{t}_t^{\text{pon}} = [\{\!\{\mathbb{m}\}\!\} \cdot [\![\mathbb{b}]\!]] \cdot \mathbf{n}^+ = [\{\!\{\mathbb{m}\}\!\} \times \mathbf{n}^+] \times [\![\mathbb{b}]\!]$	(5.136)

5.3.4 Ponderomotive force and moment

In order to gauge the influence of material magnetisation on the deformation, it may be useful to quantify the magnetic forces and torques developed within a subregion of a continuum. In Figure 5.6, a filled material system representing a magneto-mechanical continuum body is depicted.

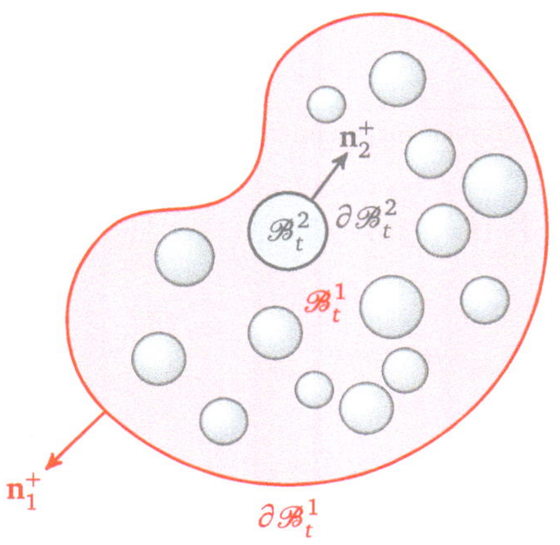

Figure 5.6: A domain composed of magnetisable particles $\mathcal{B}_t^2$ embedded inside a deformable body $\mathcal{B}_t^1$.

For a quasi-static system with no free currents, we consider the domain $\mathscr{B}_t^2$ as one of several magnetisable particles enclosed within in a carrier matrix occupying the region $\mathscr{B}_t^1$. Of interest are the resultant ponderomotive force and torque that act on the particle due to some externally applied magnetic load. In this scenario, it is known that the ponderomotive force develops not only due to volume-type forces, but also surface-type contributions at the interface $\partial\mathscr{B}_t^2$ between particle and the surrounding matrix [133, 180]. The overall ponderomotive force (as derived in Appendix C.4) is therefore decomposed into volume and interface contributions

$$\mathbf{f}_t^{\text{pon}} = \int_{\mathscr{B}_t^2} \mathbf{b}_t^{\text{pon}} \, \mathrm{dv} + \int_{\partial\mathscr{B}_t^2} \mathbf{t}_t^{\text{pon}} \, \mathrm{da} = \int_{\mathscr{B}_t^2} \nabla \cdot \boldsymbol{\sigma}^{*\text{pon}} \, \mathrm{dv} + \int_{\partial\mathscr{B}_t^2} [\![\boldsymbol{\sigma}^{*\text{pon}}]\!] \cdot \mathbf{n}_2^+ \, \mathrm{da} =: \mathbf{f}_{t,\text{vol}}^{\text{pon}} + \mathbf{f}_{t,\text{int}}^{\text{pon}} \quad , \tag{5.137}$$

wherein equation (5.127) was used in the transformation of the volume forces. Furthermore, from equation (5.128) we introduced the definition of the ponderomotive tractions

$$\mathbf{t}_t^{\text{pon}} = [\![\boldsymbol{\sigma}^{*\text{pon}}]\!] \cdot \mathbf{n}_i^+ \tag{5.138}$$

that can occur at all material discontinuities (that is to say, at both $\partial\mathscr{B}_t^1$ and $\partial\mathscr{B}_t^2$).

Similar to the general moment balance equation from which the balance of angular momentum is derived, the equivalent magnetic moments can be defined in general as

$$\mathbf{m}_t^{\text{pon}} = \int_{\mathscr{B}_t^2} [\mathbf{r} \times \mathbf{b}_t^{\text{pon}} + \mathbf{m}_t^{\text{mag}}] \, \mathrm{dv} + \int_{\partial\mathscr{B}_t^2} \mathbf{r} \times \mathbf{t}_t^{\text{pon}} \, \mathrm{da} =: \mathbf{m}_{t,\text{vol}}^{\text{pon}} + \mathbf{m}_{t,\text{int}}^{\text{pon}} \quad . \tag{5.139}$$

Here, $\mathbf{r} = \mathbf{x} - \mathbf{x}_0$ denotes the vector offset between point $\mathbf{x} \in \overline{\mathscr{B}_t^2}$ and the arbitrary reference location $\mathbf{x}_0$. It is observed in equations (5.93) and (5.139) that, due to the magnetic couple, there arises an additional magnetic moment

$$\mathbf{m}_t^{\text{mag}} = \boldsymbol{\epsilon} : [\boldsymbol{\sigma}^{*\text{mag}}]^T = \mathbb{m} \times \mathbb{b} \equiv \boldsymbol{\epsilon} : [\boldsymbol{\sigma}^{*\text{pon}}]^T \tag{5.140}$$

that consequently renders the mechanical stress $\boldsymbol{\sigma}^{\text{mech}}$ non-symmetric [125]; this is proven in Appendix C.5.2. As was shown for the ponderomotive force, the moment derived from the magnetisation of the material can also be expressed in terms of a volume and a surface contribution.

In equations (5.137) and (5.139), it is clear that several factors other than the constitution of the particle $\mathscr{B}_t^2$ determine the ratio of ponderomotive forces and torques developed in the bulk and on the surface of solid bodies. For example, the difference in relative magnetic permeability of adjacent materials influences the magnitude of the jump of the ponderomotive stress across the interface, thereby affecting the ponderomotive traction. Furthermore, the shape and size of the particle defines the surface area to volume ratio and, consequently, the weighting between the volume and surface forces and moments.

5.3.5 Formulations for magnetic potentials

To solve the governing equations for quasi-static magneto-mechano-static, a few key steps remain. Firstly one must choose a set of independent variables that, under all conditions, satisfy a subset of the balance laws. Secondly a constitutive model that links the independent variables to the remaining dependent ones must be chosen in order to provide closure of the equations. From this, the boundary value problem can be solved; for problems with complex domains, typically a numerical method involving a method of space (and time or load) discretisation is employed to find an approximate solution to the boundary value problem.

For the first point, there are a number of equally valid choices that have different repercussions in terms of their implementation and numerical qualities. Almost universally in solid mechanics (see [441] for an alternative approach), the displacement field is chosen as a primitive quantity because, among other reasons, $\boldsymbol{\varphi}$ is a potential field for $\mathbf{F}$. Therefore, the compatibility condition given in equation (5.94) is always satisfied due to the identity given in equation (A.24). In terms of the magnetic problem, formulations based on the magnetic vector potential (MVP) or the magnetic scalar potential (MSP) are commonly adopted; Figure 5.7 illustrates how the boundary conditions change for each case.

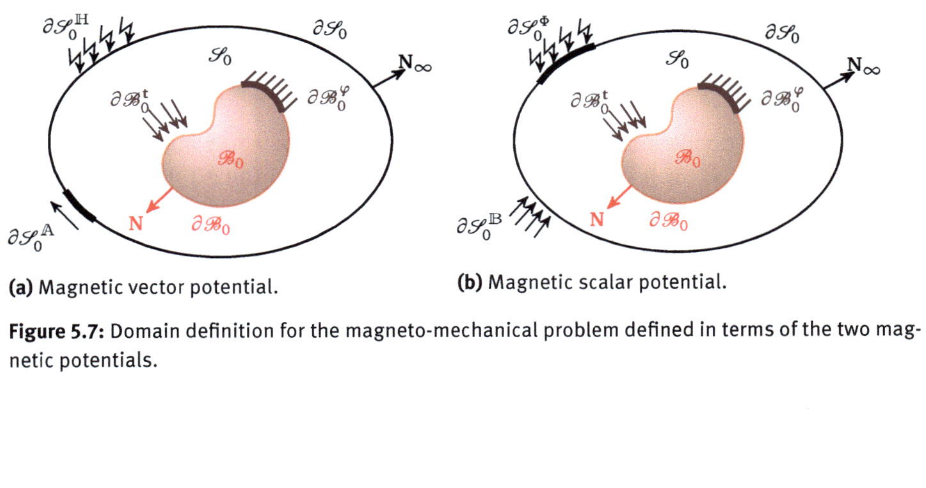

(a) Magnetic vector potential. **(b)** Magnetic scalar potential.

Figure 5.7: Domain definition for the magneto-mechanical problem defined in terms of the two magnetic potentials.

5.3.5.1 Magnetic vector potential

In this approach, it is assumed that the magnetic induction is the independent variable with a constitutive law governing the relationship between $\mathbb{B}$ and the magnetic field $\mathbb{H}(\mathbb{B})$. A fictitious vector potential field $\mathbb{A}(\mathbf{X})$ is defined [77, 277] such that

$$\mathbb{B}(\mathbf{X}) := \boldsymbol{\nabla}_0 \times \mathbb{A}(\mathbf{X}) \quad , \tag{5.141}$$

with the result that

$$\mathbf{N} \times [\![\mathbb{A}]\!] = \mathbf{0} \quad \text{on} \quad \partial\mathscr{B}_0 \quad \text{and} \tag{5.142}$$

$$\mathbf{N}_\infty \times \mathbb{A} = \mathbf{N}_\infty \times \mathbb{A}_\infty \quad \text{on} \quad \partial\mathscr{S}_0^{\mathbb{A}} \quad . \tag{5.143}$$

As is depicted in Figure 5.7a, the tangential component of the vector potential is prescribed on the essential boundary. Up to a constant vector field, $\mathbb{A}$ in equation (5.141) can be shown to satisfy equation (A.25) meaning that it satisfies equation $(5.67)_2$ regardless of the assumptions underlying the temporal dependence of the electric fields or the existence of free currents in the system.

Although in many respects this is considered the natural formulation for the magnetic problem (see [483] and the later discussions in Section 5.4.3 and Chapter 7), it presents a few challenges in terms of its numerical implementation. If the magnetic vector potential field is chosen such that its tangential components are always continuous across any interface, then equation (5.142) is always satisfied, leaving equation $(5.85)_1$ to be explicitly resolved. However, in order for $\mathbb{A}$ to be unique, is must also simultaneously satisfy either the Coulomb gauge condition [77, 99]

$$\nabla_0 \cdot \mathbb{A} = 0 \quad , \tag{5.144}$$

or the incomplete gauge condition [354, 35, 5]

$$\mathbb{A} \cdot (\bullet)^\dagger = 0 \tag{5.145}$$

for any arbitrary vector field $(\bullet)^\dagger$ that possesses no closed field lines.

5.3.5.2 Magnetic scalar potential

An alternative to the vector potential approach is the magnetic scalar potential formulation, which is often used due to its comparatively simple numerical implementation. However, its use is restricted to the case of quasi-static magnetic conditions as it is assumed that the material is electrically non-conductive and in the presence of no free currents. Under these conditions we assume that the magnetic induction $\mathbb{B}(\mathbb{H})$ is parameterised in terms of the magnetic field and that the referential magnetic field, being curl-free, is related to the fictitious magnetic scalar potential $\Phi(\mathbf{X})$ by [277, 56]

$$\mathbb{H}(\mathbf{X}) := -\nabla_0 \Phi(\mathbf{X}) \quad \text{in} \quad \mathscr{B}_0 \cup \mathscr{S}_0 \quad . \tag{5.146}$$

Such a field always satisfies equation (A.24) up to a constant value. With reference to Figure 5.7b, the continuity condition on the solid body-free space interface and the truncated far-field boundary condition are, respectively,

$$[\![\Phi]\!] = 0 \quad \text{on} \quad \partial\mathscr{B}_0 \quad , \quad \Phi = \overline{\Phi} \quad \text{on} \quad \partial\mathscr{S}_0^{\Phi} \quad . \tag{5.147}$$

5.3.6 Weak formulation of conservation laws

The foundation of any finite-element formulation is the expression of the balance laws in weak form. From the equations set out in Section 5.3.1, the following sections detail the derivation of the weak balance laws from their localised strong form. As there are no assumptions made about the nature of the underlying materials, the equations set out here are valid for any material including those for which we can describe either a total free energy function Ψ_0 (applicable to, e. g. dissipative media) or a total stored energy function Ω_0 (applicable to magneto-hyperelastic materials).

5.3.6.1 Mechanical contribution

To derive the weak form of the balance of linear momentum, we first introduce a vectorial test function $\delta\boldsymbol{\varphi}$. After multiplying equation (5.87) by the test function and integrating over the entire domain, we immediately recognise that the body force has no influence in the free space, and the result becomes

$$
\begin{aligned}
0 &= \int_{\mathscr{B}_0 \cup \mathscr{S}_0} \delta\boldsymbol{\varphi} \cdot \left[\nabla_0 \cdot \mathbf{P}^{\text{tot}}\right] \, \mathrm{dV} + \int_{\mathscr{B}_0} \delta\boldsymbol{\varphi} \cdot \mathbf{b}_0^{\text{mech}} \, \mathrm{dV} \\
&= \int_{\mathscr{B}_0} \delta\boldsymbol{\varphi} \cdot \left[\nabla_0 \cdot \mathbf{P}^{\text{tot}}\right] \, \mathrm{dV} + \int_{\mathscr{S}_0} \delta\boldsymbol{\varphi} \cdot \left[\nabla_0 \cdot \mathbf{P}^{*\max}\right] \, \mathrm{dV} + \int_{\mathscr{B}_0} \delta\boldsymbol{\varphi} \cdot \mathbf{b}_0^{\text{mech}} \, \mathrm{dV} \quad .
\end{aligned}
$$

Here, the decomposition of stresses is given by equation (5.90), wherein we recall that the free space medium is non-magnetisable. After application of integration by parts and subsequently equation (B.13), we arrive at the intermediate result that

$$
\begin{aligned}
0 = &- \int_{\mathscr{B}_0} \nabla_0 \delta\boldsymbol{\varphi} : \mathbf{P}^{\text{tot}} \, \mathrm{dV} + \int_{\mathscr{B}_0} \nabla_0 \cdot \left[\delta\boldsymbol{\varphi} \cdot \mathbf{P}^{\text{tot}}\right] \, \mathrm{dV} \\
&- \int_{\mathscr{S}_0} \nabla_0 \delta\boldsymbol{\varphi} : \mathbf{P}^{*\max} \, \mathrm{dV} + \int_{\mathscr{S}_0} \nabla_0 \cdot \left[\delta\boldsymbol{\varphi} \cdot \mathbf{P}^{*\max}\right] \, \mathrm{dV} + \int_{\mathscr{B}_0} \delta\boldsymbol{\varphi} \cdot \mathbf{b}_0^{\text{mech}} \, \mathrm{dV} \quad . \qquad (5.148)
\end{aligned}
$$

Lastly, after using the divergence theorem given in equation (B.16) and noting that any mechanical traction applied to the truncated free space boundary has no influence on the body, what remains is the result that

$$
\begin{aligned}
\int_{\mathscr{B}_0 \cup \mathscr{S}_0} \nabla_0 \delta\boldsymbol{\varphi} : \mathbf{P}^{\text{tot}} \, \mathrm{dV} &= \int_{\mathscr{B}_0} \nabla_0 \delta\boldsymbol{\varphi} : \mathbf{P}^{\text{tot}} \, \mathrm{dV} + \int_{\mathscr{S}_0} \nabla_0 \delta\boldsymbol{\varphi} : \mathbf{P}^{*\max} \, \mathrm{dV} \\
&= - \int_{\partial\mathscr{B}_0^{\mathbf{t}}} \delta\boldsymbol{\varphi} \cdot \left[\!\left[\mathbf{P}^{\text{tot}}\right]\!\right] \cdot \mathbf{N} \, \mathrm{dA} + \int_{\mathscr{B}_0} \delta\boldsymbol{\varphi} \cdot \mathbf{b}_0^{\text{mech}} \, \mathrm{dV} \\
&= \int_{\partial\mathscr{B}_0^{\mathbf{t}}} \delta\boldsymbol{\varphi} \cdot \mathbf{t}_0^{\text{mech,ext}} \, \mathrm{dA} + \int_{\mathscr{B}_0} \delta\boldsymbol{\varphi} \cdot \mathbf{b}_0^{\text{mech}} \, \mathrm{dV} \qquad (5.149)
\end{aligned}
$$

where the mechanical traction acting on the body-free space interface is derived from equation (5.98).

From equation (5.149), we can make some observations regarding the space in which the test function must exist. In particular, we note that it must derive from the differentiable (Sobolev) space

$$\delta\boldsymbol{\varphi} \in \mathscr{H}^1\left(\overline{\mathscr{B}_0 \cup \mathscr{S}_0}\right) \tag{5.150}$$

and is subject to the constraint

$$\delta\boldsymbol{\varphi} = \mathbf{0} \quad \text{on} \quad \partial\mathscr{B}_0^{\boldsymbol{\varphi}} \cup \partial\mathscr{S}_0 \quad . \tag{5.151}$$

5.3.6.2 Magnetic contribution with parameterisation in terms of $\mathbb{B}$

When considering the weak magnetic conservation laws with the magnetic induction chosen as the independent parameter, then the static form of Ampére's law is the starting point for their derivation. From equation $(5.84)_2$, again we multiply by a vector test function $\delta\mathbb{A}$ and integrate over the entire domain, rendering

$$0 = \int\limits_{\mathscr{B}_0 \cup \mathscr{S}_0} \delta\mathbb{A} \cdot [\boldsymbol{\nabla}_0 \times \mathbb{H}] \; \mathrm{d}V \quad . \tag{5.152}$$

Employing integration by parts through equation (B.12) leads to

$$0 = \int\limits_{\mathscr{B}_0 \cup \mathscr{S}_0} \boldsymbol{\nabla}_0 \cdot [\mathbb{H} \times \delta\mathbb{A}] \; \mathrm{d}V + \int\limits_{\mathscr{B}_0 \cup \mathscr{S}_0} [\boldsymbol{\nabla}_0 \times \delta\mathbb{A}] \cdot \mathbb{H} \; \mathrm{d}V \quad . \tag{5.153}$$

Thereafter, we utilise the divergence theorem (equation (B.16)) on the first term

$$\int\limits_{\mathscr{B}_0 \cup \mathscr{S}_0} \boldsymbol{\nabla}_0 \cdot [\mathbb{H} \times \delta\mathbb{A}] \; \mathrm{d}V = \int\limits_{\partial\mathscr{S}_0} \mathbf{N} \cdot [\mathbb{H} \times \delta\mathbb{A}] \; \mathrm{d}A \tag{5.154}$$

from which, with consideration of equation (A.12) for the permutation of the vectors, we attain the final result that

$$0 = \int\limits_{\mathscr{B}_0 \cup \mathscr{S}_0} [\boldsymbol{\nabla}_0 \times \delta\mathbb{A}] \cdot \mathbb{H} \; \mathrm{d}V + \int\limits_{\partial\mathscr{S}_0^{\mathbb{H}}} \delta\mathbb{A} \cdot [\mathbf{N}_\infty \times \mathbb{H}_\infty] \; \mathrm{d}A \quad . \tag{5.155}$$

The requirements from equation (5.155) on the test function is that it must conform to the Sobolev space

$$\delta\mathbb{A} \in \mathscr{H}^{\mathrm{curl}}\left(\overline{\mathscr{B}_0 \cup \mathscr{S}_0}\right) \quad \text{subject to} \quad \delta\mathbb{A} = \mathbf{0} \quad \text{on} \quad \partial\mathscr{S}_0^{\mathbb{A}} \equiv \partial\mathscr{S}_0^{\mathbb{B}} \quad . \tag{5.156}$$

5.3.6.3 Magnetic contribution with parameterisation in terms of $\mathbb{H}$

Upon the choice of the magnetic field as the independent variable, we derive the magnetic balance law with Gauss' magnetism law as the conception point. Premultiplying by $\delta\Phi$, a scalar test function, and integrating equation $(5.67)_2$ over the domain leads to

$$0 = \int_{\mathscr{B}_0 \cup \mathscr{S}_0} \delta\Phi \, [\nabla_0 \cdot \mathbb{B}] \; \mathrm{d}V \quad . \tag{5.157}$$

Thereafter, integration by parts using equation (B.9) is employed

$$0 = \int_{\mathscr{B}_0 \cup \mathscr{S}_0} \nabla_0 \cdot [\delta\Phi \, \mathbb{B}] \; \mathrm{d}V - \int_{\mathscr{B}_0 \cup \mathscr{S}_0} \nabla_0 \delta\Phi \cdot \mathbb{B} \; \mathrm{d}V \quad , \tag{5.158}$$

after which application of the divergence theorem and rearrangement of the result renders the final balance law

$$\int_{\mathscr{B}_0 \cup \mathscr{S}_0} \nabla_0 \delta\Phi \cdot \mathbb{B} \; \mathrm{d}V = \int_{\partial \mathscr{S}_0^{\mathbb{B}}} \delta\Phi \; [\mathbf{N}_\infty \cdot \mathbb{B}_\infty] \; \mathrm{d}A \quad . \tag{5.159}$$

Equation (5.159) dictates that the test function lies in the space of differentiable functions

$$\delta\Phi \in \mathscr{H}^1\left(\overline{\mathscr{B}_0 \cup \mathscr{S}_0}\right) \quad \text{subject to} \quad \delta\Phi = 0 \quad \text{on} \quad \partial\mathscr{S}_0^{\Phi} \equiv \partial\mathscr{S}_0^{\mathbb{H}} \quad . \tag{5.160}$$

5.3.7 Variational formulations

Sophisticated extended frameworks have been developed in recent years to describe the behaviour of dissipative and rate-dependent materials using variational approaches. Miehe et al. [370] has shown that the finite-strain macroscopic coupled problem can be cast in an incremental variational form, thereby accommodating materials that include dissipative magnetostrictive effects. A constitutive variational principle is applied to optimise a generalised incremental work function (describing the material response) with respect to its arguments, a set of internal state variables. By defining a quasi-hyper-magnetoelastic potential, the incremental variational problem incorporates energy storage as well as dissipative mechanisms associated with a rate-dependent material response. The existence of this potential is then exploited to formulate the incremental boundary value problem using the principle of stationary incremental energy. Mielke [378] has in fact shown that, for a large category of dissipative materials, their thermodynamically consistent coupled balance equations can be formulated in terms of General Equations for Non-Equilibrium Reversible Irreversible Coupling (more commonly known by the acronym *GENERIC*). To highlight the robustness of these generalised approaches, frameworks of this nature have also been applied to formulate or solve problems in the topics of thermomechanics [410, 565] including plasticity [378] and viscoplasticity [500], phase field models in micromagnetics [365, 138], dissipative electro-magneto-mechanics [371], gradient-extended standard dissipative solids [362], gradient plasticity [363, 375, 373], diffusive phase separation using the Cahn–Hilliard model [372, 374] and quantum mechanics [379].

To maintain a simple variational formulation, in the spirit of the work conducted by Kankanala and Triantafyllidis [248], for non-dissipative (i. e. recoverable magnetoelastic) materials one can define a total potential energy functional [204]

$$\Pi = \Pi^{\text{int}} + \Pi^{\text{ext}} \tag{5.161}$$

for which the total energy stored in the magnetoelastic body and in the free space, in general, can be expressed in terms of a total stored energy density function Ω_0 [60, 120] by

$$\Pi^{\text{int}} = \int_{\mathscr{B}_0 \cup \mathscr{S}_0} \Omega_0 \,\mathrm{dV} = \int_{\mathscr{B}_0 \cup \mathscr{S}_0} [W_0 + M_0] \,\mathrm{dV} = \int_{\mathscr{B}_0} [W_0 + M_0] \,\mathrm{dV} + \int_{\mathscr{S}_0} M_0 \,\mathrm{dV} \tag{5.162}$$

where W_0 denotes the potential energy per unit reference volume due to the magnetoelastic material's deformation and magnetisation, while M_0 quantifies the energy stored in the magnetic field. By definition, $W_0 = 0$ in $\mathscr{S}_0$.

Similar as to what was presented and discussed by Bustamante and Ogden [69] for the case involving both a solid and free space domain, in the following paragraphs we will derive the first variation of the total potential energy functional for two distinct parameterisations . Assuming a non-linear material behaviour, we will also present their linearisation which can be utilised in conjunction with a non-linear solution scheme to resolve the stationary point for a prescribed set of loading conditions.

5.3.7.1 Parameterisation in terms of $\mathbb{B}$

When the stored energy is parameterised in terms of the magnetic induction, the total internal potential energy is given by

$$\Pi^{\text{int}} = \int_{\mathscr{B}_0 \cup \mathscr{S}_0} {\Omega_0}^* (\mathbf{F}, \mathbb{B}) \,\mathrm{dV} = \int_{\mathscr{B}_0} {W_0}^* (\mathbf{F}, \mathbb{B}) \,\mathrm{dV} + \int_{\mathscr{B}_0 \cup \mathscr{S}_0} {M_0}^* (\mathbf{F}, \mathbb{B}) \,\mathrm{dV} \quad , \tag{5.163}$$

where ${\Omega_0}^*, {W_0}^*, {M_0}^*$ denote energies per unit reference volume, while the total external potential energy is [69]

$$\Pi^{\text{ext}} = -\int_{\mathscr{B}_0} \boldsymbol{\varphi} \cdot \mathbf{b}_0^{\text{mech}} \,\mathrm{dV} - \int_{\partial \mathscr{B}_0^{\text{t}}} \boldsymbol{\varphi} \cdot \mathbf{t}_0^{\text{mech,ext}} \,\mathrm{dA} + \int_{\partial \mathscr{S}_0^{\mathbb{H}}} \mathbb{A} \cdot [\mathbf{N}_\infty \times \mathbb{H}_\infty] \,\mathrm{dA} \quad . \tag{5.164}$$

The magnetic energy stored in the free field is quantified by [123, 69, 533]

$${M_0}^* (\mathbf{F}, \mathbb{B}) \coloneqq \frac{1}{2\mu_0} [\mathbb{B} \otimes \mathbb{B}] : J^{-1}\mathbf{C} \quad . \tag{5.165}$$

First variation

The equilibrium solution to equation (5.161) can be determined using the principle of stationary potential energy. This states that the stationary point at which all directional derivatives of the total potential energy disappear defines the equilibrium configuration. The expression for the Gâteaux derivative of the total potential energy given the current parameterisation is

$$\begin{aligned}\delta\Pi(\boldsymbol{\varphi},\delta\boldsymbol{\varphi},\mathbb{A},\delta\mathbb{A}) &= \delta_{\boldsymbol{\varphi}}\Pi(\boldsymbol{\varphi},\mathbb{A})+\delta_{\mathbb{A}}\Pi(\boldsymbol{\varphi},\mathbb{A})\\ &= \frac{\mathrm{d}}{\mathrm{d}\epsilon}\Pi(\boldsymbol{\varphi}+\epsilon\delta\boldsymbol{\varphi},\mathbb{A})\Big|_{\epsilon=0}+\frac{\mathrm{d}}{\mathrm{d}\gamma}\Pi(\boldsymbol{\varphi},\mathbb{A}+\gamma\delta\mathbb{A})\Big|_{\gamma=0}\quad,\end{aligned}\tag{5.166}$$

which yields the first variation $\delta\Pi = \delta\Pi^{\mathrm{int}} + \delta\Pi^{\mathrm{ext}}$. The equilibrium solution for the boundary value problem is found at the stationary point [42] of this variation; that is to say that

$$\min_{\boldsymbol{\varphi},\mathbb{A}}\Pi \quad\Rightarrow\quad \delta\Pi = \delta\Pi^{\mathrm{int}}+\delta\Pi^{\mathrm{ext}} = 0\quad.\tag{5.167}$$

From equation (5.163), and using equations (5.141) and (5.166), the variation of the internal potential energy is expressed concisely as

$$\delta\Pi^{\mathrm{int}} = \int_{\mathscr{B}_0\cup\mathscr{S}_0}\delta\mathbf{F}:\frac{\partial\Omega_0{}^*(\mathbf{F},\mathbb{B})}{\partial\mathbf{F}}\,\mathrm{dV}+\int_{\mathscr{B}_0\cup\mathscr{S}_0}\delta\mathbb{B}\cdot\frac{\partial\Omega_0{}^*(\mathbf{F},\mathbb{B})}{\partial\mathbb{B}}\,\mathrm{dV}\quad,\tag{5.168}$$

and that of the external potential energy described in equation (5.164) is

$$\delta\Pi^{\mathrm{ext}} = -\int_{\mathscr{B}_0}\delta\boldsymbol{\varphi}\cdot\mathbf{b}_0^{\mathrm{mech}}\,\mathrm{dV}-\int_{\partial\mathscr{B}_0^{\mathrm{t}}}\delta\boldsymbol{\varphi}\cdot\mathbf{t}_0^{\mathrm{mech,ext}}\,\mathrm{dA}+\int_{\partial\mathscr{S}_0^{\mathbb{H}}}\delta\mathbb{A}\cdot[\mathbf{N}_\infty\times\mathbb{H}_\infty]\,\mathrm{dA}\quad.\tag{5.169}$$

Here we define the variations, denoted by $\delta(\bullet)$, of the deformation gradient and magnetic induction respectively as

$$\delta\mathbf{F} = \nabla_0\delta\boldsymbol{\varphi}\quad,\quad\delta\mathbb{B} = \nabla_0\times\delta\mathbb{A}\quad.\tag{5.170}$$

The above collectively represents the equivalent weak form of the divergence-free condition for the stress and the curl-free condition for the magnetic field as is respectively listed in equation (5.87) and equation $(5.84)_2$ for the magneto-elastostatic case. This can be inferred by comparing equation (5.168) to equations (5.149) and (5.155), and identifying the total Piola stress and the magnetic field as

$$\mathbf{P}^{\mathrm{tot}} = \frac{\partial\Omega_0{}^*(\mathbf{F},\mathbb{B})}{\partial\mathbf{F}}\quad,\quad\mathbb{H} = \frac{\partial\Omega_0{}^*(\mathbf{F},\mathbb{B})}{\partial\mathbb{B}}\quad.\tag{5.171}$$

Furthermore, the continuity of the normal traction and tangential magnetic field (given in equation (5.98) and equation $(5.85)_1$) at the material interfaces is also embedded within this formulation. However, it remains to be ensured that the tangential continuity condition for the deformation and the continuity of normal magnetic induction (given in strong form by equation (5.95) and equation $(5.85)_2$) are satisfied.

Linearisation

As the constitutive laws governing this class of materials are typically non-linear, an iterative procedure such as the Newton–Raphson method must be utilised to determine the stationary point. The non-linear system of equations can be linearised using a first-order Taylor expansion about the current solution.

Assuming that the external loading is a dead-load, the linear Taylor expansion of equation (5.168), with the use of equation (5.173), can be expressed as

$$\begin{aligned} 0 &\doteq \delta_{\boldsymbol{\varphi}}\Pi + \Delta_{\boldsymbol{\varphi}}\delta_{\boldsymbol{\varphi}}\Pi + \Delta_{\mathbb{A}}\delta_{\boldsymbol{\varphi}}\Pi \\ 0 &\doteq \delta_{\mathbb{A}}\Pi + \Delta_{\boldsymbol{\varphi}}\delta_{\mathbb{A}}\Pi + \Delta_{\mathbb{A}}\delta_{\mathbb{A}}\Pi \quad . \end{aligned} \tag{5.172}$$

The linearisation of equation (5.168) with respect to the motion and magnetic vector potential renders the contributions

$$\Delta\delta\Pi^{\text{int}} = \Delta_{\boldsymbol{\varphi}}\left[\delta_{\boldsymbol{\varphi}}\Pi^{\text{int}} + \delta_{\mathbb{A}}\Pi^{\text{int}}\right] + \Delta_{\mathbb{A}}\left[\delta_{\boldsymbol{\varphi}}\Pi^{\text{int}} + \delta_{\mathbb{A}}\Pi^{\text{int}}\right] \quad , \tag{5.173}$$

$$\begin{aligned} &= \int_{\mathscr{B}_0\cup\mathscr{S}_0} \left[\delta\mathbf{F} : \mathscr{A}^* : \Delta\mathbf{F} + \delta\mathbb{B}\cdot\left[\mathfrak{L}^*\right]^T : \Delta\mathbf{F}\right] \,\mathrm{dV} \\ &+ \int_{\mathscr{B}_0\cup\mathscr{S}_0} \left[\delta\mathbf{F} : \mathfrak{L}^*\cdot\Delta\mathbb{B} + \delta\mathbb{B}\cdot\mathbf{D}^*\cdot\Delta\mathbb{B}\right] \,\mathrm{dV} \quad , \end{aligned} \tag{5.174}$$

with $\Delta(\bullet)$ representing value increments and

$$\Delta\mathbf{F} = \nabla_0\Delta\boldsymbol{\varphi} \quad , \quad \Delta\mathbb{B} = \nabla_0\times\Delta\mathbb{A} \tag{5.175}$$

and the material tangents, namely the fourth-order mixed elasticity tensor, the third-order mixed magnetoelastic tensor and the second-order magnetic tensor, are defined as

$$\begin{aligned} \mathscr{A}^* &= \frac{\partial^2\Omega_0{}^*(\mathbf{F},\mathbb{B})}{\partial\mathbf{F}\otimes\mathrm{d}\mathbf{F}} \quad , \quad \mathfrak{L}^* = \frac{\partial^2\Omega_0{}^*(\mathbf{F},\mathbb{B})}{\partial\mathbf{F}\otimes\mathrm{d}\mathbb{B}} \quad , \\ \left[\mathfrak{L}^*\right]^T &= \frac{\partial^2\Omega_0{}^*(\mathbf{F},\mathbb{B})}{\partial\mathbb{B}\otimes\mathrm{d}\mathbf{F}} \quad , \quad \mathbf{D}^* = \frac{\partial^2\Omega_0{}^*(\mathbf{F},\mathbb{B})}{\partial\mathbb{B}\otimes\mathrm{d}\mathbb{B}} \quad . \end{aligned} \tag{5.176}$$

Note that the usage of the differential operator $\mathrm{d}(\bullet)$ indicates the total derivative that is associated with the linearisation, rather than the partial derivative $\partial(\bullet)$ typically associated with the second variation. Note as well that the mixed elastic tangent readily incorporates both material and geometric non-linearities.

5.3.7.2 Parameterisation in terms of $\mathbb{H}$

In order to parameterise the energy in terms of the magnetic field but maintain consistency with the previously described formulation, we begin with equation (5.163) and execute a Legendre transformation of the total stored energy function

$$\Omega_0(\mathbf{F},\mathbb{H}) = \Omega_0{}^*(\mathbf{F},\mathbb{B}(\mathbb{H})) - \mathbb{H}\cdot\mathbb{B}(\mathbb{H}) = W_0(\mathbf{F},\mathbb{H}) + M_0(\mathbf{F},\mathbb{H}) \quad . \tag{5.177}$$

It is proved in equation (C.115) that the magnetic energy stored in the free field is now quantified by [531, 533]

$$M_0(\mathbf{F}, \mathbb{H}) := -\frac{\mu_0}{2}[\mathbb{H} \otimes \mathbb{H}] : J\mathbf{C}^{-1} \quad , \tag{5.178}$$

and the total internal potential energy, now defined in terms of the magnetic field, is given by

$$\Pi^{\text{int}} = \int\limits_{\mathcal{B}_0 \cup \mathcal{S}_0} \Omega_0(\mathbf{F}, \mathbb{H})\ \mathrm{dV} = \int\limits_{\mathcal{B}_0} W_0(\mathbf{F}, \mathbb{H})\ \mathrm{dV} + \int\limits_{\mathcal{B}_0 \cup \mathcal{S}_0} M_0(\mathbf{F}, \mathbb{H})\ \mathrm{dV} \quad , \tag{5.179}$$

while the total external potential energy is expressed as

$$\Pi^{\text{ext}} = -\int\limits_{\mathcal{B}_0} \boldsymbol{\varphi} \cdot \mathbf{b}_0^{\text{mech}}\ \mathrm{dV} - \int\limits_{\partial\mathcal{B}_0^{\mathrm{t}}} \boldsymbol{\varphi} \cdot \mathbf{t}_0^{\text{mech,ext}}\ \mathrm{dA} - \int\limits_{\partial\mathcal{S}_0^{\mathbb{B}}} \Phi\,[\mathbf{N}_\infty \cdot \mathbb{B}_\infty]\ \mathrm{dA} \quad . \tag{5.180}$$

First variation

The Gâteaux derivative for the transformed total potential energy is

$$\begin{aligned} \delta\Pi(\boldsymbol{\varphi}, \delta\boldsymbol{\varphi}, \Phi, \delta\Phi) &= \delta_{\boldsymbol{\varphi}}\Pi(\boldsymbol{\varphi}, \Phi) + \delta_\Phi\Pi(\boldsymbol{\varphi}, \Phi) \\ &= \frac{\mathrm{d}}{\mathrm{d}\epsilon}\Pi(\boldsymbol{\varphi} + \epsilon\delta\boldsymbol{\varphi}, \Phi)\bigg|_{\epsilon=0} + \frac{\mathrm{d}}{\mathrm{d}\gamma}\Pi(\boldsymbol{\varphi}, \Phi + \gamma\delta\Phi)\bigg|_{\gamma=0} \quad , \end{aligned} \tag{5.181}$$

Due to the use of the Legendre transformation, the equilibrium solution for the boundary value problem is now defined by the stationary (saddle-)point [42]

$$\min_{\boldsymbol{\varphi}} \max_{\Phi} \Pi \quad \Rightarrow \quad \delta\Pi = 0 \quad . \tag{5.182}$$

From equation (5.181), in conjunction with equation (5.179), the variation of the transformed internal potential energy is

$$\delta\Pi^{\text{int}} = \int\limits_{\mathcal{B}_0 \cup \mathcal{S}_0} \delta\mathbf{F} : \frac{\partial\Omega_0(\mathbf{F}, \mathbb{H})}{\partial\mathbf{F}}\ \mathrm{dV} + \int\limits_{\mathcal{B}_0 \cup \mathcal{S}_0} \delta\mathbb{H} \cdot \frac{\partial\Omega_0(\mathbf{F}, \mathbb{H})}{\partial\mathbb{H}}\ \mathrm{dV} \quad , \tag{5.183}$$

wherein the variation of the magnetic field is defined as

$$\delta\mathbb{H} = -\nabla_0\delta\Phi \quad . \tag{5.184}$$

The first variation of the external potential energy is simply

$$\delta\Pi^{\text{ext}} = -\int\limits_{\mathcal{B}_0} \delta\boldsymbol{\varphi} \cdot \mathbf{b}_0^{\text{mech}}\ \mathrm{dV} - \int\limits_{\partial\mathcal{B}_0^{\mathrm{t}}} \delta\boldsymbol{\varphi} \cdot \mathbf{t}_0^{\text{mech,ext}}\ \mathrm{dA} - \int\limits_{\partial\mathcal{S}_0^{\mathbb{B}}} \delta\Phi\,[\mathbf{N}_\infty \cdot \mathbb{B}_\infty]\ \mathrm{dA} \quad . \tag{5.185}$$

By identifying the total Piola stress and the magnetic induction as

$$\mathbf{P}^{tot} = \frac{\partial \Omega_0 (\mathbf{F}, \mathbb{H})}{\partial \mathbf{F}} \quad , \quad \mathbb{B} = -\frac{\partial \Omega_0 (\mathbf{F}, \mathbb{H})}{\partial \mathbb{H}} \tag{5.186}$$

we can observe that the above collectively represents the equivalent weak form of the divergence-free condition for both the stress and magnetic induction as are listed in equations (5.87) and (5.159). Naturally embedded within this description are the continuity conditions for the normal traction and magnetic induction (equation $(5.85)_2$) at material interfaces. The tangential continuity conditions for both the deformation and the magnetic field (listed in strong form in equation (5.95) and equation $(5.85)_1$) remain to be enforced.

Linearisation

For the coupled magnetic scalar potential variational formulation, the Taylor expansion of equation (5.183) about the current solution point is given by

$$\begin{aligned} 0 &\doteq \delta_{\boldsymbol{\varphi}}\Pi + \Delta_{\boldsymbol{\varphi}}\delta_{\boldsymbol{\varphi}}\Pi + \Delta_{\Phi}\delta_{\boldsymbol{\varphi}}\Pi \\ 0 &\doteq \delta_{\Phi}\Pi + \Delta_{\boldsymbol{\varphi}}\delta_{\Phi}\Pi + \Delta_{\Phi}\delta_{\Phi}\Pi \quad . \end{aligned} \tag{5.187}$$

The linear increments, deduced by linearisation of equation (5.183) with respect to the motion and scalar magnetic potential for a dead load, are [541, 504, 472]

$$\Delta\delta\Pi^{int} = \Delta_{\boldsymbol{\varphi}}\left[\delta_{\boldsymbol{\varphi}}\Pi^{int} + \delta_{\Phi}\Pi^{int}\right] + \Delta_{\Phi}\left[\delta_{\boldsymbol{\varphi}}\Pi^{int} + \delta_{\Phi}\Pi^{int}\right] \quad , \tag{5.188}$$

$$\begin{aligned} &= \int_{\mathscr{B}_0 \cup \mathscr{S}_0} \left[\delta\mathbf{F} : \mathscr{A} : \Delta\mathbf{F} - \delta\mathbb{H} \cdot \mathfrak{L}^T : \Delta\mathbf{F}\right] \, dV \\ &+ \int_{\mathscr{B}_0 \cup \mathscr{S}_0} \left[-\delta\mathbf{F} : \mathfrak{L} \cdot \Delta\mathbb{H} - \delta\mathbb{H} \cdot \mathbf{D} \cdot \Delta\mathbb{H}\right] \, dV \quad . \end{aligned} \tag{5.189}$$

where material tangents are given by

$$\begin{aligned} \mathscr{A} &= \frac{\partial^2 \Omega_0 (\mathbf{F}, \mathbb{H})}{\partial \mathbf{F} \otimes d\mathbf{F}} \quad , \quad \mathfrak{L} = -\frac{\partial^2 \Omega_0 (\mathbf{F}, \mathbb{H})}{\partial \mathbf{F} \otimes d\mathbb{H}} \quad , \\ \mathfrak{L}^T &= -\frac{\partial^2 \Omega_0 (\mathbf{F}, \mathbb{H})}{\partial \mathbb{H} \otimes d\mathbf{F}} \quad , \quad \mathbf{D} = -\frac{\partial^2 \Omega_0 (\mathbf{F}, \mathbb{H})}{\partial \mathbb{H} \otimes d\mathbb{H}} \quad , \end{aligned} \tag{5.190}$$

and the increment of the magnetic field vector is

$$\Delta\mathbb{H} = -\nabla_0 \Delta\Phi \quad . \tag{5.191}$$

5.4 Thermodynamics

The constitutive relationships that describe magneto-sensitive materials can be deduced from the examination of the fundamental thermodynamic balance laws that govern all underlying thermodynamic processes. Using the principles described by Truesdell and Toupin [523], and Coleman and Noll [92] for the coupled material (supplemented by details corresponding to elastic bodies given in [207, 204]), the reduced form of the balance of entropy incorporating the magnetic power can be derived from the second law of thermodynamics, local balance of energy and linear and angular momentum.

The approach adopted here to derive the material laws follows argumentations laid out by Holzapfel [204], and is based on the fundamental postulation that the Maxwell–Lorentz body force and surface traction, respectively stated in equation $(5.136)_1$ and equation (5.138), exist (see Pao [414, sec. 8]) and that there is no relative motion at material interfaces. A non-exhaustive list of literature that provides alternative derivations based on the microscopic balance laws includes [523, 520, 338, 339, 414, 133, 277, 60, 119, 125, 530].

5.4.1 First law of thermodynamics: Energy balance

The first law of thermodynamics defines the transfer of energy within a closed system. For any closed system, the total energy remains constant unless acted on by an external influence. Due to the strong coupling of the electric and magnetic fields, one must consider the full transient Maxwell equations in the initial expression of the first law of thermodynamics. Simplifications, based on the transience of the problem and strength of the electric fields, may subsequently be performed.

The rate of change of kinetic energy $\mathscr{K}(t)$ and internal energy $\mathscr{U}(t)$ is, in general, balanced by the applied external power $\mathscr{P}_{\text{ext}}(t)$ and thermal power $\mathscr{Q}(t)$ as stated by

$$\mathrm{D}_t[\mathscr{K}(t) + \mathscr{U}(t)] = \mathscr{P}_{\text{ext}}(t) + \mathscr{Q}(t) \quad , \tag{5.192}$$

where $\mathscr{P}_{\text{ext}}(t) = \mathscr{P}^{\text{mech}}_{\text{ext}}(t) + \mathscr{P}^{\text{EM}}_{\text{ext}}(t)$ consists of contributions from mechanical and electromagnetic sources. The total kinetic and internal energy, as expressed in the material setting, are respectively defined as

$$\mathscr{K}(t) = \int_{\mathscr{B}_0} \frac{1}{2}\rho_0 \mathbf{v}\cdot\mathbf{v}\,\mathrm{dV} \quad , \quad \mathscr{U}(t) = \int_{\mathscr{B}_0} U_0{}^*\,\mathrm{dV} \tag{5.193}$$

where $U_0{}^* = U_0{}^*(\mathbf{X}, t)$ is the internal energy density per unit reference volume. The external mechanical power, decomposed into contributions due to surface and volume forces, is simply

$$\mathscr{P}^{\text{mech}}_{\text{ext}}(t) = \int_{\partial\mathscr{B}_0} \mathbf{v}\cdot\mathbf{t}_0^{\text{mech,ext}}\,\mathrm{dA} + \int_{\mathscr{B}_0} \mathbf{v}\cdot\mathbf{b}_0^{\text{mech}}\,\mathrm{dV} \quad , \tag{5.194}$$

while the thermal power is

$$\mathscr{Q}(t) = \int_{\partial\mathscr{B}_0} Q_N \,\mathrm{dA} + \int_{\mathscr{B}_0} R_0 \,\mathrm{dV} \tag{5.195}$$

where $Q_N = -\mathbf{Q} \cdot \mathbf{N}$ is the normal thermal flux, with $\mathbf{Q} = \mathbf{Q}(\mathbf{X}, t)$ denoting the thermal flux vector, and $R_0 = R_0(\mathbf{X}, t)$ is a thermal source.

Accepting the existence of the Maxwell–Lorentz body force and surface traction, the electromagnetic power associated with the electromagnetic energy flux (or Poynting vector [435, 523, 262]) in matter, the magnetisation and polarisation of a body, and due to the movement of bound currents in an electric field may be derived. As is shown in Appendix C.7.1, the cumulative electromagnetic power is[2] [523, 414]

$$\mathscr{P}^{\mathrm{EM}}_{\mathrm{ext}}(t) = \int_{\partial\mathscr{B}_0} \left[\mathbf{v} \cdot \mathbf{t}^{\mathrm{pon}}_0 + [\mathbb{E} \times \mathbb{M}] \cdot \mathbf{N}\right] \mathrm{dA} + \int_{\mathscr{B}_0} \left[\mathbf{v} \cdot \mathbf{b}^{\mathrm{pon}}_0 + \mathbf{P}^{*(\mathrm{mag+pol})} : \dot{\mathbf{F}} + \mathbb{J}^{\mathrm{b}} \cdot \mathbb{E}\right] \mathrm{dV} \tag{5.196}$$

where the stress due to magnetisation and polarisation is $\mathbf{P}^{*(\mathrm{mag+pol})} = \mathbf{P}^{*\mathrm{mag}} + \mathbf{P}^{*\mathrm{pol}}$, and for which we note that the material gradient of the Lagrangian velocity vector is $\nabla_0 \mathbf{v} \equiv \dot{\mathbf{F}}$. After the appropriate transformation of the boundary terms, as is performed in Appendix C.7.2 the contribution of power can be re-expressed as

$$\mathscr{P}_{\mathrm{ext}}(t) = \int_{\mathscr{B}_0} \left[\rho_0 \mathbf{v} \cdot \mathbf{a} + \left[\mathbf{P}^{\mathrm{mech}} + \mathbf{P}^{*(\mathrm{mag+pol})}\right] : \dot{\mathbf{F}} - \mathbb{M} \cdot \dot{\mathbb{B}} + \mathbb{E} \cdot \dot{\mathbb{P}}\right] \mathrm{dV} \quad , \tag{5.197}$$

or, upon further manipulation,

$$\mathscr{P}_{\mathrm{ext}}(t) = \int_{\mathscr{B}_0} \left[\rho_0 \mathbf{v} \cdot \mathbf{a} + \mathbf{P}^{\mathrm{tot}} : \dot{\mathbf{F}} + \mathbb{H} \cdot \dot{\mathbb{B}} + \mathbb{E} \cdot \dot{\mathbb{D}} - \mathrm{D}_t M_0{}^{*}\right] \mathrm{dV} \quad . \tag{5.198}$$

where $M_0{}^{*} = M_0{}^{*(\mathrm{mag})} + M_0{}^{*(\mathrm{elec})}$ denotes the energy stored in the free space due to magnetic and electric fields.

By combining equations (5.192), (5.193), (5.195) and (5.198), the relationship between the kinetic and internal energy rates, and the power applied to the system from external sources is

$$\mathrm{D}_t \int_{\mathscr{B}_0} \left[\frac{1}{2}\rho_0 \mathbf{v} \cdot \mathbf{v} + U_0{}^{*}\right] \mathrm{dV} = \int_{\mathscr{B}_0} \Big[\rho_0 \mathbf{v} \cdot \mathbf{a} + \mathbf{P}^{\mathrm{tot}} : \dot{\mathbf{F}} + \mathbb{H} \cdot \dot{\mathbb{B}} + \mathbb{E} \cdot \dot{\mathbb{D}} - \mathrm{D}_t M_0{}^{*} - \nabla_0 \cdot \mathbf{Q} + R_0\Big] \mathrm{dV} \quad . \tag{5.199}$$

2 Due to the stationary reference frame, the magnetic power contributions as defined on the surface and volume of a body are separated into distinct parts. In contrast, the cited literature expresses this electromagnetic contribution in a spatial setting, thereby collectively expressing (due to the moving reference frame) the ponderomotive and purely electromagnetic power through quantities that capture their effect on the motion on the moving body.

The reduced material description of global balance of energy is rendered by expanding the material time derivative and rearranging the terms. Identifying the total internal energy density $E_0 = E_0(\mathbf{X}, t) = U_0{}^* + M_0{}^*$ as the sum of the energy stored due to mechanical deformation processes, magnetisation and polarisation, and that stored in the free space due to magnetic and electric fields, the internal energy rate is then

$$\mathscr{E} = \int_{\mathscr{B}_0} \dot{E}_0 \,\mathrm{dV} = \int_{\mathscr{B}_0} \left[\mathbf{P}^{\mathrm{tot}} : \dot{\mathbf{F}} + \mathbb{H} \cdot \dot{\mathbb{B}} + \mathbb{E} \cdot \dot{\mathbb{D}} - \nabla_0 \cdot \mathbf{Q} + R_0\right] \mathrm{dV} \quad . \tag{5.200}$$

As this must hold for any domain $\mathscr{B}_0$, the local balance of energy is simply

$$\dot{E}_0 = \mathbf{P}^{\mathrm{tot}} : \dot{\mathbf{F}} + \mathbb{H} \cdot \dot{\mathbb{B}} + \mathbb{E} \cdot \dot{\mathbb{D}} - \nabla_0 \cdot \mathbf{Q} + R_0 \quad . \tag{5.201}$$

From this, we note that the total internal energy includes not only the energy contribution due to electro-magneto-mechanical deformation, but also that attributed to the energy stored in the magnetic field itself, as well as that of the electric field.

5.4.2 Second law of thermodynamics: Entropy balance

The second law of thermodynamics, also known as the entropy inequality principle, dictates that certain thermodynamic processes are irreversible. It is expressed as

$$\mathscr{T}(t) = \dot{\mathscr{N}}(t) - \mathscr{R}(t) \geq 0 \tag{5.202}$$

and states that the rate of total entropy production is never negative. The total entropy, defined in the reference configuration, is

$$\mathscr{N}(t) = \int_{\mathscr{B}_0} \eta_0 \,\mathrm{dV} \tag{5.203}$$

where $\eta_0 = \eta_0(\mathbf{X}, t)$ is the entropy per unit reference volume, and the rate of entropy input is

$$\mathscr{R}(t) = \int_{\mathscr{B}_0} K_0 \,\mathrm{dV} - \int_{\partial\mathscr{B}_0} \mathbf{H} \cdot \mathbf{N} \,\mathrm{dA} \tag{5.204}$$

where $K_0 = K_0(\mathbf{X}, t)$ is a source of entropy, and $\mathbf{H} = \mathbf{H}(\mathbf{X}, t)$ is the referential entropy flux vector. It is postulated [204] that there exists a relationship between the entropy flux and heat flux, as well as entropy heat sources, as given by

$$\mathbf{H} := \frac{\mathbf{Q}}{\theta} \quad , \quad K_0 := \frac{R_0}{\theta} \quad , \tag{5.205}$$

where $\theta = \theta(\mathbf{X}, t) > 0$ is the absolute temperature.

Combining equations (5.202) to (5.205) renders the Clausius–Duhem inequality as defined in the material setting, which states that

$$\mathcal{T}(t) = \mathrm{D}_t \int\limits_{\mathcal{B}_0} \eta_0 \,\mathrm{dV} - \int\limits_{\mathcal{B}_0} \frac{R_0}{\theta} \,\mathrm{dV} + \int\limits_{\partial\mathcal{B}_0} \frac{1}{\theta} \mathbf{Q} \cdot \mathbf{N} \,\mathrm{dA} \geq 0 \quad . \tag{5.206}$$

Subsequent application of equation (B.16) to the boundary terms in equation (5.206), leads to

$$\int\limits_{\partial\mathcal{B}_0} \frac{1}{\theta} \mathbf{Q} \cdot \mathbf{N} \,\mathrm{dA} = \int\limits_{\mathcal{B}_0} \left[\frac{1}{\theta} \nabla_0 \cdot \mathbf{Q} - \frac{1}{\theta^2} \nabla_0 \theta \cdot \mathbf{Q} \right] \mathrm{dV} \tag{5.207}$$

After substitution of equation (5.207) into equation (5.206), expanding the material time derivative, localising and multiplying by θ, the local description of the Clausius–Duhem inequality is

$$\theta \dot{\eta}_0 - R_0 + \nabla_0 \cdot \mathbf{Q} - \frac{1}{\theta} \nabla_0 \theta \cdot \mathbf{Q} \geq 0 \quad . \tag{5.208}$$

A re-expression of the Clausius–Duhem inequality can be attained from equation (5.208) by using equation (5.201) to substitute for $R_0 - \nabla_0 \cdot \mathbf{Q}$. This leads to the inequality statement that

$$\mathbf{P}^{\mathrm{tot}} : \dot{\mathbf{F}} + \mathbb{H} \cdot \dot{\mathbb{B}} + \mathbb{E} \cdot \dot{\mathbb{D}} + \theta \dot{\eta}_0 - \dot{E}_0 - \frac{1}{\theta} \nabla_0 \theta \cdot \mathbf{Q} \geq 0 \quad . \tag{5.209}$$

From physical measurements of heat conduction, it is known that $-\nabla_0 \theta \cdot \mathbf{Q} \geq 0$ (consider Fourier's law), the result of which, when used with equation (5.209), leads to the local dissipation inequality

$$\mathcal{D}_{\mathrm{int}} := \mathbf{P}^{\mathrm{tot}} : \dot{\mathbf{F}} + \mathbb{H} \cdot \dot{\mathbb{B}} + \mathbb{E} \cdot \dot{\mathbb{D}} - \left[\dot{E}_0 - \theta \dot{\eta}_0 \right] \geq 0 \tag{5.210}$$

which is also known as the Clausius–Planck inequality. It is postulated [204] that there exists a valid Legendre transformation from which we can define a Helmholtz or free energy function and its material rate

$$\Psi_0{}^* = \Psi_0{}^* (\mathbf{F}, \mathbb{B}, \mathbb{D}, \theta) := E_0 - \theta \eta_0 \quad \Rightarrow \quad \dot{\Psi}_0^* = \dot{E}_0 - \dot{\theta} \eta_0 - \theta \dot{\eta}_0 \tag{5.211}$$

that, when combined with equation (5.210), results in the general expression

$$\mathcal{D}_{\mathrm{int}} = \mathbf{P}^{\mathrm{tot}} : \dot{\mathbf{F}} + \mathbb{H} \cdot \dot{\mathbb{B}} + \mathbb{E} \cdot \dot{\mathbb{D}} - \dot{\theta} \eta_0 - \dot{\Psi}_0^* (\mathbf{F}, \mathbb{B}, \mathbb{D}, \theta) \geq 0 \quad . \tag{5.212}$$

On the assumption that the material exhibits a minimal excitation to an electric field, then the expression for the dissipation further reduces to

$$\mathcal{D}_{\mathrm{int}} = \mathbf{P}^{\mathrm{tot}} : \dot{\mathbf{F}} + \mathbb{H} \cdot \dot{\mathbb{B}} - \dot{\theta} \eta_0 - \dot{\Psi}_0^* (\mathbf{F}, \mathbb{B}, \theta) \geq 0 \quad . \tag{5.213}$$

5.4.3 Parameterisation of isothermal energy functions

Now that the thermodynamics of the coupled problem has been presented, it is necessary to provide some comments on the material energy functions introduced in Section 5.3.6 in terms of the adopted nomenclature and terminology, as well as properties resulting from how they are parameterised.

Table 5.9 lists the various choices for the sets of independent parameters of the energy functions in a general thermodynamic setting. With reference to equation (5.211), in the case of where no thermal effects are considered the *free* energy essentially coincides with the *internal* energy [204]. As is mentioned in the introduction to Section 5.3.6, we will somewhat loosely refer to Ψ_0, U_0 as free *energy* functions that characterise dissipative materials, and Ω_0, W_0, M_0 as stored *energy* functions that may only capture recoverable magnetoelastic effects. We will only use two choices of parameterisation, whereby for simplicity we will not strictly characterise the material energy functions by a full classification (dissipative/recoverable free/internal energy/enthalpy). This will rather be implied by the choice of independent variables, and we will also denote energies that are yet to undergo a Legendre transformation by the superscript *.

Table 5.9: Re-parameterisation of energy functions and variables for the general magneto-mechanical problem through a Legendre transformation (LT) of the magnetoelastic variables. As a point of notation, recall that the relationship between the free energies is $\Psi_0 = U_0 + M_0$, and that for the stored energies is $\Omega_0 = W_0 + M_0$.

Variables		Number	Nature of	Energy density function	
Independent	Dependent	of LTs	functional Π	Symbols	Classification
$\mathbf{F}, \mathbb{B}$	$\mathbf{P}, \mathbb{H}$	0	"Convex"	$\Psi_0{}^*, U_0{}^*, \Omega_0{}^*, W_0{}^*, M_0{}^*$	Free energy
$\mathbf{F}, \mathbb{H}$	$\mathbf{P}, \mathbb{B}$	1	Saddle point	$\Psi_0, U_0, \Omega_0, W_0, M_0$	Mixed energy-enthalpy
$\mathbf{P}, \mathbb{B}$	$\mathbf{F}, \mathbb{H}$	1	Saddle point	–	Mixed enthalpy-energy
$\mathbf{P}, \mathbb{H}$	$\mathbf{F}, \mathbb{B}$	2	"Concave"	–	Free enthalpy

The Legendre transformed counterparts of the ponderomotive, magnetisation and Maxwell stress tensors previously presented in equations (5.91a) to (5.91c) and equations (5.131a) to (5.131c) have been derived in Appendix C.6 and conveniently tabulated in Table 5.10. Further insights into the interplay of the Legendre transformations executed on the energy functions and the resulting properties and definitions of the energies, boundary conditions and stress tensors can be found in [406, 531, 533, 125]. A detailed discussion on the choice of energy for coupled media is also provided by Miehe et al. [371].

The parametrisation of the energy functions not only has an influence on the definition of constitutive laws, but also plays a role in terms of the implementational complexity of numerical techniques used to solve boundary value problems (discussed

Table 5.10: Legendre transformed stress tensors [533, 530]. The counterpart definitions for ponderomotive, magnetisation and Maxwell stresses parameterised in terms of the magnetic induction may be found in equations (5.91a) to (5.91c) and equations (5.131a) to (5.131c).

Two-point ponderomotive, magnetisation and Maxwell stress tensors

$$\mathbf{P}^{\text{tot}} = \mathbf{P}^{\text{mech}} + \mathbf{P}^{\text{pon}} = \mathbf{P}^{\text{mech}} + \mathbf{P}^{\text{mag}} + \mathbf{P}^{\text{max}} \tag{5.214a}$$

$$\mathbf{P}^{\text{mech}} + \mathbf{P}^{\text{mag}} = \frac{\partial U_0}{\partial \mathbf{F}} \quad , \quad \mathbf{P}^{\text{max}} = \frac{\partial M_0}{\partial \mathbf{F}} \tag{5.214b}$$

$$\mathbf{P}^{\text{pon}} \equiv \mathbf{P}^{*\text{pon}} = \left[\frac{J^{-1}}{2\mu_0}\mathbf{C} : [\mathbb{B} \otimes \mathbb{B}]\right] \mathbf{F}^{-T} - [\mathbb{H} \cdot \mathbb{B}]\, \mathbf{F}^{-T} + \left[\mathbf{F}^{-T} \cdot \mathbb{H}\right] \otimes \mathbb{B} \tag{5.214c}$$

$$\mathbf{P}^{\text{mag}} = \left[\frac{1}{2}\mu_0\, [\mathbb{M} \otimes \mathbb{M}] : J\mathbf{C}^{-1}\right] \mathbf{F}^{-T} + \mu_0 J \left[\mathbf{F}^{-T} \cdot \mathbb{H}\right] \otimes \left[\mathbf{C}^{-1} \cdot \mathbb{M}\right] \tag{5.214d}$$

$$\mathbf{P}^{\text{max}} = \left[-\frac{1}{2}\mu_0\, [\mathbb{H} \otimes \mathbb{H}] : J\mathbf{C}^{-1}\right] \mathbf{F}^{-T} + \mu_0 J \left[\mathbf{F}^{-T} \cdot \mathbb{H}\right] \otimes \left[\mathbf{C}^{-1} \cdot \mathbb{H}\right] \tag{5.214e}$$

Spatial ponderomotive, magnetisation and Maxwell stress tensors

$$\boldsymbol{\sigma}^{\text{tot}} = \boldsymbol{\sigma}^{\text{mech}} + \boldsymbol{\sigma}^{\text{pon}} = \boldsymbol{\sigma}^{\text{mech}} + \boldsymbol{\sigma}^{\text{mag}} + \boldsymbol{\sigma}^{\text{max}} \tag{5.215a}$$

$$\boldsymbol{\sigma}^{\text{pon}} \equiv \boldsymbol{\sigma}^{*\text{pon}} = \frac{1}{2\mu_0}\, [\mathbb{b} \cdot \mathbb{b}]\, \mathbf{i} - [\mathbb{h} \cdot \mathbb{b}]\, \mathbf{i} + \mathbb{h} \otimes \mathbb{b} \tag{5.215b}$$

$$\boldsymbol{\sigma}^{\text{mag}} = \mu_0 \left[\frac{1}{2}\, [\mathbb{m} \cdot \mathbb{m}]\, \mathbf{i} + \mathbb{h} \otimes \mathbb{m}\right] \tag{5.215c}$$

$$\boldsymbol{\sigma}^{\text{max}} = \mu_0 \left[\mathbb{h} \otimes \mathbb{h} - \frac{1}{2}\, [\mathbb{h} \cdot \mathbb{h}]\, \mathbf{i}\right] \tag{5.215d}$$

later in Chapter 6), as well as the stability of the non-linear problem. For example, comparison between equations (5.173) and (5.188) indicates that the two variational formulations possess different stability characteristics. The magnetic vector potential formulation derives from a quasi-convex energy functional, which leads to a positive semi-definite and stable system of partial differential equations. In comparison, due to the use of the Legendre transformation a saddle point problem arises in the magnetic scalar potential formulation, and the resulting system of equations is only conditionally stable.

Depending on the choice of constitutive parameters, then under certain loading conditions the numerical problem may exhibit instabilities. These instabilities indicate that, for example, the material is no longer to support the magnetic load without collapse, or that the load associated with the onset of buckling has been exceeded. For further details on the instabilities related to coupled materials, we refer the reader to [411, 457, 228, 371, 376, 377, 169, 446].

6 General aspects of computational simulation of coupled problems

For problems defined on complex domains, it is often not easy to establish an analytical solution to the governing equations. It is therefore common to employ methods in which the continuous domain is discretised into subregions and form an approximate solution to the conservation laws as defined on these patches. In electromagnetics, both the finite volume and finite element methods are commonly utilised; however, for solid-mechanics problems only the latter option has historically been considered to be most suitable. In this chapter we focus on developing and solving the system of non-linear algebraic equations and their linearisation for the discrete multi-domain, non-linear, coupled problem based on the Galerkin FEM [577, 222, 559, 56].

6.1 Finite element discretisation

Firstly we define the discretisation of the continuous domain $\mathscr{D}_0$ into the discrete domain $\mathscr{D}_0^{\mathrm{h}}$ by

$$\mathscr{D}_0 \approx \mathscr{D}_0^{\mathrm{h}} = \bigcup_{n_{\mathrm{e}}} \mathscr{D}_0^{\mathrm{e}} = \mathscr{B}_0^{\mathrm{h}} \cup \mathscr{S}_0^{\mathrm{h}} \quad , \tag{6.1}$$

where $\mathscr{B}_0^{\mathrm{h}}$ and $\mathscr{S}_0^{\mathrm{h}}$ respectively denote the discrete representations of the body and free space. As is illustrated in Figure 6.1, the discrete domain is composed of a set of non-overlapping patches $\mathscr{D}_0^{\mathrm{e}}$, known as finite elements (FEs), each of which have a well-defined geometry.

The domain itself comprises several subregions with different magneto-mechanical characteristics. To accurately capture the discontinuities across material interfaces (and ensure that the appropriate jump conditions are enforced), we further decompose the domain into its distinct material zones

$$\mathscr{B}_0 \approx \mathscr{B}_0^{\mathrm{h}} = \bigcup_{n_{\mathrm{e}}^{\mathscr{B}}} \mathscr{B}_0^{\mathrm{e}} \quad \text{and} \quad \mathscr{S}_0 \approx \mathscr{S}_0^{\mathrm{h}} = \bigcup_{n_{\mathrm{e}}^{\mathscr{S}}} \mathscr{S}_0^{\mathrm{e}} \quad , \tag{6.2}$$

that are non-overlapping but have a conforming interface:

$$\mathscr{B}_0^{\mathrm{h}} \backslash \mathscr{S}_0^{\mathrm{h}} = \emptyset \quad \text{with} \quad \overline{\mathscr{S}_0} \cap \overline{\mathscr{B}_0} = \Gamma_0^{\mathscr{B}\mathscr{S}} \quad . \tag{6.3}$$

Indicated in Figure 6.1, and as was introduced in Section 5.3.6.1, is the boundary of the solid exposed to the free space represented by $\Gamma_0^{\mathscr{B}\mathscr{S}}$.

On each finite element $\mathscr{D}_0^{\mathrm{e}}$, the solution field will be expressed in terms of polynomial basis functions Ψ that possess the partition-of-unity property

$$\sum_{n_\Psi} \Psi(\mathbf{X}) = 1 \quad \text{on} \quad \mathscr{D}_0^{\mathrm{e}} \quad . \tag{6.4}$$

https://doi.org/10.1515/9783110418576-006

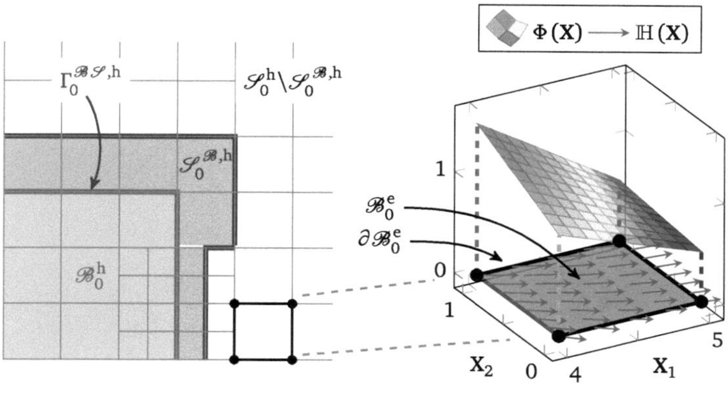

Figure 6.1: Discretisation of the continuum domain using finite elements. An example solution for the scalar potential (linear ansatz) and resulting referential magnetic field, as plotted on a single patch, is also illustrated. [417, fig. 2]

This set of basis functions is chosen based on their required properties [444, 222] for integrability of the weak form (namely their smoothness on the domain, their continuity across boundaries, and their completeness) and how well they approximate the solution space. It is also necessary to consider their numerical qualities in terms of the linearised system that is produced (namely its condition number) and the behaviour of the solution (e. g. the suppression of deformation associated with spurious zero-energy modes). Two examples of relevant basis functions are shown in Figure 6.2 and will later be discussed in more detail.

6.1.1 Displacement field

From the weak form of the momentum balance equation given in equation (5.149), the global residual contribution from the displacement field is defined as

$$r_{\boldsymbol{\varphi}} \doteq 0 = \int\limits_{\mathscr{B}_0 \cup \mathscr{S}_0} \nabla_0 \delta\boldsymbol{\varphi} : \mathbf{P}^{\text{tot}} \, \mathrm{dV} - \int\limits_{\partial \mathscr{B}_0^{\mathrm{t}}} \delta\boldsymbol{\varphi} \cdot \mathbf{t}_0^{\text{mech,ext}} \, \mathrm{dA} - \int\limits_{\mathscr{B}_0} \delta\boldsymbol{\varphi} \cdot \mathbf{b}_0^{\text{mech}} \, \mathrm{dV} \quad . \tag{6.5}$$

This quantity expresses the weak form of the governing equations or equivalently, in the case of variational formulations, the balance of virtual work. By comparison of the above to equations (5.168) and (5.183), we observe that this contribution is the first variation of the potential energy for a non-dissipative system ($U_0 \equiv W_0$).

The next step is to choose an ansatz for $\boldsymbol{\varphi}$ and form $\boldsymbol{\varphi}^{\mathrm{h}}$, a finite-dimensional subspace of $\boldsymbol{\varphi}$. We define a vector-valued basis (shape) function $\boldsymbol{\Psi}_{\boldsymbol{\varphi}}$ such that the discreti-

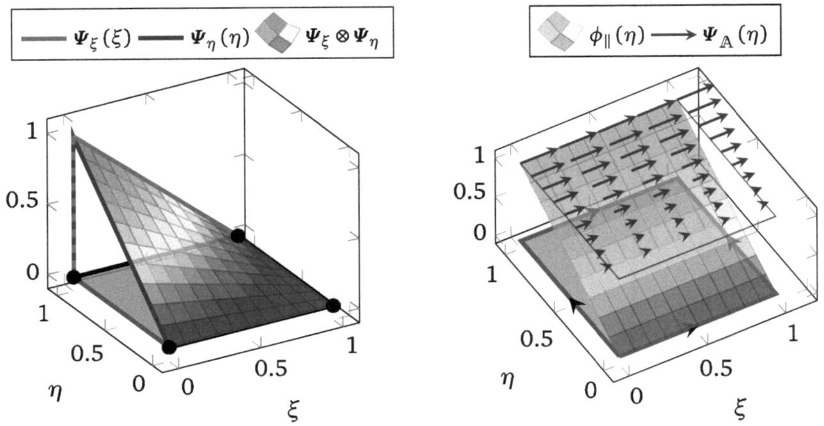

(a) A bilinear shape function (Lagrange polynomial of order 1) that forms a part of a curl-free basis. The location of the vertex-coincident support points is indicated by the black dots.

(b) A shape function (Nédélec polynomial of order 0) that forms a part of a curl-conforming basis. The orientation of edge tangent vectors are indicated by the black arrowheads.

Figure 6.2: Examples of vector-valued finite element shape functions as defined on the isoparametric domain $\Omega_\square \in [0,1]^2$. (Images based off of examples from [145].)

sation of the unknown displacement field and its gradient are expressed as

$$\boldsymbol{\varphi}(\mathbf{X}) \approx \boldsymbol{\varphi}^{\mathrm{h}}(\mathbf{X}) = \sum_I^{n_{\Psi_{\boldsymbol{\varphi}}}} \varphi^I \boldsymbol{\Psi}^I_{\boldsymbol{\varphi}}(\mathbf{X}) \quad , \quad \nabla_0 \boldsymbol{\varphi}(\mathbf{X}) \approx \nabla_0 \boldsymbol{\varphi}^{\mathrm{h}}(\mathbf{X}) = \sum_I^{n_{\Psi_{\boldsymbol{\varphi}}}} \varphi^I \nabla_0 \boldsymbol{\Psi}^I_{\boldsymbol{\varphi}}(\mathbf{X}) \quad . \tag{6.6}$$

Here specifically, the superscript I represents the Ith displacement degree-of-freedom (DoF), although in later text (when the physical coupling is reintroduced) this should be interpreted as the DoF for which the general, unconstrained basis function $\boldsymbol{\Psi}^I$ has a non-trivial value. Note also that the coefficients φ^I of the shape function $\boldsymbol{\Psi}^I_{\boldsymbol{\varphi}}$ are not spatially-dependent, although the basis vectors themselves remain so. The values and material gradients of test functions (or variations) and increments of the trial solution, corresponding to the chosen displacement ansatz, are then given by

$$\begin{aligned} \delta\boldsymbol{\varphi}(\mathbf{X}) &\approx \sum_I^{n_{\Psi_{\boldsymbol{\varphi}}}} \delta\varphi^I \boldsymbol{\Psi}^I_{\boldsymbol{\varphi}}(\mathbf{X}) \quad , \quad \nabla_0 \delta\boldsymbol{\varphi}(\mathbf{X}) \approx \sum_I^{n_{\Psi_{\boldsymbol{\varphi}}}} \delta\varphi^I \nabla_0 \boldsymbol{\Psi}^I_{\boldsymbol{\varphi}}(\mathbf{X}) \quad , \\ \Delta\boldsymbol{\varphi}(\mathbf{X}) &\approx \sum_I^{n_{\Psi_{\boldsymbol{\varphi}}}} \Delta\varphi^I \boldsymbol{\Psi}^I_{\boldsymbol{\varphi}}(\mathbf{X}) \quad , \quad \nabla_0 \Delta\boldsymbol{\varphi}(\mathbf{X}) \approx \sum_I^{n_{\Psi_{\boldsymbol{\varphi}}}} \Delta\varphi^I \nabla_0 \boldsymbol{\Psi}^I_{\boldsymbol{\varphi}}(\mathbf{X}) \quad . \end{aligned} \tag{6.7}$$

Observe that the Galerkin FEM, for which the same basis functions for $\Delta\boldsymbol{\varphi}$ as $\delta\boldsymbol{\varphi}$ are chosen, has been applied. This will ensure that the resulting system of linear equations is both square and, with an appropriate later choice of mesh update strategy, symmetric.

After application of the finite element discretisation, the residual contribution from the displacement field can be expressed in its most general form as

$$r_{\boldsymbol{\varphi}} \approx \sum_{I}^{n_{\Psi_\varphi}} \delta\varphi^I \underbrace{\left[\int_{\mathscr{B}_0 \cup \mathscr{S}_0} \nabla_0 \boldsymbol{\Psi}^I_{\boldsymbol{\varphi}} : \mathbf{P}^{\text{tot}} \, \mathrm{dV}^{\text{h}} - \int_{\partial \mathscr{B}^{\text{t}}_0} \boldsymbol{\Psi}^I_{\boldsymbol{\varphi}} \cdot \mathbf{t}_0^{\text{mech,ext}} \, \mathrm{dA}^{\text{h}} - \int_{\mathscr{B}_0} \boldsymbol{\Psi}^I_{\boldsymbol{\varphi}} \cdot \mathbf{b}_0^{\text{mech}} \, \mathrm{dV}^{\text{h}} \right]}_{r^I_{\boldsymbol{\varphi}}} \tag{6.8}$$

where $r^I_{\boldsymbol{\varphi}}$ is the contribution to the displacement-dependent part of the residual from each displacement DoF.

For application within a non-linear solution scheme, linearisation of the residual is necessary. Assuming the application of a dead load (i. e. that any applied traction is independent of the deformation and magnetic field) then, using equation $(5.175)_1$, the linearisation of this residual contribution with respect to the global displacement field alone is

$$\begin{aligned} \Delta_{\boldsymbol{\varphi}} r_{\boldsymbol{\varphi}} &\approx \sum_{I}^{n_{\Psi_\varphi}} \delta\varphi^I \frac{\mathrm{d} r^I_{\boldsymbol{\varphi}}}{\mathrm{d}\boldsymbol{\varphi}} = \sum_{I}^{n_{\Psi_\varphi}} \delta\varphi^I \left[\int_{\mathscr{B}_0 \cup \mathscr{S}_0} \nabla_0 \boldsymbol{\Psi}^I_{\boldsymbol{\varphi}} : \frac{\mathrm{d}\mathbf{P}^{\text{tot}}}{\mathrm{d}\mathbf{F}} : \Delta\mathbf{F} \, \mathrm{dV}^{\text{h}} \right] \\ &= \sum_{I}^{n_{\Psi_\varphi}} \delta\varphi^I \sum_{J}^{n_{\Psi_\varphi}} \underbrace{\left[\int_{\mathscr{B}_0 \cup \mathscr{S}_0} \nabla_0 \boldsymbol{\Psi}^I_{\boldsymbol{\varphi}} : \frac{\mathrm{d}\mathbf{P}^{\text{tot}}}{\mathrm{d}\mathbf{F}} : \nabla_0 \boldsymbol{\Psi}^J_{\boldsymbol{\varphi}} \, \mathrm{dV}^{\text{h}} \right]}_{\frac{\mathrm{d} r^I_{\boldsymbol{\varphi}}}{\mathrm{d}\varphi^J} = K^{IJ}_{\boldsymbol{\varphi}\boldsymbol{\varphi}}} \Delta\varphi^J \quad , \end{aligned} \tag{6.9}$$

where, depending on the parameterisation of the free energy function, either the linearisation $\frac{\mathrm{d}\mathbf{P}^{\text{tot}}(\mathbf{F},\mathbb{H})}{\mathrm{d}\mathbf{F}} = \mathscr{A}$ or $\frac{\mathrm{d}\mathbf{P}^{\text{tot}}(\mathbf{F},\mathbb{B})}{\mathrm{d}\mathbf{F}} = \mathscr{A}^*$ holds.

Choice of ansatz for the displacement field

What remains is to choose the basis functions for the discretisation of the displacement field. Under consideration of the discretisation shown in Figure 6.1, the restrictions on the space from which $\delta\boldsymbol{\varphi}$ must derive are no longer identical to those presented in Section 5.3.6.1 for the variation of the motion. Instead it is subject to the constraint

$$\delta\boldsymbol{\varphi} = \mathbf{0} \quad \text{on} \quad \partial\mathscr{B}^{\boldsymbol{\varphi}}_0 \cup \left[\overline{\mathscr{S}_0} \backslash \Gamma_0^{\mathscr{B}\mathscr{S}} \right] \tag{6.10}$$

where $\Gamma_0^{\mathscr{B}\mathscr{S}} = \overline{\mathscr{S}_0} \cap \overline{\mathscr{B}_0}$ indicates the surface of the solid exposed to the free space. This test function therefore vanishes not only on the displacement Dirichlet boundary surface of the elastic body, but also in the region of the free space that is not connected to the solid body. Since the free space offers no (elastic) resistance to the deformation of the solid media at low rates of deformation (assuming that viscous forces are negligible), this restriction on the cover of the test function is reasonable. If a penalty approach is to be adopted where the free space possesses a fictitious elasticity, then the restriction is removed and

$$\delta\boldsymbol{\varphi} = \mathbf{0} \quad \text{on} \quad \partial\mathscr{B}_0^{\boldsymbol{\varphi}} \cup \partial\mathscr{S}_0 \quad . \tag{6.11}$$

One compatible option for the polynomial space from which to derive $\boldsymbol{\varphi}^{\mathrm{h}}$ is that of the piecewise-defined Lagrange polynomials, often referred to as Q^n. Figure 6.2a illustrates the support of one cover function as defined on the isoparametric transformation of a local patch (element) within a domain. These polynomials reside in the Sobolev space $\mathscr{H}^1(\mathscr{B}_0 \cup \mathscr{S}_0)$ for which the boundedness of the square integral of both the function's value and gradient

$$\mathscr{H}^1((\circ), \mathscr{D}_0) := \left\{(\circ) \in \mathscr{L}^2(\mathscr{D}_0), \nabla_0(\circ) \in \mathscr{L}^2(\mathscr{D}_0)\right\}$$
$$\implies \int_{\mathscr{D}_0} |(\circ)|^2 \, \mathrm{dV} + \int_{\mathscr{D}_0} |\nabla_0(\circ)|^2 \, \mathrm{dV} < \infty$$

can be proven using functional analysis [444]; they therefore satisfy the aforementioned restrictions on the space containing the test function. However, only for a polynomial order of $n \geq 1$ do they have a non-trivial gradient, and they are at the most $\mathscr{C}^0$ continuous (the function may be continuous between adjacent elements, but the gradient will always be discontinuous between them). Regardless, the compatibility condition (equation (5.94)) is maintained due to the potential derivation of $\mathbf{F}$.

By choosing a basis that shares support points between adjacent finite elements, strong continuity of the displacement field,

$$[\![\boldsymbol{\varphi}]\!] = \mathbf{0} \quad \text{on} \quad \partial\mathscr{B}_0^{\mathrm{e}} \quad , \tag{6.12}$$

is enforced. As a result of this, both the necessary tangential compatibility between elements (specified in equation (5.95)) as well as the continuity of the tractions (as expressed in equation (5.126)) are ensured.

Low-order Q^1 (bi-/tri-linear) Lagrange elements remain the ansatz of choice in solid mechanics due to their simple computation and the low resulting bandwidth in the sparse system. They do however exhibit shear/volumetric locking in conditions of near incompressibility [449, 559], and without stabilisation may exhibit hourglass modes [40, 222, 578] when reduced-order integration schemes are employed.

Some vector-valued shape functions, such as those for the Q^n element (with polynomial order n), can be constructed from primitive, one-dimensional shape functions defined on the isoparametric domain $\xi \in [-1, 1]$. In general, the value of the base Lagrange polynomial [222] for shape function I evaluated at coordinate ξ is

$$\phi_I^n(\xi) := \frac{\Pi_{J=1, J\neq I}^n \left[\xi - \xi_J\right]}{\Pi_{J=1, J\neq I}^n \left[\xi_I - \xi_J\right]} \quad . \tag{6.13}$$

The coefficient ξ_J is the coordinate of the Jth support point in the isoparametric domain. From this base element, the multi-dimensional vector-valued shape function for a hypercube in $\mathbb{R}^{\dim}$ can be composed as

$$\boldsymbol{\Psi}^I_{\boldsymbol{\varphi}}(\boldsymbol{\xi}) = \phi_{\text{base}(I)}(\boldsymbol{\xi})\,\mathbf{e}_{\text{comp}(I)} \tag{6.14}$$

where base (I) is a function that selects the base scalar finite element associated with DoF I, and comp (I) is a function that selects its active basis direction. Two of these vector-valued shape functions are shown in Figure 6.2a to compose the cover function with support at a cell vertex.

6.1.2 Magnetic vector potential field

The global residual contribution for the magnetic vector potential, as derived from equation (5.155), is

$$r_{\mathbb{A}} \doteq 0 = \int\limits_{\mathscr{B}_0 \cup \mathscr{S}_0} [\nabla_0 \times \delta\mathbb{A}] \cdot \mathbb{H}\, \mathrm{dV} + \int\limits_{\partial\mathscr{S}_0^{\mathbb{H}}} \delta\mathbb{A} \cdot [\mathbf{N}_\infty \times \mathbb{H}_\infty]\, \mathrm{dA} \quad .$$

Again, when compared to the relevant components of equations (5.168) and (5.169) it is clear that this definition stemming from the weak-formulation of the governing equations is aligned to the first variation of the potential functional when the considered system is non-dissipative.

The unknown vector potential field and its gradient is discretised using the finite element approach, such that

$$\mathbb{A}(\mathbf{X}) \approx \sum_I^{n_{\Psi_\mathbb{A}}} \mathbb{A}^I \boldsymbol{\Psi}^I_\mathbb{A}(\mathbf{X}) \quad , \quad \nabla_0 \times \mathbb{A}(\mathbf{X}) \approx \sum_I^{n_{\Psi_\mathbb{A}}} \mathbb{A}^I \nabla_0 \times \boldsymbol{\Psi}^I_\mathbb{A}(\mathbf{X}) \quad . \tag{6.15}$$

For such an ansatz, the values and material gradients of test functions (or variations) and the increments of the trial solution are therefore expressed as

$$\begin{aligned} \delta\mathbb{A}(\mathbf{X}) &\approx \sum_I^{n_{\Psi_\mathbb{A}}} \delta\mathbb{A}^I \boldsymbol{\Psi}^I_\mathbb{A}(\mathbf{X}) \quad , \quad \nabla_0 \times \delta\mathbb{A}(\mathbf{X}) \approx \sum_I^{n_{\Psi_\mathbb{A}}} \delta\mathbb{A}^I \nabla_0 \times \boldsymbol{\Psi}^I_\mathbb{A}(\mathbf{X}) \quad , \\ \Delta\mathbb{A}(\mathbf{X}) &\approx \sum_I^{n_{\Psi_\mathbb{A}}} \Delta\mathbb{A}^I \boldsymbol{\Psi}^I_\mathbb{A}(\mathbf{X}) \quad , \quad \nabla_0 \times \Delta\mathbb{A}(\mathbf{X}) \approx \sum_I^{n_{\Psi_\mathbb{A}}} \Delta\mathbb{A}^I \nabla_0 \times \boldsymbol{\Psi}^I_\mathbb{A}(\mathbf{X}) \end{aligned} \tag{6.16}$$

when adopting the Galerkin approach. The approximation for the discrete residual contribution associated with the magnetic vector potential field is therefore

$$r_\mathbb{A} \approx \sum_I^{n_{\Psi_\mathbb{A}}} \delta\mathbb{A}^I \underbrace{\left[\int\limits_{\mathscr{B}_0 \cup \mathscr{S}_0} \left[\nabla_0 \times \boldsymbol{\Psi}^I_\mathbb{A}\right] \cdot \mathbb{H}\, \mathrm{dV}^\mathrm{h} + \int\limits_{\partial\mathscr{S}_0^{\mathbb{H}}} \boldsymbol{\Psi}^I_\mathbb{A} \cdot [\mathbf{N}_\infty \times \mathbb{H}_\infty]\, \mathrm{dA}^\mathrm{h} \right]}_{r^I_\mathbb{A}} \quad . \tag{6.17}$$

We observe that the position of the far-field boundary will always remain fixed in space and, therefore, any applied loads are deformation independent. However, this does not in general preclude the idea that applied tractions may be magnetic-field dependent. If we assume a dead magnetic load (i. e. no solution dependence), then the linearisation of the magnetic residual contribution with respect to the global potential field is

$$\begin{aligned}\Delta_{\mathbb{A}} r_{\mathbb{A}} &\approx \sum_{I}^{n_{\Psi_{\mathbb{A}}}} \delta\mathbb{A}^I \frac{\mathrm{d}r_{\mathbb{A}}^I}{\mathrm{d}\mathbb{A}} = \sum_I \delta\mathbb{A}^I \left[\int_{\mathcal{B}_0 \cup \mathcal{S}_0} \left[\nabla_0 \times \boldsymbol{\Psi}_{\mathbb{A}}^I\right] \cdot \frac{\mathrm{d}\mathbb{H}}{\mathrm{d}\mathbb{B}} \cdot \Delta\mathbb{B}\, \mathrm{dV}^{\mathrm{h}} \right] \\ &= \sum_{I}^{n_{\Psi_{\mathbb{A}}}} \delta\mathbb{A}^I \sum_{J}^{n_{\Psi_{\mathbb{A}}}} \underbrace{\left[\int_{\mathcal{B}_0 \cup \mathcal{S}_0} \left[\nabla_0 \times \boldsymbol{\Psi}_{\mathbb{A}}^I\right] \cdot \mathbf{D}^* \cdot \left[\nabla_0 \times \boldsymbol{\Psi}_{\mathbb{A}}^J\right] \mathrm{dV}^{\mathrm{h}} \right]}_{\frac{\mathrm{d}r_{\mathbb{A}}^I}{\mathrm{d}\mathbb{A}^J} = K_{\mathbb{A}\mathbb{A}}^{IJ}} \Delta\mathbb{A}^J \quad . \end{aligned} \tag{6.18}$$

using the definition for the increment of the magnetic induction provided in equation $(5.175)_2$. Furthermore, the linearisation term resulting from the coupling of the residual to the displacement field is

$$\begin{aligned}\Delta_{\boldsymbol{\varphi}} r_{\mathbb{A}} &\approx \sum_{I}^{n_{\Psi_{\mathbb{A}}}} \delta\mathbb{A}^I \frac{\mathrm{d}r_{\mathbb{A}}^I}{\mathrm{d}\boldsymbol{\varphi}} = \sum_{I}^{n_{\Psi_{\mathbb{A}}}} \delta\mathbb{A}^I \left[\int_{\mathcal{B}_0 \cup \mathcal{S}_0} \left[\nabla_0 \times \boldsymbol{\Psi}_{\mathbb{A}}^I\right] \cdot \frac{\mathrm{d}\mathbb{H}}{\mathrm{d}\mathbf{F}} : \Delta\mathbf{F}\, \mathrm{dV}^{\mathrm{h}} \right] \\ &= \sum_{I}^{n_{\Psi_{\mathbb{A}}}} \delta\mathbb{A}^I \sum_{J}^{n_{\Psi_{\boldsymbol{\varphi}}}} \underbrace{\left[\int_{\mathcal{B}_0 \cup \mathcal{S}_0} \left[\nabla_0 \times \boldsymbol{\Psi}_{\mathbb{A}}^I\right] \cdot \left[\boldsymbol{\mathfrak{L}}^*\right]^T : \nabla_0 \boldsymbol{\Psi}_{\boldsymbol{\varphi}}^J \, \mathrm{dV}^{\mathrm{h}} \right]}_{\frac{\mathrm{d}r_{\mathbb{A}}^I}{\mathrm{d}\boldsymbol{\varphi}^J} = K_{\mathbb{A}\boldsymbol{\varphi}}^{IJ}} \Delta\boldsymbol{\varphi}^J \end{aligned} \tag{6.19}$$

and, from equation (6.8), the coupling term between the displacement residual contribution and magnetic potential is

$$\begin{aligned}\Delta_{\mathbb{A}} r_{\boldsymbol{\varphi}} &\approx \sum_{I}^{n_{\Psi_{\boldsymbol{\varphi}}}} \delta\boldsymbol{\varphi}^I \frac{\mathrm{d}r_{\boldsymbol{\varphi}}^I}{\mathrm{d}\mathbb{A}} = \sum_{I}^{n_{\Psi_{\boldsymbol{\varphi}}}} \delta\boldsymbol{\varphi}^I \left[\int_{\mathcal{B}_0 \cup \mathcal{S}_0} \nabla_0 \boldsymbol{\Psi}_{\boldsymbol{\varphi}}^I : \frac{\mathrm{d}\mathbf{P}^{\mathrm{tot}}}{\mathrm{d}\mathbb{B}} \cdot \Delta\mathbb{B}\, \mathrm{dV}^{\mathrm{h}} \right] \\ &= \sum_{I}^{n_{\Psi_{\boldsymbol{\varphi}}}} \delta\boldsymbol{\varphi}^I \sum_{J}^{n_{\Psi_{\mathbb{A}}}} \underbrace{\left[\int_{\mathcal{B}_0 \cup \mathcal{S}_0} \nabla_0 \boldsymbol{\Psi}_{\boldsymbol{\varphi}}^I : \boldsymbol{\mathfrak{L}}^* \cdot \left[\nabla_0 \times \boldsymbol{\Psi}_{\mathbb{A}}^J\right] \mathrm{dV}^{\mathrm{h}} \right]}_{\frac{\mathrm{d}r_{\boldsymbol{\varphi}}^I}{\mathrm{d}\mathbb{A}^J} = K_{\boldsymbol{\varphi}\mathbb{A}}^{IJ}} \Delta\mathbb{A}^J \quad . \end{aligned} \tag{6.20}$$

Choice of ansatz for the magnetic vector potential field

From equation (6.17), it is observed that the basis functions used to discretise the magnetic vector potential field must derive from the Sobolev space

$$\begin{aligned}\mathcal{H}^{\mathrm{curl}}\left((\bullet), \mathcal{D}_0\right) &:= \left\{(\bullet) \in \mathcal{L}^2(\mathcal{D}_0), \nabla_0 \times (\bullet) \in \mathcal{L}^2(\mathcal{D}_0)\right\} \\ &\implies \int_{\mathcal{D}_0} |(\bullet)|^2 \,\mathrm{dV} + \int_{\mathcal{D}_0} |\nabla_0 \times (\bullet)|^2 \,\mathrm{dV} < \infty \quad . \end{aligned}$$

In such a function space, it is guaranteed that both the value and the curl of the field exist, and are square integrable. This aligns with the remarks made in Section 5.3.6.2, where the restrictions on the space from which $\delta\mathbb{A}$ must derive were discussed from a variational perspective.

Nédélec [396, 397] first described a suitable set of mixed basis functions for application to time-dependent electromagnetics, which has since been generalised by van Welij [527]. In contrast to the Q^1 element that has support points that are coincident with the cell vertices, the lowest order Nédélec elements have DoFs associated with cell edges that represent the tangential components of the solution. In summary, the edge-based shape functions for the lowest order element ($n = 0$) in $\mathbb{R}^2$ are [473, 530]

$$\boldsymbol{\Psi}^I_{\mathbb{A}}(\eta) = \begin{bmatrix} \phi^1_I(\eta) \\ 0 \end{bmatrix} \quad , \quad \boldsymbol{\Psi}^I_{\mathbb{A}}(\xi) = \begin{bmatrix} 0 \\ \phi^1_I(\xi) \end{bmatrix} \tag{6.21}$$

when defined on a quadrilateral with the appropriate choice of edge (unit tangent) orientation. Here the isoparametric coordinates on $\Omega_\square$ are decomposed as $\boldsymbol{\xi} = (\xi, \eta)$, and ϕ^n_I represent Lagrange polynomials, as defined in equation (6.13), of specific order. Figure 6.2b provides a visual representation of the vector-valued Nédélec shape function associated with an edge. As may be inferred from this figure, the defining feature of these basis functions is that they have a unity tangential component along one supporting edge, while the tangential contribution along all other edges is zero [571, 530]. Although the canonical element definition is still commonly utilised, recently there has been some work done by Zaglmayr [571], for example, to improve the mathematical properties and numerical formulation of the Nédélec basis.

For the choice of a continuous set of basis functions, for which the edge support points are shared between adjacent elements, the Nédélec element ensures that tangential fields conform between the elements; however, there is no enforcement of flux conservation between them. Therefore, equation $(5.134)_1$ is naturally satisfied by this ansatz but, as mentioned in Section 5.3.5.1, the divergence-free condition specified in equation $(5.134)_2$ is ensured through equation (5.141). What remains to be done is that the uniqueness of the field $\mathbb{A}$ must be enforced by gauging.

6.1.3 Magnetic scalar potential field

Lastly, the expression for the residual of the magnetic scalar potential field is

$$r_\Phi \doteq 0 = \int_{\mathscr{B}_0 \cup \mathscr{S}_0} \nabla_0 \delta\Phi \cdot \mathbb{B} \, \mathrm{dV} - \int_{\partial\mathscr{S}^{\mathbb{B}}_0} \delta\Phi \, [\mathbf{N}_\infty \cdot \mathbb{B}_\infty] \, \mathrm{dA} \quad ,$$

which is derived from equation (5.159). As seen before, it coincides with the first variation of the potential functional for non-dissipative materials as collectively provided in equations (5.183) and (5.185).

We represent the Galerkin finite element discretisation of the magnetic scalar potential field and its gradient by the ansatz

$$\Phi(\mathbf{X}) \approx \sum_{I}^{n_{\Psi_\Phi}} \Phi^I \Psi_\Phi^I(\mathbf{X}) \quad , \quad \nabla_0 \Phi(\mathbf{X}) \approx \sum_{I}^{n_{\Psi_\Phi}} \Phi^I \nabla_0 \Psi_\Phi^I(\mathbf{X}) \tag{6.22}$$

from which we derive the discretisation of the test functions (or variations) and increment of the field, as well as the material gradient of both, as

$$\begin{aligned} \delta\Phi(\mathbf{X}) &\approx \sum_{I}^{n_{\Psi_\Phi}} \delta\Phi^I \Psi_\Phi^I(\mathbf{X}) \quad , \quad \nabla_0 \delta\Phi(\mathbf{X}) \approx \sum_{I}^{n_{\Psi_\Phi}} \delta\Phi^I \nabla_0 \Psi_\Phi^I(\mathbf{X}) \quad , \\ \Delta\Phi(\mathbf{X}) &\approx \sum_{I}^{n_{\Psi_\Phi}} \Delta\Phi^I \Psi_\Phi^I(\mathbf{X}) \quad , \quad \nabla_0 \Delta\Phi(\mathbf{X}) \approx \sum_{I}^{n_{\Psi_\Phi}} \Delta\Phi^I \nabla_0 \Psi_\Phi^I(\mathbf{X}) \quad . \end{aligned} \tag{6.23}$$

Applying this discretisation leads to the result that

$$r_\Phi \approx \sum_{I}^{n_{\Psi_\Phi}} \delta\Phi^I \underbrace{\left[\int_{\mathscr{B}_0 \cup \mathscr{S}_0} \nabla_0 \Psi_\Phi^I \cdot \mathbb{B} \, dV^h - \int_{\partial \mathscr{S}_0^{\mathbb{B}}} \Psi_\Phi^I \left[\mathbf{N}_\infty \cdot \mathbb{B}_\infty \right] dA^h \right]}_{r_\Phi^I} \quad , \tag{6.24}$$

which is the discrete approximation of this residual contribution.

When considering an applied magnetic loading at the far-field boundary that is independent of the field itself, the linearisation of the magnetic contribution to the residual with respect to the potential field is

$$\begin{aligned} \Delta_\Phi r_\Phi &\approx \sum_{I}^{n_{\Psi_\Phi}} \delta\Phi^I \left[\int_{\mathscr{B}_0 \cup \mathscr{S}_0} \nabla_0 \Psi_\Phi^I \cdot \frac{d\mathbb{B}}{d\mathbb{H}} \cdot \Delta\mathbb{H} \, dV^h \right] \\ &= \sum_{I}^{n_{\Psi_\Phi}} \delta\Phi^I \sum_{J}^{n_{\Psi_\Phi}} \underbrace{\left[- \int_{\mathscr{B}_0 \cup \mathscr{S}_0} \nabla_0 \Psi_\Phi^I \cdot \mathbf{D} \cdot \nabla_0 \Psi_\Phi^J \, dV^h \right]}_{\frac{dr_\Phi^I}{d\Phi^J} = K_{\Phi\Phi}^{IJ}} \Delta\Phi^J \end{aligned} \tag{6.25}$$

with the increment of the magnetic field provided in equation (5.191).

The coupling term in the linearisation arising from sensitivity of the magnetic residual on the displacement field is

$$\begin{aligned} \Delta_{\boldsymbol{\varphi}} r_\Phi &\approx \sum_{I}^{n_{\Psi_\Phi}} \delta\Phi^I \left[\int_{\mathscr{B}_0 \cup \mathscr{S}_0} \nabla_0 \Psi_\Phi^I \cdot \frac{d\mathbb{B}}{d\mathbf{F}} : \Delta\mathbf{F} \, dV^h \right] \\ &= \sum_{I}^{n_{\Psi_\Phi}} \delta\Phi^I \sum_{J}^{n_{\Psi_{\boldsymbol{\varphi}}}} \underbrace{\left[\int_{\mathscr{B}_0 \cup \mathscr{S}_0} \nabla_0 \Psi_\Phi^I \cdot \mathfrak{L}^T : \nabla_0 \boldsymbol{\Psi}_{\boldsymbol{\varphi}}^J \, dV^h \right]}_{\frac{dr_\Phi^I}{d\varphi^J} = K_{\Phi\varphi}^{IJ}} \Delta\varphi^J \end{aligned} \tag{6.26}$$

and, from equation (6.8), the sensitivity of the displacement field with respect to the magnetic scalar potential field is captured by

$$\begin{aligned}
\Delta_{\Phi} r_{\boldsymbol{\varphi}} &\approx \sum_{I}^{n_{\Psi_{\boldsymbol{\varphi}}}} \delta\varphi^{I} \frac{\mathrm{d}r^{I}_{\boldsymbol{\varphi}}}{\mathrm{d}\Phi} = \sum_{I}^{n_{\Psi_{\boldsymbol{\varphi}}}} \delta\varphi^{I} \left[\int_{\mathscr{B}_0 \cup \mathscr{S}_0} \nabla_0 \boldsymbol{\Psi}^{I}_{\boldsymbol{\varphi}} : \frac{\mathrm{d}\mathbf{P}^{\mathrm{tot}}}{\mathrm{d}\mathbb{H}} \cdot \Delta\mathbb{H} \, \mathrm{dV}^{\mathrm{h}} \right] \\
&= \sum_{I}^{n_{\Psi_{\boldsymbol{\varphi}}}} \delta\varphi^{I} \sum_{J}^{n_{\Psi_{\Phi}}} \left[\int_{\mathscr{B}_0 \cup \mathscr{S}_0} \nabla_0 \boldsymbol{\Psi}^{I}_{\boldsymbol{\varphi}} : \left[-\frac{\mathrm{d}\mathbf{P}^{\mathrm{tot}}}{\mathrm{d}\mathbb{H}} \right] \cdot \nabla_0 \Psi^{J}_{\Phi} \, \mathrm{dV}^{\mathrm{h}} \right] \Delta\Phi^{J} \\
&= \sum_{I}^{n_{\Psi_{\boldsymbol{\varphi}}}} \delta\varphi^{I} \sum_{J}^{n_{\Psi_{\Phi}}} \underbrace{\left[\int_{\mathscr{B}_0 \cup \mathscr{S}_0} \nabla_0 \boldsymbol{\Psi}^{I}_{\boldsymbol{\varphi}} : \mathfrak{L} \cdot \nabla_0 \Psi^{J}_{\Phi} \, \mathrm{dV}^{\mathrm{h}} \right]}_{\frac{\mathrm{d}r^{I}_{\boldsymbol{\varphi}}}{\mathrm{d}\Phi^{J}} = K^{IJ}_{\varphi\Phi}} \Delta\Phi^{J} \quad .
\end{aligned} \tag{6.27}$$

Although at first glance equations (6.26) and (6.27) appear to have opposite signs, for the definition of the magnetic induction given in equation $(5.186)_2$ we conclude that this is not the case. However, as was first discussed in Section 5.4.3, comparison between equations (6.18) and (6.25) highlights the contrast in numerical qualities of the vector and scalar potentials. The former has a positive definite contribution, indicating that the solution to the vector potential problem is a minimisation problem. By comparison, the latter that has a negative definite contribution to the tangent stiffness matrix which demonstrates that a maximisation of the potential field is to be found. Boundary value problems derived from equations with such characteristics (where the equilibrium or stationary point corresponds to a mixture of maximisation and minimisation of fields) are commonly referred to as saddle point problems.

Choice of ansatz for the magnetic scalar potential field

Similar to the case the displacement field, we observe from equation (6.24) that $\delta\Phi \in \mathscr{H}^1(\mathscr{B}_0 \cup \mathscr{S}_0)$. Therefore, an appropriate choice for the basis functions for the finite element discretisation are again that derived from the Lagrange polynomials. This aligns with the restrictions noted in Section 5.3.6.3 on the spaces from which the variations $\delta\Phi$ must derive. By choosing a basis that is continuous between adjacent finite elements, the continuity of the scalar potential between material domains (equation $(5.147)_1$) is ensured, as is the continuity of the tangential magnetic field as specified in equation (5.134).

6.1.4 Summary of finite element implementation

For ease of comparison, Table 6.1 summarises the salient equations defining the FE implementation for both settings of the coupled magnetoelastic problem. Also listed are the typical choice of polynomial used for the displacement and magnetic fields

Table 6.1: Summary of finite element basis applied to the magnetoelastic coupled problem. The element types FE_Q and FE_Ned are respectively the continuous Lagrange FE and Nédélec FE. The polynomial order is denoted by n (with equal-complexity element pairs tabulated) and vector elements are highlighted by **bold** font.

Relevant equations		**Subdomain**	**Field**		
Magnetic vector potential			$\boldsymbol{\varphi}$		$\mathbb{A}$
Residual	6.8, 6.17	$\mathscr{B}_0^{h}$	**FE_Q** (n)	$\times$	**FE_Ned** $(n-1)$
Linearisation	6.9, 6.18,	$\mathscr{S}_0^{\mathscr{B},h}$	**FE_Q** (n)	$\times$	**FE_Ned** $(n-1)$
	6.19, 6.20	$\mathscr{S}_0^{h}\backslash\mathscr{S}_0^{\mathscr{B},h}$	—	$\times$	**FE_Ned** $(n-1)$
Magnetic scalar potential			$\boldsymbol{\varphi}$		Φ
Residual	6.8, 6.24	$\mathscr{B}_0^{h}$	**FE_Q** (n)	$\times$	FE_Q (n)
Linearisation	6.9, 6.25,	$\mathscr{S}_0^{\mathscr{B},h}$	**FE_Q** (n)	$\times$	FE_Q (n)
	6.26, 6.27	$\mathscr{S}_0^{h}\backslash\mathscr{S}_0^{\mathscr{B},h}$	—	$\times$	FE_Q (n)

from which to construct finite elements. For these choices, the numerical complexity for each field component (i. e. the number of DoFs on an element associated with each field component) is approximately equal.

6.1.5 Tools automating the computation of finite element linearisations and constitutive model tangent moduli

As has been shown above, the linearisations stemming from the definition of FE residuals become quite involved, especially for coupled problems and even more so for specialised FE formulations (such as one detailed in Section 11.3 for near incompressible finite-strain magnetoelasticity). Similar holds for the coupled constitutive laws that are further elaborated in Chapter 7. An accurate derivation of the FE linearisation and material tangent moduli are critical in order to achieve convergence in numerical simulations (particularly for unstable problems and when modelling materials with instabilities), as well as the optimal overall convergence rate. However, as the complexity of simulation frameworks, the boundary value problem formulation and associated constitutive laws increase, so do the number of opportunities for subtle errors to be injected into their implementation.

Nowadays there exist numerous symbolic [37, 79, 356] and numeric assistive tools that can be used to either execute and/or validate FE or constitutive model implementations. Automatic or algorithmic differentiation is one such "black box" numerical method that can be used to "automatically" compute the first, and perhaps higher-order, derivatives of function(s) with respect to one or more input variables to a very degree of high accuracy. It is often implemented as a tool to perform source code transformation [225, 459], or provided as a specialised number type that is used as a replacement to standard arithmetic numbers. The latter may take the form of so-called "taped" numbers [179], "tapeless" numbers that exploit a truncation of the Taylor series of the dependent function f,

$$f(x+\epsilon) = f(x) + \epsilon f'(x) + \frac{1}{2!}\epsilon^2 f''(x) + \frac{1}{3!}\epsilon^3 f'''(x) + \dots \quad ,$$

in conjunction with a specific choice of the perturbation parameter ϵ (forming the dual [179], complex-step [512], and hyper-dual [146] approaches), and compile-time operations using expression templates [34, 162, 423].

A comprehensive presentation on how some of these tools have been integrated into FE frameworks is given by Korelc and Wriggers [274] and Logg et al. [320]. Many modern open-source FE libraries such as FEniCS [6] and `deal.II` [10, 15], amongst others, have support for some of these assistive libraries.

6.2 Evaluation of definite integrals

Due to the complex geometries involved in the computations to which the FEM is often applied, the direct evaluation of the integrals in the equations summarised in Table 6.1 can be challenging and computationally costly. To assist with and simplify this set of calculations, we apply numerical integration [436] techniques in conjunction with a transformation of the integration domain. Consideration of the ansatz used for each field is necessary to ensure that the employed strategy can (as is most often desired) integrate the polynomial function underlying its definition.

To illustrate the steps taken to integrate any generic smooth function, we consider the continuous, smooth scalar function $f(\mathbf{X})$ integrated over the arbitrary domain $\mathscr{D}$. We first employ the finite element approximation and discretise the domain; this aligns with what has been presented thus far in Section 6.1. Next, as is shown in the last step in

$$\int_{\mathscr{D}} f(\mathbf{X})\ \mathrm{dV} \approx \int_{\mathscr{D}^{\mathrm{h}}} f(\mathbf{X})\ \mathrm{dV}^{\mathrm{h}} = \sum_{n_{\mathrm{e}}} \left[\int_{\Omega_{\mathrm{e}}} f(\mathbf{X})\ \mathrm{dV}^{\mathrm{e}} \right]_{\mathrm{e}} \quad , \tag{6.28}$$

this discrete approximation is expressed as an accumulation over patches (each of which is a finite element) by exploiting the associative property of finite elements. Here, $n_{\mathrm{e}} = |\Omega_{\mathrm{e}} \in \mathscr{D}^{\mathrm{h}}|$ is the number of elements e in the discrete domain $\mathscr{D}^{\mathrm{h}}$, and Ω_{e} represents the domain covered by any specific FE. The complexity of the calculation has thus been reduced from a domain-level integral to that of one over a patch of simplified geometry.

To further simplify the evaluation of the above, we define the transformation $\boldsymbol{\varphi}_{\square} : \mathbf{X}^{\xi} \to \mathbf{X}$ that maps points $\mathbf{X}^{\xi}$ in the isoparametric domain $\Omega_{\square}$ to $\mathbf{X}$ from the material domain (real space). In doing so, we negate the necessity to develop integration rules that accommodate the geometry of each finite element; rather a single integration rule is defined in the isoparametric space and a coordinate transformation accounts for the change in geometry. An affine transformation is typically assumed, but higher-order mappings can also be considered to better capture the geometry of curved surfaces when higher-order FEs are used. The integral on each element is therefore reduced to

$$\int_{\Omega_e} f(\mathbf{X})\ \mathrm{dV}^e = \int_{\Omega_\Box} f(\mathbf{X}) \left|\frac{\partial \mathbf{X}}{\partial \mathbf{X}^\xi}\right|_{\mathbf{X}_q} \mathrm{dV}^\Box = \sum_{n_q} f(\mathbf{X}_q) \hat{J}(\mathbf{X}_q)\, w_q\left(\mathbf{X}_q^\xi\right) \tag{6.29}$$

where $\Omega_\Box$ represents the isoparametric domain, $\mathbf{X}^\xi$ is a coordinate in the isoparametric domain, the derivative $\frac{\partial \mathbf{X}}{\partial \mathbf{X}^\xi}$ is the inverse of the transformation map $[\nabla_0 \boldsymbol{\varphi}_\Box]^{-1}$, and $\mathbf{X}_q = \mathbf{X}_q(\mathbf{X}_q^\xi)$ is the mapped location of the qth quadrature point in real space. In the last step, the numerical integration scheme has been formally applied and the integral is expressed as the weighted value of the function evaluated at predetermined points on the reference element. Here, n_q is the number of quadrature points q in the numerical integration scheme, $\hat{J}(\mathbf{X}_q) = |\frac{\partial \mathbf{X}}{\partial \mathbf{X}^\xi}|_{\mathbf{X}_q}$ is its Jacobian evaluated at quadrature point $\mathbf{X}_q$, $w_q = w_q(\mathbf{X}_q^\xi)$ is the integration weight associated with a particular quadrature point, and $\mathbf{X}_q^\xi = \mathbf{X}_q^\xi(\mathbf{X}^\xi)$ is the location of a quadrature point in the isoparametric domain. In the case that the function is discrete ($f(\mathbf{X})$ is projected to the finite element space), then

$$\int_{\mathscr{D}} f(\mathbf{X})\ \mathrm{dV} \approx \int_{\mathscr{D}^h} f^h(\mathbf{X})\ \mathrm{dV}^h = \sum_{n_e}\sum_{n_q}\sum_{n_\Psi} \Psi_f^I(\mathbf{X}_q) f^{h,I} \hat{J}(\mathbf{X}_q)\, w_q\left(\mathbf{X}_q^\xi\right) \quad . \tag{6.30}$$

Using numerical integration in conjunction with the definition of the isoparametric domain, evaluation of $r_{\boldsymbol{\varphi}}^I$, the residual at the Ith DoF in equation (6.8), can be achieved through the computation of

$$\begin{aligned}
r_{\boldsymbol{\varphi}}^I &\approx \sum_{n_e} \int_{\Omega_e} \left[\nabla_0 \boldsymbol{\Psi}_{\boldsymbol{\varphi}}^I(\mathbf{X}) : \mathbf{P}^{\mathrm{tot}}(\mathbf{X}) - \boldsymbol{\Psi}_{\boldsymbol{\varphi}}^I(\mathbf{X}) \cdot \mathbf{b}_0^{\mathrm{mech}}(\mathbf{X})\right] \mathrm{dV}^e \\
&\quad - \sum_{n_f} \int_{\partial\Omega_e^t} \boldsymbol{\Psi}_{\boldsymbol{\varphi}}^I(\mathbf{X}) \cdot \mathbf{t}_0^{\mathrm{mech,ext}}(\mathbf{X})\ \mathrm{dA}^e \\
&= \sum_{n_e} \left[\int_{\Omega_\Box} \left[\nabla_0 \boldsymbol{\Psi}_{\boldsymbol{\varphi}}^I\left(\mathbf{X}^\xi\right) : \mathbf{P}^{\mathrm{tot}}(\mathbf{X}) - \boldsymbol{\Psi}_{\boldsymbol{\varphi}}^I\left(\mathbf{X}^\xi\right) \cdot \mathbf{b}_0^{\mathrm{mech}}(\mathbf{X})\right] \left|\frac{\partial \mathbf{X}}{\partial \mathbf{X}^\xi}\right|_{\mathbf{X}_q} \mathrm{dV}^\Box\right]_e \\
&\quad - \sum_{n_f} \left[\int_{\partial\Omega_\Box} \boldsymbol{\Psi}_{\boldsymbol{\varphi}}^I\left(\mathbf{X}^\xi\right) \cdot \mathbf{t}_0^{\mathrm{mech,ext}}(\mathbf{X}) \left|\frac{\partial \mathbf{X}}{\partial \mathbf{X}^\xi}\right|_{\mathbf{X}_q} \mathrm{dv}^\Box\right]_f \\
&= \sum_{n_e} \left[\sum_{n_q} \left[\nabla_0 \boldsymbol{\Psi}_{\boldsymbol{\varphi}}^I\left(\mathbf{X}_q^\xi\right) : \mathbf{P}^{\mathrm{tot}}(\mathbf{X}_q) - \boldsymbol{\Psi}_{\boldsymbol{\varphi}}^I\left(\mathbf{X}_q^\xi\right) \cdot \mathbf{b}_0^{\mathrm{mech}}(\mathbf{X}_q)\right] \hat{J}(\mathbf{X}_q)\, w_q\left(\mathbf{X}_q^\xi\right)\right]_e \\
&\quad - \sum_{n_f} \left[\sum_{n_q} \boldsymbol{\Psi}_{\boldsymbol{\varphi}}^I\left(\mathbf{X}_q^\xi\right) \cdot \mathbf{t}_0^{\mathrm{mech,ext}}(\mathbf{X}_q) \hat{J}(\mathbf{X}_q)\, w_q\left(\mathbf{X}_q^\xi\right)\right]_f \quad ,
\end{aligned} \tag{6.31}$$

where $n_e = |\Omega_e \in \mathscr{B}_0^h \cup \mathscr{S}_0^h|$ is the total number of elements e in the discrete domain, and n_f are the number of finite element faces on the discretised traction boundary $\partial\Omega_e^t$. Using the isoparametric mapping, the gradients of the shape functions are computed by

$$\nabla_0 \boldsymbol{\Psi}^I \left(\mathbf{X}^\xi\right) = \nabla_\square \boldsymbol{\Psi}^I \left(\mathbf{X}_q^\xi\right) \cdot \frac{\partial \mathbf{X}^\xi}{\partial \mathbf{X}} \quad . \tag{6.32}$$

A similar procedure would be followed in the computation of sensitivity of the residual at the Ith DoF with respect to the motion of the Jth DoF (as formulated in equation (6.9)), leading to

$$\begin{aligned} K_{\boldsymbol{\varphi\varphi}}^{IJ} &\approx \sum_{n_e} \int_{\Omega_e} \nabla_0 \boldsymbol{\Psi}_{\boldsymbol{\varphi}}^I (\mathbf{X}) : \frac{\mathrm{d}\mathbf{P}^{\text{tot}}(\mathbf{X})}{\mathrm{d}\mathbf{F}} : \nabla_0 \boldsymbol{\Psi}_{\boldsymbol{\varphi}}^J (\mathbf{X}) \; \mathrm{d}V^{\mathrm{e}} \\ &= \sum_{n_e} \left[\int_{\Omega_\square} \nabla_0 \boldsymbol{\Psi}_{\boldsymbol{\varphi}}^I \left(\mathbf{X}^\xi\right) : \frac{\mathrm{d}\mathbf{P}^{\text{tot}}(\mathbf{X})}{\mathrm{d}\mathbf{F}} : \nabla_0 \boldsymbol{\Psi}_{\boldsymbol{\varphi}}^J \left(\mathbf{X}^\xi\right) \left| \frac{\partial \mathbf{X}}{\partial \mathbf{X}^\xi} \right|_{\mathbf{X}_q} \mathrm{d}V^\square \right]_{\mathrm{e}} \\ &= \sum_{n_e} \left[\sum_{n_q} \nabla_0 \boldsymbol{\Psi}_{\boldsymbol{\varphi}}^I \left(\mathbf{X}_q^\xi\right) : \frac{\mathrm{d}\mathbf{P}^{\text{tot}}(\mathbf{X}_q)}{\mathrm{d}\mathbf{F}} : \nabla_0 \boldsymbol{\Psi}_{\boldsymbol{\varphi}}^J \left(\mathbf{X}_q^\xi\right) \hat{J}(\mathbf{X}_q)\, w_q \left(\mathbf{X}_q^\xi\right) \right]_{\mathrm{e}} \quad . \end{aligned} \tag{6.33}$$

Note that the quadrature rules applied on $\Omega_\square$ and $\partial\Omega_\square$ are specialised to each case, due to the differing spatial dimension of each real (and isoparametric) element. Furthermore, the choice and order of quadrature formula is based on the nature of the integrand, with discretisation properties of the solution space and the location of singularities in the integrand being taken into consideration. Commonly employed quadrature rules for standard FE methods (furnishing $\mathbf{X}_q^\xi$ and $w_q(\mathbf{X}_q^\xi)$) include Gauss–Legendre quadrature [436], and Gauss–Lobatto quadrature [252].

6.3 Solution of a time/load increment

Now that the method of spatial discretisation has been discussed, what remains is its application within an algorithmic framework aimed at determining the state of the primary fields at any given time (or time step). As has been done in Section 6.1, it is assumed that all mechanical and magnetic tractions are “dead” loads. When under quasi-static conditions only non-dissipative materials are considered, time can then be interpreted as a loading parameter; however, in the case of dissipative materials, time retains its physical interpretation.

The relationship between the independent and dependent variables is non-linear (as is the coupling between fields), therefore necessitating the use of a non-linear solution scheme to resolve the problem at a fixed time. This is discussed in Section 6.3.1, while in Section 6.3.2 possible methods to solve the resulting system of linear equations are presented. For convenience, at this point we will introduce the generic magnetic potential $\mathbb{L}$ that will denote either of the magnetic quantities $\mathbb{A}$ or Φ.

6.3.1 Solution to the time-independent non-linear problem using the Newton–Raphson method

For problems that do not exhibit extreme non-linearities in the solution path, such as buckling or snap-through [559, 51], the quadratically-convergent Newton–Raphson method is often sufficiently robust to resolve the solution to a non-linear problem. At any given time step n and Newton iteration i, the first-order Taylor expansion of the residual $r(\boldsymbol{\varphi}, \mathbb{L})$ at the state $\{\boldsymbol{\varphi}, \mathbb{L}\}$ is, as expressed in continuous form,

$$0 \approx \left[r(\boldsymbol{\varphi}, \mathbb{L}) \bigg|_{\boldsymbol{\varphi}\mathbb{L}} + \frac{\mathrm{d}r(\boldsymbol{\varphi}, \mathbb{L})}{\mathrm{d}\boldsymbol{\varphi}} \bigg|_{\mathbb{L}} \Delta\boldsymbol{\varphi} + \frac{\mathrm{d}r(\boldsymbol{\varphi}, \mathbb{L})}{\mathrm{d}\mathbb{L}} \bigg|_{\boldsymbol{\varphi}} \Delta\mathbb{L} \right]_{n,i} ,$$

with the update of the primary fields performed by

$$\boldsymbol{\varphi}_{i+1} = \boldsymbol{\varphi}_i + \Delta\boldsymbol{\varphi} \quad \text{and} \quad \mathbb{L}_{i+1} = \mathbb{L}_i + \Delta\mathbb{L} \quad .$$

Expanded into component form, the former reads

$$\begin{aligned} 0 &\doteq r_{\boldsymbol{\varphi}} + \Delta_{\boldsymbol{\varphi}} r_{\boldsymbol{\varphi}} + \Delta_{\mathbb{L}} r_{\boldsymbol{\varphi}} \\ 0 &\doteq r_{\mathbb{L}} + \Delta_{\boldsymbol{\varphi}} r_{\mathbb{L}} + \Delta_{\mathbb{L}} r_{\mathbb{L}} \quad . \end{aligned} \tag{6.34}$$

Note that quadratic convergence for highly non-linear material responses is only guaranteed for "sufficiently small" load steps, that is to say when the solution $\{\boldsymbol{\varphi}, \mathbb{L}\}_n$ is "sufficiently close" to the initial linearisation point $\{\boldsymbol{\varphi}, \mathbb{L}\}_{n-1}$. If this is not the case, then backtracking (damping) algorithms such as line-search [559, 51, 436] may assist in accelerating convergence. For problems for which the solution is sensitive to the load path (i. e. the solution for a given load is non-unique), path-following or arc-length methods [559, 51] are often utilised.

6.3.2 Formation and solution of the discrete system of linear equations

Based on which potential formulation is used, we can identify each residual and linearisation contribution from those listed in Table 6.1. Applying the FE discretisation, the expansion of individual components of the linearisation from equation (6.34) are

$$\begin{aligned} 0 &\doteq \sum_{I}^{n_{\Psi_{\varphi}}} \delta\varphi^I \sum_{J}^{n_{\Psi_{\mathbb{L}}}} \left[r_{\boldsymbol{\varphi}}^I + \frac{\mathrm{d}r_{\boldsymbol{\varphi}}^I}{\mathrm{d}\varphi^J} \Delta\varphi^J + \frac{\mathrm{d}r_{\boldsymbol{\varphi}}^I}{\mathrm{d}\mathbb{L}^J} \Delta\mathbb{L}^J \right] \\ 0 &\doteq \sum_{I}^{n_{\Psi_{\mathbb{L}}}} \delta\mathbb{L}^I \sum_{J}^{n_{\Psi_{\varphi}}} \left[r_{\mathbb{L}}^I + \frac{\mathrm{d}r_{\mathbb{L}}^I}{\mathrm{d}\varphi^J} \Delta\varphi^J + \frac{\mathrm{d}r_{\mathbb{L}}^I}{\mathrm{d}\mathbb{L}^J} \Delta\mathbb{L}^J \right] \quad . \end{aligned} \tag{6.35}$$

The sparse linear system, expressed in matrix-vector form, reads

$$\begin{bmatrix} \delta\boldsymbol{\varphi} \\ \delta\mathbb{L} \end{bmatrix} \cdot \left[\underbrace{\begin{bmatrix} \boldsymbol{K}_{\varphi\varphi} & \boldsymbol{K}_{\varphi\mathbb{L}} \\ \boldsymbol{K}_{\mathbb{L}\varphi} & \boldsymbol{K}_{\mathbb{L}\mathbb{L}} \end{bmatrix}}_{K} \cdot \underbrace{\begin{bmatrix} \Delta\boldsymbol{d}_{\varphi} \\ \Delta\boldsymbol{d}_{\mathbb{L}} \end{bmatrix}}_{\Delta d} \right] = \begin{bmatrix} \delta\boldsymbol{\varphi} \\ \delta\mathbb{L} \end{bmatrix} \cdot \underbrace{\begin{bmatrix} -\boldsymbol{r}_{\varphi} \\ -\boldsymbol{r}_{\mathbb{L}} \end{bmatrix}}_{f=-r} \quad . \tag{6.36}$$

From this we identify the linear relationship

$$\boldsymbol{K} \cdot \Delta\boldsymbol{d} = \boldsymbol{f} \tag{6.37}$$

as equation (6.36) must hold for any arbitrary choice of test functions (or variations). We denote $\boldsymbol{r} \rightarrow \boldsymbol{0}$ as the global residual vector that is to be minimised, and $\Delta\boldsymbol{d}$ is the iterative increment with which we compute the global solution update for each iteration i by

$$\boldsymbol{d}_{i+1} = \boldsymbol{d}_i + \Delta\boldsymbol{d} \quad . \tag{6.38}$$

The three most practical approaches to solving the linear system of differential equations are:

1. using a direct solver to solve the global block system,
2. using an iterative solver to solve the global block system, and
3. Gaussian elimination of the full block system (using the Schur complement) and solving a condensed problem using an iterative approach.

As the first two are conceptually self-explanatory, we simply refer the reader to the works of Davis [107] and Saad [461] for reference implementations of, respectively, sparse direct and iterative solution schemes. Discussions on the development of preconditioners for global block systems, with a structure related to that discussed in item (2), can be found in [488, 461, 462].

Block Gaussian elimination

To elaborate on the third option for solving the linear system, at a fixed time and Newton iteration we expand equation (6.36) into

$$\boldsymbol{K}_{\varphi\varphi}\, \Delta\boldsymbol{d}_{\varphi} + \boldsymbol{K}_{\varphi\mathbb{L}}\, \Delta\boldsymbol{d}_{\mathbb{L}} = \boldsymbol{f}_{\varphi} \tag{6.39a}$$

$$\boldsymbol{K}_{\mathbb{L}\varphi}\, \Delta\boldsymbol{d}_{\varphi} + \boldsymbol{K}_{\mathbb{L}\mathbb{L}}\, \Delta\boldsymbol{d}_{\mathbb{L}} = \boldsymbol{f}_{\mathbb{L}} \quad . \tag{6.39b}$$

From equation (6.39b), the incremental updates for the magnetic potential can be expressed as

$$\Delta\boldsymbol{d}_{\mathbb{L}} = \boldsymbol{K}_{\mathbb{L}\mathbb{L}}^{-1} \left[\boldsymbol{f}_{\mathbb{L}} - \boldsymbol{K}_{\mathbb{L}\varphi}\, \Delta\boldsymbol{d}_{\varphi}\right] \quad . \tag{6.40}$$

Using this expression of the potential increment within a Newton iteration, Gaussian elimination of the full block system can be executed. Through substitution of equation (6.40) into equation (6.39a), and under the assumption that $\boldsymbol{K}_{\mathbb{L}\mathbb{L}}^{-1}$ is well defined, the linear problem for the displacement update is condensed to

$$\underbrace{\left[\boldsymbol{K}_{\varphi\varphi} - \boldsymbol{K}_{\varphi\mathbb{L}}\,\boldsymbol{K}_{\mathbb{L}\mathbb{L}}^{-1}\,\boldsymbol{K}_{\mathbb{L}\varphi}\right]}_{\widetilde{\boldsymbol{K}}} \Delta\boldsymbol{d}_{\varphi} = \underbrace{\boldsymbol{f}_{\varphi} - \boldsymbol{K}_{\varphi\mathbb{L}}\boldsymbol{K}_{\mathbb{L}\mathbb{L}}^{-1}\boldsymbol{f}_{\mathbb{L}}}_{\widetilde{\boldsymbol{f}}} \quad , \tag{6.41}$$

and the solution to the displacement field is then simply

$$\Delta\boldsymbol{d}_{\varphi} = \widetilde{\boldsymbol{K}}^{-1}\widetilde{\boldsymbol{f}} \quad . \tag{6.42}$$

Owing to the contribution from its second term, the Schur matrix is full rather than sparse, and is therefore expensive to compute directly; it is therefore advantageous to define its action as an operator instead. Equation (6.42) can then be computed only using an iterative solver [461] for which factorisation of the (full) matrix is not required. Once the displacement field update has been resolved, equation (6.40) is applied in order to post-process for the condensed magnetic potential field.

As a valid alternative approach, the displacement field can be condensed out and the potential solved as the primary field. Observe that each matrix-vector operation involving $\widetilde{\boldsymbol{K}}$ itself involves the inverse operation for $\boldsymbol{K}_{\mathbb{L}\mathbb{L}}$. In this alternative solution scheme, this inner solver step would involve the calculation of the action of $\boldsymbol{K}_{\varphi\varphi}^{-1}\cdot(\bullet)$ for each sweep of the Schur matrix-vector operation $[\boldsymbol{K}_{\mathbb{L}\mathbb{L}} - \boldsymbol{K}_{\mathbb{L}\varphi}\,\boldsymbol{K}_{\varphi\varphi}^{-1}\,\boldsymbol{K}_{\varphi\mathbb{L}}]\cdot(\bullet)$. Depending on the size, bandwidth and conditioning of $\boldsymbol{K}_{\varphi\varphi}$, this may be more computationally expensive than the approach detailed previously, and should be considered on a per-problem basis.

6.3.3 Special considerations when using the magnetic scalar potential formulation

Solving the saddle point problem arising from the discretisation and linearisation of the boundary value problem when using the magnetic scalar potential formulation poses several challenges [42]. One of them relates to the properties of the symmetric blocks $\boldsymbol{K}_{\varphi\varphi}$ and $\boldsymbol{K}_{\Phi\Phi}$. More specifically, their characteristics not only depend on the chosen constitutive models and their parameters, but also the magnetic field in the free space. Consider the decomposition of the global stiffness matrix presented in equation (6.36) in the following manner:

$$\boldsymbol{K}_{\varphi\varphi} = \begin{bmatrix} \boldsymbol{K}_{\varphi\varphi}^{B,B} & \boldsymbol{K}_{\varphi\varphi}^{B,T} & 0 \\ \boldsymbol{K}_{\varphi\varphi}^{T,B} & \boldsymbol{K}_{\varphi\varphi}^{T,T} & \boldsymbol{K}_{\varphi\varphi}^{T,S} \\ 0 & \boldsymbol{K}_{\varphi\varphi}^{S,T} & \boldsymbol{K}_{\varphi\varphi}^{S,S} \end{bmatrix} \quad , \quad \boldsymbol{K}_{\Phi\Phi} = \begin{bmatrix} \boldsymbol{K}_{\Phi\Phi}^{B,B} & \boldsymbol{K}_{\Phi\Phi}^{B,T} & 0 \\ \boldsymbol{K}_{\Phi\Phi}^{T,B} & \boldsymbol{K}_{\Phi\Phi}^{T,T} & \boldsymbol{K}_{\Phi\Phi}^{T,S} \\ 0 & \boldsymbol{K}_{\Phi\Phi}^{S,T} & \boldsymbol{K}_{\Phi\Phi}^{S,S} \end{bmatrix} \tag{6.43}$$

wherein we have used the notation $B = \mathscr{B}_0^{\mathrm{h}} \backslash \Gamma_0^{\mathscr{B}\mathscr{S},\mathrm{h}}$, $T = \Gamma_0^{\mathscr{B}\mathscr{S},\mathrm{h}}$ and $S = \mathscr{S}_0^{\mathrm{h}} \backslash \Gamma_0^{\mathscr{B}\mathscr{S},\mathrm{h}}$. Both of the free space tangent contributions $\boldsymbol{K}_{\boldsymbol{\varphi}\boldsymbol{\varphi}}^{S,S}$ and $\boldsymbol{K}_{\Phi\Phi}^{S,S}$, as derived from the concave transformed energy function stated in equation (5.178), are negative-definite. The magnetic tangent within the body $\boldsymbol{K}_{\Phi\Phi}^{B,B}$ may either be positive- or negative-definite depending on the constitutive law and parameters (e. g. paramagnetic versus diamagnetic materials). Ideally $\boldsymbol{K}_{\boldsymbol{\varphi}\boldsymbol{\varphi}}^{B,B}$ is always positive-definite, which would be the case if Ψ_0 is polyconvex.

However, due to the magnetic contributions (e. g. simply that given in equation (5.178)) this is not necessarily the case. They may be the source of the material instabilities (observable physical phenomena in magnetostrictive materials) referred to later in Chapter 7. At low magnetic fields and high deformation the elastic stiffness dominates, while at high magnetic fields and low deformation the contributions from the magnetic terms dominate. In the latter scenario, the material may not be capable of resisting the induced ponderomotive forces, leading to buckling or material breakdown. This effect is exacerbated by the presence of the surrounding free space. Although the tangent stiffness matrix $\boldsymbol{K}_{\boldsymbol{\varphi}\boldsymbol{\varphi}}^{S,S}$ can be ignored (because $\delta\boldsymbol{\varphi} = 0$ on $\mathscr{S}_0^{\mathrm{h}}$), that of $\boldsymbol{K}_{\boldsymbol{\varphi}\boldsymbol{\varphi}}^{T,T}$ must remain. For the DoFs on the body's boundary $\Gamma_0^{\mathscr{B}\mathscr{S}}$, $\boldsymbol{K}_{\boldsymbol{\varphi}\boldsymbol{\varphi}}^{T,T}$ includes contributions from the body (positive-definite) and free space (negative-definite). Although it is expected that at low magnetic fields the former dominates, at high magnetic fields the latter contribution is likely to be the dominant one, thereby potentially leading to numerical instabilities. In order to increase the stability of numerical simulations, it may be beneficial to neglect the material tangent of the Maxwell contribution to $\boldsymbol{K}_{\boldsymbol{\varphi}\boldsymbol{\varphi}}^{T,T}$. It has been demonstrated [417] that this does not necessarily lead to inferior convergence rates in the non-linear solution scheme for the range of stable magnetic fields.

7 Constitutive modelling

Assuming an isothermal setting ($\dot{\theta} = 0$) and introducing the total free energy function Ψ^* parameterised in terms of the referential magnetic induction vector $\mathbb{B}$, what results from equation (5.213) is the reduced Clausius–Duhem dissipation inequality from which it is determined that the dissipative power

$$\mathscr{D}_{\text{int}} = \mathscr{W}_{\text{int}} + \mathscr{M}_{\text{int}} - \dot{\Psi}_0^* = \mathbf{P}^{\text{tot}} : \dot{\mathbf{F}} + \mathbb{H} \cdot \dot{\mathbb{B}} - \dot{\Psi}_0^* \left(\mathbf{F}, \mathbf{F}_{\text{v}}^i, \mathbb{B}, \mathbb{B}_{\text{v}}^j \right) \geq 0 \quad . \tag{7.1}$$

Here it is assumed that there may exist both mechanical and magnetic dissipative mechanisms [469] in the system. The mechanical and magnetic internal power are respectively denoted by $\mathscr{W}_{\text{int}}$ and $\mathscr{M}_{\text{int}}$, while $\mathbf{F}_{\text{v}}^i = \mathbf{F}_{\text{v}}^i(t)$ and $\mathbb{B}_{\text{v}}^j = \mathbb{B}_{\text{v}}^j(t)$ are respectively the internal variables related to the ith mechanical dissipative mechanism and the jth magnetic dissipative mechanism. As a point of clarity on the nomenclature, in equation (7.1) and in the remainder of this chapter, the free energy function as it is stated is assumed to depend on all mechanical dissipative mechanisms $i \in [1, n_{\text{mech. d. m.}}]$ and magnetic dissipative mechanisms $j \in [1, n_{\text{mag. d. m.}}]$. For now, we defer the details on the viscoelasticity and viscomagnetism to Section 7.2.

Similar to what was done in Section 5.3.7.2, we introduce the Legendre transformation [71, 120, 533, 418]

$$\widetilde{\Psi_0} \left(\mathbf{F}, \mathbf{F}_{\text{v}}^i, \mathbb{H}, \mathbb{H}_{\text{v}}^j \right) = {\Psi_0}^* \left(\mathbf{F}, \mathbf{F}_{\text{v}}^i, \mathbb{B}(\mathbb{H}), \mathbb{B}_{\text{v}}^j(\mathbb{H}) \right) - \mathbb{H} \cdot \mathbb{B}(\mathbb{H}) \tag{7.2}$$

in order to express equation (7.1) with $\mathbb{H}$ as the independent variable. From this it is evident that the material rate of the total free energy is

$$\dot{\widetilde{\Psi_0}} = \dot{\Psi}_0^* - \dot{\mathbb{H}} \cdot \mathbb{B} - \mathbb{H} \cdot \dot{\mathbb{B}} \tag{7.3}$$

Using the principle of material frame indifference [80], the total free energy function can be re-parameterised in terms of, for example, the right Cauchy–Green deformation tensor

$$\Psi_0 \left(\mathbf{C}, \mathbf{C}_{\text{v}}^i, \mathbb{H}, \mathbb{H}_{\text{v}}^j \right) = \widetilde{\Psi_0} \left(\mathbf{F}, \mathbf{F}_{\text{v}}^i, \mathbb{H}, \mathbb{H}_{\text{v}}^j \right) \quad , \tag{7.4}$$

and substitution of equation (7.4) into equation (7.2) renders the transformed and re-parameterised dissipation inequality

$$\mathscr{D}_{\text{int}} = \mathbf{S}^{\text{tot}} : \frac{1}{2} \dot{\mathbf{C}} - \mathbb{B} \cdot \dot{\mathbb{H}} - \dot{\Psi}_0 \left(\mathbf{C}, \mathbf{C}_{\text{v}}^i, \mathbb{H}, \mathbb{H}_{\text{v}}^j \right) \geq 0 \quad . \tag{7.5}$$

The re-parameterised internal variable related to the ith mechanical dissipative mechanism is given by $\mathbf{C}_{\text{v}}^i = \mathbf{C}_{\text{v}}^i(t)$, and that related to the jth magnetic dissipative mechanism is given by $\mathbb{H}_{\text{v}}^j = \mathbb{H}_{\text{v}}^j(t)$.

https://doi.org/10.1515/9783110418576-007

Using the chain rule, the material rate of the Legendre transformed total free energy function given in equation (7.4) can be expanded to

$$\dot{\Psi}_0\left(\mathbf{C}, \mathbf{C}_{\mathrm{v}}^i, \mathbb{H}, \mathbb{H}_{\mathrm{v}}^j\right) = \left[2\frac{\partial \Psi_0}{\partial \mathbf{C}}\right] : \frac{1}{2}\dot{\mathbf{C}} + \sum_i \left[2\frac{\partial \Psi_0}{\partial \mathbf{C}_{\mathrm{v}}^i}\right] : \frac{1}{2}\dot{\mathbf{C}}_{\mathrm{v}}^i - \left[-\frac{\partial \Psi_0}{\partial \mathbb{H}}\right] \cdot \dot{\mathbb{H}} - \sum_j \left[-\frac{\partial \Psi_0}{\partial \mathbb{H}_{\mathrm{v}}^j}\right] \cdot \dot{\mathbb{H}}_{\mathrm{v}}^j \quad . \tag{7.6}$$

Thereafter, substitution of equation (7.6) into equation (7.5) yields

$$\mathscr{D}_{\mathrm{int}} = \left[\mathbf{S}^{\mathrm{tot}} - 2\frac{\partial \Psi_0}{\partial \mathbf{C}}\right] : \frac{1}{2}\dot{\mathbf{C}} - \sum_i \left[2\frac{\partial \Psi_0}{\partial \mathbf{C}_{\mathrm{v}}^i}\right] : \frac{1}{2}\dot{\mathbf{C}}_{\mathrm{v}}^i + \left[-\mathbb{B} - \frac{\partial \Psi_0}{\partial \mathbb{H}}\right] \cdot \dot{\mathbb{H}} + \sum_j \left[-\frac{\partial \Psi_0}{\partial \mathbb{H}_{\mathrm{v}}^j}\right] \cdot \dot{\mathbb{H}}_{\mathrm{v}}^j \geq 0 \quad . \tag{7.7}$$

Through application of the Coleman–Noll [92] (or Coleman–Gurtin [91]) procedure, we exploit the arbitrary nature of $\dot{\mathbf{C}}$ and $\dot{\mathbb{B}}$ in order to satisfy the dissipation inequality. This leads to the definitions of kinetic conjugate quantities

$$\mathbf{S}^{\mathrm{tot}} = 2\frac{\partial \Psi_0}{\partial \mathbf{C}} \quad , \quad \mathbb{B} = -\frac{\partial \Psi_0}{\partial \mathbb{H}} \tag{7.8}$$

and the final expression for the dissipative power

$$\mathscr{D}_{\mathrm{int}} = -\sum_i \left[2\frac{\partial \Psi_0}{\partial \mathbf{C}_{\mathrm{v}}^i}\right] : \frac{1}{2}\dot{\mathbf{C}}_{\mathrm{v}}^i + \sum_j \left[-\frac{\partial \Psi_0}{\partial \mathbb{H}_{\mathrm{v}}^j}\right] \cdot \dot{\mathbb{H}}_{\mathrm{v}}^j \geq 0 \quad . \tag{7.9}$$

Its requirements must be attained through a satisfactory choice of total free energy function and laws governing the evolution of $\dot{\mathbf{C}}_{\mathrm{v}}^i$ and $\dot{\mathbb{H}}_{\mathrm{v}}^j$. In the case where $\mathbf{C}_{\mathrm{v}}^i$ and $\mathbb{H}_{\mathrm{v}}^j$ are independent of one another then the dissipation condition, applicable to dissipative mechanisms that are also assumed independent, simplifies to [469]

$$\frac{\partial \Psi_0}{\partial \mathbf{C}_{\mathrm{v}}^i} : \dot{\mathbf{C}}_{\mathrm{v}}^i \leq 0 \quad , \quad -\frac{\partial \Psi_0}{\partial \mathbb{H}_{\mathrm{v}}^j} \cdot \dot{\mathbb{H}}_{\mathrm{v}}^j \geq 0 \quad \forall\, i,j \quad . \tag{7.10}$$

Analogous to equation (5.177), we can in general describe a constitutive law governing the behaviour of a magneto-sensitive material by

$$\Psi_0\left(\mathbf{C}, \mathbf{C}_{\mathrm{v}}^i, \mathbb{H}, \mathbb{H}_{\mathrm{v}}^j\right) = U_0\left(\mathbf{C}, \mathbf{C}_{\mathrm{v}}^i, \mathbb{H}, \mathbb{H}_{\mathrm{v}}^j\right) + M_0\left(\mathbf{C}, \mathbb{H}\right) \quad , \tag{7.11}$$

wherein it is assumed that the total free energy can be expressed as an additive decomposition of energy stored in the magnetic field (as given by equation (5.178)) and U_0 which represents the combined recoverable and dissipated energy stored in the material. This form of total energy function is attributable to collective works of Brigadnov and Dorfmann [60], and Dorfmann and Ogden [120, 125]. From this, paralleling the relationships stated in equation (5.214a), the additive stress decomposition that is analogous to equation (5.90) is

$$\mathbf{S}^{\mathrm{mech}} + \mathbf{S}^{\mathrm{mag}} = 2\frac{\partial U_0}{\partial \mathbf{C}} \quad , \quad \mathbf{S}^{\mathrm{max}} = 2\frac{\partial M_0}{\partial \mathbf{C}} \quad . \tag{7.12}$$

7.1 Preliminaries to magneto-mechanical energy functions

The development of properly convex energy functions is a specialised area of research, and each component of the developed constitutive law requires thorough inspection. Motivation for the formulation of phenomenological constitutive laws may be derived from one of several sources, such as experimental data [240, 402], direct micro-mechanical considerations [248, 105, 528, 529, 68, 470] and homogenisation via mean field theories (detailed later in Chapter 9). Special considerations that have to be made when developing constitutive laws include those related to material instabilities, as was mentioned in Section 5.4.3. Details regarding the construction of polyconvex energy functions can be found in, for example, [478, 52, 227] for elastic materials and [169, 228, 248] for coupled electro- and magneto-active polymers. In the remainder of this section, we will expand on how some of the fundamental characteristics of the composite material's magnetoelastic response can be captured within a phenomenological constitutive law.

7.1.1 Volumetric-isochoric split

Rubber-like materials often exhibit very different isochoric and dilatory behaviour. For this reason, following the approach of Simo et al. [495] it is often convenient to consider a multiplicative split of the deformation gradient

$$\mathbf{F} = \widehat{\mathbf{F}} \cdot \overline{\mathbf{F}} \quad , \tag{7.13}$$

where the dilatory component $\widehat{\mathbf{F}} := J^{\frac{1}{\mathrm{d}}}\mathbf{I}$ and the volume-preserving part $\overline{\mathbf{F}} := J^{-\frac{1}{\mathrm{d}}}\mathbf{F}$ (with d representing the spatial dimension), in conjunction with a further split of the magneto-mechanical stored energy into volumetric and isochoric parts

$$U_0\left(\mathbf{C}, \mathbf{C}_{\mathrm{v}}^i, \mathbb{H}, \mathbb{H}_{\mathrm{v}}^j\right) = U_0^{\mathrm{vol}}(J) + \overline{U}_0\left(\overline{\mathbf{C}}, \overline{\mathbf{C}}_{\mathrm{v}}^i, \mathbb{H}, \mathbb{H}_{\mathrm{v}}^j\right) \quad . \tag{7.14}$$

The isochoric right Cauchy–Green tensor is defined as

$$\overline{\mathbf{C}} = \overline{\mathbf{F}}^T \cdot \overline{\mathbf{F}} = J^{-\frac{2}{\mathrm{d}}}\mathbf{C} \tag{7.15}$$

and the mechanical internal variable $\overline{\mathbf{C}}_{\mathrm{v}}^i = \overline{\mathbf{C}}_{\mathrm{v}}^i(\overline{\mathbf{C}}, t)$. If we assume the material to be non-dissipative, then from this additive decomposition the total stress and its linearisation can be computed as

$$\begin{aligned}
\mathbf{S}^{\mathrm{tot}} &= 2\frac{\partial U_0^{\mathrm{vol}}(J)}{\partial \mathbf{C}} + 2\frac{\partial \overline{U}_0\left(\overline{\mathbf{C}}, \mathbb{H}\right)}{\partial \mathbf{C}} + 2\frac{\partial M_0(\mathbf{C}, \mathbb{H})}{\partial \mathbf{C}} \\
&= 2\frac{\partial U_0^{\mathrm{vol}}(J)}{\partial J}\frac{\partial J}{\partial \mathbf{C}} + 2\frac{\partial \overline{U}_0\left(\overline{\mathbf{C}}, \mathbb{H}\right)}{\partial \overline{\mathbf{C}}} : \mathscr{P} + 2\frac{\partial M_0(\mathbf{C}, \mathbb{H})}{\partial \mathbf{C}} \quad ,
\end{aligned} \tag{7.16}$$

$$\mathscr{H}^{\mathrm{tot}} := 2\frac{\mathrm{d}\mathbf{S}^{\mathrm{tot}}}{\mathrm{d}\mathbf{C}} = 4\frac{\partial^2 U_0^{\mathrm{vol}}(J)}{\partial\mathbf{C}\otimes\mathrm{d}\mathbf{C}} + 4\frac{\partial^2 \overline{U}_0\left(\overline{\mathbf{C}},\mathbb{H}\right)}{\partial\mathbf{C}\otimes\mathrm{d}\mathbf{C}} + 4\frac{\partial^2 M_0(\mathbf{C},\mathbb{H})}{\partial\mathbf{C}\otimes\mathrm{d}\mathbf{C}}$$
$$= 4\frac{\partial^2 U_0^{\mathrm{vol}}(J)}{\partial J^2}\frac{\partial J}{\partial\mathbf{C}}\otimes\frac{\mathrm{d}J}{\mathrm{d}\mathbf{C}} + 4\frac{\partial U_0^{\mathrm{vol}}(J)}{\partial J}\frac{\partial^2 J}{\partial\mathbf{C}\otimes\mathrm{d}\mathbf{C}}$$
$$+ 4\widehat{\mathscr{P}}^T : \frac{\partial^2 \overline{U}_0\left(\overline{\mathbf{C}},\mathbb{H}\right)}{\partial\overline{\mathbf{C}}\otimes\mathrm{d}\overline{\mathbf{C}}} : \widehat{\mathscr{P}} + 4\frac{\partial \overline{U}_0\left(\overline{\mathbf{C}},\mathbb{H}\right)}{\partial\overline{\mathbf{C}}} : \frac{\mathrm{d}\widehat{\mathscr{P}}}{\mathrm{d}\mathbf{C}} + 4\frac{\partial^2 M_0(\mathbf{C},\mathbb{H})}{\partial\mathbf{C}\otimes\mathrm{d}\mathbf{C}} \quad , \tag{7.17}$$

and the coupling magnetoelastic tensors are

$$\mathfrak{P}^{\mathrm{tot}} := -\frac{\mathrm{d}\mathbf{S}^{\mathrm{tot}}}{\mathrm{d}\mathbb{H}} = -2\frac{\partial^2 \overline{U}_0\left(\overline{\mathbf{C}},\mathbb{H}\right)}{\partial\mathbf{C}\otimes\mathrm{d}\mathbb{H}} - 2\frac{\partial^2 M_0(\mathbf{C},\mathbb{H})}{\partial\mathbf{C}\otimes\mathrm{d}\mathbb{H}}$$
$$= -2\widehat{\mathscr{P}}^T : \frac{\partial^2 \overline{U}_0\left(\overline{\mathbf{C}},\mathbb{H}\right)}{\partial\overline{\mathbf{C}}\otimes\mathrm{d}\mathbb{H}} - 2\frac{\partial^2 M_0(\mathbf{C},\mathbb{H})}{\partial\mathbf{C}\otimes\mathrm{d}\mathbb{H}} \quad , \tag{7.18}$$

$$\left[\mathfrak{P}^{\mathrm{tot}}\right]^T := 2\frac{\mathrm{d}\mathbb{B}}{\mathrm{d}\mathbf{C}} = -2\frac{\partial^2 \overline{U}_0\left(\overline{\mathbf{C}},\mathbb{H}\right)}{\partial\mathbb{H}\otimes\mathrm{d}\mathbf{C}} - 2\frac{\partial^2 M_0(\mathbf{C},\mathbb{H})}{\partial\mathbb{H}\otimes\mathrm{d}\mathbf{C}}$$
$$= -2\frac{\partial^2 \overline{U}_0\left(\overline{\mathbf{C}},\mathbb{H}\right)}{\partial\mathbb{H}\otimes\mathrm{d}\overline{\mathbf{C}}} : \widehat{\mathscr{P}} - 2\frac{\partial^2 M_0(\mathbf{C},\mathbb{H})}{\partial\mathbb{H}\otimes\mathrm{d}\mathbf{C}} \quad . \tag{7.19}$$

Here we have introduced the referential isochoric projection tensor,[1] which is in general defined as

$$\widehat{\mathscr{P}} := \frac{\partial\overline{\mathbf{C}}}{\partial\mathbf{C}} = J^{-\frac{2}{\mathrm{d}}}\left[\mathscr{I} - \frac{1}{\mathrm{d}}\mathbf{C}\otimes\mathbf{C}^{-1}\right] \quad . \tag{7.20}$$

The symmetric fourth-order identity tensor is

$$\mathscr{I} := \frac{1}{2}\left[\mathbf{I}\overline{\otimes}\mathbf{I} + \mathbf{I}\underline{\otimes}\mathbf{I}\right] \quad , \tag{7.21}$$

for which we define the non-standard tensor outer products $\mathscr{C} = \mathbf{A}\overline{\otimes}\mathbf{B} \rightarrow \mathscr{C}_{ijkl} = \mathrm{A}_{ik}\mathrm{B}_{jl}$ and $\mathscr{C} = \mathbf{A}\underline{\otimes}\mathbf{B} \rightarrow \mathscr{C}_{ijkl} = \mathrm{A}_{il}\mathrm{B}_{jk}$.

7.1.2 Invariants for isotropic media

For isotropic magnetoelastic materials, not only must the constitutive law remain objective under all coordinate transformations but it must also be an isotropic function of its vector and tensor valued parameters $\mathbb{H}$ and $\mathbf{F}$ [504]. Due to the requirements of objectivity and isotropy arguments, the required dependence of the free energy function is changed in the following manner:

$$U_0(\mathbf{F},\mathbb{H}) \stackrel{\mathrm{obj}}{=} U_0(\mathbf{C},\mathbb{H}) \stackrel{\mathrm{iso}}{=} U_0(I_1,\ldots,I_6) \quad . \tag{7.22}$$

1 It is crucial to note the subtle distinction between how this projection tensor is defined in equation (7.20) in comparison to the literature, such as [559, equation (3.125)] and [204, equation (6.83)], as it impacts later derivations.

In particular, a set of six irreducible invariants [497] can be determined following the representation theorem [574] that is applicable to this case. One possible compatible combination for these invariants include the three isotropic hyperelastic, one magnetostatic and two coupled magnetoelastic invariants

$$I_1(\mathbf{C}) = \mathbf{C} : \mathbf{I} \quad , \quad I_2(\mathbf{C}) = \frac{1}{2}\left[[\mathbf{C} : \mathbf{I}]^2 - \mathbf{C}^2 : \mathbf{I}\right] \quad , \quad I_3(\mathbf{C}) = \det \mathbf{C} = J^2 \quad , \tag{7.23a}$$

$$I_4(\mathbb{H}) = [\mathbb{H} \otimes \mathbb{H}] : \mathbf{I} \quad , \quad I_5(\mathbf{C}, \mathbb{H}) = [\mathbb{H} \otimes \mathbb{H}] : \mathbf{C} \quad , \quad I_6(\mathbf{C}, \mathbb{H}) = [\mathbb{H} \otimes \mathbb{H}] : \mathbf{C}^2 \quad . \tag{7.23b}$$

As a matter of convenience, one can also introduce an additional coupled invariant

$$I_7(\mathbf{C}, \mathbb{H}) = [\mathbb{H} \otimes \mathbb{H}] : \mathbf{C}^{-1} \tag{7.24}$$

that, using the Cayley–Hamilton theorem [204] in the format

$$\mathbf{C}^{-1} = \frac{1}{I_3}\left[\mathbf{C}^2 - I_1 \mathbf{C} + I_2 \mathbf{I}\right] \tag{7.25}$$

can be expressed in terms of the original irreducible invariants

$$I_7 = \frac{1}{I_3}\left[I_6 - I_1 I_5 + I_2 I_4\right] \quad . \tag{7.26}$$

Given this definition, it may be observed that $M_0(\mathbf{C}, \mathbb{H}) \equiv -\frac{\mu_0}{2} J I_7$. For completeness, we note that an alternative set of invariants solely based on the inverse of the right Cauchy–Green tensor may also be utilised [350].

7.1.3 Transverse isotropy

For transversely isotropic media, such as that which may be formed when the matrix-particle composite is cured in the present of a magnetic field, [470, 472], additional invariants can be used to account for the material anisotropy. This approach is commonly adopted in the definition of anisotropic constitutive laws for use in modelling biological [205, 416] and magneto-active [68, 105, 470] materials. Denoting the referential direction of anisotropy (orthogonal to the plane of transverse isotropy) as $\mathbf{M}$, that under the map $\boldsymbol{\varphi}$ is transformed to $\mathbf{m} = \mathbf{F} \cdot \mathbf{M}$, we can define a (referential) structure tensor

$$\mathbf{G} := \mathbf{M} \otimes \mathbf{M} \tag{7.27}$$

with which the four additional invariants [68]

$$I_8(\mathbf{C}, \mathbf{G}) = \mathbf{C} : \mathbf{G} \quad , \quad I_9(\mathbf{C}, \mathbf{G}) = \mathbf{C}^2 : \mathbf{G} \quad , \tag{7.28}$$

$$I_{10}(\mathbb{H}, \mathbf{G}) = [\mathbb{H} \otimes \mathbb{H}] : \mathbf{G} \quad , \quad I_{11}(\mathbf{C}, \mathbb{H}, \mathbf{G}) = [\mathbb{H} \otimes \mathbb{H}] : [\mathbf{C} \cdot \mathbf{G} \cdot \mathbf{C}] \tag{7.29}$$

that capture the anisotropic hyperelastic magneto-mechanical material behaviour are derived. It should be noted that in this case the transformed free energy function used in equation (7.5) now has the dependence $\Psi_0 = \Psi_0\left(\mathbf{F}, \mathbf{F}^i_{\mathrm{v}}, \mathbb{H}, \mathbb{H}^j_{\mathrm{v}}; \mathbf{G}\right)$ and that all other derived components of the constitutive laws are altered accordingly.

Care should be taken when defining Ψ_0 for anisotropic materials. Often an additive decomposition of the free energy function into an isotropic matrix component and anisotropic fibre or particle component, such as

$$\Psi_0 = \Psi_0^{\mathrm{iso}}\left(\mathbf{F}, \mathbf{F}^i_{\mathrm{v}}, \mathbb{H}, \mathbb{H}^j_{\mathrm{v}}\right) + \Psi_0^{\mathrm{aniso}}\left(\mathbf{F}, \mathbb{H}, \mathbf{G}\right) \quad , \tag{7.30}$$

is adopted, where both components are often defined in terms of invariants [223, 267, 398]. However, it is necessary that a minimum number of invariants required for completeness are used [115]. Furthermore, special consideration of the anisotropic components may be necessary when incompressible media are modelled [467].

7.2 Viscomagneto-viscoelasticity

Magneto-active polymers typically exhibit viscoelastic properties. The application of a constant deformation or body force leads to the development of a viscous stress contribution that vanishes after sufficient time has elapsed. For magneto-active materials, it is commonly assumed that, upon application of an external magnetic induction, the magnetic response is instantaneous. However, the material magnetisation is not necessarily immediate due to the motion of the underlying material microstructure. Additional dissipation therefore occurs due to the resistance of the material to a change in magnetisation. The consideration of both of these non-equilibrium responses is critical when developing technological devices based on these materials.

Saxena et al. [469] have considered both the mechanical and magnetic dissipative effects by generalising the existing magnetoelastic theory [120] and combining it with the theory of magnetic viscoelasticity. As opposed to the micro-mechanical based network modelling approach adopted by [46, 118, 43, 366], both the viscoelastic and viscomagnetic models are based on the construction of phenomenological constitutive laws with additional internal variables. It is therefore assumed that total stress is decomposed into an equilibrium stress that corresponds to the stress response at an infinitely slow rate of deformation (or, equally, the stress response when the time-dependent viscous effects are completely neglected) and a viscosity-induced overstress.

Specifically, the constitutive laws are expressed in terms of strain-type internal quantities [448, 221], as opposed to stress-type variables that take the form of convolution integrals [489, 207, 313, 245]. This approach accommodates a split of the kinematic quantities into equilibrium and non-equilibrium parts,

$$\mathcal{B}_0 \xrightarrow[\mathbf{F}^i_{\rm v},\mathbb{B}^j_{\rm v}]{} \mathcal{B}^{i,j}_{\rm v} \xrightarrow[\mathbf{F}^i_{\rm e},\mathbb{B}^j_{\rm e}]{} \mathcal{B}_t \quad (\mathbf{F},\mathbb{B}: \mathcal{B}_0 \to \mathcal{B}_t) \quad , \tag{7.31}$$

with such a decomposition defined for each mechanical dissipative mechanism $i \in [1, n_{\text{mech. d. m.}}]$ and magnetic dissipative mechanism $j \in [1, n_{\text{mag. d. m.}}]$. Shadowing the assumption made by [327, 448], the deformation gradient is multiplicatively split

$$\mathbf{F} = \mathbf{F}^i_{\rm e} \cdot \mathbf{F}^i_{\rm v} \quad \forall\, i \quad , \quad \mathbb{B} = \mathbb{B}^j_{\rm e} + \mathbb{B}^j_{\rm v} \quad \forall\, j \quad , \tag{7.32}$$

while the magnetic induction (a vectorial quantity) must have an additive decomposition. From equation $(7.32)_1$, the equilibrium and non-equilibrium right Cauchy–Green deformation tensors for each mechanical dissipation mechanism i are respectively defined as

$$\mathbf{C}_{\rm e} := [\mathbf{F}_{\rm e}]^T \cdot \mathbf{F}_{\rm e} \quad , \quad \mathbf{C}_{\rm v} := [\mathbf{F}_{\rm v}]^T \cdot \mathbf{F}_{\rm v} \quad . \tag{7.33}$$

Here, $\mathbf{C}_{\rm v}$ corresponds to a one-dimensional inelastic deformation of the dashpot in a standard rheological element [448]. The behaviour of the magnetic variable is such that on the sudden application of a constant magnetic induction then

$$\mathbb{B}_{\rm e} = \mathbb{B} \quad , \quad \mathbb{B}_{\rm v} = \mathbf{0} \quad \text{at} \quad t = 0 \quad , \tag{7.34}$$

$$\mathbb{B}_{\rm e} \to \mathbf{0} \quad , \quad \mathbb{B}_{\rm v} \to \mathbb{B} \quad \text{as} \quad t \to \infty \quad . \tag{7.35}$$

That is to say that at the instant the magnetic field accumulates in the elastic part, and is gradually transferred to the viscous component. After the application of a Legendre transformation, relationships for the magnetic field are attained, namely that

$$\mathbb{H} = \mathbb{H}^j_{\rm e} + \mathbb{H}^j_{\rm v} \quad \forall\, j \tag{7.36}$$

where, on a per dissipative mechanism basis,

$$\mathbb{H}_{\rm e} = \mathbb{H} \quad , \quad \mathbb{H}_{\rm v} = \mathbf{0} \quad \text{at} \quad t = 0 \quad , \tag{7.37}$$

$$\mathbb{H}_{\rm e} \to \mathbf{0} \quad , \quad \mathbb{H}_{\rm v} \to \mathbb{H} \quad \text{as} \quad t \to \infty \quad . \tag{7.38}$$

Like Reese and Govindjee [448], assuming an additive decomposition of the material response into equilibrium and non-equilibrium components, a general energy function for isotropic media can be expressed as

$$U_0\left(\mathbf{C}, \mathbf{C}^i_{\rm v}, \mathbb{H}, \mathbb{H}^j_{\rm v}\right) = U_0^{\rm eq}\left(\mathbf{C}, \mathbb{H}\right) + U_0^{\rm neq}\left(\mathbf{C}, \mathbf{C}^i_{\rm v}, \mathbb{H}, \mathbb{H}^j_{\rm v}\right) \quad . \tag{7.39}$$

In conjunction with equations (7.10) and (7.11), this renders the further reduced dissipation inequality

$$\frac{\partial U_0^{\rm neq}}{\partial \mathbf{C}^i_{\rm v}} : \dot{\mathbf{C}}^i_{\rm v} \le 0 \quad , \quad -\frac{\partial U_0^{\rm neq}}{\partial \mathbb{H}^j_{\rm v}} \cdot \dot{\mathbb{H}}^j_{\rm v} \le 0 \quad \forall\, i,j \quad . \tag{7.40}$$

Note that this holds only if the magnetic and mechanical non-equilibrium variables evolve independently of one another.

Assuming an additive decomposition of the magnetoelastic and magneto-inelastic energies, the energy function encapsulating the effects of material deformation and magnetisation for isotropic media could, for instance, be constructed in the general form

$$U_0\left(\mathbf{C},\mathbf{C}_\mathrm{v}^i,\mathbb{H},\mathbb{H}_\mathrm{v}^j\right) = U_0^\mathrm{eq}\left(I_1,\ldots,I_7\right) + \sum_i U_0^{\mathrm{neq},i}\left(\mathbf{C},\mathbf{C}_\mathrm{v}^i,\mathbb{H}\right) + \sum_j U_0^{\mathrm{neq},j}\left(\mathbf{C},\mathbb{H},\mathbb{H}_\mathrm{v}^j\right) \quad . \tag{7.41}$$

Analogously, and following the same reasoning, that of a transversely isotropic material could be expressed as

$$\begin{aligned} U_0\left(\mathbf{C},\mathbf{C}_\mathrm{v}^i,\mathbb{H},\mathbb{H}_\mathrm{v}^j;\mathbf{G}\right) &= U_0^\mathrm{eq}\left(I_1,\ldots,I_{11}\right) + \sum_i U_0^{\mathrm{neq},i}\left(\mathbf{C},\mathbf{C}_\mathrm{v}^i,\mathbb{H};\mathbf{G}\right) \\ &+ \sum_j U_0^{\mathrm{neq},j}\left(\mathbf{C},\mathbb{H},\mathbb{H}_\mathrm{v}^j;\mathbf{G}\right) \quad , \end{aligned} \tag{7.42}$$

with anisotropy introduced though the fixed structure tensor, expressed in equation (7.27). It remains to define the magneto-hyperelastic contribution U_0^eq and the magneto-viscoelastic components $U_0^{\mathrm{neq},i}, U_0^{\mathrm{neq},j}$, with the latter subject to the condition given in equation (7.10).

Following from the definition of the kinetic quantities given in equation (7.8), namely

$$\mathbf{S}^\mathrm{tot} = 2\frac{\partial}{\partial\mathbf{C}}\left[U_0^\mathrm{eq}\left(\mathbf{C},\mathbb{H}\right) + \sum_{i,j} U_0^\mathrm{neq}\left(\mathbf{C},\mathbf{C}_\mathrm{v}^i,\mathbb{H},\mathbb{H}_\mathrm{v}^j\right) + M_0\left(\mathbf{C},\mathbb{H}\right)\right] = \mathbf{S}_\mathrm{eq}^\mathrm{tot} + \mathbf{S}_\mathrm{neq} \quad , \tag{7.43}$$

$$\mathbb{B} = -\frac{\partial}{\partial\mathbb{H}}\left[U_0^\mathrm{eq}\left(\mathbf{C},\mathbb{H}\right) + \sum_{i,j} U_0^\mathrm{neq}\left(\mathbf{C},\mathbf{C}_\mathrm{v}^i,\mathbb{H},\mathbb{H}_\mathrm{v}^j\right) + M_0\left(\mathbf{C},\mathbb{H}\right)\right] = \mathbb{B}_\mathrm{eq} + \mathbb{B}_\mathrm{neq} \quad , \tag{7.44}$$

the non-equilibrium Piola–Kirchhoff stress and magnetic induction are

$$\mathbf{S}_\mathrm{neq} = 2\sum_{i,j}\left.\frac{\partial U_0^\mathrm{neq}\left(\mathbf{C},\mathbf{C}_\mathrm{v}^i,\mathbb{H},\mathbb{H}_\mathrm{v}^j\right)}{\partial\mathbf{C}}\right|_{\mathbf{C}_\mathrm{v}^i,\mathbb{H}_\mathrm{v}^j} \quad , \quad \mathbb{B}_\mathrm{neq} = -\sum_{i,j}\left.\frac{\partial U_0^\mathrm{neq}\left(\mathbf{C},\mathbf{C}_\mathrm{v}^i,\mathbb{H},\mathbb{H}_\mathrm{v}^j\right)}{\partial\mathbb{H}}\right|_{\mathbf{C}_\mathrm{v}^i,\mathbb{H}_\mathrm{v}^j} \quad . \tag{7.45}$$

Linearisation of the above for use with equation (5.190) renders the material tangents for non-equilibrium constituents. The direct contributions, specifically the referential elasticity tensor and referential magnetic tensor, are

$$2\frac{\mathrm{d}\mathbf{S}_\mathrm{neq}}{\mathrm{d}\mathbf{C}} = 4\left[\left.\frac{\partial^2 U_0^\mathrm{neq}}{\partial\mathbf{C}\otimes\partial\mathbf{C}}\right|_{\mathbf{C}_\mathrm{v}^i,\mathbb{H}_\mathrm{v}^j} + \sum_i\left.\frac{\partial^2 U_0^\mathrm{neq}}{\partial\mathbf{C}\otimes\partial\mathbf{C}_\mathrm{v}^i}\right|_{\mathbf{C},\mathbb{H}_\mathrm{v}^j} : \frac{\partial\mathbf{C}_\mathrm{v}^i}{\partial\mathbf{C}} + \sum_j\left.\frac{\partial^2 U_0^\mathrm{neq}}{\partial\mathbf{C}\otimes\partial\mathbb{H}_\mathrm{v}^j}\right|_{\mathbf{C},\mathbf{C}_\mathrm{v}^i} \cdot \frac{\partial\mathbb{H}_\mathrm{v}^j}{\partial\mathbf{C}}\right] \quad , \tag{7.46}$$

$$\frac{d\mathbb{B}_{neq}}{d\mathbb{H}} = -\left[\left.\frac{\partial^2 U_0^{neq}}{\partial \mathbb{H} \otimes \partial \mathbb{H}}\right|_{\mathbf{C}_v^i, \mathbb{H}_v^j} + \sum_i \left.\frac{\partial^2 U_0^{neq}}{\partial \mathbb{H} \otimes \partial \mathbf{C}_v^i}\right|_{\mathbf{C}, \mathbb{H}_v^j} : \frac{\partial \mathbf{C}_v^i}{\partial \mathbb{H}} + \sum_j \left.\frac{\partial^2 U_0^{neq}}{\partial \mathbb{H} \otimes \partial \mathbb{H}_v^j}\right|_{\mathbf{C}, \mathbf{C}_v^i} \cdot \frac{\partial \mathbb{H}_v^j}{\partial \mathbb{H}} \right] , \tag{7.47}$$

while the referential coupling magnetoelastic tensors are

$$-\frac{d\mathbf{S}_{neq}}{d\mathbb{H}} = -2\left[\left.\frac{\partial^2 U_0^{neq}}{\partial \mathbf{C} \otimes \partial \mathbb{H}}\right|_{\mathbf{C}_v^i, \mathbb{H}_v^j} + \sum_i \left.\frac{\partial^2 U_0^{neq}}{\partial \mathbf{C} \otimes \partial \mathbf{C}_v^i}\right|_{\mathbf{C}, \mathbb{H}_v^j} : \frac{\partial \mathbf{C}_v^i}{\partial \mathbb{H}} + \sum_j \left.\frac{\partial^2 U_0^{neq}}{\partial \mathbf{C} \otimes \partial \mathbb{H}_v^j}\right|_{\mathbf{C}, \mathbf{C}_v^i} \cdot \frac{\partial \mathbb{H}_v^j}{\partial \mathbb{H}} \right] , \tag{7.48}$$

$$2\frac{d\mathbb{B}_{neq}}{d\mathbf{C}} = -2\left[\left.\frac{\partial^2 U_0^{neq}}{\partial \mathbb{H} \otimes \partial \mathbf{C}}\right|_{\mathbf{C}_v^i, \mathbb{H}_v^j} + \sum_i \left.\frac{\partial^2 U_0^{neq}}{\partial \mathbb{H} \otimes \partial \mathbf{C}_v^i}\right|_{\mathbf{C}, \mathbb{H}_v^j} : \frac{\partial \mathbf{C}_v^i}{\partial \mathbf{C}} + \sum_j \left.\frac{\partial^2 U_0^{neq}}{\partial \mathbb{H} \otimes \partial \mathbb{H}_v^j}\right|_{\mathbf{C}, \mathbf{C}_v^i} \cdot \frac{\partial \mathbb{H}_v^j}{\partial \mathbf{C}} \right] . \tag{7.49}$$

Numerical examples: Viscomagneto-viscoelasticity

To demonstrate the characteristics of the viscomagnetic model, a specialised free energy function is defined, along with problem-specific kinematic conditions. In these three-dimensional examples, the constitutive law is parameterised by the magnetic induction (i. e. not under a Legendre-transformation), neglects the energy stored in the magnetic field ($M_0{}^* = 0$, rendering $\Psi_0{}^* (\mathbf{C}, \mathbf{C}_v, \mathbb{B}, \mathbb{B}_v) = U_0{}^* (\mathbf{C}, \mathbf{C}_v, \mathbb{B}, \mathbb{B}_v)$ which is the pre-transformation analogue of equation (7.11)), and assumes a single mechanical and single magnetic dissipative mechanism.

The isotropic equilibrium part of the free energy function is

$$\begin{aligned} U_0^{*,eq}(\mathbf{C}, \mathbb{H}) = \frac{\mu_e}{4}\left[1 + \alpha_e \tanh\left(\frac{I_4{}^*}{m_e}\right)\right]\left[\,[1+n]\,[I_1 - 3] + [1-n]\,[I_2 - 3]\,\right] \\ + q_e I_4{}^* + r_e I_6{}^* \quad , \end{aligned} \tag{7.50}$$

where the magnetoelastic invariants in terms of the magnetic induction, the analogue of equation (7.23b), are

$$I_4{}^*(\mathbb{B}) = [\mathbb{B} \otimes \mathbb{B}] : \mathbf{I} \quad , \quad I_5{}^*(\mathbf{C}, \mathbb{B}) = [\mathbb{B} \otimes \mathbb{B}] : \mathbf{C} \quad , \quad I_6{}^*(\mathbf{C}, \mathbb{B}) = [\mathbb{B} \otimes \mathbb{B}] : \mathbf{C}^2 \quad . \tag{7.51}$$

This Mooney–Rivlin type constitutive law exhibits stiffening due to magnetisation (as governed by the parameters α_e, m_e), as well as magnetic saturation in a component of its magneto-mechanical response. The no-field elastic shear modulus is given by μ_e, and q_e, r_e are magnetoelastic coupling parameters. The non-equilibrium component of the free energy function is

$$\begin{aligned} U_0^{*,neq}(\mathbf{C}, \mathbf{C}_v, \mathbb{B}, \mathbb{B}_v) = \frac{\mu_v}{2}\left[\mathbf{C} : \mathbf{C}_v^{-1} - 3\right] + q_v\left[\,[\mathbb{B} - \mathbb{B}_v] \otimes [\mathbb{B} - \mathbb{B}_v]\,\right] : \mathbf{I} \\ + r_v\left[\,[\mathbf{C} \cdot [\mathbb{B} - \mathbb{B}_v]] \otimes [\mathbf{C} \cdot [\mathbb{B} - \mathbb{B}_v]]\,\right] : \mathbf{I} \end{aligned} \tag{7.52}$$

with the corresponding thermodynamically consistent evolution equations for the viscous magnetic induction vector

$$\dot{\mathbb{B}}_v = \frac{2\mu_0}{\tau_m}\left[q_v\mathbf{I} + r_v\mathbf{C}^2\right][\mathbb{B} - \mathbb{B}_v] \tag{7.53}$$

and the viscous strain tensor [273]

$$\dot{\mathbf{C}}_v = \frac{1}{\tau_v}\left[\mathbf{C} - \frac{1}{3}\left[\mathbf{C} : \mathbf{C}_v^{-1}\right]\mathbf{C}_v\right] \quad . \tag{7.54}$$

Analogous to that of the elastic part, μ_v is the viscous shear modulus and q_v, r_v are visco-magnetoelastic coupling parameters. Typically τ_v is of the order of some minutes or hours, while in contrast τ_m is of the order of a few seconds or milliseconds. The evolution laws presented here have also been shown to be thermodynamically consistent with the given constitutive laws [469].

To provide some insights into the behaviour and characteristics of the viscomagnetism model, two brief examples are given and discussed below. In both cases, a pointwise computation is performed and the parameters used are (unless otherwise stated) those listed in Table 7.1. To isolate the effects of the applied magnetic induction on the history-dependent material magnetisation, no mechanical deformation is considered here ($\lambda_1 = \lambda_2 = \lambda_3 = 1$). Further examples that illustrate the influence of deformation can be found in [469].

Table 7.1: Constitutive parameters for viscomagnetism numerical examples.

(a) Equilibrium parameters.

Parameter	Value	Unit
μ_e	2.6×10^5	$\mathrm{N\,m^{-2}}$
α_e	0.3	–
m_e	1	$\mathrm{T^2}$
n	0.3	–
q_e	$1/\mu_0$	$\mathrm{A^2\,N^{-1}}$
r_e	$1/\mu_0$	$\mathrm{A^2\,N^{-1}}$

(b) Non-equilibrium parameters.

Parameter	Value	Unit
μ_v	5×10^5	$\mathrm{N\,m^{-2}}$
q_v	$5/\mu_0$	$\mathrm{A^2\,N^{-1}}$
r_v	$1/\mu_0$	$\mathrm{A^2\,N^{-1}}$
τ_m	2	s
τ_v	100	s

Stepwise magnetic induction with no deformation

Consider the scenario in which an impulse of the magnetic induction field, such as

$$\mathbb{B}_2 = \mathbb{B}_3 = 0 \quad , \quad \mathbb{B}_1 = \begin{cases} 0\,\mathrm{T} & \text{if } t < 0 \\ 0.1\mathrm{T} & \text{otherwise} \end{cases} \tag{7.55}$$

is applied to a fixed visco-magnetoelastic medium. The instantaneous application of a constant magnetic induction results in a viscous overstress and an elevated magnetic field in the direction of the applied induction, both of which reduces towards equilibrium values over an extended time duration. Figure 7.1 illustrates this effect, and the influence of the uncoupled viscomagnetic parameter q_v on the relaxation of the magnetic field and Cauchy stress.

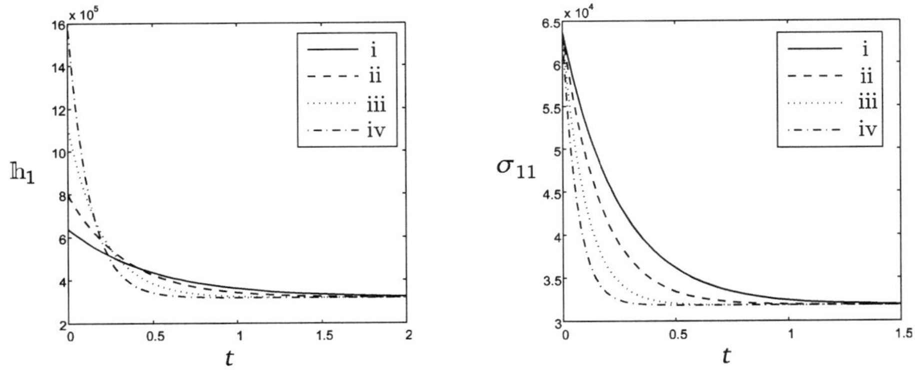

(a) Total magnetic field $\mathbb{h}_1$ [$\mathrm{A\,m^{-1}}$]. **(b)** Total Cauchy stress σ_{11} [$\mathrm{N\,m^{-2}}$].

Figure 7.1: Variation of the total magnetic field and the principal total Cauchy stress with time t [s] for no deformation in the presence of a step magnetic induction $\mathbb{B}_1 = 0.1\,\mathrm{T}$. The four curves correspond to different values of q_v: (i) $q_v = \frac{1}{\mu_0}$, (ii) $q_v = \frac{2}{\mu_0}$, (iii) $q_v = \frac{4}{\mu_0}$, (iv) $q_v = \frac{7}{\mu_0}$. [469, fig. 2 (reproduced with permission)]

It is shown that a large value of q_v leads to a high initial magnetic field, but thereafter the decay towards the equilibrium value is accelerated compared to when a lower value is used. This parameter has no influence on the initial value of the viscous overstress, but does affect the rate of decay towards its steady-state value. This is not due to a direct dependence of the stress on q_v, but rather due to its influence on the evolution of the non-equilibrium magnetic induction through equation (7.53) upon which the stress is explicitly dependent.

Time-dependent magnetic induction with no deformation

To further illustrate the magnetic history on the generated magnetic field and stresses, we consider the case where the applied induction is no longer static, but rather varies with time. The induction in the $\mathbf{e}_1$ direction is increased at a constant rate from an initial value of zero until $\mathbb{B}_1 = 0.8\,\mathrm{T}$, whereafter it is reduced at the same rate until it reaches zero again. The influence of this load history is illustrated in Figure 7.2.

For a chosen rate of applied induction, the magnetic field increases with an increasing induction and correspondingly decreases as the applied magnetic load diminishes. It is observed that at some point during unloading the magnetic field aligned

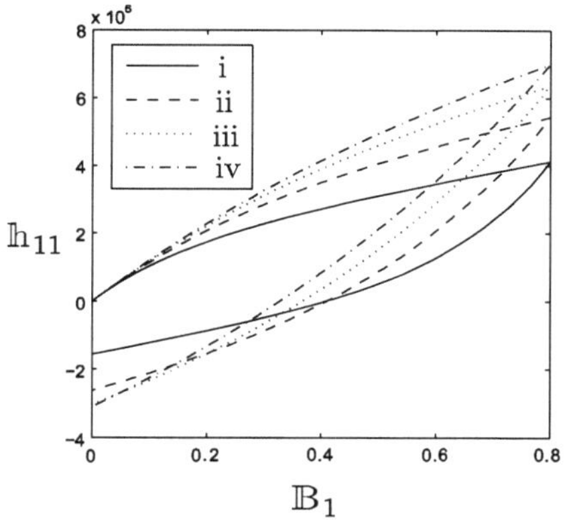

(a) Total magnetic field $\mathbb{h}_1$ [A m^{-1}].
(i) $\dot{\mathbb{B}}_1 = 1\,\mathrm{T\,s^{-1}}$, (ii) $\dot{\mathbb{B}}_1 = 2\,\mathrm{T\,s^{-1}}$,
(iii) $\dot{\mathbb{B}}_1 = 3\,\mathrm{T\,s^{-1}}$, (iv) $\dot{\mathbb{B}}_1 = 4\,\mathrm{T\,s^{-1}}$.

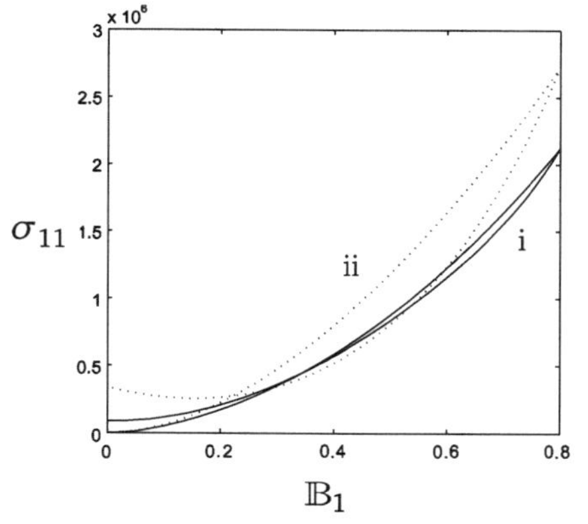

(b) Total Cauchy stress σ_{11} [N m^{-2}].
(i) $\dot{\mathbb{B}}_1 = 1\,\mathrm{T\,s^{-1}}$, (ii) $\dot{\mathbb{B}}_1 = 4\,\mathrm{T\,s^{-1}}$.

Figure 7.2: Variation of the total magnetic field and the principal total Cauchy stress with magnetic induction $\mathbb{B}_1$ [T] for a time-dependent magnetic induction. [469, fig. 10 (reproduced with permission)]

in the $\mathbf{e}_1$ direction attains a negative value. This can be explained through the relationship given in equation (5.118), and recognising that the magnetic field is generated such that it counters the material magnetisation that is aligned in the $+\mathbf{e}_1$ direction. For example, when $\mathbb{B}_1 = 0$ at the end of unloading, then $\mathbb{h}_1 < 0$ in order to offset any residual magnetisation $\mathbb{m}_1 > 0$. A higher rate of loading leads to an increase in magnitude of the generated magnetic field during loading, and a more rapid decrease and ultimate final value during unloading. The stress history differs considerably for different loading rates; and not only for the peak value that increases when the induction is increased more rapidly. The evolution of the stress during the unloading phase appears highly sensitive to the loading rate, with its value reducing towards a minimum value and subsequently increasing to a nontrivial value when $\mathbb{B}_1 = 0$.

Numerical example: Magneto-viscoelasticity

Linking the topics of experimental rheology and numerical modelling, we present an example wherein a prototype constitutive law for a magneto-viscoelastic material is evaluated for a range of mechanical and magnetic loading cases. The relevant geometry for the parallel-plate rotational rheometer is depicted in Figure 3.4, with the kinematics of the problem detailed in Section 3.1.1. From the stated kinematics, the deformation gradient (expressed in Cartesian coordinates) under harmonic torsional deformation with a compressive preload is

$$\mathbf{F}(t) = \begin{bmatrix} \frac{\cos(\alpha(t))}{\sqrt{\lambda_3}} & -\frac{\sin(\alpha(t))}{\sqrt{\lambda_3}} & -T(t)R\sqrt{\lambda_3}\sin(\Theta + \alpha(t)) \\ \frac{\sin(\alpha(t))}{\sqrt{\lambda_3}} & \frac{\cos(\alpha(t))}{\sqrt{\lambda_3}} & T(t)R\sqrt{\lambda_3}\cos(\Theta + \alpha(t)) \\ 0 & 0 & \lambda_3 \end{bmatrix} \tag{7.56}$$

and the referential magnetic field is given by

$$\mathbb{H}(t) = \begin{bmatrix} 0 & 0 & h_3(t) \end{bmatrix}^T \quad . \tag{7.57}$$

Note that, due to the parameterisation of the problem, we assume that the magnetic field applied to the material is aligned with the axis of rotation (rather than the magnetic induction as implied by Figure 3.4a).

Using the model proposed by Pelteret et al. [418], the magneto-viscoelastic component of the total free energy function that is assumed to describe the material response is defined as

$$\begin{aligned} U_0\left(\mathbf{C}, \mathbf{C}_v^i, \mathbb{H}\right) &= \frac{\mu_e}{2} f_e^{sat}\left(\mu_e, \mu_e^{\infty}, \mathbb{H}_e^{sat}, I_4\right)[I_1 - 3] + \epsilon\mu_0 J I_7 \\ &+ \sum_i \frac{\mu_v^i}{2} f_v^{sat}\left(\mu_v^i, \mu_v^{\infty,i}, \mathbb{H}_v^{sat,i}, I_4\right)\left[\left[\mathbf{C}_v^i : \mathbf{C} - 3\right] - \log(\det(\mathbf{C}_v^i))\right] \quad . \end{aligned} \tag{7.58}$$

This material model features some of the fundamental characteristics observed in experimental data (such as that shown in [545] and Chapter 4). It accommodates the multiple dissipative mechanisms that typically characterise rate-dependent materials, thereby rendering a relaxation spectrum with each component having its own characteristic relaxation time. As has been previously mentioned, materials such as that composing CIPs exhibit magnetic saturation effects, with the magnetisation of the filler plateauing after the permeating magnetic field reaches a critical strength. This is captured within the prototype constitutive model by a multiplicative factor applied to both the magnetoelastic part and each dissipative mechanism i, which effectively scales their no-field shear modulus. The saturation function is defined as

$$f^{sat}\left(\mu, \mu^{\infty}, \mathbb{H}^{sat}, I_4\right) = 1 + \left[\frac{\mu^{\infty}}{\mu} - 1\right]\tanh\left(\frac{2\, I_4}{|\mathbb{H}^{sat}|^2}\right) \quad , \tag{7.59}$$

where μ^{∞} is the saturated shear modulus, and $|\mathbb{H}^{sat}|$ is the saturation magnetic field strength; for such a function, approximately 96% of the maximum stored magnetic energy is accumulated at the saturation field strength. Similar behaviour has been previously incorporated into coupled constitutive models [72, 249, 68, 105, 498, 157, 137, 214] by means of alternatively defined saturation models. The evolution of the internal variable for each dissipative mechanism is governed by the linear evolution law

$$\dot{\mathbf{C}}_v = \frac{1}{\tau_v}\left[\mathbf{C}_v - \mathbf{C}^{-1}\right] \quad . \tag{7.60}$$

For the constitutive model to remain thermodynamically consistent, care must be taken in the choice of some of the material parameters. Following the arguments laid out by Linder et al. [312], the dissipation inequality stated in equation (7.9) is not violated if for each magnetoelastic and magneto-viscoelastic component $\mu f^{\text{sat}}(\mu, \mu^{\infty}) > 0$. The shear moduli are therefore restricted to the physically consistent range $\mu > 0$ and $\mu^{\infty} > 0$. Care must also be taken to choose a numerical time integration scheme that is sufficiently accurate and stable, and does not spuriously introduce energy into the analysed system of ordinary differential equations. For these reasons, the magneto-viscoelastic internal variable is updated using a second-order backward difference method in conjunction with the Crank–Nicholson fractional-step approach. A general framework for this category of time integrators is presented in Appendix C.10.

The chosen values of the material coefficients for a highly dissipative MAP [402] are listed in Table 7.2. As is stated in [418], the chosen parameters are fictitious but are thought to be quantitatively reasonable for the stated material. The values chosen for $\mathbb{H}_e^{\text{sat}}$, $\mathbb{H}_v^{\text{sat},1}$, $\mathbb{H}_v^{\text{sat},2}$ correspond to the magnetic induction applied to a static material with a relative permeability of 6 at $\mathbb{B}_e^{\text{sat}} = 1.6\,\text{T}$ (the saturation point for iron, although it is noted that CIP exhibits weaker magnetic properties than bulk iron [498, 256]), $\mathbb{B}_v^{\text{sat},1} = 0.7\,\text{T}$, $\mathbb{B}_v^{\text{sat},2} = 0.5\,\text{T}$.

Table 7.2: Constitutive parameters for magneto-viscoelastic numerical examples.

(a) Equilibrium parameters.

Parameter	**Value**	**Unit**
μ_e	30	kPa
μ_e^{∞}	250	kPa
$\mathbb{H}_e^{\text{sat}}$	212.2	kA m^{-1}
ϵ	−2.5	–

(b) Non-equilibrium parameters assigned to each dissipative mechanism *i*.

Parameter	**Value (diss. mech.)**		**Unit**
	1	**2**	
μ_v	20	7.5	kPa
μ_v^{∞}	35	15	kPa
$\mathbb{H}_v^{\text{sat}}$	92.84	66.31	kA m^{-1}
τ_v	0.6	4.5	s

All measurements are taken at location **P**, a point on the top outer edge of a sample with radius 10 mm and height 1 mm. The conditions under which numerical experiments were conducted are detailed in Table 7.3, and closely follow those used in Chapter 4 for the characterisation of MAPs in a laboratory setting. With respect to the mechanical conditions, an axial compression of 5 % was applied, and the amplitude of the torsion angle T was chosen such that the maximum shear strain was approximately $\gamma^{\max} = 5\,\%$. For the magnetic conditions, we wish to apply a constant average magnetic induction of between 0 T to 0.5 T as measured within the stator below the

Table 7.3: Experimental configuration for oscillatory shear with a constant perpendicularly applied magnetic field.

Parameter	Value	Unit
α^{max}	$\frac{\pi}{60}$	rad
λ_3	0.95	–
$\log(\omega)$	$\in [-1, -0.5, 0, 0.5]$	rad s^{-1}
$\mathbb{B}_3^{\text{stator}}$	$\in [0, 0.1, \ldots, 0.5]$	T

sample. From equation (5.85), we can observe that across the stator-sample interface

$$\mathbb{B}_3^{\text{stator}} \approx \mu_r \mu_0 \mathbb{H}_3^{\text{sample}} \quad \rightarrow \quad h_3 = \frac{\mathbb{B}_3^{\text{stator}}}{\mu_r \mu_0} \quad \text{with} \quad \mu_r \approx \frac{\epsilon - 0.5}{-0.5} \tag{7.61}$$

where, in this instance, μ_r represents the relative permeability of the MAP. In this approximation, it is assumed that the material behaves linearly (as would be valid at low magnetic fields), and that it can be extrapolated with fair accuracy to higher magnetic fields (which is not necessarily the case for non-linear saturating materials). From this relation, it is deduced that the relative permeability of the material (ignoring its saturation behaviour) is approximately $\mu_r = 6$, which corresponds fairly well with other approximations of the homogenised magnetic properties of CIP-based MAPs [498, 256]. Thus, $\mathbb{B}_3^{\text{stator}} = 0.5$ T corresponds to a pointwise value of the axial magnetic field inside the sample of $h_3 = 66.3\,\text{kA m}^{-1}$.

A graphical representation of the material behaviour for the range of examined experimental conditions is presented in Figure 7.3. The material clearly exhibits the qualities of a magneto-viscoelastic material, namely that (i) the effective elastic shear stiffness of the material (as given by the average slope of the curve mid-line) is increased with increasing magnetic field strength; (ii) there exists hysteretic behaviour across the spectrum of applied oscillation frequencies; and (iii) at very low excitation frequencies, the material tends to exhibit more elastic characteristics.

Further characteristics that are specific to this combination of experiment configuration and material parameters may also be observed. The stiffening response of the material due to the magnetic field is non-linear in nature. The dissipation, which is related to the enclosed area within each Lissajous curve, shows a dependence on both the oscillation frequency and the applied magnetic field; from inspection of Figures 7.3b and 7.3c it may be noted that, at a fixed excitation frequency, the dissipated energy increases as the strength of the applied magnetic field is increased. The material exhibits a broad frequency dependence due to the influence of both relaxation mechanisms, but the most dissipative behaviour is shown when both mechanisms are excited (in the range of $\omega \approx 1\,\text{rad s}^{-1}$). Typical of viscoelastic materials is that as the excitation frequency is increased above the characteristic relaxation time of a dissipative mechanism, no relaxation occurs and the viscous overstress remains. Since the dom-

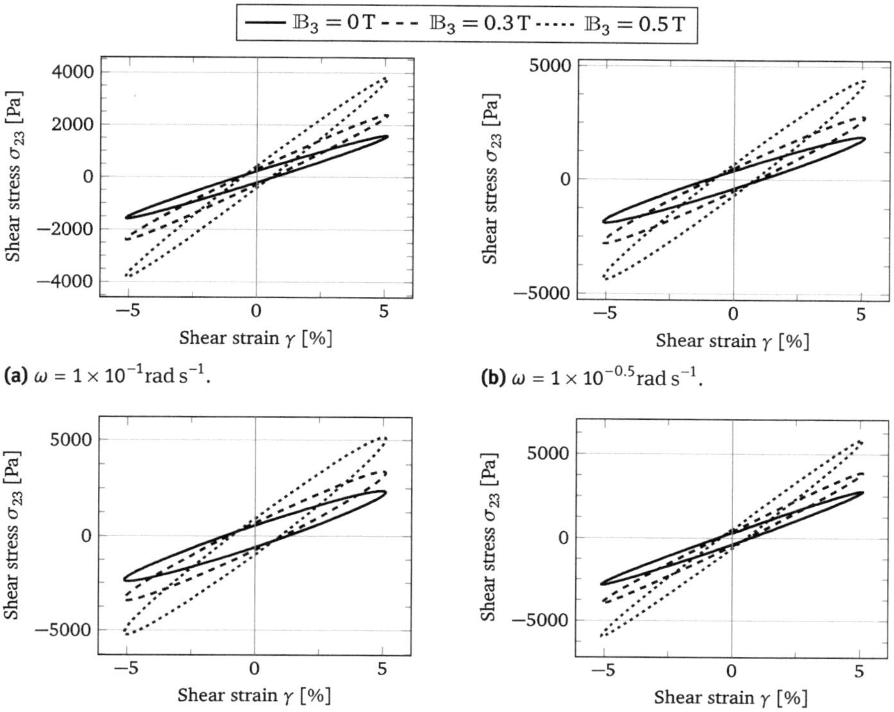

(a) $\omega = 1 \times 10^{-1} \text{rad s}^{-1}$.

(b) $\omega = 1 \times 10^{-0.5} \text{rad s}^{-1}$.

(c) $\omega = 1 \times 10^{0} \text{rad s}^{-1}$.

(d) $\omega = 1 \times 10^{0.5} \text{rad s}^{-1}$.

Figure 7.3: Steady-state Lissajous figures for a high damping magneto-active material with two magneto-mechanical dissipative mechanisms. [418, fig. 2]

inant dissipative mechanism has a relaxation time in the order of 100 ms, this effect is most readily observed at the highest tested oscillation frequency.

7.3 Transverse isotropy and particle chain dispersion

As is made clear from the literature, results and observations presented in Chapter 4, there is a significant contrast in behaviour of an MAP that is magnetically isolated during curing compared to one that has been exposed to a strong magnetic field. With reference to Figure 7.4, as the parameters that play a role in promoting particle chain formation start to dominate, the cured microstructure changes from being one that is randomly-heterogeneous (isotropic) to a predominantly transversely isotropic one. However, between these two extremes exist states in which particle migration during curing has taken place, but they do not possess a well-defined preferred directionality; the material exhibits a dispersed microstructure.

To capture this range of possible microstructures within a phenomenological model, Saxena et al. [472] utilised the methods developed to capture the anisotropy

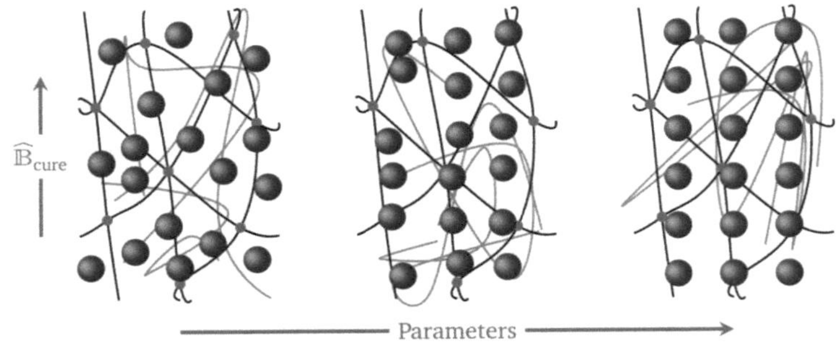

Figure 7.4: Range of particle microstructures that may arise as the curing parameters that influence particle chain formation become more significant.

of dispersed fibres in biological soft-tissues by Gasser et al. [161], Federico and Herzog [144], and Holzapfel and Ogden [206] to further develop the models presented in [470]. This approach differs substantially from that discussed in [568, 156], for which averaging methods were used to capture the microscopic behaviour of magnetisable particles in a soft carrier matrix, and [457] wherein chain-like structures were modelled using laminates.

To develop the phenomenological model that can capture anisotropy, we first introduce a spherical coordinate system with its origin located at $\mathbf{X}$ in $\mathscr{B}_0$, wherein the Eulerian angles are $\theta \in [0, \pi]$ and $\phi \in [0, 2\pi]$. In a Cartesian basis, the anisotropic direction vector can then be represented as

$$\mathbf{M}(\theta, \phi) = \sin(\theta)\cos(\phi)\,\mathbf{e}_1 + \sin(\theta)\sin(\phi)\,\mathbf{e}_2 + \cos(\theta)\,\mathbf{e}_3 \quad . \tag{7.62}$$

From this, we can immediately define the most general form of the structure tensor (valid for any local particle distribution) as

$$\mathbf{G} = \frac{1}{4\pi}\int_{\omega} \rho_{\mathrm{c}}(\mathbf{M}(\theta,\phi))\,\mathbf{M}(\theta,\phi) \otimes \mathbf{M}(\theta,\phi)\;\mathrm{d}\omega \tag{7.63}$$

where ω represents the normalised sampling volume encompassing $\mathbf{X}$ (a unit sphere), and $\rho_{\mathrm{c}}(\mathbf{M})$ is a probability density function that defines the particle chain orientation. Note that the normalisation condition requires that

$$\frac{1}{4\pi}\int_{\omega} \rho_{\mathrm{c}}(\mathbf{M}(\theta,\phi))\;\mathrm{d}\omega = 1 \quad . \tag{7.64}$$

For simplicity, we assume that the material possesses a dispersed transversally isotropic microstructure. In this case, the density function is symmetric about the normal to the plane of isotropy and thus has no dependence on the transverse angle ϕ. As is detailed by Gasser et al. [161], the distribution may for instance then be given by the π-periodic von Mises function

$$\rho_c(b,\theta) = 4\sqrt{\frac{b}{2\pi}}\frac{\exp\left(b\left[\cos(2\theta)+1\right]\right)}{\operatorname{erfi}\left(\sqrt{2b}\right)} \tag{7.65}$$

where the geometric parameter $b \in (0,\infty]$. The general structure tensor can then be expressed as

$$\mathbf{G} = \mathbf{G}(\mathbf{X}) := \kappa\mathbf{I} + [1-3\kappa]\,\overline{\mathbf{M}} \otimes \overline{\mathbf{M}} \tag{7.66}$$

where

$$\overline{\mathbf{M}} = \overline{\mathbf{M}}(\mathbf{X}) \equiv \frac{1}{4\pi}\int_{\omega} \rho_c(\mathbf{M}(\theta))\,\mathbf{M}(\theta)\ \mathrm{d}\omega \tag{7.67}$$

is the prescribed average direction of particle chain alignment at any point $\mathbf{X}$. The chain dispersion parameter, which describes the degree of material anisotropy, is defined as

$$\kappa = \kappa(\mathbf{X}) := \frac{1}{4}\int_0^{\pi} \rho_c(b,\theta)\sin^3(\theta)\ \mathrm{d}\theta \quad , \quad \kappa \in \left[0,\frac{1}{3}\right] \quad . \tag{7.68}$$

Note that when b is chosen such that $\kappa = 0$, equation (7.66) simplifies to the expression given in equation (7.27). Similarly, when $\kappa = \frac{1}{3}$, there is no dispersion and the material behaves isotropically.

A visual representation of the probability density function stated in equation (7.65) is given in Figure 7.5. One can interpret the diagram in the following way: for any chosen angle θ, the relative length of the vector $\widehat{\mathbf{M}}$ in comparison to that measured at all other angles represents the likelihood that, within the sampling volume centred at this point in the material, a particle chain with the given specific orientation will be found.

The overall microscopic viewpoint of this phenomenological model as it applies to MAPs with dispersed chain-like structures composed of magnetisable particles is illustrated in Figure 7.6. From this, it may be discerned that the phenomenological model captures anisotropy caused by microscopic particles as measured, or averaged, on the meso-scale in terms of the structural formations they are a part of. The microstructure of dispersed particle chains is wholly characterised by the prescribed values of $\kappa(\mathbf{X})$ and $\overline{\mathbf{M}}(\mathbf{X})$.

Using the above, we can also define some kinematic quantities that describe the statistical mechanics of the dispersed particle chains. The equivalent deformation gradient for the chain is defined as

$$\mathbf{F}^{c} := \frac{1}{4\pi}\int_{\omega} \rho_c\,[\mathbf{F}\cdot\mathbf{M}] \otimes \mathbf{M}\ \mathrm{d}\omega \tag{7.69}$$

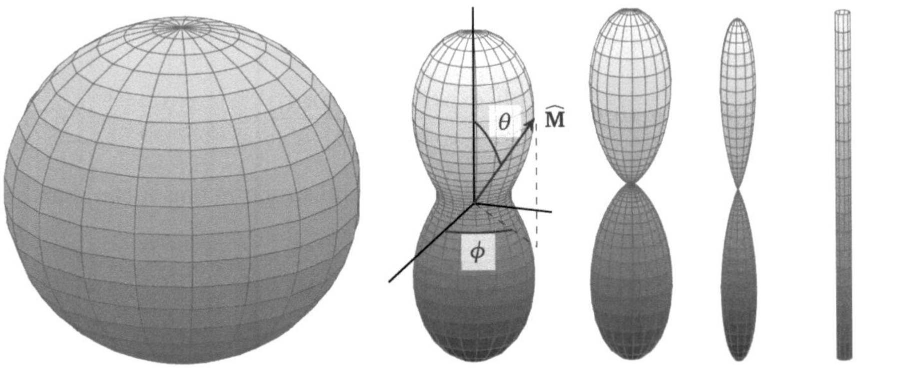

(a) $\kappa = \frac{1}{3}$. **(b)** $\kappa = \frac{1}{5}$. **(c)** $\kappa = \frac{1}{10}$. **(d)** $\kappa = \frac{1}{20}$. **(e)** $\kappa = 0$.

Figure 7.5: Probability density distributions plotted on a unit sphere according to equation (7.65). These correspond to the two limiting, and three intermediate values of the chain dispersion parameter. The figure for $\kappa = 0$ is only representative of the real geometry, which is actually a 2-D line orientated in the direction of anisotropy.

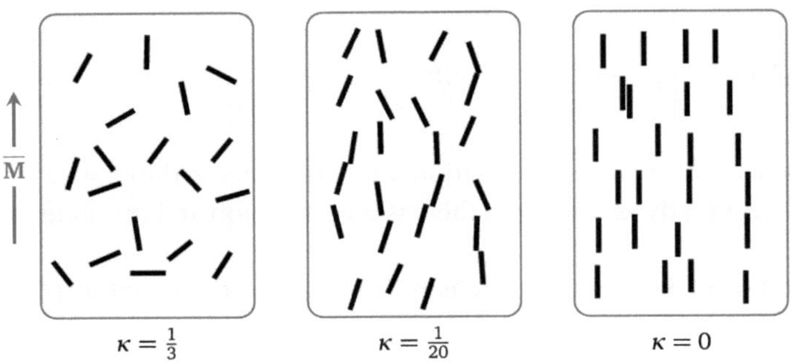

Figure 7.6: Particle microstructures are representable by the material model at any position **X** in $\mathscr{B}_0$. The phenomenological model statistically averages the influence of the distribution of particle chains, rather than individual particles themselves.

that, when using the condition given in equation (7.63), reduces to

$$\mathbf{F}^{\mathrm{c}} \equiv \mathbf{F} \cdot \mathbf{G} = \kappa \mathbf{F} + [1 - 3\kappa]\, \mathbf{F} \cdot \overline{\mathbf{G}} \quad . \tag{7.70}$$

Here, the average structure tensor is defined as $\overline{\mathbf{G}} := \overline{\mathbf{M}} \otimes \overline{\mathbf{M}}$. Similarly, one can define the chain right Cauchy–Green tensor as

$$\begin{aligned} \mathbf{C}^{\mathrm{c}} &:= \left[\mathbf{F}^{\mathrm{c}}\right]^{T} \cdot \mathbf{F}^{\mathrm{c}} = \mathbf{G} \cdot \mathbf{C} \cdot \mathbf{G} \\ &= \kappa^2 \mathbf{C} + \kappa\, [1 - 3\kappa] \left[\mathbf{C} \cdot \overline{\mathbf{G}} + \overline{\mathbf{G}} \cdot \mathbf{C}\right] + [1 - 3\kappa]^2 \left[\mathbf{C} : \overline{\mathbf{G}}\right] \overline{\mathbf{G}} \quad , \end{aligned} \tag{7.71}$$

and the referential magnetic field in the average chain direction as

$$\mathbb{H}^{\mathrm{c}} := \mathbf{G} \cdot \mathbb{H} = \kappa \mathbb{H} + [1 - 3\kappa]\, [\mathbb{H} \cdot \overline{\mathbf{M}}]\, \overline{\mathbf{M}} \quad . \tag{7.72}$$

In general, a material stored energy function can be expressed as

$$W_0 = \widetilde{W_0}(\mathbf{F}, \mathbb{H}, \mathbf{G}) \equiv W_0(\mathbf{C}, \mathbf{C}^c, \mathbb{H}, \mathbb{H}^c) \tag{7.73}$$

that, when using the arguments presented in the introduction to the chapter, is often additively decomposed into a matrix and chain component such that

$$W_0 = W_0^m(\mathbf{C}, \mathbb{H}) + W_0^c(\mathbf{C}, \mathbf{C}^c, \mathbb{H}, \mathbb{H}^c) \quad . \tag{7.74}$$

When considering equations (5.177) and (5.186), the total stress is derived as

$$\mathbf{S}^{tot} = 2\left[\left.\frac{\partial W_0(\mathbf{C}, \mathbf{C}^c, \mathbb{H}, \mathbb{H}^c)}{\partial \mathbf{C}}\right|_{\mathbf{C}^c} + \left.\frac{\partial W_0(\mathbf{C}, \mathbf{C}^c, \mathbb{H}, \mathbb{H}^c)}{\partial \mathbf{C}^c}\right|_{\mathbf{C}} : \frac{\partial \mathbf{C}^c}{\partial \mathbf{C}} + \frac{\partial M_0(\mathbf{C}, \mathbb{H})}{\partial \mathbf{C}}\right] \tag{7.75}$$

and the magnetic induction is

$$\mathbb{B} = -\left[\left.\frac{\partial W_0(\mathbf{C}, \mathbf{C}^c, \mathbb{H}, \mathbb{H}^c)}{\partial \mathbb{H}}\right|_{\mathbb{H}^c} + \left.\frac{\partial W_0(\mathbf{C}, \mathbf{C}^c, \mathbb{H}, \mathbb{H}^c)}{\partial \mathbb{H}^c}\right|_{\mathbb{H}} : \frac{\partial \mathbb{H}^c}{\partial \mathbb{H}} + \frac{\partial M_0(\mathbf{C}, \mathbb{H})}{\partial \mathbb{H}}\right] \quad . \tag{7.76}$$

In the evaluation of the above, we note that the derivatives of the kinematic chain quantities are

$$\begin{aligned}\frac{\partial \mathbf{C}^c}{\partial \mathbf{C}} &= \mathbf{G}\overline{\otimes}\mathbf{G}\\ &= \kappa^2 \mathcal{I} + \frac{\kappa}{2}[1-3\kappa]\left[\mathbf{I}\overline{\otimes}\overline{\mathbf{G}} + \mathbf{I}\underline{\otimes}\overline{\mathbf{G}} + \overline{\mathbf{G}}\overline{\otimes}\mathbf{I} + \overline{\mathbf{G}}\underline{\otimes}\mathbf{I}\right] + [1-3\kappa]^2\,\overline{\mathbf{G}}\otimes\overline{\mathbf{G}} \quad ,\end{aligned} \tag{7.77}$$

$$\frac{\partial \mathbb{H}^c}{\partial \mathbb{H}} = \mathbf{G} \quad . \tag{7.78}$$

Numerical examples: Dispersed chain-like particle structures

The following fundamental three-dimensional problems serve to demonstrate the functioning of the constitutive model for the representation of chain-like particle structures. Similar to that which was outlined for Section 7.2, contributions of the magnetic free-field energy are ignored ($M_0 = 0$) and an additive split of the energy into matrix and chain components (as expressed in equation (7.74)) is assumed. The general form of the energy functions considered here are

$$\begin{aligned}\Omega_0^m &= \frac{n_1}{\alpha}\left[\exp\left(\alpha\,[\mathbf{C}:\mathbf{I}-\mathbf{I}:\mathbf{I}]^2\right)-1\right] + n_2\,[\mathbb{H}\otimes\mathbb{H}]:\mathbf{I} + n_3\,[\mathbb{H}\otimes\mathbb{H}]:\mathbf{C}^{-1}\\ \Omega_0^c &= \frac{n_4}{\alpha}\left[\exp\left(\alpha\,[\mathbf{C}^c:\mathbf{I}-\mathbf{G}:\mathbf{I}]^2\right)-1\right] + n_5\,[\mathbb{H}^c\otimes\mathbb{H}^c]:\mathbf{I} + n_6\,[\mathbb{H}^c\otimes\mathbb{H}^c]:\mathbf{C}^{-1} \quad .\end{aligned} \tag{7.79}$$

Here, we note that both the matrix and chain contributions are themselves additively decomposed into elastic (a Fung-type law) and magnetoelastic components. Using equations (7.23a), (7.23b), (7.24) and (7.28), we can also more succinctly express the energy in terms of the invariants

$$\begin{aligned}\Omega_0^{\mathrm{m}} &= \frac{n_1}{\alpha}\left[\exp\left(\alpha\left[I_1(\mathbf{C}) - \mathbf{I}:\mathbf{I}\right]^2\right) - 1\right] + n_2 I_4(\mathbb{H}) + n_3 I_7(\mathbf{C}, \mathbb{H}) \\ \Omega_0^{\mathrm{c}} &= \frac{n_4}{\alpha}\left[\exp\left(\alpha\left[I_1(\mathbf{C}^{\mathrm{c}}) - \mathbf{G}:\mathbf{I}\right]^2\right) - 1\right] + n_5 I_4(\mathbb{H}^{\mathrm{c}}) + n_6 I_7(\mathbf{C}, \mathbb{H}^{\mathrm{c}}) \quad .\end{aligned} \tag{7.80}$$

Due to the adoption of this specific form of energy function, in particular the similarity in the energies for the matrix and particle chains, it is expected that the stress and induction developed in the chains be similar in nature to that of the matrix, but preferentially aligned with their mean orientation. The material coefficient used in these examples are listed in Table 7.4. We note that for the chosen (negative) constitutive parameters, the stress contribution that comes from the terms involving I_7 is positive when a magnetic field is applied, and subsequently reinforces the contributions arising from elastic deformation.

Table 7.4: Constitutive parameters for particle chain dispersion examples. The constant $\mu_e = 3 \times 10^4\ \mathrm{N\,m^{-2}}$.

Parameter	Value	Unit
n_1, n_4	$0.5\mu_e$	$\mathrm{N\,m^{-2}}$
n_2, n_5	$-0.5\mu_0$	$\mathrm{A^2\,N^{-1}}$
n_3, n_6	$-0.5\mu_0$	$\mathrm{A^2\,N^{-1}}$

Uniaxial deformation

We first consider the pointwise deformation of a magneto-active material with a dispersed microstructure. The mechanical and magnetic loading, as well as the average chain direction are all aligned, with

$$\mathbf{F} = \begin{bmatrix} \lambda & 0 & 0 \\ 0 & \lambda^{-1/2} & 0 \\ 0 & 0 & \lambda^{-1/2} \end{bmatrix} \quad , \quad \mathbb{H} = \begin{pmatrix} H_1 \\ 0 \\ 0 \end{pmatrix} \quad , \quad \overline{\mathbf{M}} = \begin{pmatrix} 1 \\ 0 \\ 0 \end{pmatrix} \quad ,$$

and $H_1 = 200\ \mathrm{kA\,m^{-1}}$ remains constant while the material is deformed (without volume change) by setting $\lambda \in [0.5, 2]$.

Shown in Figure 7.7 are the components of the Cauchy stress and magnetic induction aligned in the chain direction. It is observed that under conditions of large extension, the non-linear stress response is greatly increased when there is a high level of organisation of the microstructure ($\kappa = 0$) as compared to the case of an isotropic media ($\kappa = \frac{1}{3}$). Particularly noteworthy is that the material offers a greater resistance to tensile loading than compressive loading. For the given parameters the magnetic

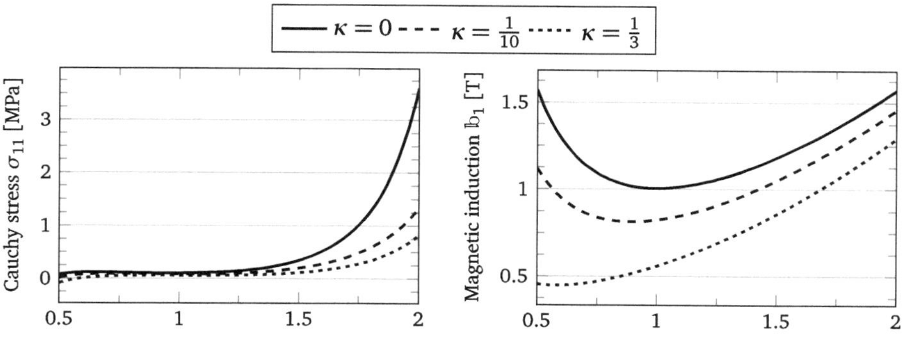

(a) Cauchy stress σ_{11}.

(b) Spatial magnetic induction $\mathbb{b}_1$.

Figure 7.7: Stress and magnetic response to a material with a dispersed chain-like microstructure subject to a uniaxial deformation and constant magnetic field. [472, fig. 6]

induction is aligned with the magnetic field, and is also significantly increased when strong particle chains are present. However, as opposed to the case for the stress, significant reinforcement during compressible loading is also observed. In the case of an isotropic material, the magnetic response is more pronounced during material extension in comparison to that of the compression case.

Inflation and extension of a magnetoelastic tube

In this second scenario, we will consider the extension and inflation of an infinitely long hollow cylinder of finite thickness. This is a commonly used example for evaluating magneto-active materials, with its analytical solution to the coupled boundary value problem first presented by Dorfmann and Ogden [123]. The cylindrical coordinate system is defined such that points $\mathbf{X} = \mathbf{X}(R, \Theta, Z)$ in $\mathscr{B}_0$ are mapped to points $\mathbf{x} = \mathbf{x}(r, \theta, z)$ in $\mathscr{B}_t$ by the non-linear deformation map $\boldsymbol{\varphi}(\mathbf{X})$. The tube is composed of an incompressible MAP with particle chains aligned circumferentially. Its initial inner and outer radius are respectively denoted by $A \leq R \leq B$ (with $\frac{B}{A} = 1.4$), and final radii are $a \leq r \leq b$. Due to the material incompressibility and symmetry of the problem, the radius of the deformed cylinder is

$$r(R) = \left[a^2 + \lambda_z^{-1}\left[R^2 - A^2\right]\right]^{\frac{1}{2}} \quad ,$$

where λ_z denotes the constant axial stretch. Expressed in the cylindrical coordinate system, the deformation gradient, magnetic field, and average chain orientation are

$$\mathbf{F} = \begin{bmatrix} \lambda_\theta^{-1}\lambda_z^{-1} & 0 & 0 \\ 0 & \lambda_\theta & 0 \\ 0 & 0 & \lambda_z \end{bmatrix} \quad , \quad \mathbb{H} = \begin{pmatrix} 0 \\ \frac{H_0}{R} \\ 0 \end{pmatrix} \quad , \quad \overline{\mathbf{M}} = \begin{pmatrix} 0 \\ 1 \\ 0 \end{pmatrix} \quad ,$$

for which the azimuthal stretch $\lambda_\theta := \frac{r}{R}$, and H_0 is a magnetic loading parameter. Note that the circumferential magnetic field (externally applied by, for example, a current-carrying wire orientated along the cylinder axis) decays through the thickness of the tube.

For this particular configuration, the magnetic field constant H_0, the axial stretch λ_z and applied radial stretch of the inner boundary $\lambda_a = \frac{a}{A}$ are prescribed. The solution to the boundary value problem is determined by solving the governing laws given by equations (5.133) and (5.135) in conjunction with the appropriate boundary conditions. From this, it is also possible to compute the internal pressure required to support the deformation. This is shown in Figure 7.8 for a fixed λ_z and $H_0 = 200\,\mathrm{kA\,m^{-1}}$, and degree of varying inflation as determined by $\lambda_a \in [0.5, 2.5]$.

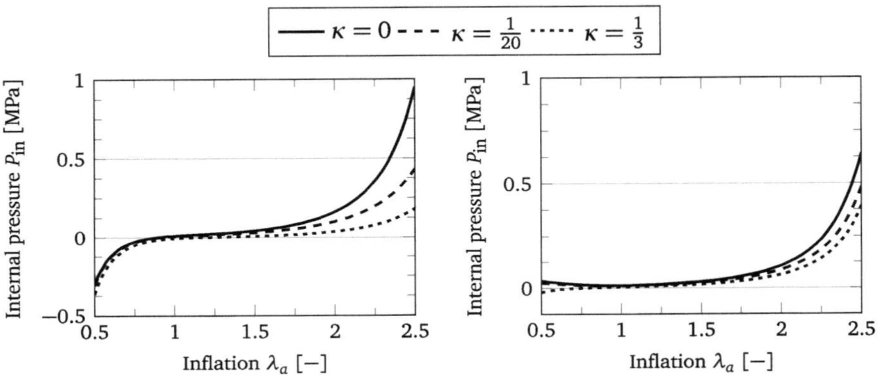

(a) Axial stretch $\lambda_z = 0.7$. **(b)** Axial stretch $\lambda_z = 1.5$.

Figure 7.8: Pressure response during inflation with constant axial deformation (compression and tension) and magnetic field. [472, fig. 8]

In the case of both axial compression and extension, the internal pressure increases exponentially as the inflation increases. When the reinforcement due to the particle chains is perfect, then the required pressure is increased considerably when compared to the isotropic case, or even that of a strong preferential direction in the material. When deflation of the tube is considered, the difference in the required (compressive) pressure on the tube surface is marginal when comparing the influence of the dispersion parameter. It is interesting to note, however, that for the case of axial extension the addition of azimuthally-aligned particle chains leads to a rise in the required pressure for both inflation and deflation. This is due to the response of the chains to the applied magnetic loading.

Considering a fixed deformation field, the influence of the magnetic loading on the internal pressure is plotted in Figure 7.9. The application of the increasing circumferential magnetic field has the tendency to shrink the tube radially, therefore requiring

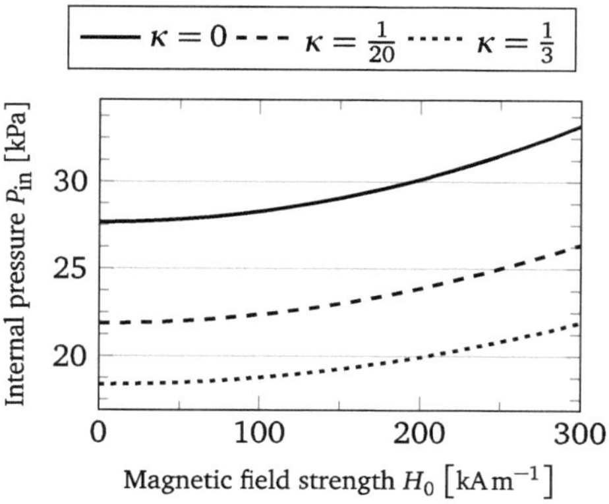

Figure 7.9: Pressure response to an applied magnetic field under conditions of constant inflation and axial stretch ($\lambda_a = 1.5$, $\lambda_z = 1.5$). [472, fig. 9]

an increase in the internal pressure. Again, this influence is greatest for the perfectly aligned particle chains that provide significant azimuthal reinforcement and, therefore, exhibit the largest response to the applied magnetic loading. As observed in the previous case, the effect of having only a slightly dispersed chain formation is quite significant. The pressure required to sustain the deformation in the isotropic case is consistently on the order of 10 kPa lower than that of $\kappa = 0$.

8 Phenomenological modelling of the curing process

Depicted in Figure 8.1 is the curing process of polymers, in which a viscous fluid is converted into a viscoelastic solid through the process of polymerisation. During this collection of exothermic chemical reactions, the short polymeric chains that comprise the low molecular weight solution bond to form long chain polymers. Until the gel point is reached, the movement of these long chain polymers is not completely constrained. Due to the migration of chains, they become interwoven and more tightly packed but there is, however, no associated change in the overall volume of the material. An increase in viscosity is observed as the friction increases between the tangled and interacting chains. Similarly, the movement of any reinforcing material, such as magnetisable particles, remains primarily unhindered.

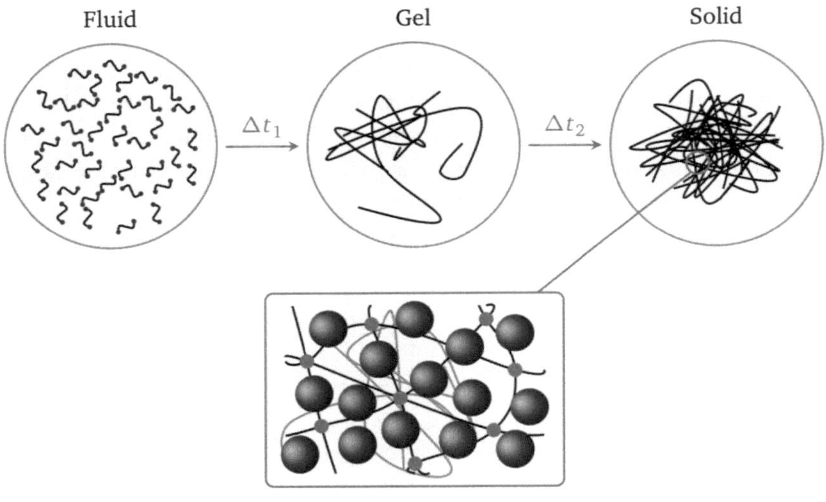

Figure 8.1: Illustration of the phase change process, from a viscous fluid to a viscoelastic solid, during curing. The inset diagram (not to scale) shows that the reinforcing particles are locked into the tightly cross-linked network. Cross linking points are depicted as red dots.

Permanent chemical bonds between the individual chains, called cross-link points, start to form after the gel point is attained. This marks the transition of the material from a gel to a macromolecular solid. Associated with the linking of the individual chains into a polymeric network is an increase in mechanical stiffness and, for a sample of constant mass, a decrease in specific volume (of the order of 1 % to 10 % [209]) of the material. The polymer is considered completely cured when the formation of cross-linking points ceases. If chemical bonding takes place between the polymer and the particles' surfaces, then the location of the reinforcing media becomes fixed within

https://doi.org/10.1515/9783110418576-008

the polymeric network. Similar can be said when the distance between cross-linking points is sufficiently small as the polymer network effectively encases the particles. For the current application to MAPs, the average distance between network points is in the order of nm while the particle diameter of in the order of µm. The latter scenario therefore applies regardless of the surface treatment of the reinforcing particles.

A simple analogue of the curing process can made by considering each individual long-chain polymer as a spring-type element. As time progresses during the cross-linking stage, the increase in composite stiffness can be represented as the sequential parallel attachment of the springs with one another. This process is illustrated in Figure 8.2.

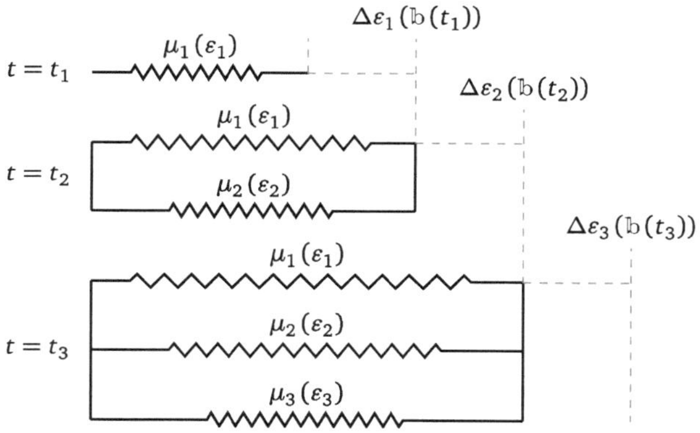

Figure 8.2: Spring-type phenomenological model [209, 191] of magnetic field-dependent curing. Mechanical deformation and magnetic loading during curing lead to the generation of residual stresses. [209, fig. 1]

Also shown is that this process can occur in the presence of mechanical deformation that interferes with the natural curing shrinkage that otherwise occurs. It has been observed that in the absence of deformation a curing material does not experience a change in stress state, even though its material properties do change [261, 171, 203]. This implies that newly formed cross-links are mechanically unstrained and stress-free. The application of a strain during curing causes the cross-linked network to stretch and a stress develops within the already cross-linked chains. During the progress of the reaction, newly added chains are not subject to the strain history that chains already fixed in the network have and, therefore, do not contribute to the stress until the deformation state is altered. The connection of varyingly prestressed elements leads to the development of residual stresses within the system. The degree to which each constituent of the network is strained may be subject to further influences, such as an imposed magnetic field or thermal effects. In composite-filled systems,

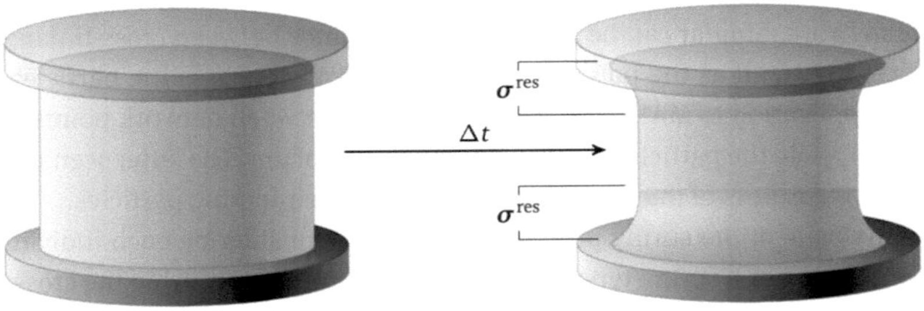

Figure 8.3: An illustration of shrinkage during curing, with an indication of the regions in the cured sample associated with high residual stresses.

such uneven curing may lead to pathological behaviours such as warping and materials cured in constrained conditions may be subject to debonding if the shrinkage-induced stresses are sufficiently large. An example result of this is seen in Figure 8.3, where a cylindrical-shaped gel is completely fixed at both ends (i. e. held at constant length) and then subject to curing. The shrinkage of the material, as well as the residual stresses throughout the specimen, are clearly visible.

As the reinforcing particles are magnetic field-sensitive, consideration of the influence of any applied magnetic field to the materials' properties that evolve during the curing process must be made. During its fluidic state, motion of the particles within the material is resisted only by the fluid's viscosity. Assuming that the uncured fluid-particle composite is sufficiently well mixed, the cured material exhibits no preferential directionality. However, as is shown in Figures 2.5 and 4.2 and stylised in Figure 8.4, exposure of the media to a magnetic field causes the particles to form

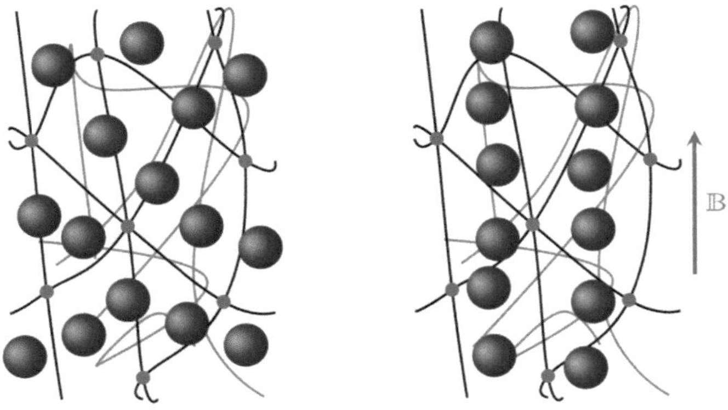

(a) No magnetic field (isotropic). **(b)** With magnetic field (anisotropic).

Figure 8.4: Particle distribution within the cross-linked network after curing with and without magnetic field. Curing under a magnetic field causes the migration of particles prior to the gel point being reached.

chain-like structures, thereby endowing it with anisotropic properties. A number of factors such as the magnetic field intensity and duration of exposure, may influence the strength of the developed anisotropy.

8.1 A continuum framework for the curing of polymers

A phenomenologically motivated hypoelastic framework that captures the evolution of mechanical properties, as well as the rate-dependence and physically observed shrinkage effects, during curing of MAPs in the presence of a magnetic field has been developed by Hossain et al. [214, 215]. The framework presented here is an extension of that expressed for the purely mechanical case for both small [209] and finite [210, 211] deformations. Similar extensions of this work have been successfully introduced for thermosetting materials with damage [353], and for multiscale modelling of curing polymers [265, 266]. Other phenomenological models of curing include those developed for thermally-sensitive curing with chemically-induced shrinkage [314, 562], for the setting of viscoelastic bio-cements [316], and for thermo-viscoplasticity with cure-dependent yield parameters [310, 311]. Similar concepts have also been used in continuum-based models that describe the chemical ageing of polymers [315, 238]. A contrasting approach has been utilised by Mahnken [331] in their macroscopic model of curing and shrinkage in temperature-dependent viscoelastomers. Further considerations have been made in [191], wherein a thermo-chemomechanical coupled polymer curing model is proposed for application to textile composites subject to damage. For a comprehensive review of curing models, we refer the reader to [208].

Starting from the second law of thermodynamics (a suitably amended form of equation (7.1)), the Clausius–Duhem inequality for an isothermal process can be expressed as

$$\mathscr{D}_{\text{int}} = \mathbf{P}^{\text{tot}} : \dot{\mathbf{F}} + \mathbb{H} \cdot \dot{\mathbb{B}} - \dot{\Psi}_0^* (t) \geq 0 \tag{8.1}$$

where the magnetoelastically coupled free energy $\Psi_0^*(t) = \Psi_0^*(t, \mathbf{F}, \mathbf{F}_{\text{v}}^i, \mathbb{B}, \mathbb{B}_{\text{v}}^j)$ captures the time-evolving material properties. The coupled energy potential, expressed in the form of a convolution integral under the presence of a magneto-mechanically coupled load, is

$$\begin{aligned}\Psi_0^* (t) = & \frac{1}{2} \int_0^t [\partial_\tau \mathscr{A}^* (\tau) : [\mathbf{F} (t) - \mathbf{F} (\tau)]] : [\mathbf{F} (t) - \mathbf{F} (\tau)] \, \mathrm{d}\tau \\ & + \frac{1}{2} \int_0^t [\partial_\tau \mathbf{D}^* (\tau) \cdot [\mathbb{B} (t) - \mathbb{B} (\tau)]] \cdot [\mathbb{B} (t) - \mathbb{B} (\tau)] \, \mathrm{d}\tau \\ & + \int_0^t [\partial_\tau \mathfrak{L}^* (\tau) \cdot [\mathbb{B} (t) - \mathbb{B} (\tau)]] : [\mathbf{F} (t) - \mathbf{F} (\tau)] \, \mathrm{d}\tau \quad ,\end{aligned} \tag{8.2}$$

where τ is the intrinsic time [314]. It captures some of the physical phenomena reported to occur during curing [171, 261], including that the stress and magnetic state remains unaltered from that induced by the past deformation history even though material properties (elastic, viscoelastic, magnetic, and viscomagnetic) continue to evolve. The time-dependent magnetoelastic tensor derivatives are denoted as

$$\partial_\tau(\diamond)(\tau) = \frac{\mathrm{d}(\diamond)}{\mathrm{d}\tau} \quad . \tag{8.3}$$

Here, $\mathscr{A}^*$, $\mathfrak{L}^*$ and $\mathbf{D}^*$ denote the fourth-, third- and second-order magnetoelastic moduli tensors. Such an energy function describes the total accumulation of energy within a system for which the magnetic and mechanical stiffness, as well as the magnetic and mechanical loading, are continuously changing. It is valid under conditions for which

$$\left.\begin{aligned} &[\partial_\tau \mathscr{A}^* : \mathbf{F}] : \mathbf{F} + [\partial_\tau \mathbf{D}^* \cdot \mathbb{B}] \cdot \mathbb{B} + \left[\partial_\tau [\mathfrak{L}^*]^T \cdot \mathbb{B}\right] : \mathbf{F} \geq 0 \\ &[\partial_\tau \mathscr{A}^* : \mathbf{F}] : \mathbf{F} \geq 0 \\ &[\partial_\tau \mathbf{D}^* \cdot \mathbb{B}] \cdot \mathbb{B} \geq 0 \end{aligned}\right\} \quad \forall\, \mathbf{F}\,,\, \mathbb{B} \quad . \tag{8.4}$$

The direct and coupling tensor moduli

$$\begin{aligned} \mathscr{A}^*(t) &= \frac{\partial^2 \Phi_0^*(t)}{\partial \mathbf{F} \otimes \mathrm{d}\mathbf{F}} \quad , \quad \mathfrak{L}^*(t) = \frac{\partial^2 \Phi_0^*(t)}{\partial \mathbf{F} \otimes \mathrm{d}\mathbb{B}} \quad , \\ [\mathfrak{L}^*]^T(t) &= \frac{\partial^2 \Phi_0^*(t)}{\partial \mathbb{B} \otimes \mathrm{d}\mathbf{F}} \quad , \quad \mathbf{D}^*(t) = \frac{\partial^2 \Phi_0^*(t)}{\partial \mathbb{B} \otimes \mathrm{d}\mathbb{B}} \end{aligned} \tag{8.5}$$

are derived from an appropriately chosen coupled auxiliary energy function $\Phi_0^*(t) = \Phi_0^*(t, \mathbf{F}, \mathbf{F}_\mathrm{v}^i, \mathbb{B}, \mathbb{B}_\mathrm{v}^j)$, thereby naturally satisfying the second and third conditions. Note that these tensors are different from equation (5.176) in the sense that they derive from the auxiliary energy function rather than a stored energy function.

The material time derivative in equation (8.1) can be evaluated using equation (8.2) and the Leibniz integral rule (stated in equation (B.20)). After some manipulation, and assuming that $\mathbf{D}^*$ is symmetric and $\mathscr{A}^*$ exhibits major symmetry (both valid since they are derived from an auxiliary energy function), this renders

$$\begin{aligned} \dot{\Psi}_0^* = &\int_0^t [\partial_\tau \mathscr{A}^*(\tau) : [\mathbf{F}(t) - \mathbf{F}(\tau)]] : \dot{\mathbf{F}}(t)\, \mathrm{d}\tau + \int_0^t [\partial_\tau \mathbf{D}^*(\tau) \cdot [\mathbb{B}(t) - \mathbb{B}(\tau)]] \cdot \dot{\mathbb{B}}(t)\, \mathrm{d}\tau \\ &+ \int_0^t [\partial_\tau \mathfrak{L}^*(\tau) \cdot [\mathbb{B}(t) - \mathbb{B}(\tau)]] : \dot{\mathbf{F}}(t)\, \mathrm{d}\tau \\ &+ \int_0^t \left[\partial_\tau [\mathfrak{L}^*]^T(\tau) : [\mathbf{F}(t) - \mathbf{F}(\tau)]\right] \cdot \dot{\mathbb{B}}(t)\, \mathrm{d}\tau \quad . \end{aligned} \tag{8.6}$$

Insertion of equation (8.6) into equation (8.1) leads to the inequality

$$\left[\mathbf{P}^{\text{tot}}(t) - \int_0^t \left[\partial_\tau \mathscr{A}^*(\tau) : [\mathbf{F}(t) - \mathbf{F}(\tau)] + \partial_\tau \mathfrak{L}^*(\tau) \cdot [\mathbb{B}(t) - \mathbb{B}(\tau)]\right] \mathrm{d}s\right] : \dot{\mathbf{F}}$$
$$+ \left[\mathbb{H}(t) - \int_0^t \left[\partial_\tau \mathbf{D}^*(\tau) \cdot [\mathbb{B}(t) - \mathbb{B}(\tau)] + \partial_\tau [\mathfrak{L}^*]^T(\tau) : [\mathbf{F}(t) - \mathbf{F}(\tau)]\right] \mathrm{d}s\right] \cdot \dot{\mathbb{B}}$$
$$\geq 0 \tag{8.7}$$

from which the form for the time-dependent total stress $\mathbf{P}^{\text{tot}}(t)$ and magnetic field $\mathbb{H}(t)$ can be straightforwardly deduced given that $\dot{\mathbf{F}}, \dot{\mathbb{B}}$ are arbitrary. Note that this result guarantees that the model is dissipation-free for arbitrary processes. Further application of equation (B.20) to the individual components of equation (8.7) renders the stress and magnetic field evolution, which are

$$\dot{\mathbf{P}}^{\text{tot}}(t) = \left[\partial_\tau \mathscr{A}^*(\tau) : [\mathbf{F}(t) - \mathbf{F}(\tau)] + \partial_\tau \mathfrak{L}^*(\tau) \cdot [\mathbb{B}(t) - \mathbb{B}(\tau)]\right]\Big|_{\tau=t} + \int_0^t \left[\partial_\tau \mathscr{A}^*(\tau) : \dot{\mathbf{F}}(t) + \partial_\tau \mathfrak{L}^*(\tau) \cdot \dot{\mathbb{B}}(t)\right] \mathrm{d}\tau \quad , \tag{8.8a}$$

$$\dot{\mathbb{H}}(t) = \left[\partial_\tau \mathbf{D}^*(\tau) \cdot [\mathbb{B}(t) - \mathbb{B}(\tau)] + \partial_\tau [\mathfrak{L}^*]^T(\tau) : [\mathbf{F}(t) - \mathbf{F}(\tau)]\right]\Big|_{\tau=t} + \int_0^t \left[\partial_\tau \mathbf{D}^*(\tau) \cdot \dot{\mathbb{B}}(t) + \partial_\tau [\mathfrak{L}^*]^T(\tau) : \dot{\mathbf{F}}(t)\right] \mathrm{d}\tau \quad , \tag{8.8b}$$

from which the final hypoelastic relations for the stress rate and rate of change of magnetic field, namely

$$\dot{\mathbf{P}}^{\text{tot}}(t) = \mathscr{A}^*(t) : \dot{\mathbf{F}}(t) + \mathfrak{L}^*(t) \cdot \dot{\mathbb{B}}(t) \quad , \tag{8.9a}$$

$$\dot{\mathbb{H}}(t) = \mathbf{D}^*(t) \cdot \dot{\mathbb{B}}(t) + [\mathfrak{L}^*]^T(t) : \dot{\mathbf{F}}(t) \quad , \tag{8.9b}$$

are deduced. Note that the last two terms in the above equations are coupling terms.

It has been reported in the literature [21, 559] that hypoelastic relations may lead to unrealistic responses (for example, in the case of elastic loading stress hysteresis may be exhibited). However, the described formulation has been thoroughly examined though numerical experimentation [214] and no significant artificial response has been observed.

8.1.1 Curing in viscomagneto-viscoelastic materials

To introduce the concept of rate-effects during curing, Figure 8.5 depicts a rheological model of curing within a linear viscoelastic setting. The model developed by Hos-

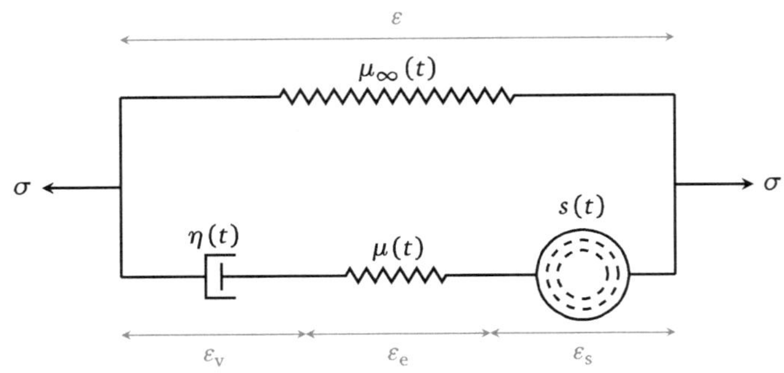

Figure 8.5: One-dimensional cure-dependent mechanical viscoelastic model with a shrinkage element of deformation [215]. In this configuration, which mimics the Zener-type viscoelastic model, the spring, the dashpot and the shrinkage element exhibit cure-dependent properties. [215, fig. 1]

sain et al. [215] and presented below is the finite-strain analogy of this idealised linear model, with extensions that incorporate field-dependent effects [214] and the non-equilibrium response [469] as introduced in Section 7.2. In this case, and in contrast to classical viscoelasticity, both equilibrium and non-equilibrium contributions will have dependence on time.

An additive decomposition of the stress and magnetic field in equilibrium and non-equilibrium components,

$$\left(\mathbf{P}^{\text{tot}}\right)^n = \left(\mathbf{P}^{\text{tot}}\right)^n_{\text{eq}} + \left(\mathbf{P}^{\text{tot}}\right)^n_{\text{neq}} \quad , \tag{8.10a}$$

$$\mathbb{H}^n = \mathbb{H}^n_{\text{eq}} + \mathbb{H}^n_{\text{neq}} \quad , \tag{8.10b}$$

comparable to that given in equations (7.43) and (7.44), is assumed at timestep n. Therein the expression for the equilibrium components is derived from the thermodynamically consistent magneto-mechanical cure-dependent relation provided by equation (8.9). In order to compute the equilibrium stress and magnetic field contributions, time discretisation must be introduced to the evolution of stress and magnetic field. Through the application of an implicit backward-Euler time integration scheme, one obtains the final result that

$$\left(\mathbf{P}^{\text{tot}}\right)^n = \underbrace{\left(\mathbf{P}^{\text{tot}}\right)^{n-1} + \mathscr{A}^* : \left[\mathbf{F}^n - \mathbf{F}^{n-1}\right] + \mathfrak{L}^* \cdot \left[\mathbb{B}^n - \mathbb{B}^{n-1}\right]}_{\left(\mathbf{P}^{\text{tot}}\right)^n_{\text{eq}}} + \left(\mathbf{P}^{\text{tot}}\right)^n_{\text{neq}} \tag{8.11a}$$

$$\mathbb{H}^n = \underbrace{\mathbb{H}^{n-1} + \mathbf{D}^* \cdot \left[\mathbb{B}^n - \mathbb{B}^{n-1}\right] + \left[\mathfrak{L}^*\right]^T : \left[\mathbf{F}^n - \mathbf{F}^{n-1}\right]}_{\mathbb{H}^n_{\text{eq}}} + \mathbb{H}^n_{\text{neq}} \tag{8.11b}$$

where the temporal function evaluation $(\bullet)^n = (\bullet)(t)$, the time update is $t = t_{n-1} + \Delta t$ and Δt is an increment in time.

For the kinematics of the viscomagnetic-viscoelastic media, the splits given in equation (7.32) are assumed. Furthermore, as was done in Section 7.2 it is again assumed that the auxiliary energy function is additively decomposed into equilibrium and non-equilibrium constituents, as expressed by

$$\Phi_0^*\left(t, \mathbf{F}, \mathbf{F}_{\mathrm{v}}^i, \mathbb{B}, \mathbb{B}_{\mathrm{v}}^j\right) = \Phi_0^{*,\mathrm{eq}}\left(t, \mathbf{F}, \mathbb{B}\right) + \Phi_0^{*,\mathrm{neq}}\left(t, \mathbf{F}, \mathbf{F}_{\mathrm{v}}^i, \mathbb{B}, \mathbb{B}_{\mathrm{v}}^j\right) \quad . \tag{8.12}$$

Herein, it is clear that the constitutive parameters in the equilibrium part can still evolve in time, as can those of the non-equilibrium part. In particular, for any time-dependent scalar parameter $\chi = \chi(t)$ it will be assumed that its evolution is governed by an exponential saturation function of the form

$$\chi(t) = \chi_0 + [\chi_\infty - \chi_0]\left[1 - \exp\left(-\kappa_\chi t\right)\right] \tag{8.13}$$

where χ_0 is the initial value, χ_∞ is the final or saturation value, and κ_χ is the curvature parameter for the material coefficient χ. This replicates the most commonly considered relationship of the curing kinetics of polymers documented within the literature [314, 261, 203]. However, currently the literature discussing the experimentally measured evolution of viscoelastic properties during curing is sparse [508, 460]. In the absence of physical evidence to the contrary, it is assumed that the evolution of the material's magnetic properties follows similar to those of the mechanical properties.

8.1.2 Curing with field-sensitive shrinkage effects

To introduce the curing-induced dilatory response to the constitutive relationship, it has been proposed that the deformation gradient can be multiplicatively split into two parts,

$$\mathbf{F} := \mathbf{F}_{\mathrm{m}} \cdot \mathbf{F}_{\mathrm{s}} \quad , \tag{8.14}$$

with $\mathbf{F}_{\mathrm{m}}$ representing the purely magneto-mechanical response. A similar approach has been adopted in [314, 562] when considering thermal and chemical influences during curing. An alternative method of decomposing the deformation gradient is presented by [331] in the context of thermo-viscoelasticity.

To incorporate the influence of the particle microstructure that have a preferential alignment (see Figure 8.4), in three-dimensions the volumetric shrinkage component of the deformation gradient tensor takes the form

$$\mathbf{F}_{\mathrm{s}} := \left[1 + \alpha(t)\; s\left(e\left(\alpha, \mathbb{B}, t\right)\right)\right]^{1/3}\left[\mathbf{I} - \beta\mathbf{G}\right] + \beta\mathbf{G} \tag{8.15}$$

where $\mathbf{G}$, which captures the anisotropy, is given by equation (7.27) and $0 \leq \beta \leq 1$ is a parameter that defines exactitude of particle alignment. Setting $\beta = 0$ recovers the isotropic shrinkage deformation gradient presented in [214].

The degree of cure $\alpha(t) \in \mathbb{R}^+[0,1]$ describes the state of transition between a gel ($\alpha = 0$) and solid ($\alpha = 1$) while the shrinkage

$$s\left(e\left(\alpha, \mathbb{B}, t\right)\right) = \frac{s_1 + s_2}{2} + \frac{s_2 - s_1}{2}\tanh\left(\eta\left[e\left(\alpha, \mathbb{B}, t\right) - \frac{e_1 + e_2}{2}\right]\right) \leq 0 \tag{8.16}$$

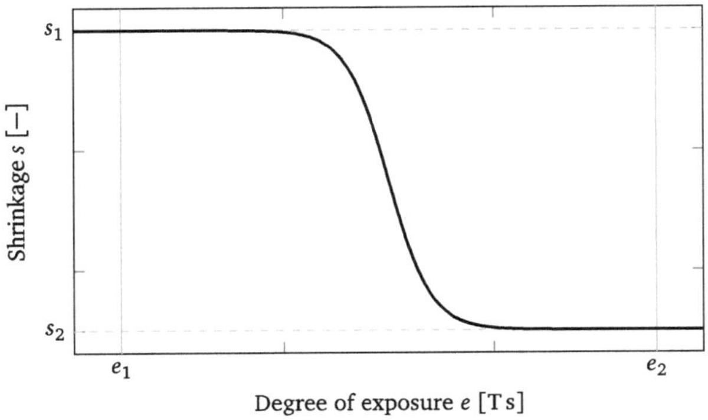

Figure 8.6: The influence of the degree of exposure (sensitive to the magnetic induction) on the shrinkage parameter.

controls the magnitude of volumetric change induced in the media. Here, η is a scaling constant and $0 < e_1 < e_2$ cut-off values of exposure. When e crosses the lower threshold e_1, it starts increasing the value of the shrinkage parameter s from an initial value of s_1. As e reaches the upper threshold e_2, the maximum prescribed value of shrinkage $s = s_2$ is reached and no more change in the shrinkage occurs. A phenomenological relationship between the shrinkage and the environmental variables to which it is sensitive is captured through the degree of exposure e, which accumulates the time and curing history for which the material has been exposed to a certain intensity of magnetic field. The degree of exposure, shown in Figure 8.6, is defined as

$$e(\alpha, \mathbb{B}, t) := \int_0^t f(\alpha(\tau)) \, |\mathbb{B}(\tau)| \, \mathrm{d}\tau \tag{8.17}$$

with the magnitude of the magnetic induction at time τ given by $|\mathbb{B}(\tau)|$ and

$$f(\alpha) = 1 - H(\alpha) \tag{8.18}$$

where $H(\circ)$ is the Heaviside step function. The definition chosen for the function $f(\alpha)$ ensures that a magnetic induction applied after the material is fully cured does not influence the degree of exposure.

From equations (5.5) and (8.14), the relation for the magneto-mechanical right Cauchy–Green tensor

$$\mathbf{C}_\mathrm{m} = \mathbf{F}_\mathrm{m}^T \cdot \mathbf{F}_\mathrm{m} = \mathbf{F}_\mathrm{s}^{-T} \cdot \mathbf{C} \cdot \mathbf{F}_\mathrm{s}^{-1} \tag{8.19}$$

can be deduced. From the usual thermodynamical argumentation (see [522] and [559, p 63]), the corresponding total Piola–Kirchhoff and Piola stress tensors are, respectively,

$$\mathbf{S}^\mathrm{tot} = \mathbf{F}_\mathrm{s}^{-1} \cdot \mathbf{S}_\mathrm{m} \cdot \mathbf{F}_\mathrm{s}^{-T} \quad , \tag{8.20}$$

$$\mathbf{P}^\mathrm{tot} = \mathbf{P}_\mathrm{m} \cdot \mathbf{F}_\mathrm{s}^{-T} \quad . \tag{8.21}$$

Therefore, the total Piola stress resulting from a magneto-mechanical load and superimposed curing shrinkage is equal to the total magneto-mechanical stress $\mathbf{P}^{\text{tot}}$ as computed from equation (8.10a) post-multiplied by $\mathbf{F}_{\text{s}}^{T}$.

Numerical examples: Curing in the presence of a magnetic field

The selective problems that follow serve to demonstrate the different interacting characteristics of the curing model. In the first problem the effect of the time-evolving constitutive parameters is isolated and highlighted, while in the second the influence of both shrinkage and material anisotropy are shown.

In both three-dimensional examples, the additive split for the auxiliary energy function defined in equation (8.12) is assumed. A compressible Mooney–Rivlin type model

$$\begin{aligned}\Phi_0^{*,\text{eq}}(t,\mathbf{C},\mathbb{B}) &= \frac{\mu_{\text{e}}(t)}{4}\left[1+\alpha_{\text{e}}\tanh\left(\frac{I_4}{m_{\text{e}}}\right)\right]\Big[\,[1+n]\,[I_1-3]+[1-n]\,[I_2-3]\,\Big]\\ &+\frac{\kappa_{\text{e}}(t)}{8}\,[\ln J]^2-\frac{\mu_{\text{e}}(t)}{2}\ln J+q(t)\,I_4^{*}+r(t)\,I_6^{*}\end{aligned} \tag{8.22}$$

for which, in general, any physical material parameter may evolve, is chosen for the equilibrium part. For the non-equilibrium component (with a single mechanical and magnetic dissipative mechanism), the constitutive relationship is

$$\begin{aligned}\Phi_0^{*,\text{neq}}(t,\mathbf{C},\mathbf{C}_{\text{v}},\mathbb{B},\mathbb{B}_{\text{v}}) &= \frac{\mu_{\text{v}}}{2}\left[\mathbf{C}_{\text{v}}^{-1}:\mathbf{C}-3\right]+\frac{\kappa_{\text{v}}}{2}\,[\ln J_{\text{e}}]^2-\mu_{\text{v}}\ln J_{\text{e}}\\ &+q_{\text{v}}\,\big[\,[\mathbb{B}-\mathbb{B}_{\text{v}}]\otimes[\mathbb{B}-\mathbb{B}_{\text{v}}]\,\big]:\mathbf{I}\\ &+r_{\text{v}}\,\big[\,[\mathbf{C}\cdot[\mathbb{B}-\mathbb{B}_{\text{v}}]]\otimes[\mathbf{C}\cdot[\mathbb{B}-\mathbb{B}_{\text{v}}]]\,\big]:\mathbf{I}\quad,\end{aligned} \tag{8.23}$$

where $J_{\text{e}}=\det\mathbf{F}_{\text{e}}$ is the magneto-mechanical volumetric Jacobian. The elastic and coupled invariants are respectively listed in equations (7.23a) and (7.51). The scaling parameter $\alpha_e>0$ induces an increase in the stiffness due to magnetisation. For the given auxiliary energy function, evolution laws that are thermodynamically consistent are

$$\dot{\mathbb{B}}_{\text{v}}=\frac{2\mu_0}{\tau_m(t)}\left[q_{\text{v}}\mathbf{I}+r_{\text{v}}\mathbf{C}^2\right]\cdot[\mathbb{B}-\mathbb{B}_{\text{v}}] \tag{8.24}$$

for the magnetic dissipative part, and

$$\dot{\mathbf{C}}_{\text{v}}=\frac{1}{\tau_v(t)}\,[\mathbf{C}-\mathbf{C}_{\text{v}}] \tag{8.25}$$

for the mechanical dissipative component. The parameters $\tau_m(t),\tau_v(t)$ are, respectively, the relaxation times for the relaxation phenomena in the viscomagnetic and viscoelastic parts. In general, they may evolve in time due to the successive linking of chains during curing.

Table 8.1: Constant constitutive parameters for curing numerical examples.

(a) Equilibrium parameters.

Parameter	Value	Unit
α_e	0.1	–
m_e	1	T^2
n	0.5	–
q_e	$1/\mu_0$	$A^2\,N^{-1}$

(b) Non-equilibrium parameters.

Parameter	Value	Unit
μ_v	5×10^5	$N\,m^{-2}$
q_v	$5/\mu_0$	$A^2\,N^{-1}$
r_v	$1/\mu_0$	$A^2\,N^{-1}$
τ_m	40	s

Table 8.2: Time evolving constitutive parameters for curing numerical examples.

Parameter	Values for equation (8.13)			Unit
	χ_0	χ_∞	κ_χ [s^{-1}]	
μ_e	1×10^{-10}	2.5×10^5	0.0255	$N\,m^{-2}$
r_e	1×10^{-10}	$1/\mu_0$	0.0255	$A^2\,N^{-1}$
τ_v	1×10^{-10}	10	0.0255	s

For the described examples, the elastic and viscous material parameters that are chosen to remain fixed are listed in Table 8.1. All time-dependent material parameters given in Table 8.2 evolve according to equation (8.13). Both the elastic and viscous bulk moduli are calculated using the relationship $\frac{\mu}{\kappa} = 0.1$ that, in the case of small deformations, represents a material with a Poisson's ratio $\nu = 0.45$.

All presented numerical examples correspond to uniaxial deformation of an MAP with the magnetic field, applied during curing, in the direction of deformation. The deformation gradient can thus be expressed as $\mathbf{F} = \mathrm{diag}\,(\lambda_1, \lambda_2, \lambda_3)$, with $\lambda_2 = \lambda_3$, and the applied magnetic induction $\mathbb{B} = \{\mathbb{B}_1, 0, 0\}^T$. Plane stress conditions are assumed, rendering only a single component of the Piola stress tensor non-trivial. The magnetic and mechanical loading conditions that will be applied in these examples are described visually in Figure 8.7. Three phases of loading are assumed: In the first 20 s the load is linearly increased to a predefined value. After this, the load is held at a constant value for a further 160 s. Thereafter, the load increases linearly again until a maximum value is attained at $t = 200$ s.

Curing without shrinkage and anisotropy

Assuming an isotropic material response ($\beta = 0$) and that no shrinkage occurs ($s = 0$), then under pure mechanical loading ($\mathbb{B} = 0$) the stress-stretch response is given in Figure 8.8. Up until the holding phase the stress response to the deformation is nearly linear. Within the holding phase, during which the stretch remains constant but curing still occurs, stress relaxation occurs. The stress increases again when additional mechanical deformation is applied. However, due to the evolution of the material pa-

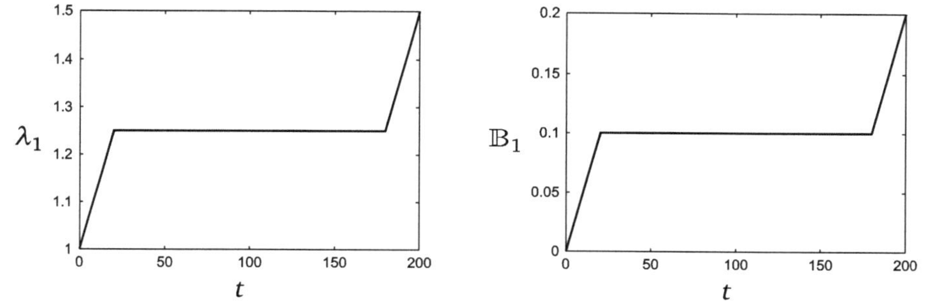

(a) Mechanical stretch λ_1 [–]. **(b)** Magnetic induction $\mathbb{B}_1$ [T].

Figure 8.7: Three-phase loading conditions (*load-hold-load*) for mechanical stretch and magnetic induction versus time t [s]. [215, fig. 4 (reproduced with permission)]

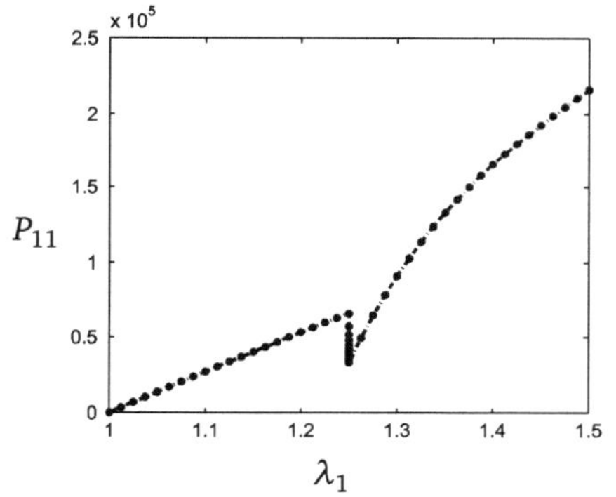

Figure 8.8: Piola stress P_{11} [MPa] versus stretch λ_1 [–] for three-phase mechanical loading of a curing viscomagneto-viscoelastic material. Clearly visible in both plots is a discontinuity that marks the evolving material response during the hold-phase of curing. During the phase, relaxation occurs but the material response becomes stiffer upon reapplication of a load. [215, fig. 7b (reproduced with permission)]

rameters due to cross-linking during the elongated holding period, the mechanical stiffness of the material has increased considerably.

In the scenario where the deformation is set to remain constant ($\lambda_1 = \lambda_2 = \lambda_3 = 1$), and magnetic loading occurs alone, similar behaviour can be observed for the magnetic field-induction response. In Figure 8.9a, the significant and rapid relaxation of the magnetic field after the initial response during the first phase is clearly visible. During the second loading stage, again a stiffer response due to the increased material parameters can be noted. The induced axial stress, plotted in Figure 8.9b, also shows a more stiff response as the coupled magnetoelastic parameters evolve during curing. Slight stress relaxation, due to viscomagnetic effects, is visible during the holding phase.

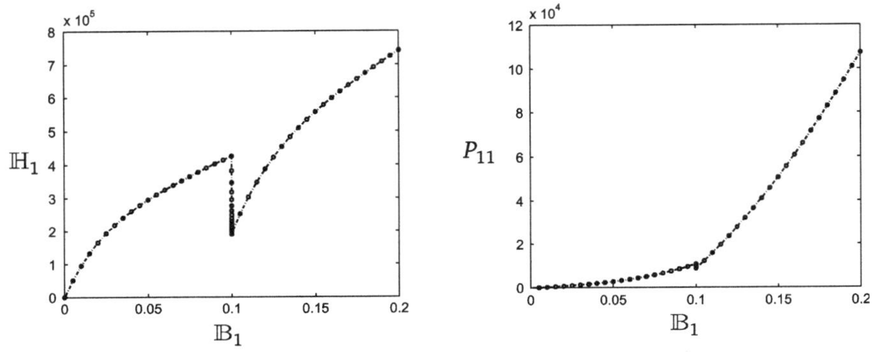

(a) Magnetic field $\mathbb{H}_1$ $[\mathrm{A\,m^{-1}}]$. **(b)** Piola stress P_{11} $[\mathrm{MPa}]$.

Figure 8.9: Stress and magnetic field response versus magnetic induction $\mathbb{B}_1$ [T] for three-phase magnetic loading of a curing viscomagneto-viscoelastic material. [215, figs. 9b,8b (reproduced with permission)]

Curing with shrinkage and anisotropy

As a first example, we consider a case of pure mechanical shrinkage. For the curing conditions that influence the shrinkage deformation gradient given in equation (8.15), a prescribed exponentially saturating α history, defined using equation (8.13) is assumed. Since there are no coupled effects considered for the shrinkage mechanism, the shrinkage parameter s is considered constant. A summary of the shrinkage conditions for this case is given in Table 8.3a.

Table 8.3: Shrinkage parameters for numerical examples of curing with shrinkage.

(a) Pure mechanical shrinkage.

Parameter	Value	Unit
α_0	1×10^{-8}	–
α_∞	1	–
κ_α	0.0225	$\mathrm{s^{-1}}$
β	0	–
$s = \dfrac{s_1+s_2}{2}$	−0.05	–

(b) Coupled magneto-mechanical shrinkage.

Parameter	Value	Unit
s_1	0	–
s_2	−0.05	–
e_1	0	T s
e_2	15	T s
η	0.5	–

Figure 8.10a depicts the assumed curing profile, while Figure 8.10b provides comparison of the stress evolution during three-phase mechanical loading when mechanical shrinkage and no shrinkage are considered. Due to the additional hydrostatic loading generated during curing, the shrinkage-induced total stress is greater than that generated during pure loading. Notably, the stress continues to increase during the holding phase as the degree of cure continues to evolve.

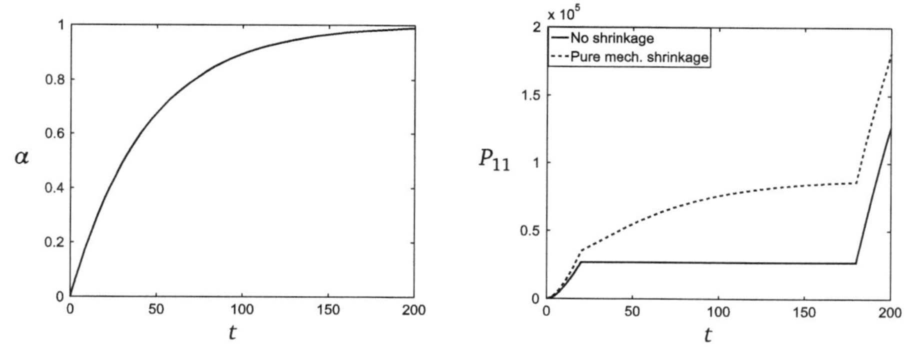

(a) Degree of cure α $[-]$.

(b) Piola stress P_{11} [MPa].

Figure 8.10: Curing history and shrinkage-induced stress development during three-phase mechanical loading conditions. [215, fig. 15 (reproduced with permission)]

In a second scenario, we consider the application of coupled magneto-mechanical loading conditions as well as a coupled magneto-mechanical shrinkage response for both isotropic ($\beta = 0$) and anisotropic ($\beta = 0.5; \mathbf{M} = [1, 0, 0]^T$) media. With the shrinkage parameters for the general case provided in Table 8.3b and the curing profile shown in Figure 8.10a, the shrinkage history governed by equation (8.16) evolves with the material's exposure to the magnetic field in the manner shown in Figure 8.11a for the anisotropic case.

The evolution of the stress response for the magneto-mechanical problem is illustrated in Figure 8.11b. As a baseline with which to compare the coupled shrinkage model's response, the coupling of the shrinkage model to the magnetic induction is removed reducing the problem to one of pure mechanical shrinkage; however, the coupling in the constitutive model remains. Comparing this case to the fully cou-

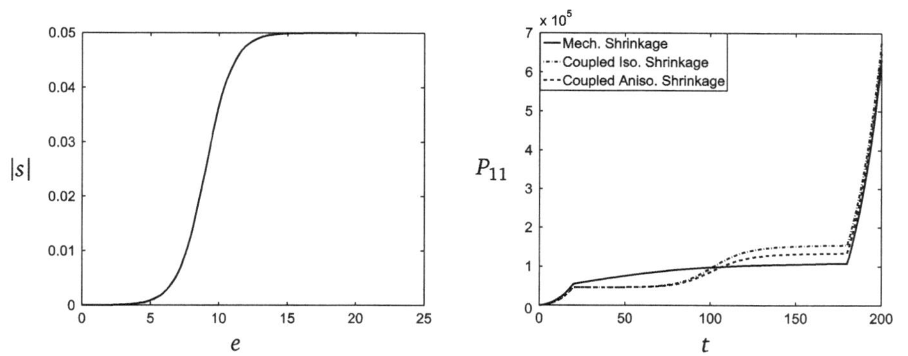

(a) Magnitude of the shrinkage s $[-]$ versus degree of exposure e [T s].

(b) Piola stress P_{11} [MPa] versus time t [s].

Figure 8.11: Shrinkage history and shrinkage-induced stress development during three-phase magneto-mechanical coupled loading conditions. [215, fig. 16 (reproduced with permission)] pled cases, the influence of the material's exposure to the magnetic field is clearly visible during the holding phase. As the degree of exposure increases, the shrinkage increases which consequently increases the material stiffness and hydrostatic loading. When the case of anisotropy is considered, the particle alignment hinders the shrinkage process and the shrinkage-generated stress is reduced when compared to the isotropic case.

9 Homogenisation

In this chapter, we will explore a number of approaches that can be taken to perform homogenisation of MAPs with particulate-filled microstructures. As this method bridges the gap between two extremes, we first consider the farthest ends of the spectrum in terms of modelling heterogeneous materials.

Sketched in Figure 9.1a is the detail to be captured within a full resolution model (also known as direct numerical simulation) of a heterogeneous media. Such models are very computationally expensive, as the fine-grained features and phenomena must be resolved using a single geometry encompassing all significant length-scales. It is therefore challenging to describe and mesh the relevant microstructure, especially given that the size and position of the embedded particles is random. The arduous task of developing a representative geometry must be repeated for every permutation of the macro-scale geometry. Local mesh refinement can be used as a method of reducing computational expense, but it is necessary to preserve curved surfaces and other significant features of the microstructure. This requires that the complex micro-scale geometry be known a priori and its features should be parametrically describable. Further complexities are introduced when the microstructure exhibits complex nonlinear behaviour; this may limit the extent to which the geometry can be loaded before numerical issues arise.

Phenomenological constitutive laws, such as that discussed in Section 7.3, are commonly applied in order to alleviate some of the complexities arising from the direct representation of material substructures. By viewing the microstructure as a smoothed continuum (as is illustrated in Figure 9.1b), they capture its influence at a macro-scale level and, therefore, significantly reduce the material complexity of a macro-scale geometry. However, the problem of defining phenomenological constitutive laws becomes considerably more difficult as the number of significant micro-scale material and geometric non-linearities increases. It is therefore not necessarily easy or even possible to define a phenomenological material law that captures the behaviour and response of the underlying media in all scenarios.

Analytical and computational homogenisation thus offer a balance between the ends of the spectra of incorporating feature details into mathematical models. In a broad sense, the idea (as it illustrated in Figure 9.1c) is to provide or describe an analogue for the average micro-scale material behaviour that is then utilised at the macro-scale to pointwise describe the material response to a given load. These methods require that consistency between the two length scales, in the form of the stored and dissipated energy as well as the balance laws, is maintained. Homogenisation techniques may be computationally expensive because a micro-scale model must be evaluated at each computation point in the homogenised material. However, in the context of the FE^2 approach, methods such as model order reduction via proper orthogonal decomposition [13, 102] have been regularly and successfully employed to offset this computational expense.

https://doi.org/10.1515/9783110418576-009

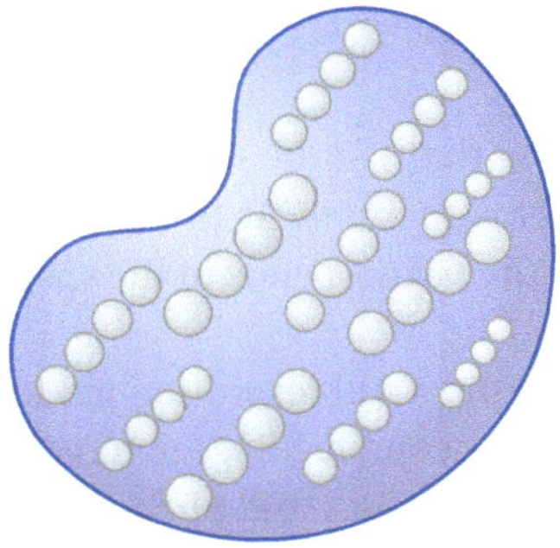

(a) Full resolution model.

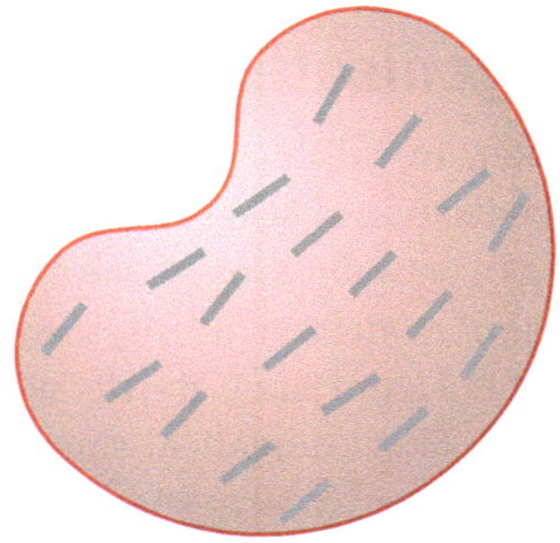

(b) Phenomenological constitutive law.

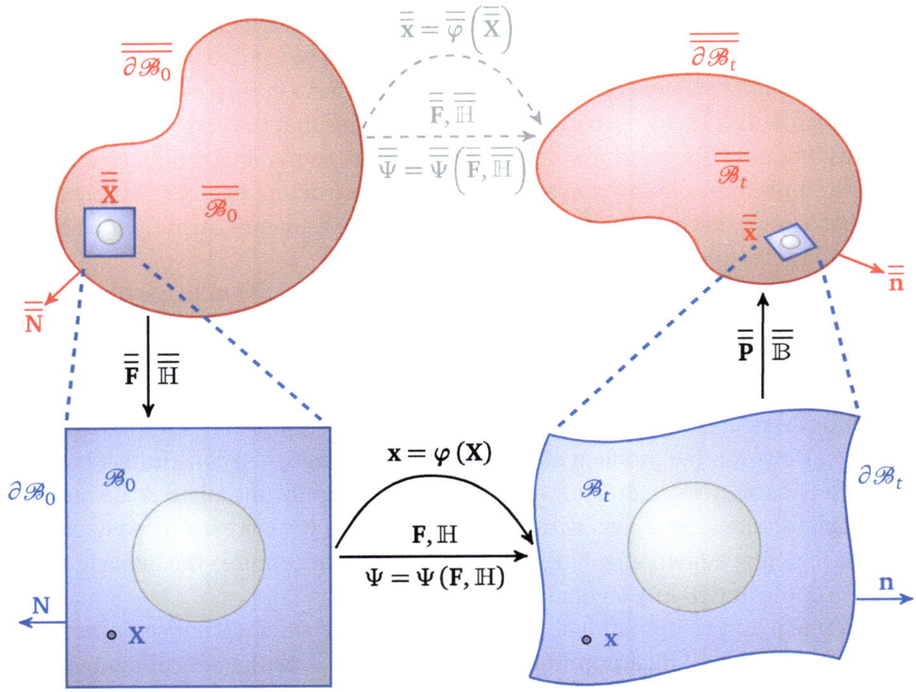

(c) Computational homogenisation via a micro-macro scale transition (FE^2). The red bodies represent the macro-scale domain, while the blue bodies are the micro-scale domain (with a single inclusion depicted). The left images represent the reference configuration and the right are the current configuration, attained after magnetic and mechanical loading.

Figure 9.1: Modelling approaches for materials with sub-scale material features.

9.1 First-order homogenisation of magneto-coupled materials

Topics related to the homogenisation of composite materials cover a wide expanse of theory, models, and computational methods related to their analytical evaluation and numerical implementation. Kanouté et al. [250], Pindera et al. [424], Charalambakis [81], and Geers et al. [164] provide overviews of the field in the context of finite (in-)elasticity, which is supplemented by a more recent review conducted by Saeb et al. [463], as well as a general review on the field of multiscale modelling of heterogeneous materials by Matouš et al. [335]. Fundamentals on the theory for solid mechanics are discussed in detail by Geers et al. [165], as well as [433, 572], while Zohdi and Wriggers [580], and Schröder [476] cover various aspects of the numerical problem related to the two-scale approach.

Analytical and computational homogenisation has seen application in a broad spectrum of topics directly and tangentially related to magnetoelasticity. Below we provide an extended literature review on the subject in general, after which details on how Hill-type averaging can be derived for—and implemented within—a magneto-mechanical framework is provided in Sections 9.1.1 and 9.1.2. Later, an example using the Mori–Tanaka approach will be presented in Section 9.1.4.

Hill-type averaging

Self-consistent kinematic quantities and tangent moduli are commonly computed for macro–micro scale transition models using Hill-type averaging [196, 197]. This method hinges on the equality of power between the micro- and macro-scales. It has been applied to heterogeneous meso- and micro-structures comprising hyperelastic materials [369, 201, 432, 515, 526], with special application to problems in configurational mechanics [453] and surface elasticity [235, 236]. It is also suitable to capture dissipative and coupled effects, and has been used to this effect to model viscoelastic [243], elastoplastic [368, 361, 360, 399, 357, 276], visco-elastoplastic [319], thermoelastic, [164, 516, 335] and thermo-elastoplastic [412] media, among many others. Linked to the multi-level modelling concept is the “generalised method of cells”, which is a higher-order theory considering periodically repeating representative volume elements (RVEs) used to construct tensors of effective material moduli. It has been utilised in the homogenisation of electro-magneto-thermoelastic multiphase composites [3], elastic media with non-linearly elastic, elastoplastic and viscoplastic constituents [187], and elastic media with damage [188].

FE^2 methods

Hill-type averaging is often used as the cornerstone of the FE^2 approach since a FE model is constructed to represent both the macro-scale model and to pointwise capture micro-scale phenomena, and there is some mechanism in place to transfer pertinent information between the scales. Regarding specific application of the

FE^2 method, Coquelle et al. [95] conducted FE analyses of particle systems and compared them to models derived using Hashin–Shtrikman homogenisation in the small-strain regime. Chatzigeorgiou et al. [82] present a unified framework for magneto-mechanical homogenisation; they derive an array of choices for boundary conditions for the micromechanical problem based on various formulations for the macro- and micro-scale potential functions, all of which satisfy the Hill–Mandel condition. This work is followed up by Javili et al. [234] who conducted a comprehensive characterisation of a magneto-sensitive matrix-particle (or, in two dimensions, matrix-fibre) composite microstructure. The sensitivity of the RVEs response to fixed loading conditions (uniaxial deformation and shear) for different material parameters was established. This unit-cell was subsequently used in a fully-coupled two-scale simulation of a fibre-reinforced magneto-active composite beam, for which an analysis of both numerical qualities of the problem as well as the multiscale material response was evaluated. More recently, Zabihyan et al. [570] have performed a comprehensive analysis of the same RVE, as well as examples exhibiting other particle dispersion and size properties. They focussed specifically on the influence of the boundary conditions on the predicted homogenised energy, stress and magnetic fields. Kuznetsov [281] examined a collection of numerical examples in the context of non-linear electromechanical coupling, comparing the result of mathematical homogenisation to highly refined direct numerical simulation. As an addendum to their work deriving a model for MAPs, Danas et al. [105] assess a simplified periodic non-linear elastic microstructure under conditions of isochoric uniaxial stretch. It was determined that even small applied strains resulted in the exhibition of non-linear material behaviour, with tension-compression asymmetry being displayed by the transversely isotropic media. Very recently, Keip and Rambausek [255] applied the FE^2 approach to a two-dimensional soft MAP immersed in free space. At the micro-scale, periodic boundary conditions were applied to the RVE that had circular inclusions. They reported on a comprehensive numerical study, including the magnetostrictive response of the macro- and micro-scale bodies under different loading conditions. Also of particular interest is the work of Schröder and Keip [477], who present the two-scale homogenisation of piezocomposites comprising ellipsoidal inclusions.

We note that there exist few robust analytical alternatives [86, 439, 380] to the computationally intensive FEM- or boundary element method (BEM)-based FE^2 method. An approach based on a direct Fourier analysis of microstructural images using FFT was developed by Moulinec and Suquet [388]. It was subsequently used in [389, 357] to model the small-strain response of heterogeneous elastic and elastoplastic composites, respectively; in both of these articles, comparison to a FEM analysis of an equivalent microstructure was performed and analysed. This method has since been extended to a more general variational framework that can account for a greater contrast between material heterogeneities [62].

Analytical methods and the application of mean field theory

In comparison to the approach mentioned above, Hill [196], and Christensen and Lo [87] derive analytical solutions for isotropic material parameters of linear elastic composites; the latter gives rise to the generalised self-consistent model [87, 86]. A more flexible analytical solution [439, 330] can be determined using the generalised Mori–Tanaka [387, 41] approach, wherein concentration tensors for each phase are computed by means of the Eshelby tensor [134], which itself captures the geometric non-linearities of the considered microstructure. It has been used to great effect in the computation of effective, anisotropic material properties of composite materials, and it has also been extended to coupled problems. For example, it has been utilised in the determination of the mean properties of elastic multiphase materials [218, 269], piezoelectric [128, 129, 220] and (piezo-)electro-magnetoelastic composites [304, 305, 303, 299, 573, 450], and electro-magneto-thermoelastic fibre-reinforced media [511]. Klusemann et al. [269] also considers the upper and lower limit on the predicted material stiffness tensor, which for the case of elastic materials are given by the Hashin–Shtrikman bounds [190]. Collectively, the aforementioned papers consider both spherical and ellipsoidal inclusions; however, the theory (although accounting for material anisotropy) is only linear, and is thus not entirely suitable to model soft elastomer-based MAPs. The incremental Mori–Tanaka method extends the traditional approach in order to capture material non-linearities along with the geometric non-linearities of the microstructure. This approach has been used successfully to model finite strain thermo-elastoplastic composites [420, 421], and in crystal plasticity [111]. The Eshelby tensor has also been used in conjunction with an incremental formulation to directly compute the effective non-linear material response ("mean field Eshelby-based homogenisation"); for example, in the homogenisation of non-linear elastoplastic [117], and electro-magneto-thermomechanical composites with damage [484]. In contrast, the self-consistent approach was used in Taya [513] to homogenise piezo-electric media. Note that although not explicitly stated, some of the above literature compares results obtained by one or more of the aforementioned analytical homogenisation techniques.

With a more specific focus on the application of mean field theories in magneto-mechanics, Jolly et al. [240] developed and validated a one-dimensional, quasi-static model for MAPs based on the dipole interaction of embedded particles. The model, which accounts for non-linearities such as magnetic saturation, makes an assumption about the magnetic flux distribution within the media and relies on the computation of the average magnetisation of the magnetisable particles. Similarly, Borcea and Bruno [53] derive a three-dimensional homogenised small-strain model for an MAP using a dipole model under the assumption of uniformly magnetised particles. Yin et al. [567] applied an Eshelby formulation to produce a hyperelastic constitutive law of an MAP based on the assumption of a chain network existing between magnetostrictive particles (but governing only its effective mechanical response). Both magnetic and mechanical loads were considered, and the former was taken into consideration by imposing its influence on the particle's magnetostrictive strain response. In later work, Yin et al. performed homogenisation in the linear material regime using Green's function to derive the local magnetic and elastic fields, and Eshelby's equivalent inclusion method [134, 135] to compute the effective mechanical strain and material properties. This was achieved for materials consisting of either chain-like particle structures [568] and randomly dispersed particles [566]. For both microstructures, the mechanical response was compared to that derived using a Mori–Tanaka approach; it was found that the two methods presented a consistent response which differs only due to the presence of an additional particle-particle interaction term in Yin et al.'s model. Liu et al. [318] have also applied the Eshelby dilute approach to perform homogenisation on magnetostrictive composites. More recently, Ponte Castañeda and Galipeau [428] put forward a homogenisation framework for MAPs, employing a modified Eshelby formulation in conjunction with Hill-type averaging, that accounts for both finite strains and magnetic dipole interactions. By introducing a partial decomposition of the magnetoelastic energy, the theory requires that only a purely mechanical homogenisation problem together with an estimate of particle rotations within the media be solved first, followed by a purely magnetic problem. This framework can be used to homogenise isotropic and anisotropic composites with both a linear and non-linear (saturating) magnetic particle response, and is used further in [156, 155, 157, 158] to investigate the influence of MAP microstructure at finite strains.

9.1.1 Hill's condition on the equality of micro- and macro-scale virtual powers

To begin, we must distinguish between the two boundary value problems defined at the macro- and micro-scales as illustrated in Figure 9.1c. The characteristic length scale of the latter is typically several orders of magnitude smaller than the former, but is not so small that continuum-related assumptions no longer hold. Another notable feature of the micro-scale problem is that it is assumed that an appropriate constitutive law can be defined pointwise, while it is assumed that the micro-scale response is too complex to be represented directly by a material law at the macro-scale.

For the micro-scale problem, we define the referential micro-scale body $\mathscr{B}_0$ and describe its continuum setting exactly in the manner documented in Section 5.1. The kinematics and kinetic quantities, as well as the balance laws specified in Section 5.3 (and, for the quasi-static problem, Table 5.8 in particular) apply. From the local constitutive law, the increments in energy (the virtual power) can be inferred from the dissipation inequality associated with each potential formulation; this approach follows Hill's incremental model [198, 199]. For example, from equation (7.1) we can deduce that the virtual power density associated with the MVP formulation

$$\wp_0^* (\mathbf{F}, \mathbb{B}) := \mathbf{P}^{\text{tot}} : \delta\mathbf{F} + \mathbb{H} \cdot \delta\mathbb{B} \quad , \tag{9.1}$$

while the virtual power density associated with the MSP formulation, namely

$$\wp_0(\mathbf{F}, \mathbb{H}) := \mathbf{P}^{\text{tot}} : \delta\mathbf{F} - \mathbb{B} \cdot \delta\mathbb{H} \quad , \tag{9.2}$$

is derived from equation (7.5).

For the macro-scale problem, the material macro-scale body is represented by $\overline{\overline{\mathscr{B}_0}}$, with reference position vector $\overline{\overline{\mathbf{X}}}$ mapped to the position $\overline{\overline{\mathbf{x}}} = \overline{\overline{\boldsymbol{\varphi}}}(\overline{\overline{\mathbf{X}}})$. Quantities associated with the macro-scale problem will, in general, be annotated as $\overline{\overline{(\circ)}}$. We further define volume and surface averaging operators for material and spatial quantities by

$$\begin{aligned} \langle(\circ)\rangle_0 &= \frac{1}{V_0} \int\limits_{\mathscr{B}_0} (\circ)\ \mathrm{dV} \quad , \quad \langle(\circ)\rangle_t = \frac{1}{v_t} \int\limits_{\mathscr{B}_t} (\circ)\ \mathrm{dv} \quad , \\ \lceil(\circ)\rfloor_0 &= \frac{1}{V_0} \int\limits_{\partial\mathscr{B}_0} (\circ)\ \mathrm{dA} \quad , \quad \lceil(\circ)\rfloor_t = \frac{1}{v_t} \int\limits_{\partial\mathscr{B}_t} (\circ)\ \mathrm{da} \quad . \end{aligned} \tag{9.3}$$

From their fundamental definitions (arising from the extension of the atomistic Irving–Kirkwood–Noll theory [226, 401, 300] to continuum mechanics [332]), with the use of the averaging operators it is possible to derive definitions of the macro-scale kinematic and kinetic quantities in terms of their micro-scale counterparts. Given the potential field for the motion, $\boldsymbol{\varphi}(\mathbf{X}, t)$, the macro-scale deformation gradient and Piola stress tensor are respectively defined as [82]

$$\overline{\overline{\mathbf{F}}}\left(\overline{\overline{\mathbf{X}}}\right) := \langle\mathbf{F}\rangle_0 = \lceil\boldsymbol{\varphi} \otimes \mathbf{N}\rfloor_0 \quad \text{and} \tag{9.4}$$

$$\overline{\overline{\mathbf{P}^{\text{tot}}}}\left(\overline{\overline{\mathbf{X}}}\right) := \left\langle\mathbf{P}^{\text{tot}}\right\rangle_0 = \left\lceil\mathbf{t}_0^{\text{tot}} \otimes \mathbf{X}\right\rfloor_0 \quad \text{with} \quad \mathbf{t}_0^{\text{tot}} = \mathbf{P}^{\text{tot}} \cdot \mathbf{N} \quad . \tag{9.5}$$

Similarly, the macro-scale magnetic field and induction are, in general,

$$\overline{\overline{\mathbb{H}}}\left(\overline{\overline{\mathbf{X}}}\right) := \langle\mathbb{H}\rangle_0 \quad \text{and} \quad \overline{\overline{\mathbb{B}}}\left(\overline{\overline{\mathbf{X}}}\right) := \langle\mathbb{B}\rangle_0 , \tag{9.6}$$

respectively. When the MVP is considered, the magnetic induction can also be computed by

$$\overline{\overline{\mathbb{B}}}\left(\overline{\overline{\mathbf{X}}}\right) = -\lceil\mathbb{A} \times \mathbf{N}\rfloor_0 \quad \text{with} \quad \mathbb{B} = \nabla_0 \times \mathbb{A} \quad \text{and} \quad \overline{\overline{\mathbb{H}}}\left(\overline{\overline{\mathbf{X}}}\right) = \lceil[\mathbb{H} \times \mathbf{N}] \times \mathbf{X}\rfloor_0 \quad , \tag{9.7}$$

while in the case of the MSP, the magnetic field is also described by

$$\overline{\overline{\mathbb{H}}}\left(\overline{\overline{\mathbf{X}}}\right) = -\lceil\Phi\mathbf{N}\rfloor_0 \quad \text{with} \quad \mathbb{H} = -\nabla_0\Phi \quad \text{and} \quad \overline{\overline{\mathbb{B}}}\left(\overline{\overline{\mathbf{X}}}\right) = \lceil[\mathbb{B} \cdot \mathbf{N}]\,\mathbf{X}\rfloor_0 \quad . \tag{9.8}$$

We can observe that equation (9.8) compares formally to equations (9.4) and (9.5). A setting for the mechanical quantities similar to equation (9.7) is conceptually possible based on a tensor potential $\mathbf{A}$ for the stress, so that $\mathbf{P} = \nabla_0 \times \mathbf{A}$ and, consequentially, $\nabla_0 \cdot \mathbf{P} = \mathbf{0}$. This, however, would be an unusual parameterisation and, in the finite strain setting, might have consequences related to the possible non-uniqueness between $\mathbf{P}$ and $\mathbf{F}(\mathbf{P})$.

The macro-scale balance laws for the quasi-static problem,

$$\begin{aligned} \overline{\overline{\nabla_0 \cdot}}\, \overline{\overline{\mathbf{P}^{\mathrm{tot}}}} = \mathbf{0} \quad &, \quad \overline{\overline{\nabla_0 \times}}\, \overline{\overline{\mathbf{F}}} = \mathbf{0} \quad , \\ \overline{\overline{\nabla_0 \cdot}}\, \overline{\overline{\mathbb{B}}} = 0 \quad &, \quad \overline{\overline{\nabla_0 \times}}\, \overline{\overline{\mathbb{H}}} = \mathbf{0} \quad , \end{aligned} \tag{9.9}$$

are entirely analogous to those given in Table 5.8. Similarly, the macroscopic virtual power density for the MVP and MSP formulations are respectively expressed as

$$\overline{\overline{\wp_0^*}}\left(\overline{\overline{\mathbf{F}}}, \overline{\overline{\mathbb{B}}}\right) := \overline{\overline{\mathbf{P}^{\mathrm{tot}}}} : \delta\overline{\overline{\mathbf{F}}} + \overline{\overline{\mathbb{H}}} \cdot \delta\overline{\overline{\mathbb{B}}} \quad \text{and} \tag{9.10a}$$

$$\overline{\overline{\wp_0}}\left(\overline{\overline{\mathbf{F}}}, \overline{\overline{\mathbb{H}}}\right) := \overline{\overline{\mathbf{P}^{\mathrm{tot}}}} : \delta\overline{\overline{\mathbf{F}}} - \overline{\overline{\mathbb{B}}} \cdot \delta\overline{\overline{\mathbb{H}}} \quad , \tag{9.10b}$$

and can be seen to have a similar structure to those stated for the micro-scale problem.

9.1.2 Consistent boundary conditions arising from the equality of virtual power

To establish consistency across the different length scales, it is necessary to maintain power equality between them. This is expressed through the equality of macro- and micro-scale virtual power, as either

$$\overline{\overline{\wp_0^*}}\left(\overline{\overline{\mathbf{X}}}\right) = \langle \wp_0^*(\mathbf{X}) \rangle_0 \quad \text{or} \quad \overline{\overline{\wp_0}}\left(\overline{\overline{\mathbf{X}}}\right) = \langle \wp_0(\mathbf{X}) \rangle_0 \quad , \tag{9.11}$$

which simply states the equivalence of the macro-scale magnetoelastic energy density function and the volume average of the micro-scale magnetoelastic energy density function. This is principally known as the Hill–Mandel condition for virtual power increments.

There are several possible cases for boundary conditions (uniform or periodic fields and tractions), and in the following paragraphs we will state the various micro-scale boundary conditions required to satisfy equation (9.11). The mechanical and magnetic problems will be considered individually as it is perfectly valid to mix boundary conditions for the various fields in the coupled problem. In both scenarios, there exist two cases from which we will determine legitimate Dirichlet, Neumann and periodic boundary conditions. That is to say that we will determine the conditions under which

$$\overline{\overline{\mathbf{P}^{\mathrm{tot}}}} : \delta\overline{\overline{\mathbf{F}}} = \left\langle \mathbf{P}^{\mathrm{tot}} : \delta\mathbf{F} \right\rangle_0 \quad \text{and} \tag{9.12}$$

$$\overline{\overline{\mathbb{H}}} \cdot \delta\overline{\overline{\mathbb{B}}} = \langle \mathbb{H} \cdot \delta\mathbb{B} \rangle_0 \quad \text{or} \quad \overline{\overline{\mathbb{B}}} \cdot \delta\overline{\overline{\mathbb{H}}} = \langle \mathbb{B} \cdot \delta\mathbb{H} \rangle_0 \tag{9.13}$$

are individually satisfied. The derivation for the following equations are given in [82], and is repeated in full in Appendix C.8. In summary, it can be shown that for equation (9.12), equation $(9.13)_1$ and equation $(9.13)_2$ to be fulfilled,

$$\overline{\overline{\mathbf{P}^{\text{tot}}}} : \delta\overline{\overline{\mathbf{F}}} - \left\langle \mathbf{P}^{\text{tot}} : \delta\mathbf{F} \right\rangle_0 = \int\limits_{\partial\mathscr{B}_0} \left[\boldsymbol{\varphi}(\mathbf{X}) - \overline{\overline{\mathbf{F}}} \cdot \mathbf{X} \right] \cdot \left[\mathbf{t}_0^{\text{tot}} - \overline{\overline{\mathbf{P}^{\text{tot}}}} \cdot \mathbf{N} \right] \mathrm{dA} = 0 \quad , \tag{9.14}$$

$$\overline{\overline{\mathbb{H}}} \cdot \delta\overline{\overline{\mathbb{B}}} - \langle \mathbb{H} \cdot \delta\mathbb{B} \rangle_0 = \int\limits_{\partial\mathscr{B}_0} \left[\mathbb{A}(\mathbf{X}) - \frac{1}{2}\overline{\overline{\mathbb{B}}} \times \mathbf{X} \right] \cdot \left[\mathbb{H} \times \mathbf{N} - \overline{\overline{\mathbb{H}}} \times \mathbf{N} \right] \mathrm{dA} \quad , \tag{9.15}$$

$$\overline{\overline{\mathbb{B}}} \cdot \delta\overline{\overline{\mathbb{H}}} - \langle \mathbb{B} \cdot \delta\mathbb{H} \rangle_0 = \int\limits_{\partial\mathscr{B}_0} \left[\Phi(\mathbf{X}) + \overline{\overline{\mathbb{H}}} \cdot \mathbf{X} \right] \left[\mathbb{B} \cdot \mathbf{N} - \overline{\overline{\mathbb{B}}} \cdot \mathbf{N} \right] \mathrm{dA} = 0 \quad . \tag{9.16}$$

9.1.2.1 Mechanical contribution to virtual power

Fluctuation in potential field

To extract the first two sets of boundary conditions, we assume that the micro-scale motion (displacement field) is the result of the superposition of two components, as given by

$$\boldsymbol{\varphi}(\mathbf{X}) = \overline{\overline{\mathbf{F}}} \cdot \mathbf{X} + \widetilde{\boldsymbol{\varphi}}(\mathbf{X}) \tag{9.17}$$

that, after consideration of equation (5.1), implies

$$\mathbf{u}(\mathbf{X}) = \left[\overline{\overline{\mathbf{F}}} - \mathbf{I} \right] \cdot \mathbf{X} + \widetilde{\mathbf{u}}(\mathbf{X}) \equiv \overline{\overline{\nabla_0 \overline{\mathbf{u}}}} \cdot \mathbf{X} + \widetilde{\mathbf{u}}(\mathbf{X}) \quad . \tag{9.18}$$

The first component has a linear relation with the macro-scale deformation gradient and the second describes the displacement fluctuation field. Note that this is the first-order relation between the micro-scale and macro-scale motion from which the theory gets its name; the corresponding second-order Taylor expansion of the micro-scale motion would be [165]

$$\boldsymbol{\varphi}(\mathbf{X}) = \overline{\overline{\mathbf{F}}} \cdot \mathbf{X} + \frac{1}{2}\mathbf{X} \cdot \overline{\overline{\mathfrak{G}}} \cdot \mathbf{X} + \widetilde{\boldsymbol{\varphi}}(\mathbf{X})$$

where $\overline{\overline{\mathfrak{G}}}(\overline{\overline{\mathbf{X}}}) = \overline{\overline{\nabla_0 \mathbf{F}}}(\overline{\overline{\mathbf{X}}})$ is a rank-3 tensor representing the second material gradient of the macro-scale motion field. Further information on second-order theory and FEM computations therewith can be found in [429, 430, 431, 275, 276, 517].

The micro-scale deformation gradient is then expressed as

$$\mathbf{F} = \overline{\overline{\mathbf{F}}} + \nabla_0 \widetilde{\boldsymbol{\varphi}} = \overline{\overline{\mathbf{F}}} + \widetilde{\mathbf{F}} \tag{9.19}$$

using equations (5.3) and (9.18). Inserting equation (9.19) into equation (9.12), the resulting expression for the mechanical contribution to the Hill–Mandel condition is

$$\left\langle \mathbf{P}^{\text{tot}} : \delta\mathbf{F} \right\rangle_0 - \overline{\overline{\mathbf{P}^{\text{tot}}}} : \delta\overline{\overline{\mathbf{F}}} = \left[\delta\widetilde{\boldsymbol{\varphi}} \cdot \left[\mathbf{P}^{\text{tot}} \cdot \mathbf{N} \right] \right]_0 \quad . \tag{9.20}$$

The first of two plausible boundary conditions required to make the surface integral in equation (9.20) vanish is

$$\widetilde{\boldsymbol{\varphi}}(\mathbf{X}) = \mathbf{0} \quad \text{on} \quad \partial\mathscr{B}_0 \quad \Rightarrow \quad \boldsymbol{\varphi}(\mathbf{X}) = \overline{\overline{\mathbf{F}}} \cdot \mathbf{X} \quad \text{on} \quad \partial\mathscr{B}_0 \tag{9.21}$$

which implies that the displacement field is linear on $\partial\mathscr{B}_0$ and $\widetilde{\boldsymbol{\varphi}}$ holds everywhere in $\mathscr{B}_0$. This condition is imposed as a Dirichlet constraint on the micro-scale RVE. In the case of a non-trivial fluctuation field,

$$\begin{aligned} &[\![\widetilde{\boldsymbol{\varphi}}]\!] = \mathbf{0} \quad \text{and} \quad \{\!\{\mathbf{P}^{\text{tot}} \cdot \mathbf{N}\}\!\} = \mathbf{0} \quad \text{between} \quad \partial\mathscr{B}_0^+ \text{ and } \partial\mathscr{B}_0^- \\ \Rightarrow \quad &\widetilde{\boldsymbol{\varphi}}(\mathbf{X}) \text{ is periodic} \quad \text{and} \quad \mathbf{P}^{\text{tot}} \cdot \mathbf{N} \text{ is anti-periodic} \quad \text{on} \quad \partial\mathscr{B}_0^\pm \quad ; \end{aligned} \tag{9.22}$$

this encapsulates a set of periodic conditions for a geometrically periodic RVE, with opposing surface pairs $\{\partial\mathscr{B}_0^+, \partial\mathscr{B}_0^-\}$ denoted by $\partial\mathscr{B}_0^\pm$. Note that both the Dirichlet and periodic conditions guarantee that $\langle \mathbf{F} \rangle = \overline{\overline{\mathbf{F}}}$.

Spatially piecewise constant applied traction

If we alternately assume that the mechanical traction acting on $\mathscr{B}_0$ is prescribed and piecewise spatially constant, that is,

$$\mathbf{P}^{\text{tot}}(\mathbf{X}) \cdot \mathbf{N} = \overline{\overline{\mathbf{P}^{\text{tot}}}} \cdot \mathbf{N} \quad \text{on} \quad \partial\mathscr{B}_0 \quad , \tag{9.23}$$

then it can be shown that

$$\left\lceil \delta\widetilde{\boldsymbol{\varphi}} \cdot \left[\mathbf{P}^{\text{tot}} \cdot \mathbf{N}\right] \right\rfloor_0 = \left\lceil \delta\widetilde{\boldsymbol{\varphi}} \cdot \left[\overline{\overline{\mathbf{P}^{\text{tot}}}} \cdot \mathbf{N}\right] \right\rfloor_0 = \overline{\overline{\mathbf{P}^{\text{tot}}}} : \lceil \delta\widetilde{\boldsymbol{\varphi}} \otimes \mathbf{N} \rfloor_0 = \overline{\overline{\mathbf{P}^{\text{tot}}}} : \left\langle \widetilde{\delta\mathbf{F}} \right\rangle_0 = 0 \tag{9.24}$$

with the consideration of equation (9.4), and noting that there exists no average fluctuation field over the microscopic body. This immediately confirms that the constant traction (Neumann) boundary condition given in equation (9.23) satisfies the Hill–Mandel condition. For the two magnetic potentials, we will evaluate similar set of restrictions on the fields that will result in an analogous set of constraints.

9.1.2.2 Magnetic contribution to virtual power: Magnetic vector potential

Fluctuation in potential field

Presupposing the existence of a magnetic vector potential fluctuation field, we consider a scenario in which the magnetic potential has a linear relation with the macro-scale magnetic induction and its superposition with the fluctuation field. The result is that we choose

$$\mathbb{A}(\mathbf{X}) = \frac{1}{2}\overline{\overline{\mathbb{B}}} \times \mathbf{X} + \widetilde{\mathbb{A}}(\mathbf{X}) \quad . \tag{9.25}$$

Using equations (5.141) and (9.25), the micro-scale magnetic induction is therefore composed as

$$\mathbb{B} = \overline{\overline{\mathbb{B}}} + \nabla_0 \times \widetilde{\mathbb{A}} = \overline{\overline{\mathbb{B}}} + \widetilde{\mathbb{B}} \tag{9.26}$$

and the resulting expression for the Hill–Mandel condition is

$$\langle \mathbb{H} \cdot \delta\mathbb{B} \rangle_0 - \overline{\overline{\mathbb{H}}} \cdot \delta\overline{\overline{\mathbb{B}}} = \lceil \delta\widetilde{\mathbb{A}} \cdot [\mathbb{H} \times \mathbf{N}] \rfloor_0 \quad . \tag{9.27}$$

Once again, what remains is to determine the appropriate conditions under which the surface integral on the right-hand side of equation (9.27) is zero. The first option is to negate the presence of the fluctuation field on the boundary

$$\widetilde{\mathbb{A}}(\mathbf{X}) = \mathbf{0} \quad \text{on} \quad \partial\mathscr{B}_0 \quad \Rightarrow \quad \mathbb{A}(\mathbf{X}) = \frac{1}{2}\overline{\overline{\mathbb{B}}} \times \mathbf{X} \quad \text{on} \quad \partial\mathscr{B}_0 \tag{9.28}$$

thereby implying that the magnetic vector potential field is linear on $\partial\mathscr{B}_0$. The second option is to enforce periodicity of the fields, that is,

$$\begin{aligned} & [\![\widetilde{\mathbb{A}}]\!] = \mathbf{0} \quad \text{and} \quad \{\!\{\mathbb{H} \times \mathbf{N}\}\!\} = \mathbf{0} \quad \text{between} \quad \partial\mathscr{B}_0^+ \text{ and } \partial\mathscr{B}_0^- \\ \Rightarrow \quad & \widetilde{\mathbb{A}}(\mathbf{X}) \text{ is periodic} \quad \text{and} \quad \mathbb{H} \times \mathbf{N} \text{ is anti-periodic} \quad \text{on} \quad \partial\mathscr{B}_0^\pm \quad , \end{aligned} \tag{9.29}$$

which are sensible choices in the context of a geometrically periodic RVE. The Dirichlet and periodic conditions both ensure that $\langle \mathbb{B} \rangle = \overline{\overline{\mathbb{B}}}$.

Spatially piecewise constant applied "traction"

If it is instead assumed that a spatially piecewise constant magnetic traction is applied to $\partial\mathscr{B}_0$, then

$$\mathbb{H}(\mathbf{X}) \times \mathbf{N} = \overline{\overline{\mathbb{H}}} \times \mathbf{N} \quad \text{on} \quad \partial\mathscr{B}_0 \quad . \tag{9.30}$$

As a consequence of this choice, it is determined that

$$\lceil \delta\widetilde{\mathbb{A}} \cdot [\mathbb{H} \times \mathbf{N}] \rfloor_0 = \left\lceil \delta\widetilde{\mathbb{A}} \cdot \left[\overline{\overline{\mathbb{H}}} \times \mathbf{N}\right] \right\rfloor_0 = \overline{\overline{\mathbb{H}}} \cdot \lceil \mathbf{N} \times \delta\widetilde{\mathbb{A}} \rfloor_0 = \overline{\overline{\mathbb{H}}} \cdot \langle \delta\widetilde{\mathbb{B}} \rangle_0 = 0 \quad , \tag{9.31}$$

which verifies that the Hill–Mandel condition is satisfied by the constant magnetic field (Neumann) boundary condition given in equation (9.30) if the average fluctuation $\langle \delta\widetilde{\mathbb{B}} \rangle_0$ vanishes.

9.1.2.3 Magnetic contribution to virtual power: Magnetic scalar potential

Fluctuation in potential field

For the final case of the MSP, we first assume that the potential has a linear relation with the macro-scale magnetic field and that there is a magnetic scalar potential fluctuation field superimposed on the macro-scale field. Consequently, we can state that

$$\Phi(\mathbf{X}) = -\overline{\overline{\mathbb{H}}} \cdot \mathbf{X} + \widetilde{\Phi}(\mathbf{X}) \quad , \tag{9.32}$$

from which we express the micro-scale magnetic field as

$$\mathbb{H} = \overline{\overline{\mathbb{H}}} - \nabla_0 \widetilde{\Phi} = \overline{\overline{\mathbb{H}}} + \widetilde{\mathbb{H}} \tag{9.33}$$

using equations (5.146) and (9.32). The resolved expression for the Hill–Mandel condition is

$$-\langle \mathbb{B} \cdot \delta \mathbb{H} \rangle_0 + \overline{\overline{\mathbb{B}}} \cdot \delta \overline{\overline{\mathbb{H}}} = \lceil \delta \widetilde{\Phi} \left[\mathbb{B} \cdot \mathbf{N} \right] \rfloor_0 \quad . \tag{9.34}$$

In order to make the right-hand side term in equation (9.34) vanish, we are again presented with two options. The first is to presume that the micro-scale fluctuation field is non-existent on the boundary, which renders the Dirichlet condition

$$\widetilde{\Phi}(\mathbf{X}) = 0 \quad \text{on} \quad \partial \mathscr{B}_0 \quad \Rightarrow \quad \Phi(\mathbf{X}) = -\overline{\overline{\mathbb{H}}} \cdot \mathbf{X} \quad \text{on} \quad \partial \mathscr{B}_0 \tag{9.35}$$

and insinuates that the magnetic scalar potential field is linear on $\partial \mathscr{B}_0$. The second is to assume that for a geometrically periodic RVE the set of periodic conditions, namely

$$\begin{aligned} &[\![\widetilde{\Phi}]\!] = 0 \quad \text{and} \quad \{\!\!\{ \mathbb{B} \cdot \mathbf{N} \}\!\!\} = 0 \quad \text{between} \quad \partial \mathscr{B}_0^+ \text{ and } \partial \mathscr{B}_0^- \\ \Rightarrow \quad &\widetilde{\Phi}(\mathbf{X}) \text{ is periodic} \quad \text{and} \quad \mathbb{B} \cdot \mathbf{N} \text{ is anti-periodic} \quad \text{on} \quad \partial \mathscr{B}_0^\pm \quad , \end{aligned} \tag{9.36}$$

appropriately describe the micro-scale field. Both choices for the constraints on the fields have the ancillary result that $\langle \mathbb{H} \rangle = \overline{\overline{\mathbb{H}}}$.

Spatially piecewise constant applied “traction”

Alternately, assuming that a spatially piecewise constant magnetic traction

$$\mathbb{B}(\mathbf{X}) \cdot \mathbf{N} = \overline{\overline{\mathbb{B}}} \cdot \mathbf{N} \quad \text{on} \quad \partial \mathscr{B}_0 \tag{9.37}$$

is imposed on the microstructure, then it can be deduced that

$$\lceil \delta \widetilde{\Phi} \left[\mathbb{B} \cdot \mathbf{N} \right] \rfloor_0 = \left\lceil \delta \widetilde{\Phi} \left[\overline{\overline{\mathbb{B}}} \cdot \mathbf{N} \right] \right\rfloor_0 = \overline{\overline{\mathbb{B}}} \cdot \lceil \delta \widetilde{\Phi} \mathbf{N} \rfloor_0 = -\overline{\overline{\mathbb{B}}} \cdot \langle \delta \widetilde{\mathbb{H}} \rangle_0 = 0 \quad . \tag{9.38}$$

The piecewise constant magnetic induction (Neumann) boundary condition given in equation (9.37) thus fulfils the Hill–Mandel condition if the average fluctuation $\langle \delta \mathbb{H} \rangle_0$ vanishes.

9.1.2.4 Upper and lower bounds on the averaged material response

The boundary conditions presented in Sections 9.1.2.1 to 9.1.2.3 can be mixed in different combinations, with the result that a range of material responses may be predicted. It is, however, expected that as the size of the RVE is increased, the homogenised material properties converge to a consistent set of values. The Hill–Voigt–Reuss bounds [534, 451, 196, 580] are a set of conditions that present the least and most conservative cases for the predicted micro-scale response. These bounds are presented and discussed by Javili et al. [236] in the context of elastic materials, as well as Zabihyan et al. [570] for coupled magnetoelastic materials.

Voigt [534] considered a case where the primary fields are considered entirely linear inside the body. This is also known as the Taylor assumption [412, 236], and typically elicits the most "stiff" response from the RVE. For the coupled problem, this condition can be summarised by

$$\begin{aligned} \boldsymbol{\varphi}(\mathbf{X}) &= \overline{\overline{\mathbf{F}}} \cdot \mathbf{X} \quad \text{in} \quad \overline{\mathscr{B}_0} \quad \Leftrightarrow \quad \mathbf{F} = \overline{\overline{\mathbf{F}}} \quad \text{in} \quad \overline{\mathscr{B}_0} \quad \text{and} \\ \mathbb{A}(\mathbf{X}) &= \frac{1}{2} \overline{\overline{\mathbb{B}}} \times \mathbf{X} \quad \text{in} \quad \overline{\mathscr{B}_0} \quad \Leftrightarrow \quad \mathbb{B} = \overline{\overline{\mathbb{B}}} \quad \text{in} \quad \overline{\mathscr{B}_0} \quad \text{or} \\ \Phi(\mathbf{X}) &= -\overline{\overline{\mathbb{H}}} \cdot \mathbf{X} \quad \text{in} \quad \overline{\mathscr{B}_0} \quad \Leftrightarrow \quad \mathbb{H} = \overline{\overline{\mathbb{H}}} \quad \text{in} \quad \overline{\mathscr{B}_0} \quad . \end{aligned} \tag{9.39}$$

In contrast, Reuss [451] evaluated the scenario wherein the tractions developed within any cut in the micro-scale domain are spatially constant (also known as the Sachs assumption [412, 236]). For a magnetoelastic material, this is expressed as

$$\begin{aligned} \mathbf{t}_0^{\text{mech}} &= \overline{\overline{\mathbf{P}^{\text{tot}}}} \cdot \mathbf{N} \quad \text{in} \quad \overline{\mathscr{B}_0} \quad \Leftrightarrow \quad \mathbf{P}^{\text{tot}} = \overline{\overline{\mathbf{P}^{\text{tot}}}} \quad \text{in} \quad \overline{\mathscr{B}_0} \quad \text{and} \\ \mathbf{t}_0^{\text{mag}} &= \overline{\overline{\mathbb{H}}} \times \mathbf{N} \quad \text{in} \quad \overline{\mathscr{B}_0} \quad \Leftrightarrow \quad \mathbb{H} = \overline{\overline{\mathbb{H}}} \quad \text{in} \quad \overline{\mathscr{B}_0} \quad \text{or} \\ t_0^{\text{mag}} &= \overline{\overline{\mathbb{B}}} \cdot \mathbf{N} \quad \text{in} \quad \overline{\mathscr{B}_0} \quad \Leftrightarrow \quad \mathbb{B} = \overline{\overline{\mathbb{B}}} \quad \text{in} \quad \overline{\mathscr{B}_0} \end{aligned} \tag{9.40}$$

and typically renders the most "compliant" material response. For the above, recall that $\overline{\mathscr{B}_0} = \mathscr{B}_0 \cup \overline{\partial \mathscr{B}_0}$.

9.1.2.5 Implementation of constraints for periodically repeating Rves

It is necessary to adopt different strategies in order to enforce the different types of boundary conditions on the microstructural problem. Application of the pure Dirichlet constraints (or appraisal of the more restrictive spatially constant Voigt bound) relies on the same mechanisms that one would utilise to apply Dirichlet conditions to any traditional FEM problem, namely row-column elimination or through an affine constraints matrix [485, 28, 165]. The enforcement of the spatially piecewise constant boundary traction condition (or evaluation of the more restrictive spatially constant Reuss bound) is more challenging, involving the iterative solution of a non-linear problem; one algorithm that can be used to examine this condition is detailed by Zabihyan et al. [570]. Alternatively, Lagrange multipliers [357, 361] or penalty methods [515] may also be used to apply constraints.

Miehe [361] details a method to prescribe the boundary constraints to a parallelepiped-shaped RVE of a heterogeneous, geometrically periodic material. A Lagrangian periodicity frame, centred at the origin $\mathbf{X}_0$, is defined by a set of vectors

$$\{\mathbf{L}_\alpha\} \quad \text{with} \quad \alpha \in \{1, \ldots, \mathrm{d}\} \tag{9.41}$$

from which the periodic repetition of the microstructural geometry can be described as the admissible translations

$$\mathbf{L} = \sum_{\alpha=1}^{d} m_\alpha \mathbf{L}_\alpha \quad \text{with} \quad m_\alpha \in \{-\infty, \ldots, -1, 0, 1, \ldots, \infty\} \tag{9.42}$$

of the unit-cell or RVE. An example of a regular RVE is illustrated in Figure 9.2, with shape of the geometry defined by the periodic frame. More specifically, for this simple geometry we consider a straight-edged parallelepiped with vertices coincident with $\mathbf{L}_\alpha$ and opposing surfaces denoted as $[\partial\mathscr{B}_0^\pm]_\alpha$. It should be noted, however, that neither the RVE nor the periodicity frame itself is unique [12, 361].

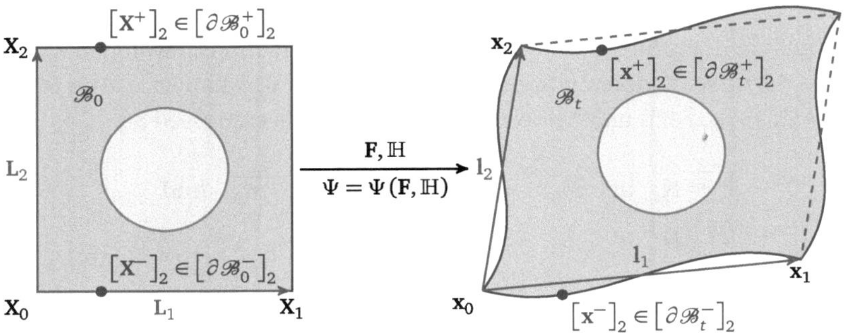

Figure 9.2: Application of boundary conditions to a geometrically periodic, heterogeneous RVE / unit-cell.

Considering this periodicity frame, the offset between the reference surfaces $[\partial\mathscr{B}_0^+]_\alpha$ and $[\partial\mathscr{B}_0^-]_\alpha$ is given exactly by $\mathbf{L}_\alpha$. Therefore, the distance between adjacent corners of the unit-cell can be calculated from

$$\mathbf{X}_\alpha^+ - \mathbf{X}_\alpha^- = [\![\mathbf{X}]\!]_\alpha \equiv \mathbf{L}_\alpha \quad , \tag{9.43}$$

and, upon deformation of the material,

$$\mathbf{x}_\alpha^+ - \mathbf{x}_\alpha^- = [\![\boldsymbol{\varphi}]\!]_\alpha \equiv \mathbf{l}_\alpha \tag{9.44a}$$

$$= \overline{\overline{\mathbf{F}}} \cdot \mathbf{L}_\alpha = \overline{\overline{\mathbf{F}}} \cdot [\![\mathbf{X}_\alpha]\!] \quad . \tag{9.44b}$$

It therefore follows upon application of equation (5.1) that

$$[\![\mathbf{u}]\!]_\alpha = \overline{\overline{\nabla_0 \overline{\overline{\mathbf{u}}}}} \cdot \mathbf{L}_\alpha \quad \text{with} \quad \mathbf{u}(\mathbf{X}_0) = \mathbf{0} \quad . \tag{9.45}$$

The above requires that the microstructure, mesh topology and resultant FE discretisation itself is periodic. It has also been proposed that approximate periodic boundary conditions can be imposed in a weak sense, or using surface-to-surface constraints [475].

Inspection of equations (9.17), (9.25) and (9.32) reveal that the periodic problem corresponding to the mechanical and magnetic fields are defined up to a constant field; that is, they are not unique. By prescribing equation (9.45), the translational DoFs associated with equation (9.18) become fixed, thereby removing the mechanical rigid body modes of $\mathscr{B}_0$. Similarly, the macro-scale fields associated with equations (9.25) and (9.32) become unique when the magnetic equivalent of the above, namely

$$[\![\mathbb{A}]\!]_\alpha = \frac{1}{2}\overline{\overline{\mathbb{B}}} \times \mathbf{L}_\alpha \quad \text{with} \quad \mathbb{A}(\mathbf{X}_0) = \mathbf{0} \quad , \qquad \text{or} \tag{9.46}$$

$$[\![\Phi]\!]_\alpha = \overline{\overline{\nabla_0 \overline{\Phi}}} \cdot \mathbf{L}_\alpha = -\overline{\overline{\mathbb{H}}} \cdot \mathbf{L}_\alpha \quad \text{with} \quad \Phi(\mathbf{X}_0) = 0 \tag{9.47}$$

are prescribed. For all other mechanical or magnetic DoFs, periodic constraints are applied by imposing

$$[\![\mathbf{u}]\!] = \overline{\overline{\nabla_0 \overline{\mathbf{u}}}} \cdot [\![\mathbf{X}]\!] \quad \text{on} \quad \partial\mathscr{B}_0^\pm \quad \text{and/or} \tag{9.48}$$

$$[\![\mathbb{A}]\!] = \frac{1}{2}\overline{\overline{\mathbb{B}}} \times [\![\mathbf{X}]\!] \quad \text{on} \quad \partial\mathscr{B}_0^\pm \quad \text{or} \tag{9.49}$$

$$[\![\Phi]\!] = -\overline{\overline{\mathbb{H}}} \cdot [\![\mathbf{X}]\!] \quad \text{on} \quad \partial\mathscr{B}_0^\pm \tag{9.50}$$

following from equations (9.17), (9.25) and (9.32).

For unit-cells with a more exotic structure, such as hexagonal geometries, alternate strategies may be required. Michel et al. [357] propose a method of accommodating such a geometry by translating the hexagonal RVE into one of several composite parallelepipeds. In the case that the applied load aligns with the axis of a stacked 3-D hexagonal microstructure, Gărăjeu et al. [160] suggest that the problem may be approximated by one that is 2-D axisymmetric; this approach has been used in the context of MAPs by Danas et al. [105]. Lastly, we note that the Lagrange multiplier and penalty approaches are particularly helpful when the discretisation is not consistent across $\partial\mathscr{B}_0^\pm$ and periodic conditions are to be considered.

Numerical example: Boundary conditions and the homogenised material response

To demonstrate the influence of the boundary conditions on both the micro-scale and the effective material response, Zabihyan et al. [570] considered a simple two-dimensional composite RVE of unit length with a stiff cylindrical inclusion of radius $R = 0.282$ (fibre $\phi = 25$vol.-%). The micro-scale (coupled, Neo–Hookean type) constitutive law for both the matrix and inclusion, differentiated by the subscripts m and p, respectively, was

$$\Psi(\mathbf{F}, \mathbb{H}) = \frac{1}{2}\mu\left[\mathbf{F} : \mathbf{F} - \mathrm{d} - 2\ln(J)\right] + \frac{1}{2}\lambda \ln^2(J) - \frac{1}{2}\upsilon J I_7 \tag{9.51}$$

Table 9.1: Fictitious constitutive parameters for numerical examples demonstrating influence of boundary conditions on a magnetoelastic RVE response.

Matrix		Inclusion	
Parameter	**Value**	**Parameter**	**Value**
μ_m	8	μ_p	80
λ_m	12	λ_p	120
υ_m	0.001	υ_p	0.01

where μ is the elastic shear modulus, λ is the Lamé parameter, and υ is the material's magnetic permeability. The chosen material parameters for this illustrative example are listed in Table 9.1. For the purpose of brevity, we will consider only the case of macro-scale uniaxial tension with magnetic loading in the direction of stretch, that is,

$$\overline{\overline{\mathbf{F}}} = \begin{bmatrix} \overline{\overline{\mathbf{F}}}_{11} & 0 \\ 0 & 1 \end{bmatrix} \quad \text{and} \quad \overline{\overline{\mathbb{H}}} = \begin{bmatrix} \overline{\overline{\mathbb{H}}}_1 \\ 0 \end{bmatrix} \quad . \tag{9.52}$$

Figure 9.3 presents the normalised micro-scale component of the Piola stress and referential magnetic induction aligned with the $\mathbf{e}_1$ direction for three different cases

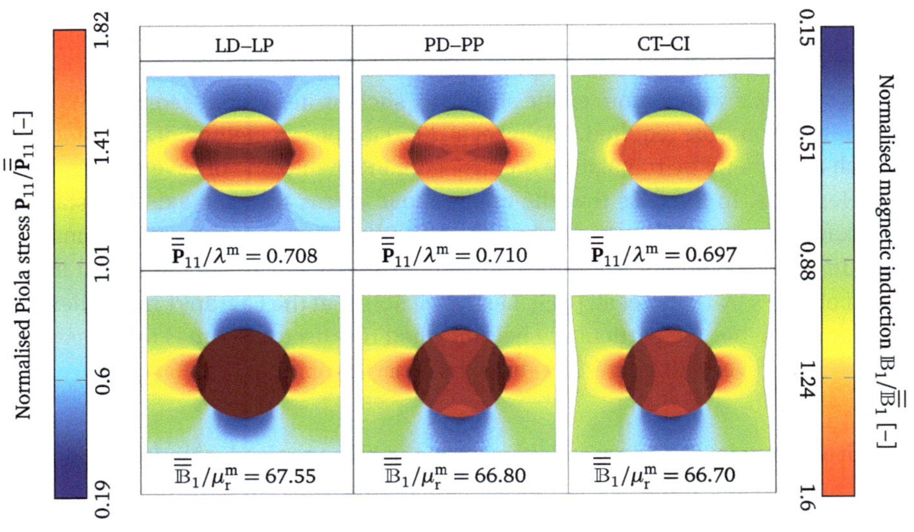

Figure 9.3: Distribution of the normalised Piola stress and magnetic induction components $\mathbf{P}_{11}$ (top row) and $\mathbb{B}_1$ (bottom row) for three combinations of boundary conditions. The applied macro-scale stretch and magnetic field are $\overline{\overline{\mathbf{F}}}_{11} = 1.1$ and $\overline{\overline{\mathbb{H}}}_1 = 50$, respectively. The boundary conditions showcased in these three cases are respectively linear displacement and magnetic scalar potential (LD–LP, equations (9.21) and (9.35)), periodic displacement and magnetic scalar potential (PD–PP, equations (9.22) and (9.36)) and constant mechanical and magnetic tractions (CT–CI, equations (9.23) and (9.37)). [570, fig. 4 (reproduced with permission)]

of applied boundary conditions. Both the qualitative and quantitative micro-scale responses are very similar, with the pure Neumann condition (CT–CI) rendering the most compliant response. The prediction for the homogenised macro-scale response is also consistent between all cases, with under 2 % difference between them. Note also that in the first two cases the RVE retains its geometric periodicity under loading, while in the third case this is not enforced.

As is shown in Figure 9.4, the size of the RVE is also a factor that, along with the applied constraints, should be considered. To generate a larger RVE, the simple unit cell is repeated translationally, thereby retaining the fibre-to-matrix ratio constant. Under the aforementioned loading conditions, the pure Dirichlet constraints consistently over-predict the averaged enthalpy, stress and magnetic induction, while the pure Neumann constraints similar under-predict them. The averaged fields predicted when pure periodic constraints are applied consistently lie between these two limits, and are hardly affected by the RVE size. For all but the induction, the LD–LP condition tends to converge more rapidly to the periodic solution as the RVE size increases than

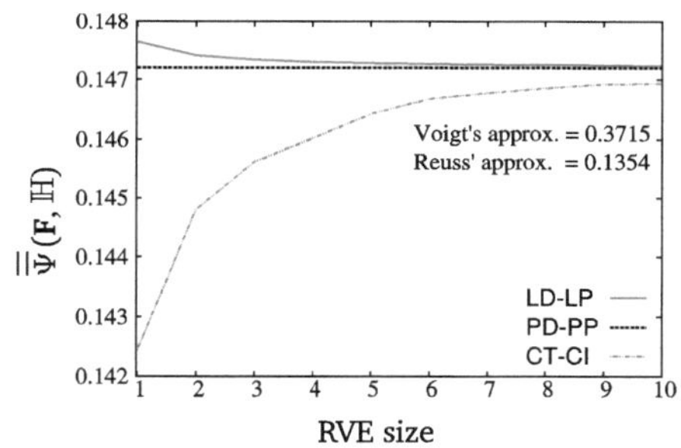

(a) Averaged mixed energy-enthalpy.

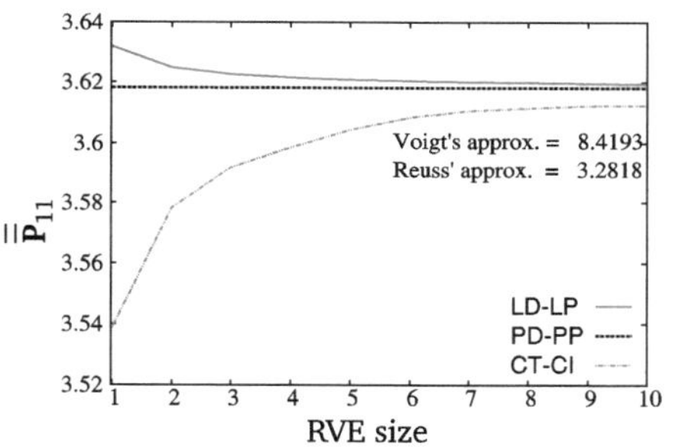

(b) Averaged Piola stress.

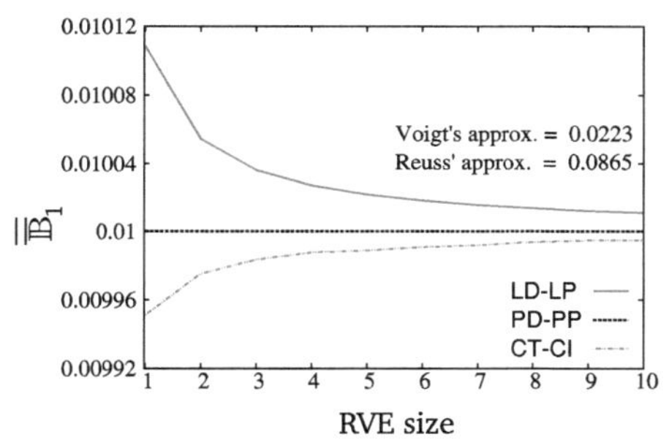

(c) Averaged magnetic induction.

Figure 9.4: Components of averaged quantities versus the RVE size (number of repetitions of the periodic structure). The size of the RVE increases from that of a unit cell to a 10×10 geometrically periodic microstructure that contains all underlying lower size RVEs. The applied macro-scale stretch and magnetic field are $\overline{\overline{\mathbf{F}}}_{11} = 1.1$ and $\overline{\overline{\mathbb{H}}}_1 = 7.6$, respectively. Note that the Voigt and Reuss bounds are listed but not plotted on the graphs themselves. [570, fig. 10 (reproduced with permission)]

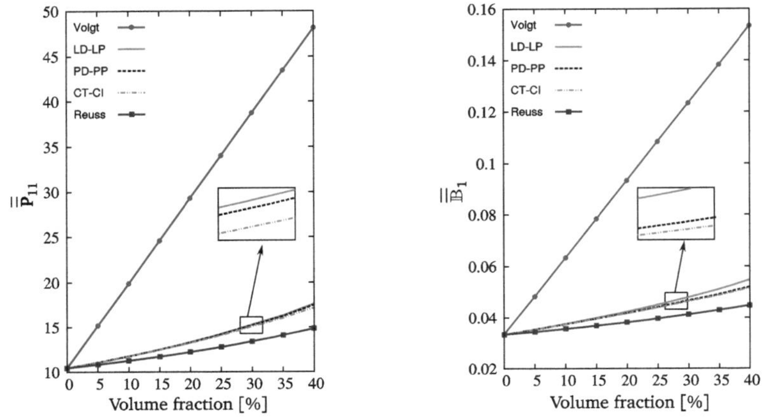

(a) Averaged Piola stress. **(b)** Averaged magnetic induction.

Figure 9.5: Influence of the inclusion size on the macro-scale field response. Only a single unit cell was considered, and the applied macro-scale stretch and magnetic field are $\overline{\overline{\mathbf{F}}}_{11} = 1.5$ and $\overline{\overline{\mathbb{H}}}_1 = 50$, respectively. [570, fig. 14 (reproduced with permission)]

the CT–CI boundary conditions do. In these cases, more of the unit cells (within the core of the RVE) experience periodic-like conditions as the size of the RVE increases.

Lastly, the influence of the fibre volume fraction on the macro-scale response is presented in Figure 9.5. Intuitively, increasing the volume of the fibre component renders both a stiffer mechanical and magnetic response than at a lower volume fraction. The increasing response is, however, non-linear as the influence of the geometric nonlinearity introduced by the presence of the inclusion becomes more dominant. The Dirichlet-type boundary conditions again result in a stiffer response than the periodic and Neumann-type constraints, while the Voigt and Reuss bounds offer greatly over- and under-predicted responses. The discrepancy between all of the evaluated cases consistently increases as ϕ becomes greater; between the LD-LP, PD-PP, and CT-CI cases, the increase is only marginal.

9.1.3 Computation of consistent tangent moduli for the macro-scale problem

In the numerical implementation of the FE^2 approach, at quadrature points within the discretised FE macroscopic domain the constitutive response of the coupled material is provided by RVEs (i. e. micro-scale boundary value problems). The consistent linearisation of the macro-scale stress and magnetic induction for the parameterisation given in equation (9.2) are respectively given by

$$\Delta\overline{\overline{\mathbf{P}}} = \frac{\partial\overline{\overline{\mathbf{P}}}}{\partial\overline{\overline{\mathbf{F}}}} : \Delta\overline{\overline{\mathbf{F}}} + \frac{\partial\overline{\overline{\mathbf{P}}}}{\partial\overline{\overline{\mathbb{H}}}} \cdot \Delta\overline{\overline{\mathbb{H}}} \equiv \overline{\overline{\mathscr{A}}} : \Delta\overline{\overline{\mathbf{F}}} - \overline{\overline{\mathfrak{L}}} \cdot \Delta\overline{\overline{\mathbb{H}}} \quad , \tag{9.53a}$$

$$-\Delta\overline{\overline{\mathbb{B}}} = -\frac{\partial\overline{\overline{\mathbb{B}}}}{\partial\overline{\overline{\mathbf{F}}}} : \Delta\overline{\overline{\mathbf{F}}} - \frac{\partial\overline{\overline{\mathbb{B}}}}{\partial\overline{\overline{\mathbb{H}}}} \cdot \Delta\overline{\overline{\mathbb{H}}} \equiv -\overline{\overline{\mathfrak{L}}}^{T} : \Delta\overline{\overline{\mathbf{F}}} - \overline{\overline{\mathbf{D}}} \cdot \Delta\overline{\overline{\mathbb{H}}} \quad . \tag{9.53b}$$

This is analogous to the linearisation presented in Section 5.3.7.2 that is applicable to the micro-scale problem. To assemble the system of linear equations at the macro-scale level, it is therefore necessary to compute the sensitivities of the macro-scale dependent variables with respect to the macro-scale independent fields. This, however, is not completely straightforward as the primary micro-scale fields depend on both the averaged macro-scale fields as well as the fluctuation fields (as can be observed in equations (9.19) and (9.33)). Thus, the microstructural response (and, therefore, its homogenised macro-scale counterparts) have additional components related to the fluctuation fields that must be accounted for.

There are several methods that one can use to form the consistent macro-scale tangent moduli, or compute a reasonable approximation to it. This includes (i) condensation of the micro-scale stiffness matrix by considering a decomposition of the micro-scale DoFs into the boundary and internal components [361, 412, 400, 165, 515, 516]; (ii) local homogenisation of the tangent moduli [519], where the tangent is computed by exploiting the incremental formulation in order to approximate partial derivatives of the micro-scale fields with respect to their macro-scale counterparts; (iii) application of FFT-based methods [389, 357, 176]; (iv) perturbation approaches; and (v) calculation of algorithmically correct tangent moduli. In the following sections, we will describe in more detail the latter two approaches.

9.1.3.1 Perturbation method

Perturbation approaches, or application of numerical differentiation techniques, offer a simple method by which to compute effective homogenised tangent moduli for the microstructure. They have been employed for the purpose of computational homogenisation in the context of finite strain elasticity [359, 514], elasto-plasticity [518, 368], and contact mechanics [515], and magneto-mechanics [234, 38].

Using a condensed indicial format applying, for example, either Voigt or Kelvin notation [393, 112], the compact expression for the linear increment of the coupled MSP problem is

$$\Delta\overline{\overline{\Xi}} = \overline{\overline{\mathscr{K}}} \cdot \Delta\overline{\overline{\mathbf{Y}}} \quad \Longrightarrow \quad \begin{bmatrix} \Delta\overline{\overline{\mathbf{P}}} \\ -\Delta\overline{\overline{\mathbb{B}}} \end{bmatrix} = \begin{bmatrix} \overline{\overline{\mathscr{A}}} & -\overline{\overline{\mathfrak{L}}} \\ -\overline{\overline{\mathfrak{L}}}^{T} & -\overline{\overline{\mathbf{D}}} \end{bmatrix} \cdot \begin{bmatrix} \Delta\overline{\overline{\mathbf{F}}} \\ \Delta\overline{\overline{\mathbb{H}}} \end{bmatrix} \tag{9.54}$$

where $\overline{\overline{\Xi}}(\overline{\overline{\mathbf{Y}}}(\overline{\overline{\mathbf{F}}}, \overline{\overline{\mathbb{H}}}))$ and $\overline{\overline{\mathbf{Y}}}(\overline{\overline{\mathbf{F}}}, \overline{\overline{\mathbb{H}}})$ are ordered vectors respectively containing the macro-scale kinetic and kinematic quantities, and $\overline{\overline{\mathscr{K}}}(\overline{\overline{\mathbf{F}}}, \overline{\overline{\mathbb{H}}})$ is the consistently ordered matrix of macro-scale tangent moduli that should be computed. For convenience, we will denote the perturbation of the jth component of $\overline{\overline{\mathbf{Y}}}$ as

$$\delta_j^{+}\overline{\overline{\mathrm{Y}}}_i = \begin{cases} \overline{\overline{\mathrm{Y}}}_i + \varepsilon & \text{if} \quad j = i \\ \overline{\overline{\mathrm{Y}}}_i & \text{otherwise} \end{cases} \tag{9.55}$$

where the superscript + indicates that an incremental (rather than decremental) perturbation is performed. One may apply numerical differentiation to approximate the macro-scale tangent moduli by rearranging the above as

$$\overline{\overline{\mathscr{K}}}_{ij} = \frac{\Delta \overline{\overline{\Xi}}_i}{\Delta \overline{\overline{\mathbf{Y}}}_j} \quad . \tag{9.56}$$

The first-order forward difference approximation for each component of $\overline{\overline{\mathscr{K}}}$ is then given by

$$\overline{\overline{\mathscr{K}}}_{ij} = \frac{\left[\overline{\overline{\Xi}}\left(\delta_j^+ \overline{\overline{\mathbf{Y}}}\right) - \overline{\overline{\Xi}}\left(\overline{\overline{\mathbf{Y}}}\right)\right]_i}{\left[\delta_j^+ \overline{\overline{\mathbf{Y}}} - \overline{\overline{\mathbf{Y}}}\right]_j} = \frac{\overline{\overline{\Xi}}_i\left(\delta_j^+ \overline{\overline{\mathbf{Y}}}\right) - \overline{\overline{\Xi}}_i\left(\overline{\overline{\mathbf{Y}}}\right)}{\varepsilon} \quad \text{(no sum over } j) \quad . \tag{9.57}$$

One may observe that from the perturbation of each row j of $\overline{\overline{\mathbf{Y}}}$ one may extract an entire i^{th} column of $\overline{\overline{\mathscr{K}}}$, thereby only necessitating the solution of $n_{\mathbf{F}} + n_{\mathbb{H}}$ linear systems of equations for this coupled problem. For details on the construction of the appropriate linear problem to be solved at the equilibrium state, refer to [518, 234]. The choice of the perturbation parameter is critical, as too large a value renders the approximation insufficiently accurate and too small a value introduces substantial numerical round-off errors. Albeit in a slightly different context, Tanaka et al. [512] gives further demonstration of the consequences of the inappropriate choice of the perturbation size, while Miehe [359] provides insight on how to sensibly choose this parameter. To maintain numerical accuracy it may be necessary to define $\varepsilon = \varepsilon(\overline{\overline{\mathbf{Y}}}_j)$ for each perturbed component j as opposed to it being considered a constant value chosen a priori. A more robust alternative would be to use the second-order central-difference approximation

$$\overline{\overline{\mathscr{K}}}_{ij} = \frac{\left[\overline{\overline{\Xi}}\left(\delta_j^+ \overline{\overline{\mathbf{Y}}}\right) - \overline{\overline{\Xi}}\left(\delta_j^- \overline{\overline{\mathbf{Y}}}\right)\right]_i}{\left[\delta_j^+ \overline{\overline{\mathbf{Y}}} - \delta_j^- \overline{\overline{\mathbf{Y}}}\right]_j} = \frac{\overline{\overline{\Xi}}_i\left(\delta_j^+ \overline{\overline{\mathbf{Y}}}\right) - \overline{\overline{\Xi}}_i\left(\delta_j^- \overline{\overline{\mathbf{Y}}}\right)}{2\varepsilon} \quad \text{(no sum over } j) \tag{9.58}$$

which has the disadvantage of being more computationally costly, as it requires double the number of perturbation steps to be performed. Coefficients for even higher-order finite difference stencils are listed by Fornberg [149]. Another option to the strategy to that described above is that presented by Bayat and Gordaninejad [38], who reported an application of the perturbation method as used to resolve only the sensitivity of the fluctuating fields with respect to their macro-scale counterparts.

9.1.3.2 Algorithmically consistent tangent moduli

A more rigorous approach to computing macro-scale moduli, which specifically considers the treatment of both the constant and macro-scale fluctuation fields, was discussed by Miehe et al. [367] in the context of finite strain elasto-plasticity. It has since been extended to the areas of thermoplasticity [27]; infinitesimal [477] and finite strain electroelasticity [257]; large deformation magneto-mechanics [255]; and electro-magneto-mechanics [480, 377]. To compute the algorithmically correct tangent moduli, it is necessary to consider the incremental form of the constitutive relations as stated in equation (9.53), and account for the sensitivity with respect to both the macro-scale fields, as well as the superimposed fluctuation fields.

In summary, the macro-scale tangent moduli related to a free energy function parameterised as $\overline{\overline{\Psi}} = \overline{\overline{\Psi}}(\overline{\overline{\mathbf{F}}}, \overline{\overline{\mathbb{H}}})$ can be determined from equation (9.53), by considering the decomposition of the macro-scale fields into their two superimposed components (namely the linear and fluctuating parts), exploiting the FE discretisation stated in equations (6.7) and (6.23), and defining an equivalent ansatz for the fluctuation fields. Considering the above, the collection of algorithmically correct decoupled and coupled macro-scale material tangents are concisely expressed as

$$\begin{bmatrix} \overline{\overline{\mathscr{A}}} & -\overline{\overline{\mathfrak{L}}} \\ -\overline{\overline{\mathfrak{L}}}^T & -\overline{\overline{\mathbf{D}}} \end{bmatrix} = \left\langle \begin{bmatrix} \mathscr{A} & -\mathfrak{L} \\ -\mathfrak{L}^T & -\mathbf{D} \end{bmatrix} \right\rangle_0 - \frac{1}{V_0} \boldsymbol{L}^T \cdot \boldsymbol{K}^{-1} \cdot \boldsymbol{L} \quad . \tag{9.59}$$

The derivation of this result is provided in Appendix C.8.3. We observe that the first term on the right-hand side is simply the volume average of the microstructural tangent moduli, while the second is a correction term that accounts for the sensitivity of the microstructure to periodic fluctuations. The elements of the matrix $\boldsymbol{L}$ are

$$\begin{aligned} \boldsymbol{L}_{\varphi\varphi} &= \int_{\mathscr{D}_0} \nabla_0 \boldsymbol{\Psi}^I_{\varphi} : \mathscr{A} \, dV \quad , \quad \boldsymbol{L}_{\varphi\Phi} = - \int_{\mathscr{D}_0} \nabla_0 \boldsymbol{\Psi}^I_{\varphi} : \mathfrak{L} \, dV \quad , \\ \boldsymbol{L}_{\Phi\varphi} &= \int_{\mathscr{D}_0} \nabla_0 \Psi^I_{\Phi} \cdot \mathfrak{L}^T \, dV \quad , \quad \boldsymbol{L}_{\Phi\Phi} = \int_{\mathscr{D}_0} \nabla_0 \Psi^I_{\Phi} \cdot \mathbf{D} \, dV \end{aligned} \tag{9.60}$$

and those of the coupled algorithmic stiffness matrix $\boldsymbol{K}$ are

$$\begin{aligned} \boldsymbol{K}_{\varphi\varphi} &= \int_{\mathscr{D}_0} \nabla_0 \boldsymbol{\Psi}^I_{\varphi} : \mathscr{A} : \nabla_0 \boldsymbol{\Psi}^J_{\varphi} \, dV \quad , \quad \boldsymbol{K}_{\varphi\Phi} = \int_{\mathscr{D}_0} \nabla_0 \boldsymbol{\Psi}^I_{\varphi} : \mathfrak{L} \cdot \nabla_0 \Psi^J_{\Phi} \, dV \quad , \\ \boldsymbol{K}_{\Phi\varphi} &= \int_{\mathscr{D}_0} \nabla_0 \Psi^I_{\Phi} \cdot \mathfrak{L}^T : \nabla_0 \boldsymbol{\Psi}^J_{\varphi} \, dV \quad , \quad \boldsymbol{K}_{\Phi\Phi} = - \int_{\mathscr{D}_0} \nabla_0 \Psi^I_{\Phi} \cdot \mathbf{D} \cdot \nabla_0 \Psi^J_{\Phi} \, dV \quad , \end{aligned} \tag{9.61}$$

the latter aligning with the standard tangent matrix $\boldsymbol{K}$ for the original boundary value problem. Note that the correction term, which requires the inverse of $\boldsymbol{K}$, is itself computed by solving a boundary value problem; the product

$$-\boldsymbol{K}^{-1} \cdot \boldsymbol{L} \equiv \begin{bmatrix} \dfrac{\partial \Delta \tilde{\boldsymbol{d}}_{\varphi}}{\partial \overline{\overline{\mathbf{F}}}} & \dfrac{\partial \Delta \tilde{\boldsymbol{d}}_{\varphi}}{\partial \overline{\overline{\mathbb{H}}}} \\ \dfrac{\partial \Delta \tilde{\boldsymbol{d}}_{\Phi}}{\partial \overline{\overline{\mathbf{F}}}} & \dfrac{\partial \Delta \tilde{\boldsymbol{d}}_{\Phi}}{\partial \overline{\overline{\mathbb{H}}}} \end{bmatrix} \tag{9.62}$$

represents the sensitivity of the increments of the DoFs used to discretise the fluctuation fields with respect to the incremental change in the macro-scale gradient fields.

It is important to note that for some alternative parameterisations of the microscopic free energy function, for example $\Psi = \Psi(\mathbf{C}, \mathbb{H})$ or $\Psi = \Psi(\mathbf{E}, \mathbb{H})$ that have symmetric strain-like tensors as arguments, it is still necessary to use aforementioned expressions for computing the macro-scale tangent moduli. This is because the number of independent components of $\Delta\mathbf{C}$ and $\Delta\mathbf{E}$ is less than $\Delta\mathbf{F}$ and, therefore, there exists no unique link between the constraints imposed on $\widetilde{\mathbf{F}}$ and $\widetilde{\mathbf{C}}$ or $\widetilde{\mathbf{E}}$. Using the fundamental relationship $\mathbf{P} = \mathbf{F} \cdot \mathbf{S}$, it is possible to determine the appropriate transformation between the fully-referential and two-point descriptions of the material elasticity and magnetoelasticity tensors. Accordingly, the push-forward of these quantities are [492]

$$\mathscr{A} = \mathbf{F} \cdot \mathscr{H} : \left[\mathbf{F}^T \overline{\otimes} \mathbf{I}\right] + \mathbf{I} \overline{\otimes} \mathbf{S} \quad , \quad \mathfrak{L} = \mathbf{F} \cdot \mathfrak{P} \quad , \tag{9.63}$$

where the Piola–Kirchhoff stress, material elasticity tangent moduli and magnetoelastic tangent moduli are respectively given by equations (7.16) to (7.18).

9.1.4 Homogenisation of curing using the Mori–Tanaka method

Returning to the curing of MAPs, Hossain et al. [213] extended the curing and shrinkage model detailed in Chapter 8 to account for the materials' microstructure using homogenisation as opposed to a phenomenological constitutive law. Derived for the small strain case, the dissipation inequality now reads

$$\mathscr{D}_{\text{int}} = \boldsymbol{\sigma} : \dot{\boldsymbol{\varepsilon}} + \mathbb{h} \cdot \dot{\mathbb{b}} - \dot{\Psi}_0^* (t) \geq 0 \tag{9.64}$$

where $\boldsymbol{\varepsilon} = \frac{1}{2}[\nabla_0 \mathbf{u} + [\nabla_0 \mathbf{u}]]$ is the infinitesimal strain tensor. As this potential has the same structure as equation (8.2), by reinterpreting the kinetic terms $\{\mathbf{F}, \mathbb{B}\} \rightarrow \{\boldsymbol{\varepsilon}, \mathbb{b}\}$ we can follow the same procedure as is detailed in Section 8.1 to deduce the stress and magnetic field evolution. After executing a Legendre transformation on the magnetic variables [82], the time-evolution of the stress and magnetic induction are

$$\dot{\boldsymbol{\sigma}}^{\text{tot}} (t) = \hat{\mathscr{A}} (t) : \dot{\boldsymbol{\varepsilon}} (t) + \hat{\mathfrak{L}} (t) \cdot \left[-\dot{\mathbb{h}} (t)\right] \quad , \tag{9.65a}$$

$$\dot{\mathbb{b}} (t) = -\hat{\mathbf{D}} (t) \cdot \left[-\dot{\mathbb{h}} (t)\right] + \hat{\mathfrak{L}}^T (t) : \dot{\boldsymbol{\varepsilon}} (t) \quad , \tag{9.65b}$$

where the linearisation tensors are related to the small-strain equivalent of those in equation (8.9) by

$$\hat{\mathscr{A}} = \mathscr{A} + \hat{\mathfrak{L}} \cdot \mathfrak{L}^T \quad , \quad \hat{\mathfrak{L}} = -\mathfrak{L} \cdot \mathbf{D}^{-1} \quad , \quad \hat{\mathbf{D}} = \mathbf{D}^{-1} \quad . \tag{9.66}$$

Consistent with the infinitesimal strain assumption, the isotropic small strain shrinkage model

$$\boldsymbol{\varepsilon} := \boldsymbol{\varepsilon}_{\text{m}} + \boldsymbol{\varepsilon}_{\text{s}} = \boldsymbol{\varepsilon}_{\text{m}} + s(t, \mathbb{b})\,\mathbf{I} \quad , \tag{9.67}$$

similar to that in [209] is adopted. The evolution of the shrinkage s is governed by equation (8.13), and the field-dependent total volume shrinkage s_∞ is collectively defined by equations (8.16) to (8.18).

To capture the behaviour of the particulate microstructure, Hossain et al. utilised the Mori–Tanaka method [128] extended to magneto-mechanical materials to consider a composite comprising $n_{\text{ph}} + 1$ material phases. The micromechanical small strain constitutive law for each phase can be written in a general manner (using a condensed indicial notation) as

$$\dot{\Sigma}_{ij} = \mathscr{L}_{ijmn}(\mathbf{E}, t)\left[\dot{\mathrm{E}}_{mn} - [\dot{\mathrm{E}}_{\mathrm{s}}]_{mn}\right] \tag{9.68}$$

where

$$\Sigma_{ij} = \begin{cases} \sigma_{ij} & \text{if} \quad i \in \mathscr{M}_{\text{mech}} \\ \mathbb{b}_j & \text{if} \quad i \in \mathscr{M}_{\text{mag}} \end{cases} \qquad \forall\, j \in \mathscr{M}_{\text{d}} \tag{9.69a}$$

$$E_{mn} = \begin{cases} \varepsilon_{mn} & \text{if} \quad m \in \mathscr{M}_{\text{mech}} \\ -\mathbb{h}_n & \text{if} \quad m \in \mathscr{M}_{\text{mag}} \end{cases} \qquad \forall\, n \in \mathscr{M}_{\text{d}} \tag{9.69b}$$

$$[E_{\mathrm{s}}]_{mn} = \begin{cases} [\varepsilon_{\mathrm{s}}]_{mn} := s(t, \mathbb{b})\,\delta_{mn} & \text{if} \quad m \in \mathscr{M}_{\text{mech}} \\ 0 & \text{if} \quad m \in \mathscr{M}_{\text{mag}} \end{cases} \qquad \forall\, n \in \mathscr{M}_{\text{d}} \tag{9.69c}$$

$$\mathscr{L}_{ijmn} = \begin{cases} \alpha(t)\,\hat{\mathscr{A}}_{ijmn} & \text{if} \quad i, m \in \mathscr{M}_{\text{mech}} \\ \hat{\mathfrak{L}}_{nij} & \text{if} \quad i \in \mathscr{M}_{\text{mech}},\ m \in \mathscr{M}_{\text{mag}} \\ \hat{\mathfrak{L}}_{jmn} & \text{if} \quad i \in \mathscr{M}_{\text{mag}},\ m \in \mathscr{M}_{\text{mech}} \\ -\hat{\mathbb{D}}_{jn} & \text{if} \quad i, m \in \mathscr{M}_{\text{mag}} \end{cases} \qquad \forall\, j, n \in \mathscr{M}_{\text{d}} \tag{9.69d}$$

for the tensor component index sets

$$\mathscr{M}_{\text{d}} = \{1, \ldots, \mathrm{d}\} \quad , \quad \mathscr{M}_{\text{mech}} = \{1, \ldots, \mathrm{d}\} \quad \text{and} \quad \mathscr{M}_{\text{mag}} = \{\mathrm{d}+1\} \quad . \tag{9.70}$$

Note that, although similar, the condensed notation presented here does indeed differ to that used in Section 9.1.3.1; the left-hand side of equation (9.54) is single indexed, while that of equation (9.68) is multi-indexed.

The purpose of homogenisation is to determine the equivalent macro-scale response for the composite, namely

$$\dot{\overline{\overline{\boldsymbol{\Sigma}}}} = \overline{\overline{\mathscr{L}}} : \left[\dot{\overline{\overline{\mathbf{E}}}} - \dot{\overline{\overline{\mathbf{E}}}}_{\mathrm{s}}\right] \quad , \tag{9.71}$$

which follows a similar constitutive law to its micro-scale constituents. Following the typical assumption in micromechanics, the macro-scale kinematic and kinetic response is the volume-weighted sum of its constituent's components response

$$\overline{\overline{\mathbf{E}}} = \sum_{r=0}^{n_{\text{ph}}} \phi_r \mathbf{E}_r \quad , \quad \overline{\overline{\boldsymbol{\Sigma}}} = \sum_{r=0}^{n_{\text{ph}}} \phi_r \boldsymbol{\Sigma}_r \tag{9.72}$$

where ϕ_r is the volume fraction of the rth phase ($r = 0$ signifying the matrix, $r \in \mathbb{Z}[1, n_{\text{ph}}]$ the inclusions that may be of a varying nature). As a simplification for these non-linear materials [283], it is also assumed that the relation expressed in equation (9.68) holds for each phase individually (and independent of one another); that is to say that

$$\boldsymbol{\Sigma}_r = \mathscr{L}_r(\mathbf{E}_r, t) : \mathbf{E}_r \quad . \tag{9.73}$$

Given these assumptions, it can be shown that the effective homogenised magneto-mechanical tangent modulus is given by

$$\overline{\overline{\mathscr{L}}} = \sum_{r=0}^{n_{\text{ph}}} \phi_r \mathscr{L}_r : \mathbf{T}_r^{\text{MT}} \quad . \tag{9.74}$$

Therein the generalised Mori–Tanaka strain concentration tensor for any constituent phase r is

$$\mathbf{T}^{\text{MT}} = \mathbf{T}^{\text{dil}} : \left[\sum_{s=0}^{n_{\text{ph}}} \phi_s \mathbf{T}_s^{\text{dil}} \right]^{-1} \quad , \tag{9.75}$$

for which the dilute strain concentration tensor of each phase r reads as

$$\mathbf{T}^{\text{dil}} = \left[\mathscr{I} + \mathscr{E} : \mathscr{L}_0^{-1} : [\mathscr{L} - \mathscr{L}_0] \right]^{-1} \quad . \tag{9.76}$$

The tensor $\mathscr{L}$ are phase-dependent material moduli given by equation (9.69d), and $\mathscr{L}_0$ is the effective magneto-mechanical tangent modulus for the matrix phase. It may be observed that for the matrix phase itself ($r = 0$), the last term is zero and $\mathbf{T}_0^{\text{dil}} \equiv \mathscr{I}$.

The Eshelby geometric tensor $\mathscr{E}$ depends on the geometry of the phase, and is computed either analytically or, as is done in this example, numerically using Gaussian quadrature. Assuming a three-dimensional problem and that all magnetisable particles are elliptically-shaped with major and minor axes aligned with the Cartesian bases, then $\mathscr{E}$ for the particle phases is determined from

$$\mathscr{E}_{ijkl} = \begin{cases} \frac{1}{8\pi} [\mathscr{L}_0]_{mnkl} \int_{-1}^{1} \int_{0}^{2\pi} \left[\hat{\zeta}_j \hat{\zeta}_n Z_{im}^{-1} + \hat{\zeta}_i \hat{\zeta}_n Z_{jm}^{-1} \right] \mathrm{d}\omega \, \mathrm{d}\zeta_3 & \text{if} \quad i \in \mathscr{M}_{\text{mech}} \\ \frac{1}{4\pi} [\mathscr{L}_0]_{mnkl} \int_{-1}^{1} \int_{0}^{2\pi} \left[\hat{\zeta}_j \hat{\zeta}_n Z_{im}^{-1} \right] \mathrm{d}\omega \, \mathrm{d}\zeta_3 & \text{if} \quad i \in \mathscr{M}_{\text{mag}} \end{cases} \tag{9.77}$$

with

$$\hat{\zeta}_1 = \sqrt{1 - \zeta_3} \frac{\cos\omega}{a_1} \quad , \quad \hat{\zeta}_2 = \sqrt{1 - \zeta_3} \frac{\sin\omega}{a_2} \quad ,$$

$$\hat{\zeta}_3 = \frac{\zeta_3}{a_3} \quad , \quad Z_{ij} = [\mathscr{L}_0]_{ikjl} \hat{\zeta}_k \hat{\zeta}_l \quad . \tag{9.78}$$

The parameters $\{a_1, a_2, a_3\}$ denote the half-lengths of the ellipse axes respectively aligned in the $\{\mathbf{e}_1, \mathbf{e}_2, \mathbf{e}_3\}$ directions.

Lastly, it is necessary to consider the shrinkage strains from a microstructural viewpoint. Following the approach used in [302], the macro-scale generalised shrinkage tensor is determined from

$$\overline{\overline{\mathbf{E}}}_{\mathrm{s}} = \sum_{r=0}^{n_{\mathrm{ph}}} \phi_r \left[\overline{\overline{\mathbf{E}}}_{\mathrm{s}}\right]_r + \sum_{r=1}^{n_{\mathrm{ph}}} \phi_r \left[\left[\mathscr{I} - \overline{\overline{\mathscr{L}}}^{-1} : \mathscr{L}_r\right] : \mathscr{R}_r : \left[\left[\overline{\overline{\mathbf{E}}}_{\mathrm{s}}\right]_r - \left[\overline{\overline{\mathbf{E}}}_{\mathrm{s}}\right]_0\right]\right] \tag{9.79}$$

where $[\overline{\overline{\mathbf{E}}}_{\mathrm{s}}]_r$ is the generalised shrinkage tensor for each constituent of the composite and, for each phase r

$$\mathscr{R} = \mathscr{E} : [[\mathscr{L} - \mathscr{L}_0] : \mathscr{E} + \mathscr{L}_0]^{-1} : \mathscr{L} - \mathscr{I} \quad . \tag{9.80}$$

Note that, with specific application to MAPs, metallic inclusions do not undergo curing; thus $[\overline{\overline{\mathbf{E}}}_{\mathrm{s}}]_r = \mathbf{0}$ for $r > 0$, which reduces the complexity of equation (9.79) significantly.

Like for equation (8.11), a backward-Euler time discretisation scheme for the constitutive relation given by equation (9.68) is adopted for each phase r, namely

$$\boldsymbol{\Sigma}^n = \boldsymbol{\Sigma}^{n-1} + \mathscr{L}^n : \left[[\mathbf{E} - \mathbf{E}_{\mathrm{s}}]^n - [\mathbf{E} - \mathbf{E}_{\mathrm{s}}]^{n-1}\right] \quad . \tag{9.81}$$

Although it might appear that the material response is instantaneously linear due to the small strain setting, this is in fact not the case due to the dependence of the moduli $\mathscr{L}^n = \mathscr{L}^n(\mathbf{E}^n, t)$, and the strong non-linear coupling of the shrinkage strains $\boldsymbol{\varepsilon}_{\mathrm{s}}^n = \boldsymbol{\varepsilon}_{\mathrm{s}}^n(\mathbb{b}^n)$ to the magnetic field. It is thus typically necessary to employ the incremental Mori–Tanaka approach [420] to update the solution at a fixed time step. As an alternative, Hossain et al. [213] detail a relaxation scheme for the computation of the moduli and the shrinkage strains in order to reduce the problem to one that is linear in time.

Numerical examples: Micromechanical model of curing

To illustrate some of the characteristics of the micromechanical model, the curing of a composite comprising cobalt ferrite ($CoFe_2O_4$) particles embedded in an epoxy matrix is considered. The particles are assumed to be spherical in shape ($\{a_1, a_2, a_3\} = 1$), and migration of the particles when cured under a magnetic field (leading to preferential directionality) is not taken into account. The material, curing and shrinkage parameters for the scenario are given in Table 9.2, while the ratio of material components is varied.

Figure 9.6 provides an indication of how the material tangent moduli increases as the material cures (under conditions wherein shrinkage is not considered). The overall stiffness of the material evolves exponentially with time, and when compared to the

Table 9.2: Parameters for numerical examples of homogenised model of curing with shrinkage [213, 220, 413]. The chosen value for κ dictates that the curing and shrinkage occurs over a timespan of 50 s.

(a) Material parameters (expressed in Voigt notation). Note that $\hat{\mathscr{A}}_{66} = \frac{1}{2}\left[\hat{\mathscr{A}}_{11} - \hat{\mathscr{A}}_{12}\right]$ and $\hat{\mathbb{D}}_{11} = \hat{\mathbb{D}}_{22}$.

Material	$\hat{\mathscr{A}}_{11}$	$\hat{\mathscr{A}}_{12}$	$\hat{\mathscr{A}}_{13}$	$\hat{\mathscr{A}}_{33}$	$\hat{\mathscr{A}}_{44}$	$\hat{\mathfrak{L}}_{15}$	$\hat{\mathfrak{L}}_{31}$	$\hat{\mathfrak{L}}_{33}$	$\hat{\mathbb{D}}_{11}$	$\hat{\mathbb{D}}_{33}$
	GPa					**N A^{-1} m^{-1}**			**μN A^{-2}**	
Epoxy	5.53	2.97	2.97	5.53	1.28	0	0	0	1.257	1.257
Cobalt ferrite	286	173	170	269	45	580	580	700	−590	157

(b) Curing parameters.

Parameter	**Value**	**Unit**
α_0	1×10^{-4}	–
α_∞	1	–
κ_α	0.0925	s^{-1}

(c) Shrinkage parameters.

Parameter	**Value**	**Unit**
s_0	0	–
κ_s	0.0925	s^{-1}
s^1_∞	0	–
s^2_∞	−0.0005	–
η_∞	5	–
$\frac{e^1_\infty + e^2_\infty}{2}$	0.055	–

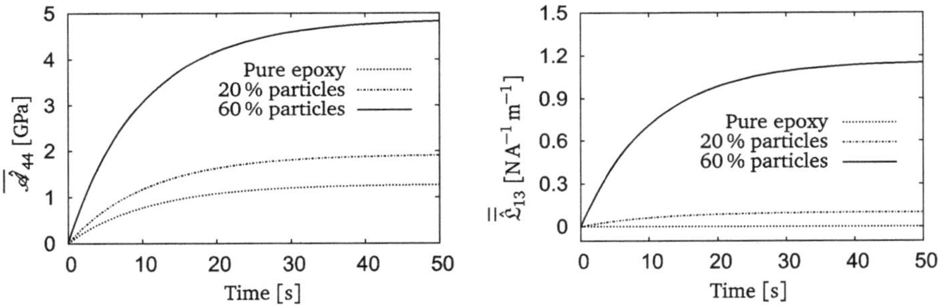

(a) Shear modulus.

(b) Coupled term.

Figure 9.6: Evolution of macro-scale quantities during the progression of matrix curing, excluding shrinkage effects. [213, figs. 2a,2c (reproduced with permission)]

case of pure epoxy it is clear that the addition of the stiff particles invokes a marked increase in material rigidity. The coupled term similarly evolves in an exponential manner when magnetostrictive particles are added to the epoxy. This demonstrates that their presence introduce a piezomagnetic response to the composite which is not observable in the matrix alone.

We now consider the shrinkage phenomenon while assuming the presence of no macro-scale strain ($\overline{\overline{\boldsymbol{\varepsilon}}} = \mathbf{0}$), and cure under a magnetic field that is linearly increasing from $\mathbb{h}_3$ =0 A m^{-1} at t =0 s to $\mathbb{h}_3$ =100 kA m^{-1} at t =50 s. Figure 9.7 presents the time-

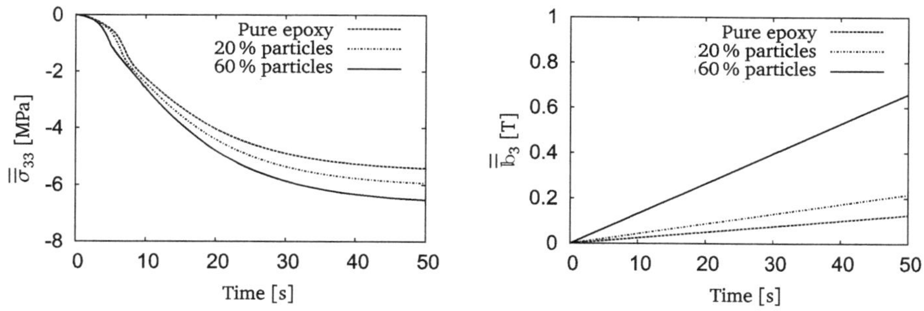

(a) Normal stress.

(b) Magnetic induction.

Figure 9.7: Evolution of macro-scale quantities during the progression of matrix curing, accounting for the shrinkage effects. [213, fig. 10 (reproduced with permission)]

dependent normal stress and magnetic components aligned with the magnetic field for different ratios of the constituents. Compared to when no shrinkage is considered, accounting for this effect causes the stress to increase by two orders of magnitude without affecting the magnetic induction. The presence of the magnetisable particles produces both an increase in the compressive stress generated within the media and a proportional increase in its magnetic response.

Lastly, in order to provide an indication as to the significance of the magnetic effect with a high volume fraction of particles, Figure 9.8 shows the normal stress generated within the material of $\phi = 0.6$ particles with and without the presence of a linearly increasing magnetic field. Under the magnetic field, the increased material stiffness hinders the material's ability to shrink due to its increased stiffness, resulting in the generation of 50 % increase in the contractile stress.

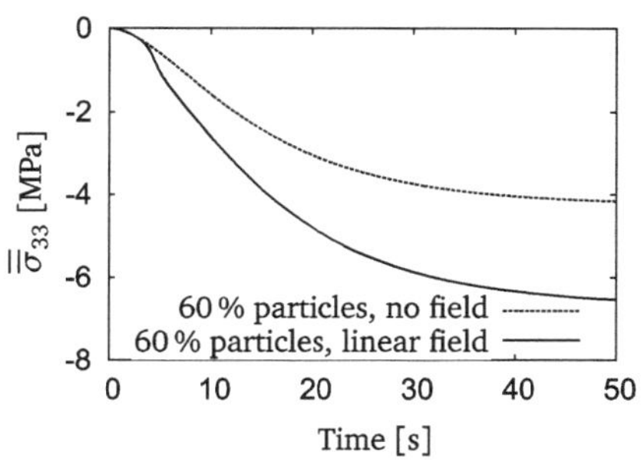

Figure 9.8: Influence of the magnetic field on the evolution of the macro-scale normal stress with time and as shrinkage strains increase. [213, fig. 11 (reproduced with permission)]

9.2 The stochastic finite element method

As has been discussed in Chapters 2 to 4, there exists a vast number of factors that influence the behaviour of MAPs, both in terms of its fundamental behaviour and elicited response during testing. Obvious determinants, such as the gross material type and loading scenario, influence the bulk response of the material. However, more subtle uncertainties are widely prevalent throughout both physical and numerical experimentation. From the computational perspective, input data such as material properties, the domain in terms of modelled internal material structure, and the applied boundary conditions and loading parameters all have some degree of uncertainty associated with them. Without further examination, it is sometimes difficult to quantify which of these nondeterministic parameters have a significant influence on the predicted and examined quantities of interest.

One method that can be used to quantify the influence of the spectrum of all input parameters is the Monte Carlo method. In this “brute force” approach, separate simulations are conducted to solve a problem individually for a set (ensemble) of model configurations (samples). Although it offers a guaranteed convergence on the mean values and variation of any measured quantity, when high accuracy is required it is prohibitively expensive. This issue is compounded when higher-order accuracy of a measured quantity is required, and any errors associated with the numerical approach itself must also be considered in a probabilistic manner [110]. Perturbation methods, which forward propagate uncertainties by determining the sensitivity of the solution and post-processed data with respect to some variables, have also found success in analysing the stochastic properties of homogenised media [264, 11, 465]. They are, however, limited to scenarios in which the configuration of the primary system is close to the mean state of the system, and in which the perturbations are small. Additionally, there exist some higher-order formulations of the perturbation method [464, 247, 317].

An alternative to these approaches is the use of the stochastic Galerkin based stochastic finite element method (SFEM), which was pioneered by Ghanem and Spanos [166]. This approach is motivated by the fact that numerous simulation parameters are non-deterministic and are thus treated this way from the outset. With SFEM, it is possible to use only a single numerical simulation to interrogate the entire solution space that encompasses the stochastic parameters. However, both the problem complexity (in terms of the number of DoFs required to simulate the additional stochastic dimensions) and the sophistication of the numerical implementation are increased. Nevertheless, it has found application in the quantification of uncertainties in elastic fibre composites [298], defective elastic material that undergo fracture [116], two phase media [501], polycrystalline materials [328] and coupled media with material discontinuities [425, 426]. Methods to couple stochastic and deterministic models have also been developed [96].

Although computational homogenisation of random media can be achieved by analysing a large sample that collectively exhibits the average material properties [8], both the Monte Carlo method and SFEM allow for the consideration of ergodicity [96, 298, 116, 501, 328, 426], thereby allowing a prohibitively large sample to be substituted by a stochastic one. The averaged material response is considered equal to the ensemble average and, therefore, can be computed from the analysis of a large variation of representative samples with differing characteristics. It is with this in mind that we present in the following sections the formulation of the SFEM as used by Pivovarov and Steinmann [425, 426] in the analysis of both elastic and magnetoelastic media with statistically quantifiable heterogeneities.

9.2.1 Extension into stochastic dimensions

Before describing notation for the stochastic variables, we first introduce a similar description of the notation for the variables that we are by this point most familiar with. The Euclidean space $\mathbb{E}$ defines the set of points $\mathbf{X} \in \mathbb{R}^{\mathrm{d}}$ and has the Euclidean norm as its metric. We can thus define the Hilbert space $\mathscr{H}$ for the physical domain as that containing the inner product of functions over $\mathscr{B}_0 \subset \mathbb{E}$ with values lying on the real number line $\mathbb{R}$. Therefore, in $\mathscr{H}$ the scalar product of two position-dependent (physical) functions g_1, g_2 is

$$\mathscr{L}^2\left(g_1\left(\mathbf{X}\right), g_2\left(\mathbf{X}\right), \mathscr{B}_0\right) = \int_{\mathscr{B}_0} g_1\left(\mathbf{X}\right)\, g_2\left(\mathbf{X}\right)\, \mathrm{dV} \quad . \tag{9.82}$$

Taking this into account, and consulting Figure 9.9 as a navigational aid, we describe the stochastic domain $\mathscr{S}_\Theta$ in the stochastic space $\mathbb{S}$ that contains the set of independent random variables $\boldsymbol{\Theta} = \boldsymbol{\Theta}\left(\omega\right) \in \mathbb{R}^{n_\Theta}$. The parameter ω is known as an elementary event, with which we parameterise a collection of random or stochastic variables $\boldsymbol{\Theta} = \boldsymbol{\Theta}\left(\omega\right)$ that vary with some distribution. In practice, each stochastic variable $\Theta\left(\omega\right)$ will be used to govern the sampling value

$$\chi\left(\Theta\left(\omega\right)\right) = \mu_\chi + \sigma_\chi \Theta\left(\omega\right) \tag{9.83}$$

of some parameter $\chi\left(\omega\right)$, be it a material or geometric quantity, with a given mean value μ_χ and standard deviation σ_χ. In this manner, the parameters themselves are endowed with a probabilistic quality. For practical purposes, we will restrict $\Theta\left(\omega\right) \in [-3, 3]$ to truncate the Gaussian distribution. This truncation assists in the stabilisation of the computational implementation, and furthermore has a basis in physical observations of the variable parameters that will be examined later. For example, when describing the size of particles in a microstructure, this value has definite upper bounds due to the screening process of particulates during manufacture. Additionally, particles that are less than a certain size require a special mathematical treatment and so cannot be directly included in the model.

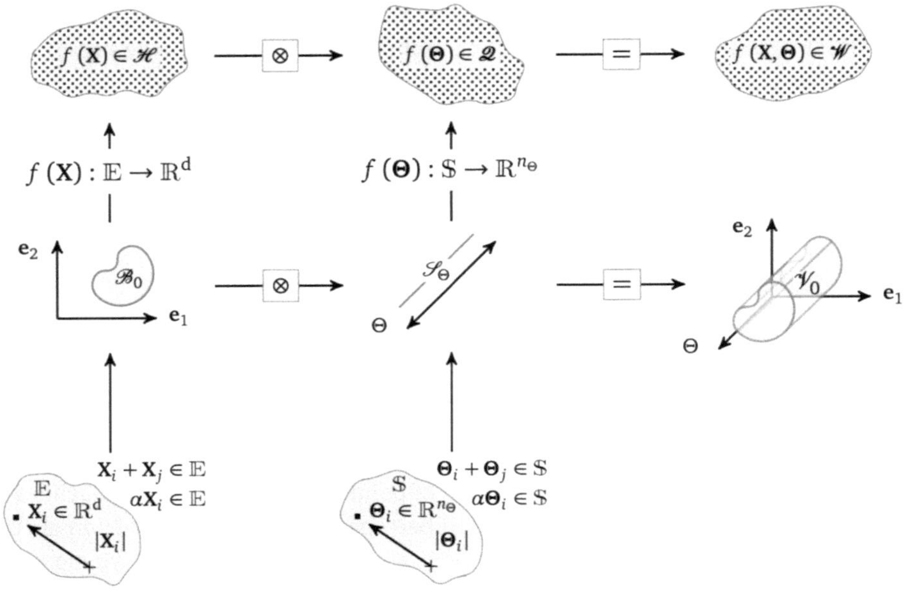

Figure 9.9: The physical, stochastic and product spaces and domains. Note that the stochastic space $\mathbb{S}$ of random variables $\boldsymbol{\Theta} = \boldsymbol{\Theta}(\omega)$ is itself a mapping from the underlying probability space of elementary events ω. [426, fig. 1]

We can then define $\mathscr{Q}$ as the inner product of functions over the stochastic domain $\mathscr{S}_\Theta \subset \mathbb{S}$ with values also found on $\mathbb{R}$. With some manipulation [425], it can be shown that in the Hilbert space $\mathscr{Q}$ the inner product of two probabilistic functions, once reparameterised in terms of the random variables $\boldsymbol{\Theta}$ into a stochastic function over the stochastic domain is

$$\mathscr{L}^2\left(g_1(\omega), g_2(\omega), \mathscr{S}_\Theta\right) = \int_{\mathscr{S}_\Theta} g_1(\boldsymbol{\Theta})\, g_2(\boldsymbol{\Theta}) f_\Theta \mathrm{d}\boldsymbol{\Theta} \quad . \tag{9.84}$$

Here, the Lebesgue integral over the probability space has been transformed to a Riemann integral over the stochastic domain, and $\boldsymbol{\Theta}$ are now truly considered independent of ω (that is to say, they are primary variables). The weighting function f_Θ is known as the joint probability density function of the independent basis variables $\boldsymbol{\Theta}$ and is illustrated in Figure 9.10. It is the mapping Jacobian between the stochastic and probability spaces. Due to the truncation of the Gaussian probability distributions, it is scaled according to

$$f_\Theta \to \frac{1}{p} f_\Theta \quad \text{where} \quad p := \int_{-3}^{3} f_\Theta \mathrm{d}\Theta \approx 0.9973 \tag{9.85}$$

and outside of the range $[-3, 3]$ has a value of zero.

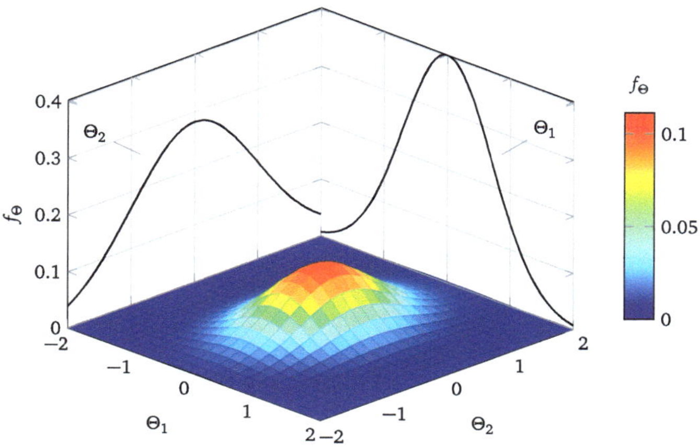

Figure 9.10: Joint probability density function for two independent Gaussian random variables. Although not illustrated here, the support of the Gaussian variables is limited to $\Theta_i(\omega) \in [-3, 3]$.

Combining these concepts, we now define the product domain $\mathscr{V}_0 := \mathscr{B}_0 \otimes \mathscr{S}_\Theta$ as product space of the physical and stochastic domains, with the corresponding Hilbert spaces $\mathscr{W} = \mathscr{H} \times \mathscr{Q}$. From this definition, and those given before, in $\mathscr{W}$ the inner product of two functions parameterised in terms of the physical and stochastic variables is

$$\mathscr{L}^2\left(g_1(\mathbf{X},\boldsymbol{\Theta}), g_2(\mathbf{X},\boldsymbol{\Theta}), \mathscr{V}_0\right) = \int_{\mathscr{B}_0}\int_{\mathscr{S}_\Theta} g_1(\mathbf{X},\boldsymbol{\Theta})\, g_2(\mathbf{X},\boldsymbol{\Theta}) f_\Theta \mathrm{d}\boldsymbol{\Theta}\ \mathrm{dV} \quad . \tag{9.86}$$

By extension, for a magneto-coupled problem this simply reads

$$\mathscr{L}^2\left(g_1(\mathbf{X},\Phi,\boldsymbol{\Theta}), g_2(\mathbf{X},\Phi,\boldsymbol{\Theta}), \mathscr{V}_0\right) = \int_{\mathscr{B}_0}\int_{\mathscr{S}_\Theta} g_1(\mathbf{X},\Phi,\boldsymbol{\Theta})\, g_2(\mathbf{X},\Phi,\boldsymbol{\Theta}) f_\Theta \mathrm{d}\boldsymbol{\Theta}\ \mathrm{dV} \quad . \tag{9.87}$$

In many respects, and as is depicted in Figure 9.9, the addition of the stochastic variables can be interpreted as an "extrusion" of the traditional spatial coordinate system into a higher dimensional space containing the random variables. This expansion of space leads the problem from having d number of spatial dimensions in $\mathscr{H}$ to $\mathrm{d} + n_\Theta$ dimensions in $\mathscr{W}$. By describing the governing equations in a general manner, we can more easily relate how the boundary value problem is posed when stochastic variables are also to be considered. In view of the structure to equations (5.149) and (5.159) (and equation (5.155) for the vector potential formulation which is not discussed further in this section), we can introduce a general differential operator $\mathbf{D}(\mathbf{X})$ and forcing function $\mathrm{f}(\mathbf{X},\Phi)$ such that

$$\mathbf{D}(\mathbf{X})\,\mathbf{u}(\mathbf{X},\Phi) = \mathrm{f}(\mathbf{X},\Phi) \quad . \tag{9.88}$$

where $\mathbf{u}(\mathbf{X}, \Phi)$ is the solution to the coupled boundary value problem. Collecting the FE discretisation in terms of the basis functions (equations (6.6), (6.15) and (6.22)) into a global vector

$$\mathbf{u}(\mathbf{X}, \Phi) \approx \sum_{I}^{n_{\Psi_\mathbf{u}}} \mathbf{u}^I \Psi_\mathbf{u}^I(\mathbf{X}, \Phi) \quad , \tag{9.89}$$

the discrete projection of the differential equation onto the finite element basis is

$$\mathscr{L}^2\left(\Psi_\mathbf{u}^I, \mathbf{D}(\mathbf{X})\,\mathbf{u}(\mathbf{X}, \Phi) - \mathrm{f}(\mathbf{X}, \Phi), \mathscr{B}_0\right) = 0 \quad \forall\, I \in \{1, \ldots, n_\Psi\} \quad . \tag{9.90}$$

When considering the solution to this general problem when numerous stochastic parameters are also considered, the above is modified such that

$$\mathbf{D}(\mathbf{X}, \boldsymbol{\Theta})\,\mathbf{u}(\mathbf{X}, \Phi, \boldsymbol{\Theta}) = \mathrm{f}(\mathbf{X}, \Phi, \boldsymbol{\Theta}) \quad \text{and} \quad \mathbf{u}(\mathbf{X}, \Phi, \boldsymbol{\Theta}) \approx \sum_{I}^{n_{\Psi_\mathbf{u}}} \mathbf{u}^I \Psi_\mathbf{u}^I(\mathbf{X}, \Phi, \boldsymbol{\Theta}) \quad , \tag{9.91}$$

and the solution vector itself is now also considered a random field. It therefore follows that the solution to the discrete problem satisfies

$$\mathscr{L}^2\left(\Psi_\mathbf{u}^I, \mathbf{D}(\mathbf{X}, \boldsymbol{\Theta})\,\mathbf{u}(\mathbf{X}, \Phi, \boldsymbol{\Theta}) - \mathrm{f}(\mathbf{X}, \Phi, \boldsymbol{\Theta}), \mathscr{V}_0\right) = 0 \quad \forall\, I \in \{1, \ldots, n_\Psi\} \quad . \tag{9.92}$$

What remains is to choose the basis functions $\Psi_\mathbf{u}^I(\mathbf{X}, \Phi, \boldsymbol{\Theta})$ in the product space $\mathscr{W}$. By making different selections for the discretisation, various formulations for the SFEM with different qualities can be developed and assessed; this is presented in Section 9.2.3, after a brief diversion into how the stochastic variables can be used to define material uncertainties.

9.2.2 Defining material discontinuities using level set functions

Throughout the remainder of the section, we will refer to level set functions when describing the location of material interfaces. These multivariate functions describe hypersurfaces in multidimensional space (isocontours in three dimensions), and are used to distinguish the material properties associated with any coordinate $\mathbf{X}$ for a choice of $\boldsymbol{\Theta}$.

Figure 9.11a depicts a single level set function that is used to impress geometric information on a two-dimensional geometry. As the stochastic parameter Θ is varied, the area encapsulated in the level set function $z(\mathbf{X}, \Theta)$ changes. It can be inferred from the illustration that, in this case

$$z(\mathbf{X}, \Theta) > 0: \quad \text{matrix} \quad \text{and} \quad z(\mathbf{X}, \Theta) < 0: \quad \text{inclusion}$$

holds as a definition for any fixed value of Θ. By varying Θ, the sampling value χ can be interpreted as the radius $r \in [r^{\min}, r^{\max}]$ of a centred circular inclusion. However, as

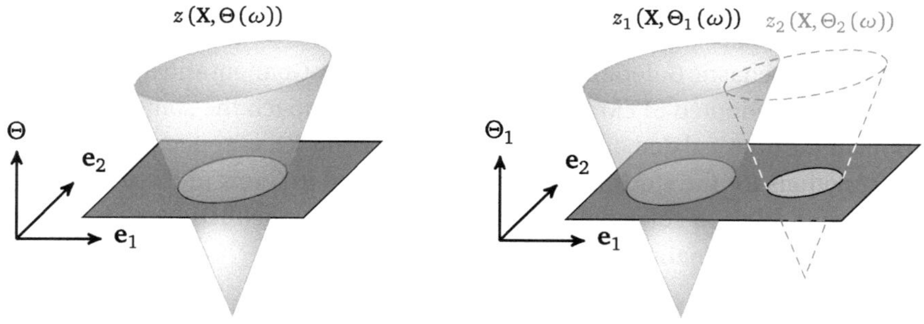

(a) A single linear level set function versus spatial coordinates.

(b) Merged level set function versus spatial coordinates.

Figure 9.11: Level set approach for material discontinuities. [426, fig. 8]

will be detailed later, we will instead choose $z(\mathbf{X}, \Theta)$ to govern some material property that varies throughout the physical domain, through which the presence of a geometric heterogeneity will be implied.

The case where more than one level set function is used to delineate multiple material zones in the physical space (a multivariate uncertainty) is presented in Figure 9.11b. Typically the functions that are used to define material properties can only be expressed in terms of a single level set function and, therefore, the two functions used to define the geometry in Figure 9.11b could not be directly substituted into them. It can, however, be shown that an arbitrary number of level set functions that satisfy the necessary criteria for the presented later framework can be concatenated into a single governing level set function in the following manner:

$$z(\mathbf{X}, \boldsymbol{\Theta}) := \min_{\forall \mathbf{X} \in \mathscr{B}_0} \left(z_1(\mathbf{X}, \Theta_1), \ldots, z_p(\mathbf{X}, \Theta_p) \right) \equiv z(\mathbf{X}, \omega) \quad . \tag{9.93}$$

Note, however, that although in the graphic the level set functions no longer appear to be completely independent (they are parameterised in terms of a single elementary event ω), in the computational framework all combinations of parameters $\boldsymbol{\Theta}$ are indeed considered.

9.2.3 Basis function selection

In this section, we will provide a brief introduction into several useful and different implementations of the SFEM [425, 426] as applied to the modelling of uncertain material microstructures of MAPs.

Global stochastic basis

The spectral (or stochastic Galerkin) FEM, first proposed by Ghanem and Spanos [166], is the most widely adopted SFEM formulation. In this approach the standard FE discretisation for the physical domain, as presented in Section 6.1, is applied. In contrast, the basis functions for the stochastic domain are defined in terms of the polynomial chaos expansion [100], a collection of continuous, globally defined basis functions that possess the orthogonally property given by

$$\mathscr{L}^2\left(\overline{\boldsymbol{\Psi}}^I_{\boldsymbol{\Theta}}(\boldsymbol{\Theta}), \overline{\boldsymbol{\Psi}}^J_{\boldsymbol{\Theta}}(\boldsymbol{\Theta}), \mathscr{V}_0\right) = \mathscr{L}^2\left(\left[\overline{\boldsymbol{\Psi}}^I_{\boldsymbol{\Theta}}(\boldsymbol{\Theta})\right]^2, \mathscr{V}_0\right)\delta^{IJ} \quad . \tag{9.94}$$

The overall approximation space is therefore

$$\boldsymbol{\Psi}(\mathbf{X}, \Phi, \boldsymbol{\Theta}) \equiv \boldsymbol{\Psi}_{\mathbf{u}}(\mathbf{X}, \Phi, \boldsymbol{\Theta}) = \boldsymbol{\Psi}_{\varphi}(\mathbf{X}) \otimes \boldsymbol{\Psi}_{\Phi}(\mathbf{X}) \otimes \overline{\boldsymbol{\Psi}}_{\boldsymbol{\Theta}}(\boldsymbol{\Theta}) \quad . \tag{9.95}$$

that is, the tensor product of the basis functions discretising the physical and magnetic problem (each defined in the physical domain and having only local support) with the global basis functions with which the stochastic domain is approximated.

The primary benefit to this approach is that the integration over the physical and stochastic domains can be split. Therefore, numerically efficient and effective schemes can be independently chosen for both sub-problems. As a FE discretisation is only applied to the physical space, the usual restrictions on the choice of ansatz and numerical methods apply to this part of the problem. As the random variables $\boldsymbol{\Theta}$ are restricted to being Gaussian in nature, Hermitian polynomials (shown in Figure 9.12a) can be used as the global basis functions in the stochastic domain. When combined with a suitably truncated integration scheme (e. g. Gauss–Hermite with modifications applied due to the truncation of the stochastic variables), this renders significantly increased solution accuracy. A drawback that was noted is that for problems which included discontinuities, such as the inclusions that represent CIP particles, the method failed. This was overcome by regularising the discontinuity through the introduction of a smoothed material domain transition region. However, Pivovarov and Steinmann [425] observed oscillations in the solution field at the extents (and particularly the "corners") of the stochastic space, which are attributed to Runge's phenomenon. Consequently, Pivovarov and Steinmann [425] proposed using trigonometric basis functions, namely the quasi-Fourier basis (shown in Figure 9.12b) derived from the fundamental sequence

$$\overline{\Psi}^*_{\boldsymbol{\Theta}}(\Theta) = \left\{1, \sin\left(\frac{2n-1}{2}\frac{\pi}{3}\Theta\right), \sin\left(\frac{2n-1}{2}\frac{\pi}{3}\Theta\right)\Theta\right\} \quad , \tag{9.96}$$

where $n \in \mathbb{Z}^+$. The developed basis was orthonormalised using the Gram–Schmidt process. By employing these stochastic basis functions, exponential convergence of this method was observed with respect to the reference solution, and the approximation on the stochastic boundary was improved.

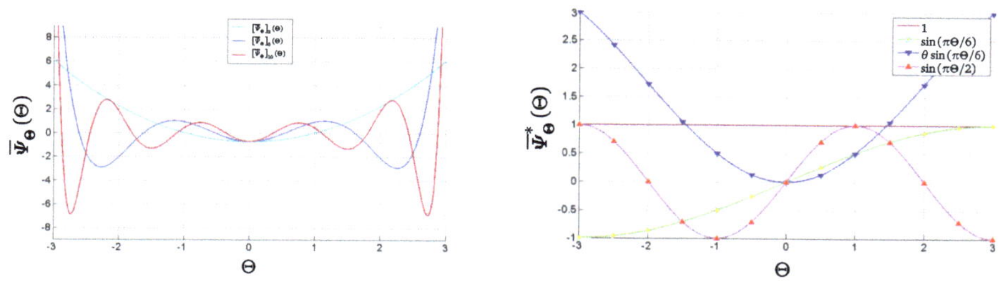

(a) A selection of Hermite polynomials limited to the sampling range of the Gaussian variables.

(b) A selection of quasi-Fourier function sequence limited to the sampling range of the Gaussian variables.

Figure 9.12: A selection of global shape functions specialised for the truncated stochastic domain. [425, figs. 14c,29e (reproduced with permission)]

Local stochastic basis

An alternative to the use of global basis function for the stochastic domain was suggested by Deb [109], and Babuška et al. [25, 26]. Therein, it was proposed that a FE discretisation be applied to both the physical and stochastic domains, which would admit the use of advanced FEM methods, such as mesh and polynomial adaptivity, in both spaces. There are two manners in which this can be implemented, namely using the tensor product of basis functions as discussed in [404, 403], or using isoparametric shape functions as explored in [110].

Each have their own benefits and disadvantages, but the latter will be the focus of further discussion. It is easily accommodated by an extension to the approach described in Section 6.1 and its evaluation discussed in Section 6.2. To do so, it is necessary to introduce a FE discretisation to the stochastic variables thereby extending the isoparametric domain into the stochastic space. Now the ansatz for the physical and stochastic dimensions are

$$\mathbf{X} \approx \mathbf{X}^{\mathrm{h}} = \sum_{I}^{n_{\Psi_{\mathbf{X}}}} X^{I} \boldsymbol{\Psi}_{\boldsymbol{\varphi}}^{I}\left(\mathbf{X}^{\xi}\right) \quad \text{and} \quad \boldsymbol{\Theta} \approx \boldsymbol{\Theta}^{\mathrm{h}} = \sum_{I}^{n_{\Psi_{\boldsymbol{\Theta}}}} \Theta^{I} \boldsymbol{\Psi}_{\boldsymbol{\Theta}}^{I}\left(\mathbf{X}^{\xi}\right) \tag{9.97}$$

such that each coordinate $\mathbf{X}^{\xi} \in \Omega_{\square}$ now has dimension

$$\dim\left(\mathbf{X}^{\xi}\right) = \dim\left(\mathbf{X}\right) + \dim\left(\boldsymbol{\Theta}\right) \quad . \tag{9.98}$$

The similarity between this method and traditional FE methods hints at an obvious advantage from an implementational perspective. Furthermore, efficient treatment of scenarios involving material discontinuities and stress concentrations can be performed by generating a conforming domain throughout $\mathscr{V}_0$ and by using h-adaptivity. The computational cost of such an implementation can be reduced through the application of n-dimensional serendipity elements [16], which only require lower-order quadrature rules for their evaluation [222, 426]. It is important to note, however, that during integration shape function derivatives are only taken with respect to the physical coordinates. Additionally, all integrals must include the (truncated) joint probability density function $f_{\Theta}(\boldsymbol{\Theta})$ as a weighting factor.

Extended FEM framework

As was previously mentioned, the introduction of material discontinuities provides some challenges for modelling within an SFEM framework. So far two approaches to incorporating these sharp feature changes have been mentioned, namely the use of a blending function to regularise the description of spatially-dependent material property, or the creation of a finite element mesh that is conformal in both the physical and stochastic spaces. These two methods trade off between the ease of meshing the (high-dimensional) domains and the accuracy to which material discontinuities can be resolved. The extended stochastic FE method, developed by Nouy et al. [404] and used further in [403, 498], introduces a third alternative wherein the higher-dimensional Cartesian mesh can still be used while resolving the location of material discontinuities, and their influence on the solution, with high precision.

The extended SFEM involves the local enrichment of the FE basis functions with a set of non-smooth, and potentially discontinuous, enrichment functions. Introducing a single enrichment function $F = F\left(\mathbf{X}^{\xi}\right)$ localised on an individual finite element, the approximation space on that element e is

$$\mathbf{u}^{\mathrm{e}}(\mathbf{X}, \Phi, \boldsymbol{\Theta}) = \sum_{I}^{n_{\Psi_{\mathbf{u}}}} \left[\mathbf{u}^{I} + \mathbf{a}^{I} F\left(\mathbf{X}^{\xi}\right)\right] \Psi_{\mathbf{u}}^{I}\left(\mathbf{X}^{\xi}\right) \tag{9.99}$$

where **a** are additional solution coefficients associated with the enrichment of **u**. Using the definition put forward in [385], the enrichment function is defined as

$$F\left(\mathbf{X}^{\xi}\right) = \sum_{J}^{n_{\Psi_{\mathbf{u}}}} |z^{J}| \Psi_{\mathbf{u}}^{I}\left(\mathbf{X}^{\xi}\right) - \left|\sum_{J}^{n_{\Psi_{\mathbf{u}}}} z^{J} \Psi_{\mathbf{u}}^{I}\left(\mathbf{X}^{\xi}\right)\right| \tag{9.100}$$

where z^{J} represents the values of level set functions evaluated at the support points. As mentioned in Section 9.2.2, these level set functions will be defined in such a way that they describe the location of the material discontinuities throughout the physical and stochastic spaces.

We note that this choice of enrichment function [151] is non-unique, with alternatives based on Heaviside step functions being employed in [286, 287]. Due the local properties of the enrichment function, only elements containing the material discontinuities have additional DoFs associated with them, leading to a minimal computational cost. That said, the introduction of the discontinuous basis requires that numerical integration occurs on each side of the discontinuity, therefore necessitating the subdivision of the Cartesian-aligned computational grid along the material interface. Two approaches to achieve this are division into simplexes (a higher-dimensional Delaunay triangulation) and regular hypercube decomposition. The computational cost attached to these methods depends on the required resolution of the material interface and the number of stochastic dimensions. In later results only the simplex-splitting method is applied, and a generalised Gauss quadrature rule is used to integrate over these simplexes.

Numerical examples: SFEM with uni- and multi-variate uncertainties

By way of an introduction to the configuration of an SFEM problem, we first examine a uni- and multi-variate problem in the context of finite strain elasticity using global basis functions. Therein it will be demonstrated how, in a single numerical experiment, an entire space of possible material responses can be captured. This is followed by an application in magneto-mechanics wherein a comparison is made between results obtained using an assortment of basis function definitions. In all examples, the physical domain of the RVE is wholly encapsulated within the two-dimensional unit square, and periodic boundary conditions are always enforced.

Elasticity

As was mentioned in Section 9.2.2, we can employ a level set function to define a material property that varies throughout a domain, thereby introducing a geometric heterogeneity (and also uncertainty) into the problem. In the spirit of equation (8.13), we will therefore define the shear modulus for the elastic material by

$$\mu(\mathbf{X}, \boldsymbol{\Theta}) = \mu_m + \frac{1}{2}\left[\mu_p - \mu_m\right]\left[1 - \tanh\left(kz\left(\mathbf{X}, \boldsymbol{\Theta}\right)\right)\right] \tag{9.101}$$

where μ_m and μ_p represent the matrix and particle stiffness, and k is a parameter that controls the sharpness of the transition between the matrix and inclusion. In this example, the choices for the different material parameters, valid for the Neo–Hookean type coupled constitutive law state in equation (9.51), are listed in Table 9.3.

Table 9.3: Fictitious constitutive parameters for numerical examples demonstrating the influence of uni- and multi-variate uncertainties on an elastic RVE response.

Matrix		**Inclusion**	
Parameter	**Value**	**Parameter**	**Value**
μ_m	0.4	μ_p	2
ν_m	0.25	ν_p	0.25

The two geometric uncertainties that will be examined, which only have two physical space dimensions represented, are illustrated in Figure 9.13. For both examples, the macro-scale deformation gradient that is applied is given by

$$\overline{\overline{\mathbf{F}}} = \begin{bmatrix} 1.25 & 0 \\ 0 & 1 \end{bmatrix} .$$

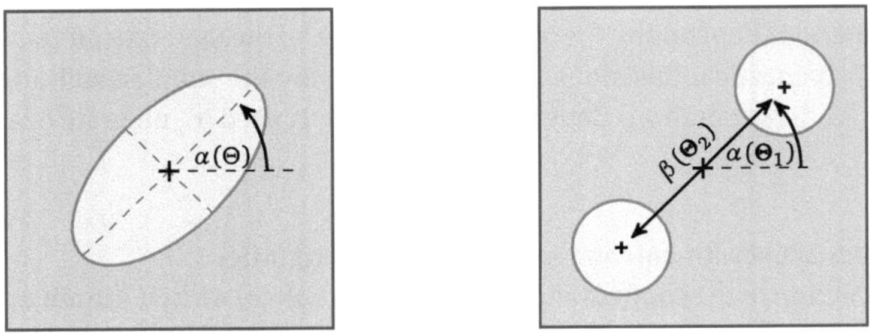

(a) A univariate uncertainty (orientation of major axis).

(b) A multivariate uncertainty (inclusion distance and orientation).

Figure 9.13: Examples of geometric uncertainties in materials with inclusions.

Comparison of the predicted material response will be made to results obtained by Monte Carlo simulation; that is, by exhaustively sampling the parameter space by means of individual numerical simulations.

As can be inferred from Figure 9.13a, the univariate uncertainty to be modelled is that of the orientation angle of the major axis of an elliptical inclusion (with its major and minor radii given as $a_1 = 0.5, a_2 = 0.32$). We will consider the case when the particle's orientation angle with respect to the $\mathbf{e}_1$ basis $\alpha(\Theta) \in [0, \frac{\pi}{2}]$ is uncertain, so with reference to equation (9.83) its mean value and standard deviation are $\mu_\alpha = \frac{\pi}{4}$ and $\sigma_\alpha = \frac{\pi}{12}$, respectively. The level set function that defines whether a point $\mathbf{X}$ lies within or outside of the inclusion is

$$z(\mathbf{X}, \Theta) = \sqrt{a_1 a_2} \left[\sqrt{\frac{\widetilde{X}_1^2}{a_1^2} + \frac{\widetilde{X}_2^2}{a_2^2}} - 1 \right] \quad , \tag{9.102}$$

where the pseudo-position vector $\widetilde{\mathbf{X}}$ is

$$\widetilde{X}_1 = X_1 \cos(\alpha(\Theta)) + X_2 \sin(\alpha(\Theta)) \quad , \quad \widetilde{X}_2 = -X_1 \sin(\alpha(\Theta)) + X_2 \cos(\alpha(\Theta)) \quad .$$

Figure 9.14 shows the predicted volume-average von Mises stress for the RVE as computed using the Monte Carlo method and SFEM with global quasi-Fourier basis functions. It can be observed that the trend for the average stress over the range of the parameter space is, for all intents and purposes, identically predicted by both approaches. Note that from the graphs one can discern both the stress state of the underlying geometry as well as the statistical probability that a RVE with such a stress condition will be measured; the probability distribution differs only due to the number of sampling points. Not only do the results demonstrate that the SFEM provides a valid approach to modelling statistically varying characteristics of materials but, since the probabilities are embedded within the stochastic space, it also highlights how rich a dataset can be extracted from a single numerical simulation. By comparison, the Monte Carlo method not only requires that numerous simulations be conducted over

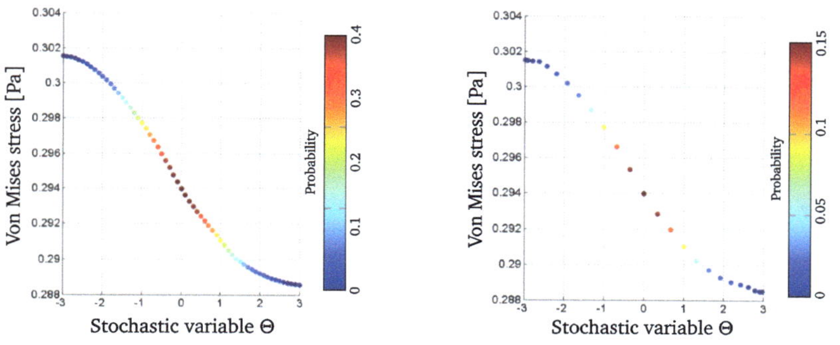

(a) Monte-Carlo. **(b)** SFEM.

Figure 9.14: Homogenised von Mises stress versus the stochastic variable Θ for a univariate geometric uncertainty that dictates the orientation angle of an elliptical inclusion. The colour of the points corresponds to the probability that the interrogated state is observed in the RVE. [425, fig. 43 (reproduced with permission)]

the entire set of permutations for the stochastic variables, but the numerical data also need be collated and subsequently post-processed to compute the probability distribution.

To increase the complexity of the examined problem we next consider a case where there exists a multivariate uncertainty. We will consider a situation in which the distance from the origin $\beta(\Theta_p)$ and orientation angle $\alpha(\Theta_p)$ of a symmetric pair of circular inclusions of constant radius $R = 0.2$ is unknown. The level set function governing the location of each inclusion p is

$$z_p(\mathbf{X}, \Theta_p) = R\left[\sqrt{\frac{[\widetilde{X}_p]_1^2}{R^2} + \frac{[\widetilde{X}_p]_2^2}{R^2}} - \sqrt{\frac{1}{2}}\right] \tag{9.103}$$

where the pseudo-position vector $\widetilde{\mathbf{X}}$ is

$$[\widetilde{X}_p]_1 = [X_p]_1 + s_p\beta(\Theta_p)\cos(\alpha(\Theta_p)) \quad , \quad [\widetilde{X}_p]_2 = [X_p]_2 + s_p\beta(\Theta_p)\sin(\alpha(\Theta_p)) \tag{9.104}$$

and the factor that determines the quadrant in which the inclusion lies is $s_p = \{-1, +1\}$. The particles' orientation angle is $\alpha(\Theta) \in [0, \frac{\pi}{2}]$, with the same mean value and standard deviation as in the previous example. The distance between the particles $\beta(\Theta) \in [0.125, 0.575]$ with a respective mean value and standard deviation of $\mu_p = 0.35$ and $\sigma_p = 0.075$.

The volume average von Mises stress distribution for the range of parameter distribution is presented in Figure 9.15. The complex non-linear landscape of the both the RVE stress response and its statistical probability of occurrence was, again, accurately rendered with the SFEM when compared to the exhaustive and computationally expensive Monte-Carlo approach.

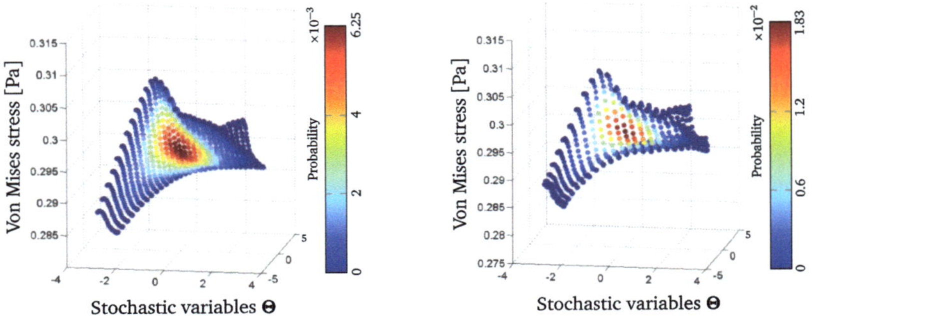

(a) Monte-Carlo. **(b)** SFEM.

Figure 9.15: Homogenised von Mises stress versus the stochastic variables **Θ** for a multivariate geometric uncertainty that describes the orientation of two particles and the distance between them. The colour of the points corresponds to the probability that the interrogated state is observed in the RVE. [425, fig. 45 (reproduced with permission)]

Magneto-mechanics

For the magneto-mechanical problem, we will consider an RVE that is more closely aligned to the material as described in Section 2.2. In such a case, not only is the particle size given by the experimentally-consistent log-normal distribution, illustrated in Figure 2.4b, but so is the volume of its enclosing Voronoi bounding box when the presence of multiple particles in a sufficiently large isotropic sample are taken into account. Transforming the geometrically consistent problem to one defined on the unit square, the analytical expression for the radius is not given by equation (9.83), but rather by

$$R(\Theta) = \exp\left[\mu_R + \sigma_R \Theta(\omega)\right] \quad . \tag{9.105}$$

Limiting the $3 - \sigma$ particle radius to $R(\Theta) \in [0.25, 0.6]$, the mean value and standard deviation are then $\mu_R = \frac{1}{2}\log(0.15)$ and $\sigma_R = \frac{1}{6}\log(2.4)$.

As the solution of a periodic RVE with one circular inclusion is independent of the inclusion's location, the position of the particle may remain fixed. A single particle centred at the origin with varying radius $R(\Theta)$ can be described by the level set function

$$z(\mathbf{X}, \Theta) = R(\Theta)\left[\sqrt{\frac{X_1^2}{R(\Theta)^2} + \frac{X_2^2}{R(\Theta)^2}} - 1\right] \quad . \tag{9.106}$$

The area enclosed within the particle, as superimposed on a conformal and Cartesian mesh, or the given range of possible particle radius is thus illustrated in Figure 9.16.

In addition to equation (9.101), the spatially- and stochastically-dependent magnetic permeability is defined by

$$\upsilon(\mathbf{X}, \boldsymbol{\Theta}) = \upsilon_{\mathrm{m}} + \frac{1}{2}\left[\upsilon_{\mathrm{p}} - \upsilon_{\mathrm{m}}\right]\left[1 - \tanh\left(kz(\mathbf{X}, \boldsymbol{\Theta})\right)\right] \tag{9.107}$$

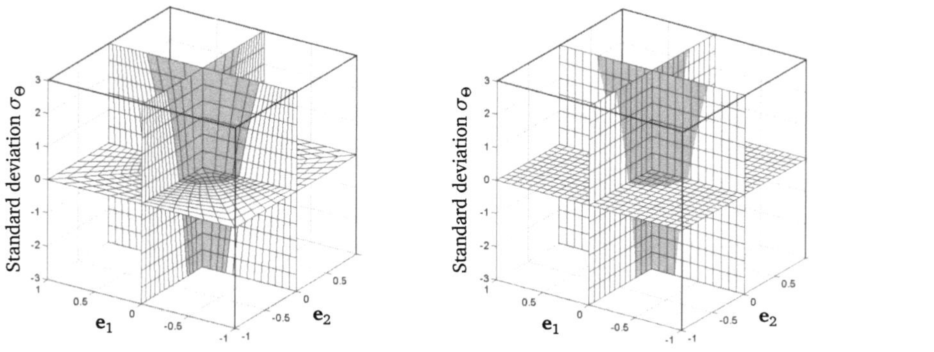

(a) SLFEM sharp (conforming). **(b)** SXFEM; SLFEM smooth (uniform).

Figure 9.16: Three-dimensional (two physical and one stochastic dimension) meshes used with different SFEM methods. SLFEM refers to SFEM with local basis functions and SXFEM to the extended SFEM. [426, figs. 17, 18 (reproduced with permission)]

where the matrix and particle magnetic permeabilities are governed by υ_m and υ_p, and k is the transition parameter. The material parameters that will be utilised in the following example are listed in Table 9.4, while the mechanical and magnetic loading is fixed as

$$\overline{\overline{\mathbf{F}}} = \begin{bmatrix} 1.1 & 0 & 0 \\ 0 & 1 & 0 \\ 0 & 0 & 1 \end{bmatrix} \quad , \quad \overline{\overline{\mathbb{H}}} = \begin{Bmatrix} 40 \\ 0 \\ 0 \end{Bmatrix} \ .$$

Note that the definition of the basis functions incorporates an extra component accounting for the extra stochastic dimension.

Table 9.4: Fictitious constitutive parameters for numerical examples demonstrating the influence of uncertainties on the response of an magnetoelastic RVE. For conformal meshes, the material transition parameter in equation (9.101) is $k = \infty$, while for fully Cartesian meshes demanding a smooth transition $k = 37$.

Matrix		**Inclusion**	
Parameter	**Value**	**Parameter**	**Value**
μ_m	8	μ_p	80
ν_m	0.3	ν_p	0.3
υ_m	0.001	υ_p	0.01

Figure 9.17a presents the reference solution as computed by the Monte Carlo method for the range of the particle size distribution. As is expected, associated with an increasing inclusion volume fraction is an increase in the average stress magnitude for

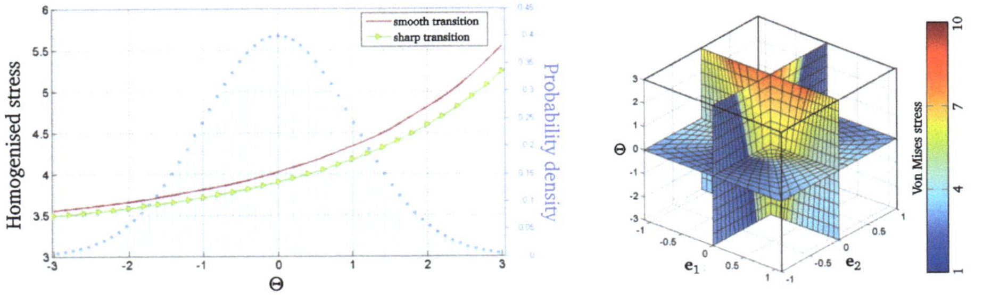

(a) Homogenised stress response predicted using Monte Carlo simulations for material parameters described using sharp and smooth transitions.

(b) Spatial von Mises stress plotted along the coordinate planes for SLFEM with a conformal mesh.

Figure 9.17: Comparison of the stress response for a simulations of an MAP with a stochastic inclusion radius. [426, figs. 21, 24b (reproduced with permission)]

the fixed deformation and magnetic fields. This is observable in Figure 9.17b, where the local stress generated as the particle size increases ($\Theta \to 3$) increases considerably.

With the level set function chosen such that $\phi = 25\,\%$, Figure 9.18 shows the contours of a stress component plotted on a cross-section of the physical domain. Each approach renders a qualitatively similar solution, although one may observe fluctuations in the post-processed stress when a gradual transition between the matrix and inclusion is modelled. It is clear that by using a conformal mesh (and a sharp transition parameter) leads to a solution with highest fidelity at the material interface, although this is nearly matched in accuracy by the extended SFEM even though the mesh is Cartesian aligned. Although dependent on which quantities were being compared, it was determined [426] that the best convergence characteristics for the magnetoelastic

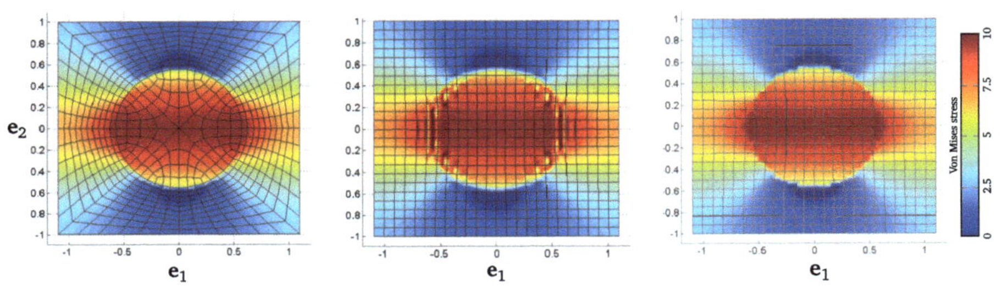

(a) SLFEM sharp transition (conforming).

(b) SLFEM smooth transition (Cartesian).

(c) SXFEM (Cartesian).

Figure 9.18: Normalised micro-scale Piola stress distribution ($\mathbf{P}_{11}$) over the physical domain, as predicted by the various SFEM methods. Although perturbations in the visualisation exist due to the visualisation process, the oscillations visible in Figure 9.18b are primarily a result of the use of a smooth transition. A comparative result can be found in [234]. The radius of the inclusion is fixed at 0.564. [426, figs. 25, 26, 27 (reproduced with permission)] problem (with respect to the reference solution) were typically exhibited by the SFEM with local basis functions and a conformal mesh. For a similar simulation complexity, it predicted the mean value of the homogenised stress with several orders higher accuracy than SFEM with a local basis and a smooth transition. Although the extended SFEM provides a lesser accuracy than when the local basis functions are employed, its performance is generally superior to that of SFEM with global basis functions.

10 Modelling and computational simulation at the micro-scale

There are numerous numerical and practical challenges to overcome when simulating complex composites at the micro-scale using FE analysis. As an introduction to these topics, within this chapter we will demonstrate the validation of a numerical implementation of the magnetostatic problem. Thereafter, we will explore what might be some of the good practices to employ when conducting FEM simulations of these MAP composite microstructures; these will then be considered in the studies that follow. The predominant theme will be to bring together some of the core concepts presented in previous chapters, for example, the constitutive modelling of different material components in the composite, understanding its force generation properties, as well as its homogenised material properties and effective response when considering interesting and relevant microstructural arrangements.

10.1 Single particle representative volume element problem

As a precursor to conducting complex microstructural simulations, the analysis of a RVE with a single inclusion can provide significant insight in terms of both the physical and numerical behaviour of the coupled problem. We therefore first consider the generic configuration of a two-dimensional RVE encapsulating a single particle, as is shown in Figure 10.1. The square RVE has side length L and, for later convenience, has the $\mathbf{e}_1$-aligned surfaces divided into sections of equal length. It is predominantly composed of matrix material that is denoted as $\mathscr{B}_0^1$. The elliptical inclusion $\mathscr{B}_0^2$, with axes of dimension a_1 and a_2, is offset from the centre of its bounding media by length l_0 and rotated by an angle θ with respect to the $\mathbf{e}_1$ basis.

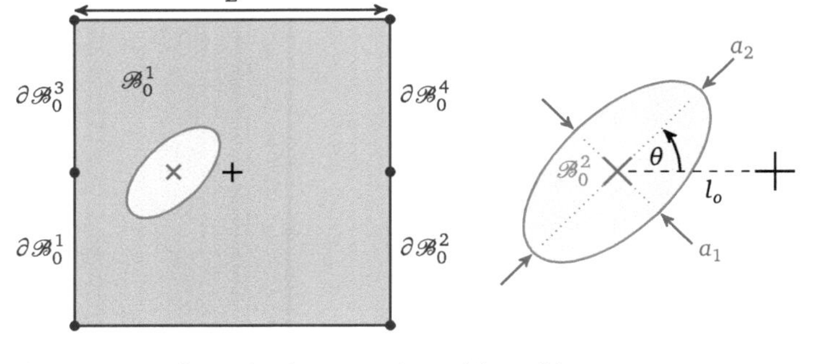

Figure 10.1: Configuration for a generic particle problem.

https://doi.org/10.1515/9783110418576-010

10.1.1 Solution accuracy and finite element discretisation

The most meaningful simple scenario that can be assessed is that of a circular or spherical particle with radius $a_1 = a_2 = R$ centred within an infinite medium. More particularly, we will present the analytical solution in two dimensions (equally valid in three dimensions when the particle is an extrusion of circular cross-section) to the problem of a magnetostatic, linearly magnetisable particle immersed in a matrix of infinite size with a uniform magnetic field aligned in the $\mathbf{e}_1$ direction. The fundamentals of the potential theory used to derive the solution have been presented in detail by Griffiths [180], and Durst [130], among others.

10.1.1.1 Analytical solution

The analytical solution is derived from the characteristic solutions to the two-dimensional, polar expression of Laplace's equation

$$\frac{1}{r}\frac{\partial}{\partial r}\left[r\frac{\partial \Phi}{\partial r}\right] + \frac{1}{r^2}\frac{\partial^2 \Phi}{\partial \theta^2} = 0 \tag{10.1}$$

with $\theta \in [0, 2\pi]$ being periodic and where $r(\mathbf{X}) = |\mathbf{X}|$ is the distance from the particle centre (i. e. the origin), and $\theta(\mathbf{X}) = \arctan\left(\frac{X_2}{X_1}\right)$ is the inclination of the coordinate vector from the $\mathbf{e}_1$ basis. From the general solutions to this equation, the scalar potential both inside and outside the origin-centred particle can be determined when considering the kinematic and continuity conditions

$$\Phi^{\mathrm{m}} = \Phi^{p} \quad \text{at} \quad r = R \quad , \tag{10.2a}$$

$$\mu_{\mathrm{r}}^{\mathrm{m}}\frac{\partial \Phi^{\mathrm{m}}}{\partial r} = \mu_{\mathrm{r}}^{p}\frac{\partial \Phi^{p}}{\partial r} \quad \text{at} \quad r = R \quad , \tag{10.2b}$$

the far-field condition

$$\Phi^{\mathrm{m}} \to -H_{\infty} r \cos(\theta) \quad \text{if} \quad r \gg R \tag{10.2c}$$

that ensures that the magnetic field far away from the particle is uniform, and by assuming the symmetry condition

$$\Phi\left(r, \frac{\pi}{2}\right) = \Phi\left(r, \frac{3\pi}{2}\right) = 0 \quad . \tag{10.2d}$$

Here, the strength of the magnetic field infinitely far away from the particle is denoted by $H_{\infty} = |\mathbb{H}_{\infty}|$, and we assume that $\mathbb{H}_{\infty}$ is aligned with $\mathbf{e}_1$. Considering the above, the solution for a cylindrical particle is

$$\Phi(\mathbf{X}) = \begin{cases} \Phi^{p} = C_1 r \cos(\theta) & \text{if} \quad r \leq R \\ \Phi^{\mathrm{m}} = \left[-H_{\infty} + \frac{C_2}{r^2}\right] r \cos(\theta) & \text{otherwise} \end{cases} \quad , \tag{10.3}$$

for which the two coefficients are

$$C_1 = -H_\infty \frac{2}{1+\widehat{\mu_\mathrm{r}}} \quad \text{and} \quad C_2 = R^2\left[-H_\infty - C_1\widehat{\mu_\mathrm{r}}\right] \quad ,$$

and $\widehat{\mu_\mathrm{r}} = \mu_\mathrm{r}^p/\mu_\mathrm{r}^\mathrm{m}$ is the ratio between the magnetic permeability of the particle and the surrounding matrix. Applying the chain rule, the gradient of the analytical solution, given in Cartesian coordinates, is therefore

$$\nabla_0\Phi = \begin{cases} \left[\dfrac{\partial\Phi^p}{\partial r}\dfrac{\partial r}{\partial X_1} + \dfrac{\partial\Phi^p}{\partial\theta}\dfrac{\partial\theta}{\partial X_1} \quad , \quad \dfrac{\partial\Phi^p}{\partial r}\dfrac{\partial r}{\partial X_2} + \dfrac{\partial\Phi^p}{\partial\theta}\dfrac{\partial\theta}{\partial X_2}\right]^T & \text{if} \quad r \le R \\ \left[\dfrac{\partial\Phi^\mathrm{m}}{\partial r}\dfrac{\partial r}{\partial X_1} + \dfrac{\partial\Phi^\mathrm{m}}{\partial\theta}\dfrac{\partial\theta}{\partial X_1} \quad , \quad \dfrac{\partial\Phi^\mathrm{m}}{\partial r}\dfrac{\partial r}{\partial X_2} + \dfrac{\partial\Phi^\mathrm{m}}{\partial\theta}\dfrac{\partial\theta}{\partial X_2}\right]^T & \text{otherwise} \quad . \end{cases} \tag{10.4}$$

The derivatives of the potential field inside and outside of the particle with respect to the polar coordinates are

$$\frac{\partial\Phi^p}{\partial r} = C_1\cos(\theta) \quad , \quad \frac{\partial\Phi^p}{\partial\theta} = -C_1 r\sin(\theta) \quad ,$$
$$\frac{\partial\Phi^\mathrm{m}}{\partial r} = \left[-H_\infty - \frac{C_2}{r^2}\right]\cos(\theta) \quad , \quad \frac{\partial\Phi^\mathrm{m}}{\partial\theta} = \left[H_\infty - \frac{C_2}{r^2}\right] r\sin(\theta) \quad ,$$

and the partial derivatives of the radial coordinates with respect for the Cartesian directions are

$$\frac{\partial r}{\partial X_1} = \frac{X_1}{r} \quad , \quad \frac{\partial\theta}{\partial X_1} = \frac{-X_2}{X_1^2[1+\frac{X_2^2}{X_1^2}]} \quad \text{and} \quad \frac{\partial r}{\partial X_2} = \frac{X_2}{r} \quad , \quad \frac{\partial\theta}{\partial X_2} = \frac{1}{X_1[1+\frac{X_2^2}{X_1^2}]} \quad .$$

An example of the two-dimensional analytical solution to the differential equations can be seen in Figure 10.2. A similar derivation can be performed for the three-dimensional case. Through application of the constitutive law, the resulting ponderomotive force and torque can be computed.

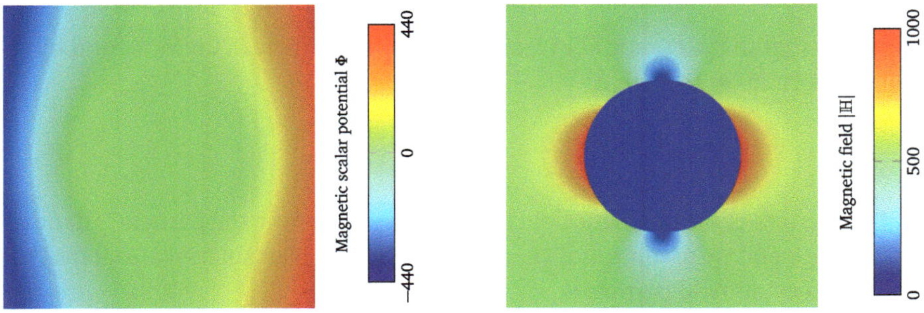

(a) Magnetic scalar potential field. **(b)** Magnetic field.

Figure 10.2: Truncated view of the analytical solution to the 2-D problem. The parameters to the problem were chosen as follows: $H_\infty = 500\ \mathrm{A\,mm^{-1}}$, $R = 0.5$ mm, $\mu_\mathrm{r}^\mathrm{m} = 1$, and $\mu_\mathrm{r}^p = 5000$.

Having an analytical solution allows one to (i) verify the correctness of the implementation of the FE problem (and post-processing routines, such as the computation of magnetic forces and torques as is discussed in [533]), and (ii) validate the implementation through analysis of the error associated from the FE discretisation. Due to the symmetry of the problem, the total magnetic force generated by the particle is zero, and so to quantify the force-generation aspects of a magnetisable particle we will, in Section 10.1.2, consider more complex scenarios. A brief digression into the subject of FE analysis is therefore made in order to investigate some points that should be considered when performing micro-structural studies.

10.1.1.2 Error analysis

Comparison of the solution to a boundary value problem, as computed using FEM, with its analytical solution provides the opportunity to probe the errors associated with the FE discretisation and chosen ansatz. For this purpose, we will consider two measures of importance [579, 444], namely the energy error associated with the discrete solution

$$\epsilon = \left[\int_{\mathscr{D}_0^{\mathrm{h}}} \left[\mathbb{H} - \mathbb{H}^{\mathrm{h}}\right] \mu_{\mathrm{r}} \mu_0 \left[\mathbb{H} - \mathbb{H}^{\mathrm{h}}\right] \, \mathrm{dV} \right]^{\frac{1}{2}} \tag{10.5}$$

with $\mathbb{H} = -\nabla_0 \Phi$ where $\Phi \in \mathscr{H}^m(\mathscr{D}_0)$ is the exact solution that lies in the Sobolev space[1] of order $m \geq 1$ [22, 444]. As such, the $\mathscr{H}^1$-error of the discrete solution is

$$\|\Phi - \Phi^{\mathrm{h}}\|_{\mathscr{H}^1} = \left[\int_{\mathscr{D}_0^{\mathrm{h}}} \left[|\Phi - \Phi^{\mathrm{h}}|^2 + |\nabla_0 [\Phi - \Phi^{\mathrm{h}}]|^2 \right] \, \mathrm{dV} \right]^{\frac{1}{2}} . \tag{10.6}$$

Babuška and Suri [23, 24] have shown that for linear elliptic problems solved on a regular grid with a uniform choice of h and p, respectively representing the maximum characteristic mesh size and the polynomial order of the FE discretisation, and under some additional assumptions, the a priori $\mathscr{H}^1$-error bound is

$$\|\Phi - \Phi^{\mathrm{h}}\|_{\mathscr{H}^1} \leq C \frac{h^{\eta-1}}{p^{m-1}} \|\Phi\|_{\mathscr{H}^m} \tag{10.7}$$

1 The smallest parameter m for which $\Phi \in \mathscr{H}^m(\mathscr{D}_0)$ holds is known as the Sobolev regularity of the function Φ [382]. For the general heterogeneous Poisson problem $-\nabla_0 \cdot [\mu(\mathbf{X}) \nabla_0 \Phi] = f(\mathbf{X})$ in $\mathscr{D}_0$, with a non-constant coefficient $\mu(\mathbf{X})$, the regularity of the solution can be reduced by imposing jumps in the material coefficient within the domain [258] thereby rendering $\mu(\mathbf{X}) \nabla_0 \Phi$ not continuously differentiable. The regularity of the solution is further affected by the smoothness of both the boundary of the geometry and the right-hand side function $f(\mathbf{X})$.

where $\eta = \min(m, p+1)$, and C is a constant that is independent of h, p and η. Guo and Babuška [182] have proven that under certain conditions, including that of a suitably selected hp discretisation, then the $\mathscr{H}^1$-error bound is [383]

$$\|\Phi - \Phi^{\mathrm{h}}\|_{\mathscr{H}^1} \le C_1 \exp\left(-C_2 \left[n_{\mathrm{DOF}}\right]^{\frac{1}{3}}\right) \tag{10.8}$$

where n_{DOF} is the number of DoFs, and C_1 and C_2 are positive scalars that are independent of n_{DOF}. This indicates that an exponential convergence may be achieved with the hp-FEM. In a similar manner, it can be shown that for Poisson problems the energy error is bounded by [132, 444]

$$\epsilon \le C \sum_{\mathrm{e} \in \mathscr{D}_0^{\mathrm{h}}} h_{\mathrm{e}}^{p} \|\nabla^{p+1}\Phi\|_{\mathscr{L}^2(\mathrm{e})} \quad . \tag{10.9}$$

This, in a sense, gives a measure of the next gradient of the solution which cannot be exactly represented by the FE space.

The outlined conclusions of an error analysis indicate that, for the ideal conditions (with the possibility for geometric singularities), it is possible to achieve polynomial or even an exponential convergence rate with hp-refinement. We reiterate that the above represents errors bounds for homogeneous problems, so equations (10.7) to (10.9) could only act as a guideline for problems involving heterogeneous materials such as those considered when modelling RVEs. It should also be noted that it is the energy error stated in equation (10.9) that is minimised by the total energy functional, so a decrease in this measure upon h- or p-refinement is expected. Since error estimators are typically driven by the jump in the solution gradient between elements the $\mathscr{H}^1$ semi-norm of the error

$$|\Phi - \Phi^{\mathrm{h}}|_{\mathscr{H}^1} = \left[\int_{\mathscr{D}_0^{\mathrm{h}}} \left|\nabla_0[\Phi - \Phi^{\mathrm{h}}]\right|^2 \mathrm{dV}\right]^{\frac{1}{2}} \tag{10.10}$$

is also useful to investigate (although any decrease in this metric upon adaptivity is somewhat coincidental).

Numerical example: Error analysis of a magnetostatic RVE

The magnetostatic formulation outlined in Section 6.1.3 was implemented within the `deal.II` [29, 30, 10] FEM framework, and a number of permutations of the geometry, ansatz and mesh refinement methods were evaluated. To mitigate the effect of domain truncation, we utilise the *method of manufactured solutions* and apply the analytical solution to the stationary problem as the Dirichlet boundary condition on $\partial\mathscr{B}_0 \equiv \partial\mathscr{B}_0^{\boldsymbol{\varphi}}$. In particular, in a dimensionless setting we choose for $H_\infty = -\frac{\Delta\Phi}{L}$ with $\Delta\Phi = 1000$, the geometric parameters $L = 2$ and $R = \frac{L}{4}$ and the material parameters $\mu_{\mathrm{r}}^{\mathrm{m}} = 1$ and $\mu_{\mathrm{r}}^{p} = 5000$. Two methods of producing a computational geometry are investigated. In the first, the FEM discretisation is fully aligned with the Cartesian coordinate system, and the material subdomains are defined according the position of FE cell centres and the expected location of the spherical inclusion. This is a convenient approach from which to define complex geometries, but has the disadvantage that the curvature of the particles is not captured accurately. The second follows the particle geometry exactly (in the discrete setting), and the surface geometry continues to be accounted for during subsequent global (GMR) or adaptive (AMR) mesh refinement. Where AMR is applied, the Kelly error estimator [259] is used as the marking strategy for cells to be refined and coarsened.

A visual summary of the computational domain and cell-average $\mathscr{H}^1$ semi-norm of the errors is shown in Figure 10.3 for the two geometries after several iterations of AMR. It is observed that the cells near the material discontinuity are preferentially refined in order to accurately capture the jump in the magnetic field across the discontinuity. However, the coarse description of the inclusion by the Cartesian-aligned grid inadequately resolves the solution around material discontinuity. It results in the development of artificial singularities at the material subdomain boundary, leading to large local error in magnetic field in this vicinity.

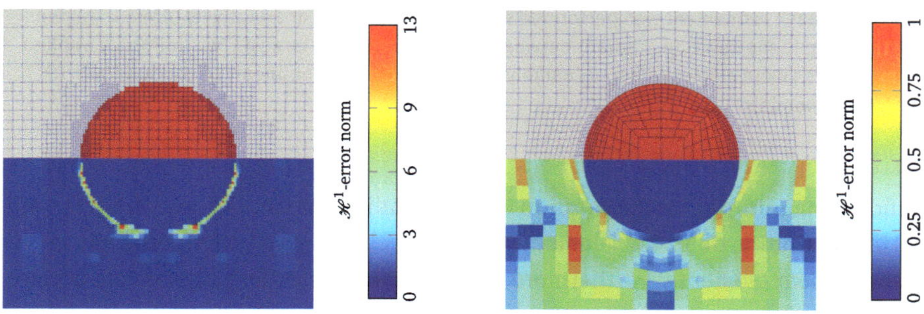

(a) Cartesian-aligned mesh.

(b) Exact geometry.

Figure 10.3: Depiction of the adaptively refined mesh with material regions (upper half) and cell $\mathscr{H}^1$ semi-norms of the error (lower half). Bilinear (Q^1) elements were used in the discretisation of the domain. Note the order of magnitude difference in the norms of the two examined meshes.

As a supplement to the preliminary visual analysis, Figure 10.4 graphs the convergence characteristics of the aforementioned solution norms when increasing the refinement level of the discretisation. Additionally, it is clear that AMR performs better than GMR in terms of computational efficiency. Through the marking strategy the additional DoFs are located in the region of highest solution gradients, namely at the interface where there is the solution discontinuity. Note though that the rate of convergence of GMR is roughly equal to that of AMR; this is to be expected as rate of convergence depends on the mesh size, and only the error constant is increased (in this instance, by a

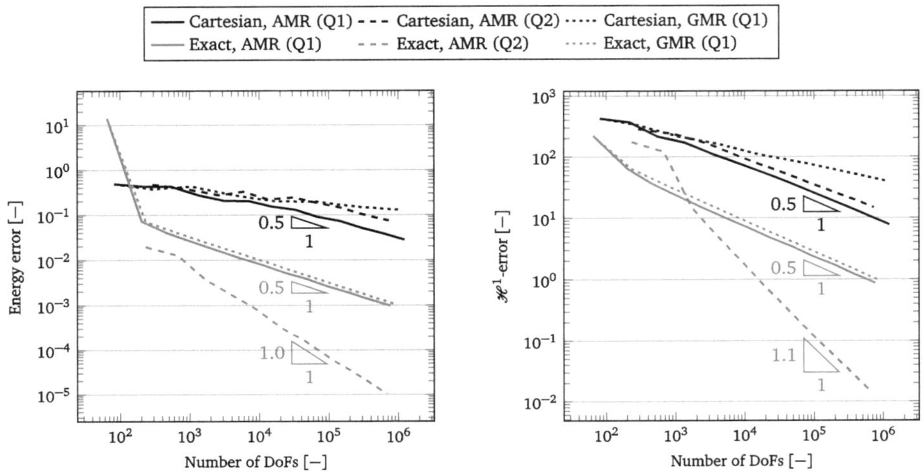

(a) Energy error norm.

(b) $\mathscr{H}^1$ semi-norm of the error.

Figure 10.4: Plot of error norm measures versus the total number of DoFs. For brevity, we omit the $\mathscr{H}^1$-error norm, which closely follows that of the $\mathscr{H}^1$ semi-norm of the error since $\|\Phi - \Phi^h\|_{\mathscr{H}^1} = \|\Phi - \Phi^h\|_{\mathscr{L}^2} + |\Phi - \Phi^h|_{\mathscr{H}^1}$ and $|\Phi - \Phi^h|_{\mathscr{H}^1} \gg \|\Phi - \Phi^h\|_{\mathscr{L}^2}$. The two examined meshes were either Cartesian-aligned or matching the exact geometry of the particle (in a discrete sense). Mesh refinement was performed either globally (GMR) or adaptively (AMR) in regions where the jump in the solution gradient between adjacent cells was largest. Two uniformly applied FE element types were also considered, namely those based on either linear (Q^1) or quadratic (Q^2) Lagrange polynomials.

significant magnitude). As the solution (and its gradient) is smooth away from the material discontinuity, quadratic elements perform well and give better performance per DoF than linear elements. Furthermore, the use of higher-order polynomials accelerates the rate of convergence when the geometry is exact, but for the Cartesian-aligned mesh the discretisation error dominates the examined error measures.

10.1.2 Computation of magnetic forces and torques

When its equilibrium state for a given load is computed, an analysis of the force- and torque-generation properties of MAPs can be conducted. This may provide a deeper insight into the mechanics of the microstructure, and the influence of the geometry on the force-generation properties of the composite. From the analytical expressions given in equations (5.137) and (5.139), it is possible to express the ponderomotive force and moment acting on an arbitrary subdomain in two ways: (i) directly from the strong form, as given in the previously mentioned equations, or (ii) through its re-expression in the weak form.

10.1.2.1 Computations derived from the strong form

In the case of computations derived directly from the strong form, there are two points that require consideration. When considering the arbitrary subdomain $\mathscr{B}_t^i \subset \mathscr{B}_t$, the expression for the ponderomotive force and torque strictly under magnetostatic conditions is

$$\mathbf{f}_t^{\text{pon}} = \int_{\mathscr{B}_t^i} \mathbb{m} \cdot [\nabla \mathbb{b}] \, \mathrm{dv} + \int_{\partial\mathscr{B}_t^i} [\{\!\{\mathbb{m}\}\!\} \cdot [\![\mathbb{b}]\!]] \, \mathbf{n}^+ \, \mathrm{da} \quad , \tag{10.11a}$$

$$\mathbf{m}_t^{\text{pon}} = \int_{\mathscr{B}_t^i} [\mathbf{r} \times [\mathbb{m} \cdot [\nabla \mathbb{b}]] + \mathbb{m} \times \mathbb{b}] \, \mathrm{dv} + \int_{\partial\mathscr{B}_t^i} \mathbf{r} \times [\{\!\{\mathbb{m}\}\!\} \cdot [\![\mathbb{b}]\!]] \, \mathbf{n}^+ \, \mathrm{da} \quad . \tag{10.11b}$$

From this, two key points should be observed. Firstly, when employing the MSP formulation the gradient of the magnetic induction implies that the second derivative of the magnetic scalar potential must be non-trivially calculable. This condition is, however, not fulfilled when linear shape functions are used for the discretisation of the MSP field. The second point of consideration is that for the jump in the ponderomotive stress across a material interface to be captured, it is necessary that magnetic fields remain discontinuous across these interfaces.

Vogel et al. [533] account for both points using a $\mathscr{L}^2$ smoothing procedure [222] applied to all magnetic fields but localised within a continuous material subdomain $\mathscr{B}_t^i$. The magnetic fields are initially computed at the quadrature points using the prescribed constitutive laws, and are projected to the FE support points on subdomain $\mathscr{B}_t^i$ through the solution of the linear system given by

$$\boldsymbol{M} \cdot \tilde{\mathbb{I}} = \mathbb{I} \quad \text{in} \quad \mathscr{B}_t^i \quad . \tag{10.12}$$

Here, $\tilde{\mathbb{I}}$ represents the $\mathscr{L}^2\left(\mathscr{B}_t^i\right)$ smoothed counterpart of the spatial magnetic field $\mathbb{I} \in \{\mathbb{h}, \mathbb{m}, \mathbb{b}\} \in \mathscr{B}_t^i$. Discretising the magnetic field by

$$\mathbb{I}(\mathbf{x}) \approx \sum_{I}^{n_{\Psi_{\mathbb{I}}}} \mathbb{I}^I \boldsymbol{\Psi}_{\mathbb{I}}^I(\mathbf{x}) \quad \text{on} \quad \mathscr{B}_t^i \quad , \tag{10.13}$$

the expression for the contributions to the consistent mass matrix and right-hand side are

$$\boldsymbol{M}^{IJ} = \int_{\mathscr{B}_t^i} \boldsymbol{\Psi}_{\mathbb{I}}^I(\mathbf{x}) \, \boldsymbol{\Psi}_{\mathbb{I}}^J(\mathbf{X}) \, \mathrm{dv} \quad , \quad \mathbb{I}^I = \int_{\mathscr{B}_t^i} \boldsymbol{\Psi}_{\mathbb{I}}^I(\mathbf{x}) \cdot \mathbb{I}(\mathbf{x}) \, \mathrm{dv} \quad \text{in} \quad \mathscr{B}_t^i \quad . \tag{10.14}$$

As this process is conducted for each material subdomain individually, duplicate values for each field are generated along all material discontinuities. This attends to both aforementioned requirements, as the reasonable expectation that the magnetic field quantities are smooth on a per-material basis is maintained, while there is no restriction on their smoothness properties across the interface.

10.1.2.2 Computations derived from the weak form

To express the weak expression of the magnetic forces generated on a material domain, it is necessary to apply the principle of virtual work [538]. Concisely stated, the virtual work performed by the ponderomotive force an arbitrary domain $\mathscr{D}_t$ is

$$\delta\Pi^{\text{pon}} = \int_{\mathscr{D}_t} \delta\boldsymbol{\varphi} \cdot \mathbf{b}_t^{\text{pon}} \,\text{dv} + \int_{\partial\mathscr{D}_t} \delta\boldsymbol{\varphi} \cdot \mathbf{t}_t^{\text{pon}} \,\text{da} \tag{10.15}$$

where $\delta\boldsymbol{\varphi}$ is the virtual displacement. Application of equations (B.16) and (5.136) leads to the result [533] that

$$\delta\Pi^{\text{pon}} = \int_{\partial\mathscr{D}_t} \delta\boldsymbol{\varphi} \cdot [\boldsymbol{\sigma}^{*\text{pon}} \cdot \mathbf{n}^+] \,\text{da} - \int_{\mathscr{D}_t} \nabla\delta\boldsymbol{\varphi} : \boldsymbol{\sigma}^{*\text{pon}} \,\text{dv} \equiv \sum_{I\in\mathscr{D}_t}^{n_{\Psi_\varphi}} \delta\boldsymbol{\varphi}^I \cdot [\mathbf{f}_t^{\text{pon}}]^I \tag{10.16}$$

when considering the FE discretisation of the computational domain, namely equation (6.7). Given the arbitrary nature of $\delta\boldsymbol{\varphi}$, it can therefore be discerned from this that

$$\sum_I^{n_{\Psi_\varphi}} \delta\boldsymbol{\varphi}^I \cdot [\mathbf{f}_t^{\text{pon}}]^I = \sum_I^{n_{\Psi_\varphi}} \delta\varphi^I \left[\int_{\partial\mathscr{D}_t} \boldsymbol{\Psi}_{\boldsymbol{\varphi}}^I \cdot [\boldsymbol{\sigma}^{*\text{pon}} \cdot \mathbf{n}^+] \,\text{da} - \int_{\mathscr{D}_t} \nabla\boldsymbol{\Psi}_{\boldsymbol{\varphi}}^I : \boldsymbol{\sigma}^{*\text{pon}} \,\text{dv} \right] , \tag{10.17}$$

with the ponderomotive stress tensor defined in equation (5.131a). Finally, the resultant force of the computational subdomain $\mathscr{B}_t^i \subset \mathscr{D}_t$ can be calculated from the global vector of ponderomotive forces by

$$[\mathbf{f}_t^{\text{pon}}]_{\mathscr{B}_t^i} = \sum_{I\in\mathscr{B}_t^i}^{n_{\Psi_\varphi}} [\mathbf{f}_t^{\text{pon}}]^I \quad . \tag{10.18}$$

Compared to the approach detailed in Section 10.1.2.1, computing the magnetically-induced forces from the weak form is straight-forward. The main caveat of this approach is that it is not possible to distinguish and localise the forces acting, for example, solely on an individual particle from those acting on the matrix. This is because the nodal force vector is distributed over the entire computational domain and, at a material interface, the assignment of the nodal forces to the adjacent subdomains is ambiguous. However, the issue is minimised if the contrast in the relative magnetic permeabilities between the two media is large, as the ponderomotive force computed at interface DoFs will be dominated by contributions stemming from the medium with the larger relative permeability.

Numerical examples: Force and torque measurements on magnetostatic RVEs

Vogel et al. [533] advocate (in preference to other examined techniques) the use of the strong form calculations in conjunction with $\mathscr{L}^2$ smoothing for all fields. Although it is not strictly necessary to smooth all of the magnetic fields (in some instances, values at calculation points can be validly extrapolated from the quadrature points), they found that the numerical results to be more favourable in comparison to those produced by the other approaches. The use of smoothed values is expected to generate superior results since smoothing procedures lead to better numerical approximations; this property is exploited in the development of error estimators [443, 579]. Furthermore, they established that the use of a linear FE ansatz in conjunction with the smoothing method had no impact on the quality of the computations, both in terms of the final computed values as well as that of the convergence characteristics measured upon refinement of the computational grid. The results presented below are thus restricted to using these preferential options.

With reference again to Figure 10.1, a 2.5-D representation of the $L = 10$ mm geometry was constructed through an extrusion of 0.4 mm into the third dimension. To generate a non-uniform magnetic field, the Dirichlet boundary conditions were defined such that $\Delta\Phi = 2000$ A between $\partial\mathscr{B}_0^1$ and $\partial\mathscr{B}_0^2$, while on $\partial\mathscr{B}_0^3$ and $\partial\mathscr{B}_0^4$ $\Phi = 0$ A. The material parameters were chosen to be $\mu_r^m = 10$ for the lightly magnetisable matrix and $\mu_r^p = 5000$ for the particle.

By adjusting the offset distance l_0 of a $R = 1$ mm radius particle the influence of the asymmetry of the surrounding magnetic field can be established. Figure 10.5 illustrates the magnetic field surrounding the particle as well as the ponderomotive forces acting on the particle. It is evident that the non-trivial magnetic field acts to deform and translate the particle. The dominant volumetric force is aligned with the gradient of the magnetic potential field, while in contrast the resultant interface force acts almost orthogonally to the line of action of the volumetric force.

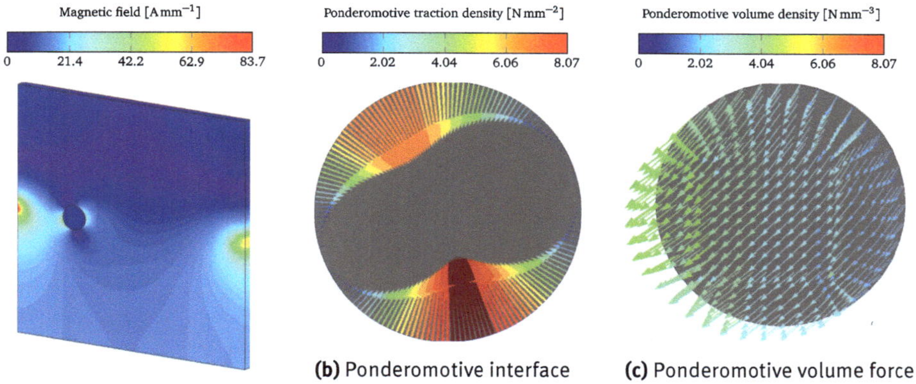

(a) Magnetic field.

(b) Ponderomotive interface force density.

(c) Ponderomotive volume force density.

Figure 10.5: Cylindrical particle, within an elastomer offset by $l_0 = 2.5$ mm to the left side along the $\mathbf{e}_1$-axis, in asymmetric magnetic field. Note that Figures 10.5b and 10.5c represent the nodal representation of the ponderomotive force densities. [533, figs. 13, 14 (reproduced with permission)]

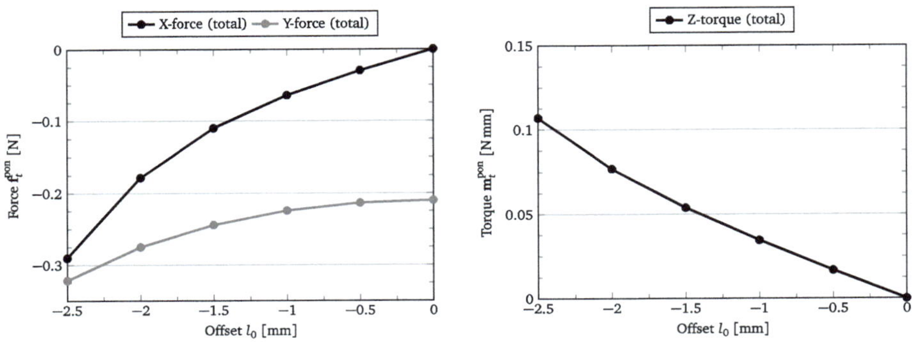

(a) Total ponderomotive force.

(b) Total ponderomotive torque.

Figure 10.6: Interface and volume forces and magnetic torque produced in the magnetised circular particle offset in an asymmetric field. [533, fig. 16]

As the particle is moved in the magnetic field, the ponderomotive force and traction acting on it changes accordingly; this is quantified in Figure 10.6. When the particle is centred within the domain, the magnetic field induces only an $\mathbf{e}_2$-aligned ponderomotive force. When the particle is offset, it becomes attracted to the source of the potential and the effective ponderomotive moment, measured about its centre of mass, increases non-linearly. For a non-static problem, this implies that not only would the cylindrical particle exhibit translation but rotation as well.

The forces and torques that a magnetisable particle generates are not only affected by the surrounding magnetic field, but their geometry as well. As an illustration of this point, the cross-section of the particle is changed from cylindrical to elliptical, with $a_1 = 1$ mm and $a_2 = 2$ mm while the offset was fixed at $l_0 = 1.5$ mm. The resulting distribution of traction forces shown in Figure 10.7 highlights the measurable influence that the attitude of a non-symmetric particle within a non-uniform magnetic field has on the ponderomotive traction force exerted on it.

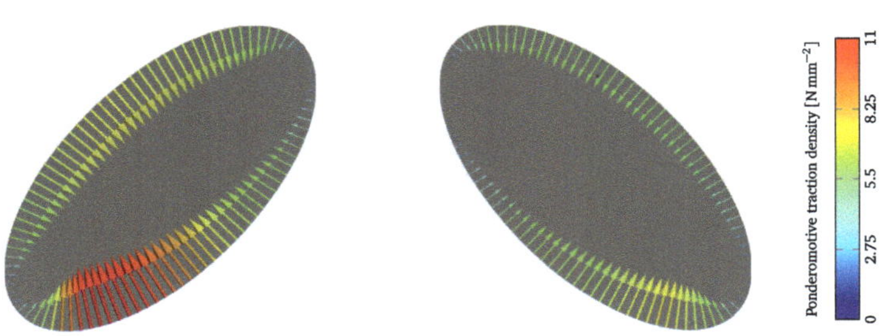

(a) Rotation of $\theta = 45°$.

(b) Rotation of $\theta = -45°$.

Figure 10.7: Nodal representation of ponderomotive interface force density arising from change in orientation of an elliptical particle offset in an asymmetric magnetic field. The fixed geometric parameters were $a_1 = 1$, $a_2 = 2$ and $l_0 = 1.5$. [533, fig. 16 (reproduced with permission)]

As a result of the non-uniformity of the magnetic field at the fixed position, the particle experiences a net downwards ponderomotive force, regardless of its angle of orientation and with the volume and traction forces always opposing one another. Although the volume force is predominantly greater in magnitude than the traction force, for certain orientations of the particle the force components in the $\mathbf{e}_1$ direction equilibrate one another. In this scenario, the particle instantaneously experiences no attraction to the potential source, but will rather only undergo translation in the $\mathbf{e}_2$ direction in conjunction with a magnetically-induced rotation. This is concluded from Figure 10.8, and a similar observation can be made regarding the equilibrium orientation for the torque induced in the inclusion. However, in contrast to the force decomposition, the asymmetry of the interface force causes its contribution to the torque to dominate that of the volume force.

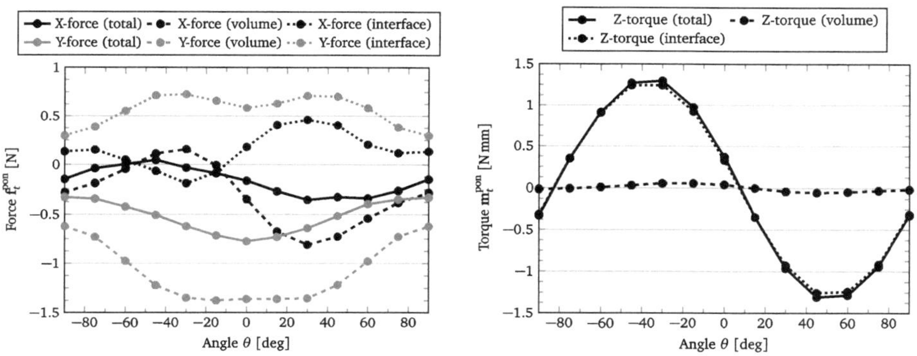

(a) Total ponderomotive force.

(b) Total ponderomotive torque.

Figure 10.8: Interface and volume forces and magnetic torque produced in the magnetised elliptical particle offset and rotated in an asymmetric field. [533, fig. 18]

10.2 Micro-structural studies of MAP compositions

Using the knowledge gained from the initial study presented in Section 10.1.1, what follows is a few further studies that illustrate some of the challenges in the simulation of geometrically detailed microstructures of MAPs, and their complex mechanical response. This supplements the numerous numerical studies of MAP microstructures that may be found in the recent literature, which includes parametric studies of the mechanical and magnetic response of (i) a unit cell, with a focus on particle size [105], shape [256] and orientation [155]; (ii) particle arrangements and chain distributions leading to isotropic [355], orthotropic [355], and perfect and dispersed transversely isotropic [189, 355] behaviour; and (iii) complex microstructures, with investigations on the effect of (randomly distributed) particle sizes [355], (random and structured) particle distributions and orientations [355, 244, 256], and particulate volume fraction [355]. Work has also been conducted on the influence of particle geometries in the mechanical response of RVEs using analytical techniques [156, 487, 158]. More focussed analytical and numerical studies related to MAP microstructures include those investigating forces and torques acting on magnetised subdomains [157, 533], ferromagnetic subdomains in the context of micro-magnetoelasticity [137], and coupling effects in magneto-electro-elastomers [442]. To complement the studies conducted using numerical approaches on the topic of preprocessing, Schneider et al. [474] discuss mechanisms to automate the generation and meshing of randomised, highly filled microstructures; a recent review by Bargmann et al. [31] presents the state-of-the-art techniques for generating heterogeneous microstructures for RVEs. Furthermore, Miehe and Bayreuther [364] introduce some numerical methods (including multiscale multigrid techniques) in order to alleviate some of the computational effort needed to simulate finely-meshed heterogeneous geometries.

10.2.1 Full resolution simulation of a prototype magnetostatic microstructural model

To illustrate the challenges related to producing full resolution models of MAPs, a 2-D magnetostatic idealised representation of a 0.25 mm^2 section of such a material was implemented. The isotropic composite has a particle density of 25vol.-% (520 particles in total) and the size distribution given in Figure 2.4b. The iron particles were given a relative magnetic permeability of $\mu_r^{\text{particle}} = 5000$, while for the matrix $\mu_r^{\text{matrix}} = 1$. Boundary conditions were applied on all surfaces of the material such that the vertically aligned magnetic field $|\overline{\overline{\mathbb{H}}}| = 200\,\text{kA}\,\text{m}^{-1}$, with both Dirichlet and periodic condition evaluated.

The first issue is the generation of the computational mesh itself. Since both the size and location of the particles in an isotropic medium must conform to a preset distribution, it is sensible to make use of a random number generator (e. g. the high-quality pseudo-random number generator MT19937 [63]) to assist in the stipulation of these properties. Random selection of particle sizes from the given log-normal size distribution was conducted until the prescribed volume fraction was applied. In this example, a fair isotropic distribution of the particles was subsequently obtained by randomly positioning the particles, largest to smallest, such that they remained at a minimum distance from one another (i. e. they are non-overlapping). Particles overlapping the boundary were moved if the arc subtending the overlap chord was outside of a certain range. Both of these limitations were stipulated to ensure that a high-quality mesh could be generated. Given the domain size and inclusion properties, the multi-region geometry was then automatically constructed using the Open CASCADE [408] geometry engine and subsequently meshed using Cubit [282]. To assist in this process, an automated mesh size scheme was developed to define the mesh size relative to the individual particle sizes, and the mesh was then entirely generated using an unstructured quadrilateral meshing algorithm. In three dimensions, this final step would be difficult to achieve autonomously, which would result in a large effort being necessary to generate a single microstructure.

This model was assessed using the `deal.II` [29, 30, 10] FEM framework in conjunction with the *Trilinos* [193] parallel linear algebra library. Lagrange polynomials were used for the ansatz within the standard Galerkin FEM; more specifically, bilinear elements were used within the particles, while both bilinear and biquadratic elements were considered in the discretisation of the surrounding matrix. Figure 10.9a provides an illustration of the result of the numerical study. A large potential gradient is developed between the magnetised particles, which leads to large interparticle magnetic fields. Adaptive mesh refinement was employed to capture these inhomogeneously distributed steep gradients, the location of which is unknown, and not easily predicted a priori. An example of the mesh after the refinement process is shown in Figure 10.9b. DoFs associated with the hanging nodes are eliminated from the linear system through the prescription of additional affine constraints [28]. Note that the smooth surface geometry of the particles is maintained upon refinement. This helps to ensure optimal convergence of the solution, as otherwise artificial magnetic singularities may be generated at meshing-induced sharp corners of material interfaces.

Table 10.1 highlights some of the numerical data derived from the simulation of the linearly magnetisable MAP. A large number of DoFs are necessary in order to resolve the interparticle magnetic field and accurately capture the magnetic properties of the variably sized particles. Importantly, and as discussed in Section 9.1.2, it may be observed that the applied boundary condition has a very large effect on the predicted homogenised value for the magnetic induction. Comparison of this value based on the solutions generated on the coarsest and finest discretisations for a given set of boundary conditions and for both FE ansatz revealed a difference of less than 1 %. Overall, the application of Dirichlet boundary conditions leads to a large over-prediction in the induction strength. This is due to the strong enforcement of the potential field within particles that cut the boundary, which leads to a very large (and physically inaccurate) magnetic field being generated within these few particles. Associated with the use of (h-adaptive) refinement is a large increase in solver time, as the size of the problem increases. The use of a solver is necessary in the primary problem to compute the solution to the magnetic scalar potential problem, while the post-processing steps utilises a similar solver to solve for the $\mathscr{L}^2$-smoothed magnetic fields on each individual subdomain. Although parallel computing along with a highly efficient iterative solver (conjugate gradient [192, 461] with an algebraic multigrid preconditioner [163]) was employed for both the primary and post-processing solver steps, both took increasingly more time to converge to the prescribed tolerance (the residual norm $|\boldsymbol{r}| < 1 \times 10^{-6}$). Lastly, as a confirmation of that the material is indeed isotropic, the volume-average magnetisation in the direction perpendicular to the applied magnetic field was determined to be three orders of magnitude less than that aligned with it.

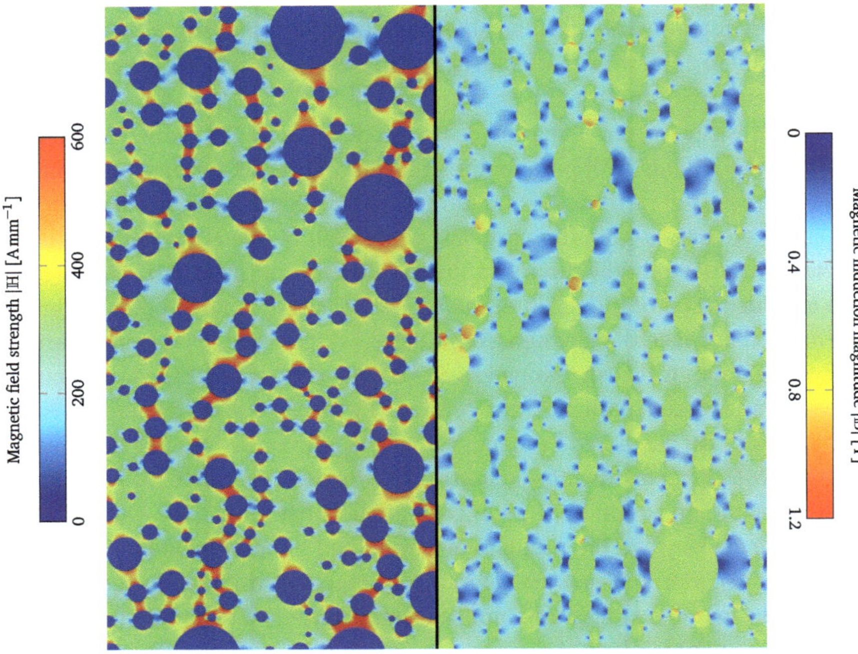

(a) Magnitude of the magnetic field and induction through the composite's microstructure. Periodic boundary conditions have been applied to the RVE.

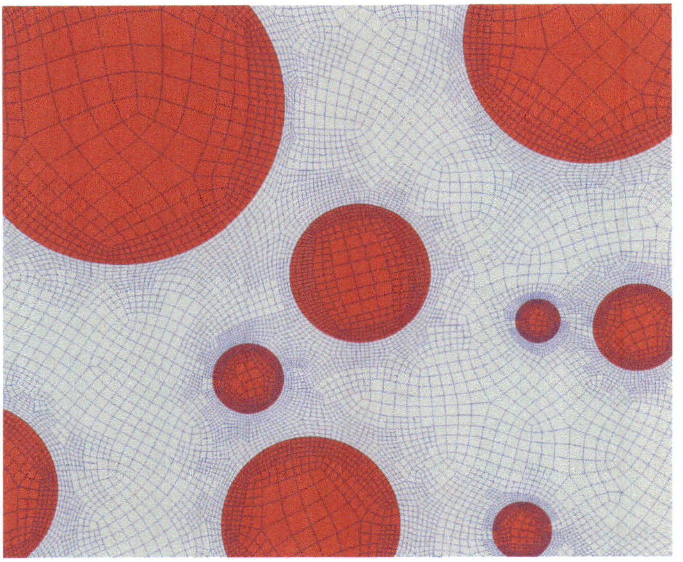

(b) Discretisation around a cluster of particles after two adaptive refinement cycles.

Figure 10.9: A 2-D full resolution model of a $0.25\,\mathrm{mm}^2$ section of an idealised isotropic MAP.

Table 10.1: Computational details of full resolution magnetostatic simulation. We denote the macro-scale magnetic induction measured when applying Dirichlet and periodic boundary conditions as $|\overline{\overline{\mathbb{B}}}|^{\text{dir}}$ and $|\overline{\overline{\mathbb{B}}}|^{\text{per}}$, respectively. All timings are computed relative to the non-refined FE_Q (1) – FE_Q (1) case.

Refinement cycle	Number of cells (approx.)	Number of DoFs (approx.)	$\lvert\overline{\overline{\mathbb{B}}}\rvert^{\text{dir}}$ [T]	$\lvert\overline{\overline{\mathbb{B}}}\rvert^{\text{per}}$ [T]	Solver time (rel.; approx)	Post-processing time (rel.; approx)
		FE_Q (1) on $\mathscr{B}_0^{\text{m}}$, FE_Q (1) on $\mathscr{B}_0^{\text{p}}$				
0	156700	159000	1.118259	0.437075	1	1
1	302500	322100	1.117627	0.437957	2.7	2.1
2	771800	822500	1.116755	0.438173	9.7	5.1
3	1848100	1964800	1.116260	0.438199	35.9	12.1
4	4343700	4583000	1.115958	0.438191	64	30.2
		FE_Q (2) on $\mathscr{B}_0^{\text{m}}$, FE_Q (1) on $\mathscr{B}_0^{\text{p}}$				
0	156700	533800	1.117175	0.435989	7	3.4
1	300700	1047400	1.116617	0.437617	11.2	6.4
2	763200	2459500	1.116119	0.438039	37	14.7
3	1876800	6245100	1.115894	0.438139	102.9	38.6

The final problem size roughly represents the limiting problem size that can be solved on a standard modern desktop computer. For the 3-D problem, significantly fewer DoFs can be used due to the increased bandwidth of the sparse system. Modern technologies such as matrix-free methods [279] and/or massive parallelism on supercomputer clusters [30] in conjunction with the *hp*-FEM can be employed [28] in order to investigate significantly larger boundary value problems.

When considering the coupled magnetoelastic problem, further considerations of the problem size and geometric complexity are required. The elastic problem has more DoFs than the magnetic scalar potential problem, and the size of the sparse system becomes significantly larger due to the coupling of all elastic DoFs in addition to the magneto-mechanical coupling. Large deformations due to mechanical or magnetic loads, or a combination thereof, may lead to significant grid distortion and render the problem indeterminant when $\det(\mathbf{F}) \leq 0$. This problem may be alleviated though the use of advanced techniques such as adaptive remeshing. Such computational challenges highlight the core advantage to using the Eshelby dilute approach applied in Section 9.1.4.

10.2.2 Influence of microstructural organisation in a representative volume element on the response characteristics of a prototype magnetoelastic material

Subjecting RVEs with different microstructural formations to various loading conditions allows the influence of both the load and the microstructure itself on the homogenised material response to be probed. In this section, we will present the outcome of two parametric studies. The first is aimed at comparing the effective material response of various CIP arrangements under different loading conditions, while the second gives more insight into the physical mechanism of particle attraction in dispersed chain-like microstructures.

Periodic boundary conditions are considered in all cases presented below, thereby rendering, according to the results presented in Chapter 9, an accurate microstructural stress and inductive response. Noting the decomposition given in equation (7.14), the volumetric material law that was considered for both the matrix and particles is given by [204]

$$U_0^{\text{vol}}(J) = \frac{\kappa}{4}\left[J^2 - 1 - 2\ln J\right] \quad , \tag{10.19a}$$

while the prototype isochoric free energy for the matrix and inclusion are respectively given by

$$\overline{U}_0^{\text{m}}\left(\overline{\mathbf{C}}, \mathbb{H}\right) = \frac{\mu_{\text{e}}}{K}\sum_{i=1}^{5}\gamma_i\,\omega^{[i-1]}\left[\overline{I}_1^{\,i} - \text{d}^i\right] - \frac{\mu_0}{2}\left[\mu_{\text{r}} - 1\right]J\overline{I}_7 \quad \text{and} \tag{10.19b}$$

$$\overline{U}_0^{\text{p}}\left(\overline{\mathbf{C}}, \mathbb{H}\right) = \frac{\mu_{\text{e}}}{2}\left[\overline{I}_1 - \text{d}\right] - \frac{\mu_0}{2}\left[\mu_{\text{r}} - 1\right]J\overline{I}_7 \quad . \tag{10.19c}$$

Note further that as a part of the total free energy (as stated in equation (7.11)) the above is used in conjunction with equation (5.178), thereby taking into consideration the energy stored in the magnetic field itself. The Arruda–Boyce 2^{d}-chain model[2] [17] is used to represent the elastic part of the matrix, and the constants in the model are [57, 85]

$$\omega = \frac{1}{N} \quad , \quad K = \left[1 + \frac{\text{d}}{5}\omega + \frac{11\text{d}^2}{175}\omega^2 + \frac{19\text{d}^3}{875}\omega^3 + \frac{519\text{d}^4}{67375}\omega^4\right] \quad , \tag{10.20a}$$

$$\gamma_1 = \frac{1}{2} \quad , \quad \gamma_2 = \frac{1}{20} \quad , \quad \gamma_3 = \frac{11}{1050} \quad , \quad \gamma_4 = \frac{19}{7000} \quad , \quad \gamma_5 = \frac{519}{673750} \quad . \tag{10.20b}$$

(The parameter K is defined such that the Arruda–Boyce model satisfies the linear-elastic consistency condition that $\frac{\partial U_0^{\text{AB}}(\mathbf{C})}{\partial I_1}\Big|_{I_1=\text{d}} = \frac{\mu}{2}$.) For the constitutive parameter representing the number of chain segments, we choose $N = 125$ which falls within the physically appropriate range $N \in [100, 200]$ for highly cross-linked elastomers [17, 554, 535]. The material parameters common to both constitutive laws, and the accompanying chosen FE ansatz spaces, are listed in Table 10.2. In all cases, the mesh was refined several times around the particles in order to ensure that the high magnetic field and stress gradients in the vicinity of the material discontinuities were accurately resolved.

2 This is a dimension-independent expression of the 4-chain and 8-chain model that are valid in two- and three-dimensions, respectively.

Table 10.2: Constitutive parameters and FE discretisation applied to the parameter study identifying the RVE effective response under different loading conditions. Note that the matrix is near-incompressible in nature and that the iron particles, although not taking the true shear modulus of iron (in the order of GPa), remain several orders of magnitude stiffer than the surrounding matrix.

Material	Constitutive parameter			FE ansatz space
	μ_e [Pa]	ν [–]	μ_r [–]	
Matrix	30×10^3	0.499	1.5	FE_Q(2) on $\mathscr{B}_0^{\mathrm{m}}$
Particles	30×10^6	0.3	5000	FE_Q(1) on $\mathscr{B}_0^{\mathrm{p}}$

10.2.2.1 Effective stiffness moduli in isotropic, orthotropic and transversely isotropic microstructures

By using, for instance, the perturbation approach outlined in Section 9.1.3 it is possible to determine the effective stiffness of any RVE to the applied loading. Through careful selection of this loading condition, the sensitivities can be directly related to homogenised mechanical or magnetic properties of the complex microstructure. With reference to the examples presented below, the effective shear modulus and an effective elastic stiffness modulus can be respectively computed by

$$\bar{\bar{\mu}}_{12} := \left.\frac{\Delta\bar{\bar{\sigma}}_{12}}{\Delta\bar{\bar{\gamma}}_{12}}\right|_{\bar{\bar{\mathbb{H}}}} \quad \text{and} \quad \bar{\bar{E}}_{22} := \left.\frac{\Delta\bar{\bar{\sigma}}_{22}}{\Delta\bar{\bar{\varepsilon}}_{22}}\right|_{\bar{\bar{\mathbb{H}}}} \tag{10.21}$$

under conditions of simple shear and uniaxial stress, respectively. Note that in the latter scenario the condition of zero transverse stress does not arise naturally, and it is necessary to evolve this state through a non-linear solution method. By linearisation of the transverse stress, this implies that

$$\left[\bar{\bar{\sigma}}_{11} - \bar{\bar{\sigma}}_{11}^{*}\right] + \frac{\partial\bar{\bar{\sigma}}_{11}}{\partial\bar{\bar{F}}_{11}}\Delta\bar{\bar{F}}_{11} \equiv 0 \tag{10.22}$$

with the other components of $\bar{\bar{\mathbf{F}}}$, as well as those of $\bar{\bar{\mathbb{H}}}$ fixed, and where we have denoted the fixed goal value of the transverse stress $\bar{\bar{\sigma}}_{11}^{*} \equiv 0$.

The magnetic and mechanical loads were applied in a staggered manner, the macro-scale magnetic field (aligned in the $\mathbf{e}_2$ direction) first being applied and then held constant, after which the mechanical deformation was enforced. The deformation (specifically the macro-scale shear strain $\bar{\bar{\gamma}}_{12}$, or macro-scale tensile stretch $\bar{\bar{\lambda}}_2$ aligned with the applied magnetic field) is increased logarithmically in order to distinguish between the linear and non-linear response regimes. The maximum applied magnetic field strength $|\bar{\bar{\mathbb{H}}}| = 225\,\mathrm{A\,mm^{-1}}$ corresponds to a magnetic induction of $|\bar{\bar{\mathbb{B}}}| \approx 0.5\,\mathrm{T}$ being generated in a linear material with $\mu_r \approx 6$. Under conditions of no (macro-scale) deformation, the range of recorded magnetic response $|\bar{\bar{\mathbb{B}}}|$ for the examined $\phi_{\mathrm{CIP}} = 5\,\%$ microstructures (subject to discretisation errors) was 0.4670 T to 0.4754 T, while that of the $\phi_{\mathrm{CIP}} = 10\,\%$ microstructures was 0.5125 T to 0.5584 T.

In Figures 10.10 to 10.12, the effective response of three different two-dimensional, geometrically periodic microstructural arrangements are presented while considering various combinations of mechanical and magnetic loads, as well as two volume fractions of the particulates. Note that the two-dimensional isotropic material has a hexagonal arrangement of the CIPs, while the orthotropic material has a regular grid-like spacing of the particles. The transverse isotropic microstructure has the particles arranged in a chain-like structure. The bounding box dimensions are not constant for the different microstructures but, assuming a fixed particle radius of $R = 1.3\,\mu\text{m}$, the characteristic dimensions of the RVE are in the range of $L = 21\,\mu\text{m}$ to $26\,\mu\text{m}$ for $\phi_{\text{CIP}} =$ for 5 %, and 15 μm to 18.5 μm for $\phi_{\text{CIP}} =$ for 10 %.

As can be observed in Figures 10.10 and 10.11, the homogenised shear stiffness for the tested particle volume fractions is linear until the applied shear strain is in the order of 5 % to 10 %. Considering no magnetic load, the small strain shear stiffness modulus is greatest for the isotropic material, followed by the orthotropic and then the anisotropic materials. This is due to the interparticle spacing, with the isotropic material having the most reinforcement while the transversely isotropic material is weaker due to the reduction in the amount of the elastomer between the particles. For each geometry, the overall stiffness increases with an increase in ϕ_{CIP}. As the mechanical deformation is increased into the finite strain regime, two noteworthy effects can be distinguished. The stiffening behaviour is dominated by the non-linear hyperelastic nature of the matrix, with the isotropic material exhibiting an exponential increase in the shear modulus. However, geometric effects due to the particle arrangement are also clearly visible, as the response of the orthotropic and transverse isotropic microstructures do not present the same response.

Upon application of a strong magnetic field, the homogenised shear stiffness measured at small strains increases measurably. For the dispersed microstructures the increase is in the order of only a few percent for both examined volume fractions. However, the organised nature of the chain-like microstructures leads to very strong interparticle interactions and, therefore, a considerable increase in recorded stiffness. This effect is highly amplified as the volume fraction of CIP is increased, with a 100 % increase in $\overline{\overline{\mu_{12}}}$ for $\phi_{\text{CIP}} = 10\,\%$. At finite strains, the microstructure tends to exhibit a weakening behaviour due to the rearrangement of particles and the associated change in interparticle spacing (in particular, that aligned with the magnetic field). This is particularly notable in the chain-like microstructure, for which the interparticle spacing always increases upon the application of a simple shear deformation. At some critical point, the reduction in stiffness due to weakening dipole-dipole interactions is again superseded by the matrix elasticity, with superimposed geometric effects adding some additional non-linearities to the complex response. Lastly, it should be noted that, for the tested loading conditions, the examined microstructures remained stable under pure shear loads even though they (i) did not always exhibit shear stiffening behaviour, and (ii) the stiffening response was neither monotonically convex nor concave.

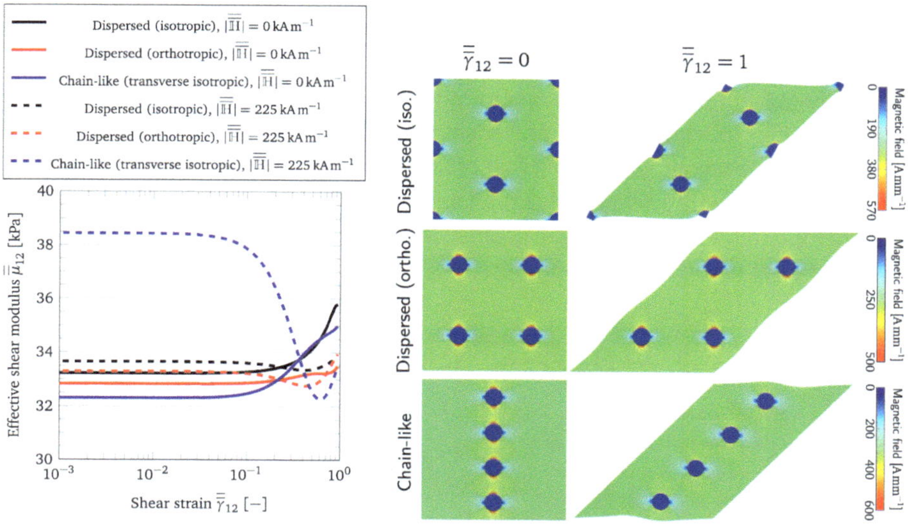

(a) Effective shear modulus. **(b)** Deformation and magnetic field strength.

Figure 10.10: Measurement of the shear response of three microstructures with $\phi_{\mathrm{CIP}} = 5\,\%$. The RVEs experience both mechanical and magnetic loading. Simple shear deformation has been applied, while the referential magnetic field is aligned in the $\mathbf{e}_2$ direction.

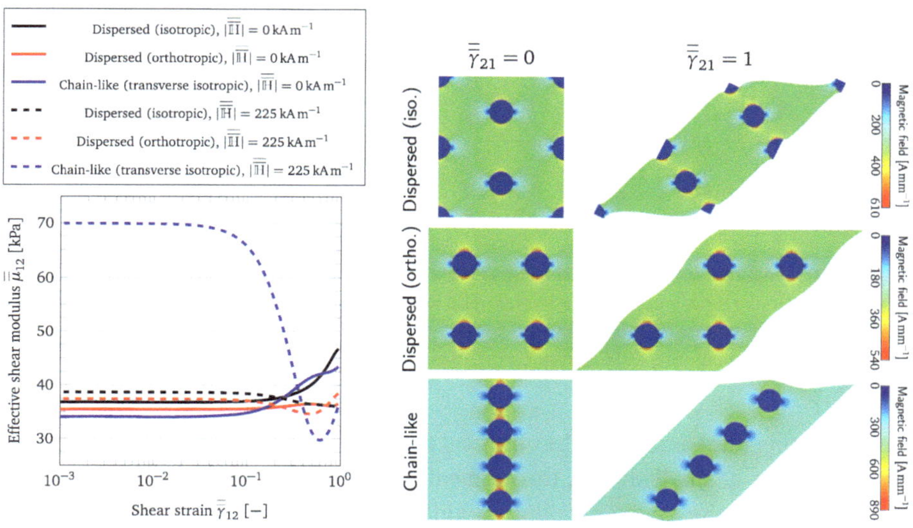

(a) Effective shear modulus. **(b)** Deformation and magnetic field strength.

Figure 10.11: Measurement of the shear response of three microstructures with $\phi_{\mathrm{CIP}} = 10\,\%$. The RVEs experience both mechanical and magnetic loading. Simple shear deformation has been applied, while the referential magnetic field is aligned in the $\mathbf{e}_2$ direction.

The same microstructures were examined under tensile deformations (specifically, that resulting in uniaxial stress) and the response is depicted in Figure 10.12. Compared to the shear case, under tensile load these microstructures do not exhibit an extended linear response regime and the transition from a linear to non-linear response in terms of the stiffness modulus is at around 1 % axial strain. As opposed to the shear case, the stiffest axial response is presented for the transversely isotropic microstructure, with the most compliant being that of the isotropic RVE. For a purely mechanical load, the effective stiffness modulus demonstrates a slight weakening at large deformations (due to the presence of the geometrically non-linear particles) until the non-linear response of the matrix dominates.

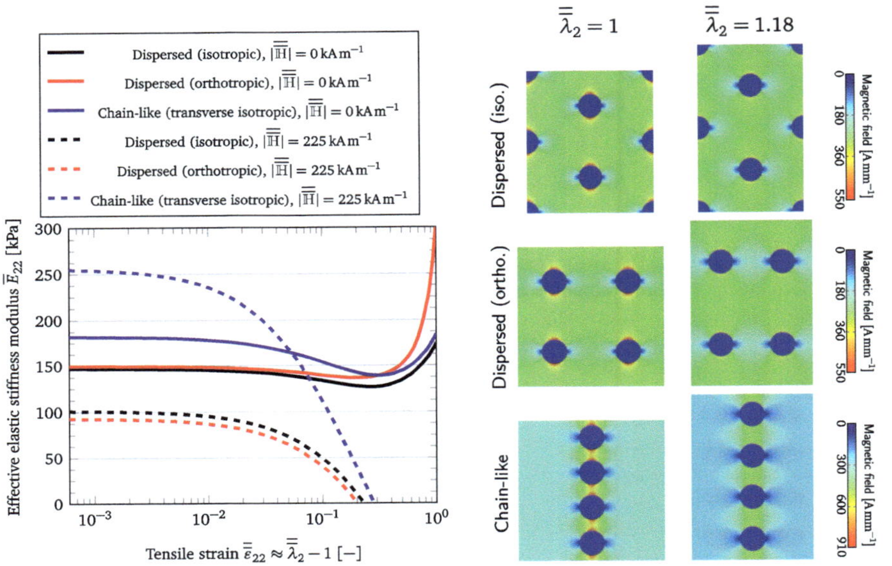

(a) Effective elastic stiffness modulus.

(b) Deformation and magnetic field strength.

Figure 10.12: Measurement of the uniaxial stress response of three microstructures with $\phi_{\mathrm{CIP}} = 10\,\%$. The RVEs experience both mechanical and magnetic loading. A tensile strain deformation has been applied in the $\mathbf{e}_2$ direction, with the transverse strain prescribed such that the transverse stress is zero. The referential magnetic field is aligned in the $\mathbf{e}_2$ direction.

Two interesting phenomena can be observed upon the application of a magnetic field. The first is that, at small strains, the permeating magnetic field does not necessarily reinforce the effective Young's modulus (while it did for the effective shear modulus for all examined microstructures). Both the isotropic and orthotropic microstructures are weakened in axial stiffness, whereas that of the chain-like microstructure demonstrate a stiffer response due to the strong interparticle interactions being aligned with the applied deformation. Secondly, the effective stiffness modulus for all of the microstructures exposed to a fixed, strong axial magnetic field decreased monotonically from their small strain value. For an applied stretch in the range of 10 % to 20 %, all examined microstructures exhibited a material instability as they could no longer sustain the applied tensile load. This is expected due to both the saddle point nature of the FE formulation and the associated non-convexity of the constitutive model (the mixed energy-enthalpy parametrised in terms of the magnetic field $\mathbb{H}$), as is discussed in Section 5.4.3.

10.2.2.2 Magnetic forces generated in dispersed chain-like structures

The study presented in Section 10.2.2.1 illustrates the clear influence of the arrangement of the microstructure on its mechanical behaviour. Through application of equation (10.11a), we may gain further understanding of the interparticle interactions in terms of their ponderomotive attractive forces. We consider a two-dimensional microstructure with particles arranged in a "wavy" (dispersed) chain-like structure, where the average alignment of the particles $\overline{\mathbf{M}}$ is parallel to the $\mathbf{e}_2$ direction, but alternating particles are offset in the $\mathbf{e}_1$ direction by a constant value $\pm\delta^p$ with respect to the RVE mid-line. Denoting the RVE characteristic length as L, one may then define a geometric chain dispersion parameter by

$$\kappa^* := \frac{|\delta^p|}{L/4} \quad . \tag{10.23}$$

Note that $\kappa^* = 0$ represents a perfect chain-like structure while $\kappa^* = 1$ indicates fully dispersed orthotropic microstructure, and that geometric periodicity of the microstructure is always maintained.

Figure 10.13 illustrates the magnetic field strength generated throughout the microstructure for various arrangements of particles with varying κ^*. Here, the square bounding box has dimension $L = 15\,\mu\text{m}$, and the volume of the cylindrical magnetic inclusions remained constant with $\phi_{\text{CIP}} = 10\,\%$. The material properties and the method in which the boundary conditions are applied remain the same as introduced in Section 10.2.2.1. The applied macro-scale magnetic field was aligned in the $\mathbf{e}_2$ direction with $|\overline{\overline{\mathbb{H}}}| = 225\,\text{kA m}^{-1}$ and the imposed macro-scale deformation gradient $\overline{\overline{\mathbf{F}}} \approx \mathbf{I}$. When

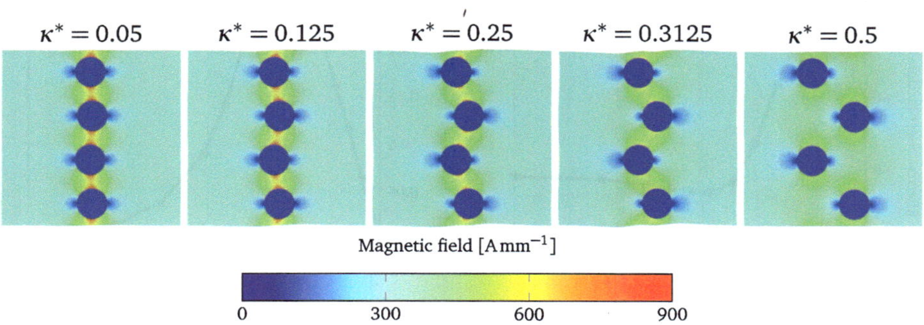

Figure 10.13: Deformation and magnetic field strength of dispersed chain-like structures with $\phi_{\text{CIP}} = 10\,\%$.

the particles are highly organised ($\kappa^* \rightarrow 0$), a very large local magnetic field can be observed in the intraparticle spaces. As the dispersion of the chain increases ($\kappa^* \rightarrow 1$), the intraparticle magnetic field weakens until the matrix has a magnetic field of near-constant strength permeating it. Also noticeable is the influence of the attraction of the particles to one another; as can be observed by the deformation of the boundary of the RVE, the matrix in the intraparticle space is disturbed as the particles move to align themselves in the $\mathbf{e}_2$ direction. If the imposition of the macro-scale deformation were not enforced, it may be expected that the RVE exhibits magnetostrictive behaviour (see [105, figure 5]).

By mechanically perturbing the microstructure within the small strain regime, a measurement of its effective shear modulus with and without an applied magnetic load can be made. The result of this procedure for the aforementioned geometries and load cases is presented Figure 10.14a. When strongly anisotropic, the non-magnetised RVE is most mechanically compliant for the range of examined κ^*, but becomes mechanically stiffest when highly magnetised. As the chain dispersion increases this stark contrast reduces, and at maximum dispersion the difference in effective shear modulus for the magnetised versus non-magnetised RVE is in the order of 5 %. When a strong magnetic field is introduced to the microstructure, its effective response is highly sensitive to the structure of the particle chains when they are in a near organised configuration. There is also some intermediate arrangement ($\kappa^* \approx 0.43$) for which the microstructure's homogenised shear behaviour appears to be independent of the applied magnetic field.
Figure 10.14b plots the strength of the ponderomotive force generated in a single particle for the range of tested geometries. Due to their symmetric arrangement, this in fact measures the $\mathbf{e}_1$ aligned magnetic force that attracts each particle towards one another and, therefore, the RVE mid-line. At the two extremes, the particles are either perfectly

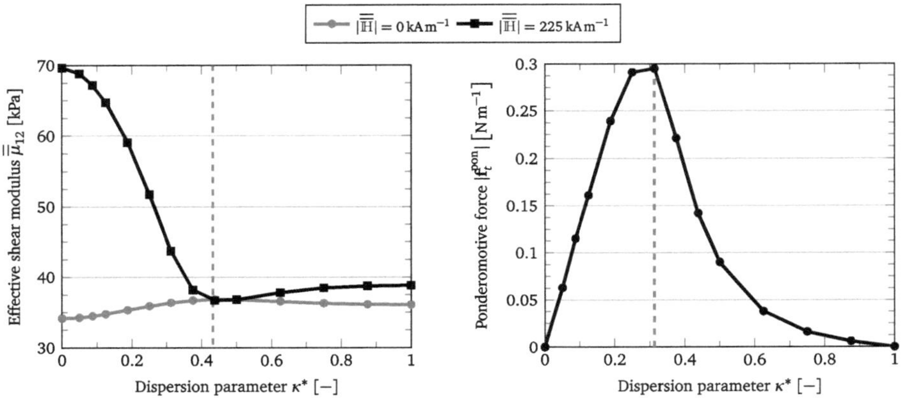

(a) Effective shear modulus. **(b)** Particle ponderomotive force magnitude.

Figure 10.14: Measurement of the small strain ($\bar{\bar{\gamma}}_{12} = 1 \times 10^{-3}$) response of dispersed chain-like microstructures with $\phi_{\mathrm{CIP}} = 10\,\%$.

aligned or evenly distributed such that there is no overall attractive force between the particles. There exists some optimal value of $\kappa^* \approx 0.32$ for which the force-generating capacity of the homogeneously sized particles is maximised. This suggests that the microstructure can be optimised to generate a maximal displacement of the particles, and distortion of the RVE as a whole, under application of a magnetic field. It is also interesting to note that with κ^* below this optimal value the force-generating capacity of the particles tends to decrease linearly towards the minimum, while above it the magnitude of the ponderomotive force decays exponentially towards zero.

11 Modelling and computational simulation at the macro-scale

Modelling MAPs at the macro-scale presents some interesting challenges, primarily related to the inclusion of the free space into which the composite polymer is immersed. In this chapter, we will cover aspects of the implementation of boundary value problems considering the free space domain, and will demonstrate the application of the previously discussed phenomenological constitutive laws that can be used to induce interesting macroscopic material responses.

In the first numerical example presented in Section 11.1, we link together the various topics discussed in Chapter 9 in order to characterise the macroscopic response of a bi-material MAP using the FE^2 method. Mechanisms to model magnetisable bodies immersed in free space are presented in Section 11.2; this expands on the topic as it was first discussed in the context of the FE discretisation in Chapter 6. Therein we also present a macroscopic boundary value problem that incorporates the dispersed particle chain model presented in Section 7.3. To complete the chapter, we introduce a mixed FE formulation that is suitable to model quasi-incompressible MAPs in Section 11.3; this extends the base description given in Section 5.3.7.

In addition to the numerical examples presented below, there have been several FE implementations of scenarios involving the free space in the recent literature. These include the study of magnetoelastic films [32], isotropic magnetoelastic samples under magnetic and shear loading [74, 255], a circular [370] and spherical [137] MAP sample in free space, a magnetostrictive linear micro-motor [370], and materials with anisotropic microstructures [256]. To supplement this, analytical solutions to coupled magnetoelastic boundary value problems have been explored in the context of MAP membranes [502], and the extension and inflation of magnetoelastic tubes [72, 406, 472, 468], among others.

11.1 Macro- to micro-scale transition using the FE^2 approach

To illustrate the influence of the microstructure on the macroscopic behaviour of an MAP, we will model a two-dimensional bi-material actuator using the FE^2 approach, with a configuration similar to that presented by Ethiraj [137]. The geometry, discretisation, boundary conditions, and microstructure of the composite are depicted in Figure 11.1. The laminate beam-like actuator has dimensions 1×4 mm^2, with the discretisation being 2×8 bi-quadratic elements (the same ansatz was used for both displacement and MSP DoFs). Concerning the macroscopic boundary conditions, a condition of zero displacement in $\mathbf{e}_2$ direction on the $-\mathbf{e}_2$ surface was allowed and the central support point on the same surface was pinned. The magnetic boundary conditions

https://doi.org/10.1515/9783110418576-011

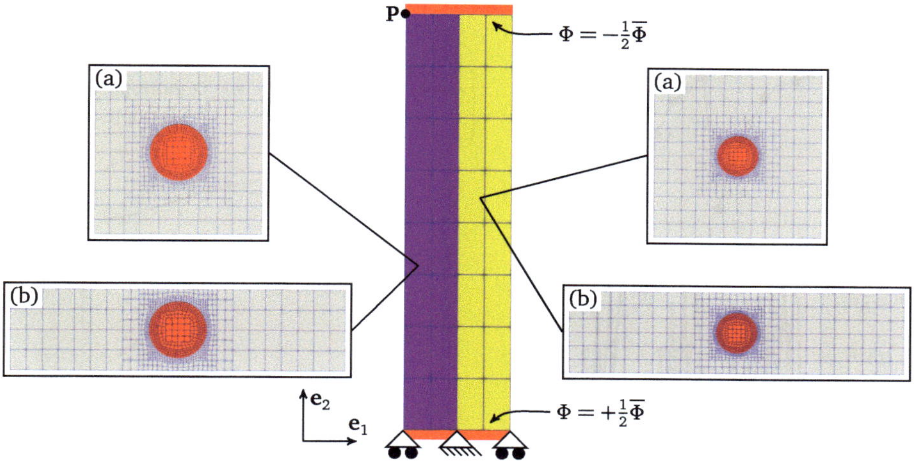

Figure 11.1: Problem setup, FE discretisation and material subdomains of the macroscopic problem and underlying RVEs. For the macroscopic problem, the laminate beam is split into two subdomains; the purple region comprises a microstructure with $\phi_{\text{CIP}} = 10\,\%$, while the yellow subdomain has $\phi_{\text{CIP}} = 5\,\%$.

are such as to suggest that the magnetic load is applied by means of a pliable, current-conducting wire wrapped tightly around the actuator. To this end, the MSP was incrementally increased on the upper and lower $\mathbf{e}_2$ surfaces; the final attainable magnitude was dependent on the microstructure that was evaluated.

The left half of the macroscopic composite is composed of an $\phi_{\text{CIP}} = 10\,\%$ MAP, while the right half has $\phi_{\text{CIP}} = 5\,\%$ microstructure with the same symmetries. In the two considered cases, the microstructure either corresponds to the (a) orthotropic or (b) strongly anisotropic geometry, for which the boundary conditions are always assumed to be periodic for both the elastic deformation and MSP fields. Accordingly, the geometries (CIP dispersion and volume fraction), phenomenological constitutive laws, constitutive properties, FE discretisation and (periodic) boundary conditions utilised for the microstructures are identical to those previously presented and evaluated in Section 10.2.2. As a reminder, the information flow between the two length scales is as shown in Figure 9.1c. To conduct the micro- to- macro-scale translation of the effective tangent moduli, the algorithmically consistent tangent approach discussed in Section 9.1.3.2 was employed.

Inspection of the convergence rates of the macro- and micro-scale problems reveals that the implementation of the algorithmically consistent tangent moduli along with the consistent tangents for the micro-scale phenomenological models, was conducted precisely. Shown in Figure 11.2a is a representative extract of the convergence history (on both length scales) for three time steps in which the numerical problem is well-conditioned and well-posed. The quadratic convergence at the macro-scale, as

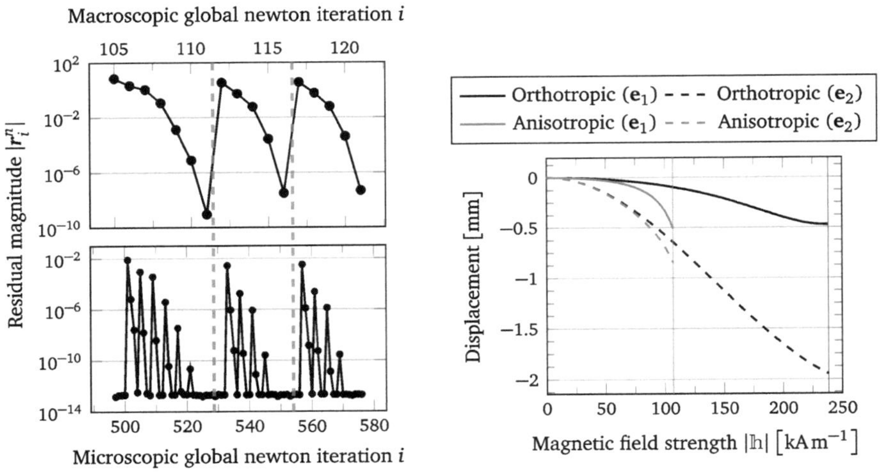

(a) Representative convergence rates for three macroscopic time steps (anisotropic).

(b) Displacement solution at point **P** versus the spatial magnetic field magnitude.

Figure 11.2: Results derived from the simulation of two MAPs using the FE^2 approach.

promised by the algorithmically consistent tangent moduli is clearly observable, while the fully manually implemented tangent moduli (formally verified using the tools discussed in Section 6.1.5) render similarly excellent convergence characteristics. However, at very strong magnetic fields, the large displacements of the matrix material in the RVE leads to distorted elements, and thus a poorly conditioned isoparametric transformation map; in such a scenario, quadratic convergence is no longer observed and divergence of the solution follows soon after. To delay this issue and to achieve the high magnetic loads applied here, an adaptive time stepping (or load sizing) approach was required.

Examining the displacement of point **P** against the spatial magnetic field experienced at the same point, in Figure 11.2b it may be observed that both the orthotropic and anisotropic geometries exhibit approximately 5 mm of displacement in the $-\mathbf{e}_1$ direction at the maximal applied magnetic load. Furthermore, both microstructures generate a contraction of the material as **P** displaces in the $-\mathbf{e}_2$ as well. Although, the contractile magnetostriction of the orthotropic microstructure is significantly greater than that of the anisotropic one, the latter exhibits enhanced bending characteristics at lower applied magnetic loads.

This is further fortified by visualising the macroscopic and microscopic configurations side by side in Figure 11.3. The differential properties through the material thickness leads to the two zones with differing particulate volume fractions generating different magnetic induction fields and, therefore, ponderomotive stresses. Due to the near-incompressibility of the matrix (and rigidity of the inclusions), the orthotropic

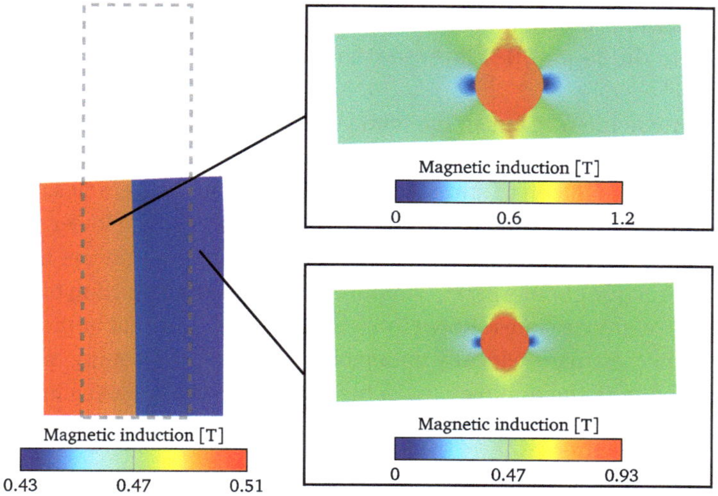

(a) Orthotropic microstructure (final $\Delta\overline{\Phi} = 476.1$ A, corresponding to $|\overline{\mathbb{H}}| = 119.0$ kA m^{-1}).

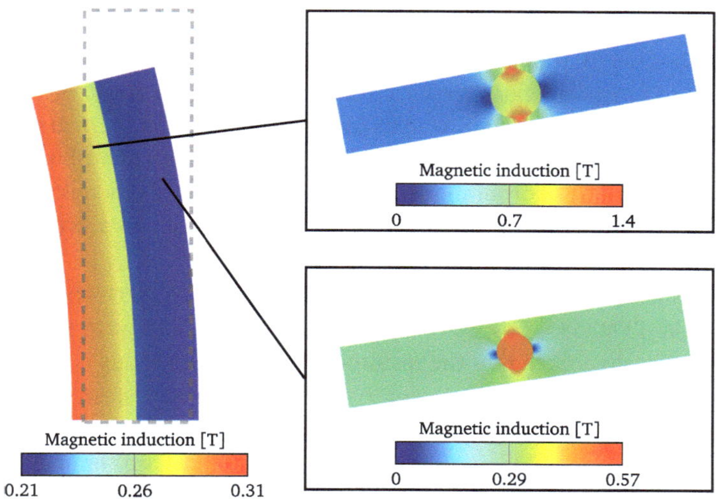

(b) Anisotropic microstructure (final $\Delta\overline{\Phi} = 341.6$ A, corresponding to $|\overline{\mathbb{H}}| = 85.4$ kA m^{-1}).

Figure 11.3: Displaced solution showing the magnetic induction magnitude. The inset images provide a view of the microstructure located at the indicated positions. The original configuration of the macroscopic body is indicated by the dotted outline.

microstructure renders a compression-dominated macroscopic response with a significant lateral extension along the displacement-constrained surface. In contrast, the chain-like anisotropic microstructure induces large interparticle forces in the high volume fraction region that dominate those produced in the other laminate layer, leading to bending in the macroscopic domain. With $|\mathbb{h}| \approx 110\,\text{kA}\,\text{m}^{-1}$, the microstructural deformation is such that in the $\phi_{\text{CIP}} = 10\,\%$ layer the particle network is near a state of percolation.

The strong interparticle interactions in the high volume fraction microstructures can be deduced by examining the isocontours of the magnetic induction. Particularly in the chain-like RVE, the generation of strong magnetic dipoles in the particles that have very little matrix material between them is readily observed. For the lower volume fraction media, the magnetic induction at the boundaries is more uniform than for their higher volume fraction counterparts. Comparing the final configuration of the microstructures to their reference states shown in Figure 11.1, it is clear that at the examined positions each microstructure locally undergoes significant shortening in the $\mathbf{e}_2$ direction, and elongation in the $\mathbf{e}_1$ direction. In the case of the chain-like composite, a finite rotation associated with the bending of the macroscopic geometry is also noticeable.

11.2 Immersion of magnetic bodies in free space

It is well understood that forces arising from the Maxwell stress [414, 338] in the free space are often central to functioning of MAP-based devices. As was discussed in Section 6.3.3, one of the key issues that arises due to its inclusion is that of numerical stability; for this reason, it may be preferential to neglect the Maxwell stress contribution [466]. However, for the most part incorporation of its influences into macroscopically-focussed numerical models, especially simulations of industrial components, is a critical feature that cannot often be ignored.

There exist several methods whereby the contribution of the free space can be captured. The coupled BEM–FEM [539, 540, 504] has a relatively straightforward implementation, but has the disadvantage that the resulting system of equations is dense and non-symmetric. Inspiration for both free space-solid coupling and mesh movement algorithms can also be taken from the realm of fluid-structure interaction, where monolithic approaches [39, 108] and immersed methods [419, 50, 195] are commonly used. In the following section, we will focus on another option, namely that whereby the FEM is used in conjunction with a truncation of the surrounding free space. With this approach, the FE mesh is extended considerably beyond the magnetoelastic body in order to capture the correct response of the system (that being the deformable body and surrounding free space collectively). This allows for the Maxwell stress to be accounted for in a practical manner. Furthermore, is admits the decoupling of the coupled problem and the mesh update scheme, thereby facilitating the use of a robust scheme for describing the movement of the free space mesh during large deformations of the magnetoelastic body.

11.2.1 Mesh motion in the free space

It is often anticipated that at the solid body-free space interface, and at sharp corners of the geometry where magnetic singularities arise, there will exist a steep magnetic field gradient due to the differing magnetic properties across this interface. In order to capture these phenomena accurately, typically a fine FE discretisation would be employed in this locale. With reference to the notation presented in Figure 6.1, the large deformations that are simulated may lead to impingement by the body on the fine adjacent cells, as is illustrated in Figure 11.4a; if sufficiently large, this deformation renders the displacement map in $\mathscr{S}_0^{\mathscr{B},\mathrm{h}}$ (and perhaps regions of $\mathscr{S}_0^{\mathrm{h}} \backslash \mathscr{S}_0^{\mathscr{B},\mathrm{h}}$) non-invertible. One method of circumventing this issue would be to impose of a fictitious motion to the free space mesh, such as is illustrated in Figure 11.4b, which would facilitate modelling the scenario at higher loads and deformations. It is important to note that, with the use of such an update procedure, all spatial quantities (such as those listed in Section 5.3.2) computed in $\mathscr{S}_0$ must consider the fictitious map.

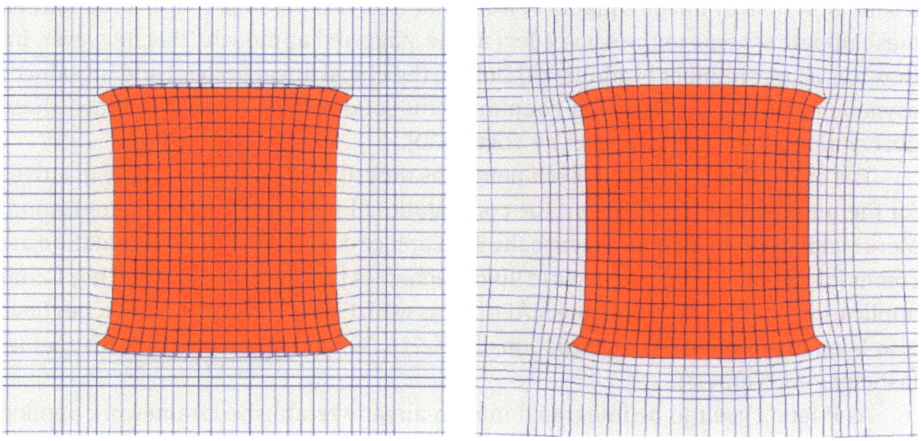

(a) No update of mesh in free space. (b) With update of mesh in free space.

Figure 11.4: Illustration of mesh update in the free space. The deformable and magnetisable quasi-incompressible body (shown in red) is pinned at its corners and exposed to a magnetic field that permeates it and the free space (shown in grey). If the deformation is sufficiently large, the cells in the free space and adjacent to the body may become inverted, leading to computational issues.

There are a number of approaches to incorporate an update to the free space deformation map within a numerical framework. The free space can be directly modelled as a compliant elastic medium with the magnetic properties of a vacuum, as was done in [255, 256]. This method, however, has several drawbacks. The first is that there is a marginal penalty (mechanical) traction applied to the magnetoelastic body $\mathscr{B}_0$. The second, and most problematic for geometries incorporating large free space volume, is that the number of elastic DoFs increases significantly. Thirdly, the condition number for the resulting linear system of equations may be poor due to the compliant properties chosen for the fictitious elastomer in $\mathscr{S}_0$.

Alternatively, as rationalised in Section 6.1.1 and elucidated in Table 6.1, the elastic DoFs in the free space may be removed from the primary coupled problem. By employing this approach, the free space update is completely decoupled from the primary problem, and it is necessary to compose a secondary problem to compute a suitable fictitious deformation map $\boldsymbol{\varphi}$ in $\mathscr{S}_0$. In fact, there is no limitation on the movement of DoFs within the free space domain except that material domains with different magnetic permeabilities must retain their original topology. The primary and auxiliary problems are solved using a staggered approach, with the mesh update being necessary only if the displacement of the magnetoelastic body is larger than a threshold value (this being related to the mesh size). If the secondary problem is not evaluated, then only the displacement DoFs on $\Gamma_0^{\mathscr{B}\mathscr{S}}$ need be incrementally updated to ensure that $\boldsymbol{\varphi}$ in $\mathscr{S}_0^{\mathscr{B},\mathrm{h}}$ remains synchronised between the primary and auxiliary problems. The resulting continuous map is then used in the computation of equation (6.35) or a similar system of linear equations. Note that it is also possible to use a monolithic mesh update scheme to the same effect as the staggered approach. This, however, introduces a large number of additional DoFs to the global linear system, and introduces asymmetry as a result of one-way coupling at the solid body-free space interface; these observations are detailed further by Vogel [530].

The optimal movement of mesh points, as described by Knupp et al. [271], strives for the retention or improvement the computational properties of the discretisation. More simplified approaches that use the 3-D Laplace equations as their foundation are commonly employed for problems in fluid mechanics [232, 321]; this method has found application in scenarios involving MAPs [74] as well. Elasticity-based approaches have also been explored in fluid simulations [239, 217, 39] and applications in magnetoelasticity [472, 530, 417].

As it lends itself to be integrated into an already established framework applied to FEM solutions of MAPs, we will concentrate further on the general physics-based approach for the free space update that has been presented by Pelteret et al. [417]. Therein it was proposed that the equation governing the mesh motion problem and the associated linear constitutive law be defined as

$$\nabla_0 \cdot [\mathscr{K}\, [\nabla_0 \boldsymbol{\varphi}]^{\mathrm{S}}] = \mathbf{0} \quad \text{on} \quad \mathscr{S}_0 \quad \text{with} \quad \mathscr{K}(\mathbf{X}) := k(\mathbf{X})\, [[1-\alpha]\, \mathscr{I} + \alpha \mathbf{I} \otimes \mathbf{I}] \quad , \tag{11.1}$$

where $k(\mathbf{X}) > 0$ is a parameter that governs the effective stiffness magnitude at the given coordinate position. The balance law stated in equation $(11.1)_1$ can be identified as that for quasi-static linear elasticity, but its properties are dependent on the choice of the blend parameter $-1 \leq \alpha \leq 1$ in equation $(11.1)_2$. The characteristics of the algorithm based on the prescribed blend parameter are summarised in Table 11.1. Since the displacement on the solid body-free space interface is prescribed by the solution

Table 11.1: Noteworthy choices of the blending parameter and their influence on the physical interpretation of the mesh update scheme. These can be derived from the relationship between the Poisson ratio, shear modulus and Lamé parameter, and by considering the governing equation written in the form of the static Navier–Cauchy displacement equations, $[\lambda + \mu]\, \nabla [\nabla \cdot \boldsymbol{\varphi}] + \mu \nabla \cdot [\nabla \boldsymbol{\varphi}] = \mathbf{0}$.

Blend parameter α	Effective Poisson ratio ν^*	Characteristics of update algorithm
−1	−0.5	Laplace smoothing
0	0	Highly compressible material
$\to 1$	$\to 0.5$	Increasing incompressibility
1	0.5	Fully "incompressible" media; $\mu = 0$, λ is finitely bounded

of the primary coupled problem (denoted as $\overline{\boldsymbol{\varphi}}$), the boundary conditions associated with the auxiliary problem are given by

$$\boldsymbol{\varphi} = \overline{\boldsymbol{\varphi}} \quad \text{on} \quad \Gamma_0^{\mathscr{B}\mathscr{S}} \quad \text{and} \quad \boldsymbol{\varphi} = \overline{\overline{\boldsymbol{\varphi}}} \quad \text{on} \quad \partial \mathscr{S}_0 \quad . \tag{11.2}$$

We further note that mesh sliding is allowed on the far-field boundary $\partial \mathscr{S}_0$, but the shape of the geometry must remain the same while a magnetic load is applied.

Through the function $k(\mathbf{X})$, it is possible to dictate the effective local stiffness of the fictitious free space material. Reasonable choices for the stiffness function include [232] the homogeneous distribution, power law or exponential law

$$k = \overline{k} = 1 \quad \text{or} \quad k(d) = d^{-p} \quad \text{or} \quad k(d) = \exp(-d\,p) \quad , \tag{11.3}$$

where p is a fixed parameter and $d(\mathbf{X})$ is a metric based on the discretisation or geometry. Two sensible choices for the aforementioned metric are for it to describe the referential FE cell diameter, or for it to measure the distance from $\mathbf{X}$ to the nearest solid-body vertex. Both of these measurements may take place in either the reference or spatial configurations, although using the former has the benefit of implementation simplicity when using a total Lagrangian formulation. Little relative motion between the vertices of small diameter cells can be accommodated before they become significantly deformed. Thus, given the first option, the most deformable cells would be those of largest diameter, while the smaller cells would tend to convect with the solid body. The effectiveness of the second option relies on the construction of a mesh to suit its characteristics, in particular that the smallest cells should be located in the near vicinity of the solid body. As it stiffens cells adjacent to the solid, their mapping should be translation dominated, while those in the far-field deform more freely. Both options are suitable for application to modelling MAPs with the surrounding free space, as the largest magnetic field gradients are expected to be found at the interface of the solid and free space. It is therefore likely that the discretisation will be finely concentrated

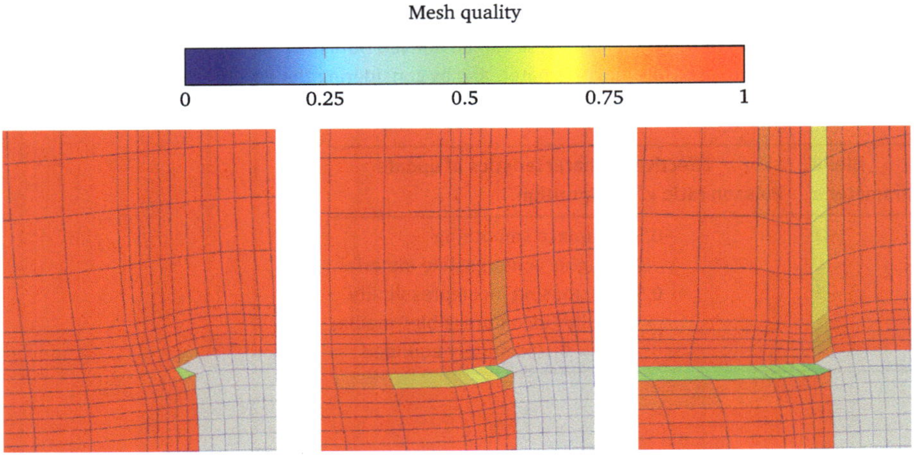

(a) $\alpha = -1$. **(b)** $\alpha = 0.98$. **(c)** $\alpha = 1$.

Figure 11.5: Comparison of a selection of choices for the blend parameter. The stiffness function $k(\mathbf{X})$ is given by equation $(11.3)_2$, with $d(\mathbf{X})$ measuring the distance to the nearest solid vertex and $p = 2$. The mesh quality measure is the scaled Jacobian [507]. The boundary layer mesh surrounding the elastic body is clearly visible. [417, fig. 5 (reproduced with permission)]

in this region and not in the far-field where a more uniform magnetic field is expected. Figure 11.5 depicts the resulting update for a finite deformation of the representative mesh first shown in Figure 11.4.

Numerical example: MAP with dispersed chain-like particle structures immersed in free space

To further illustrate the application of the solid body-free space discretisation and the associated mesh update procedure, we return to a more exotic application of the dispersed particle chain phenomenological model introduced in Section 7.3. The problem described graphically in Figure 11.6a is one of a thick-walled tube comprising a magnetisable material that is placed within an axially aligned magnetic field. In particular, it is assumed that the microstructure has an helical arrangement of its average preferred direction. The dimensions of the solid and surrounding free space geometries are listed in Table 11.2; it is assumed that the geometry, material structure and loading is symmetric around the mid-plane, therefore necessitating the modelling of only half of the domain. A section through the geometry is shown in Figure 11.6b to clarify the assignment of the various material to the different finite element subdomains defined in Figure 6.1. Observations are made at $\mathbf{P}(0, 0.75, 2.25)$ mm, a measurement point near the end of the tube exposed to the free space and halfway between the inner and outer radii. In this example, a trilinear ansatz is applied to approximate each individual component of the solution field.

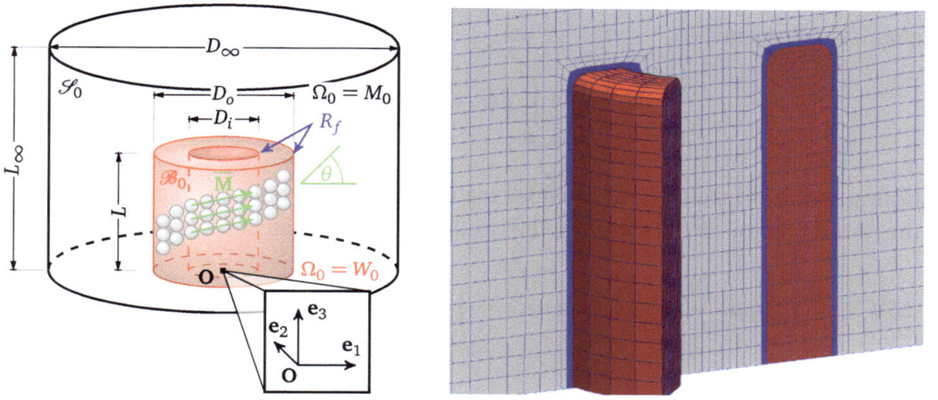

(a) Geometry and domain decomposition. **(b)** Section of geometry displaying the mesh.

Figure 11.6: Problem configuration with microstructure orientation and FE discretiation. Both the displacement and magnetic fields were discretised using linear FEs. In Figure 11.6b, the coloured regions align with those depicted in Figure 6.1. [472, figs. 16,17]

Table 11.2: Dimensions of the geometry for a magnetisable tube immersed in free space.

Parameter	Value	Unit	Parameter	Value	Unit
D_i	1	mm	R_f	0.25	mm
D_o	2	mm	D_∞	6	mm
L	2.5	mm	L_∞	5	mm

In terms of the boundary conditions applied to the discrete problem, the maximum potential difference between the mid-plane and far-field axial boundary is $\Delta\Phi^{\max} = 250$ A; this induces an axially-aligned magnetic field in the free space. Along the mid-plane, symmetry conditions are applied to the mechanical problem. In addition, it is assumed that the outer radius is completely fixed mechanically. Lastly, the spatial traction condition that is applied on the radial surfaces of the tube is

$$\mathbf{t}_t^{\text{mech,ext}} = -\lambda\,[\boldsymbol{\varphi} \cdot \mathbf{e}_r]\,\mathbf{e}_r \quad , \tag{11.4}$$

with $\lambda = 250\,\text{N}\,\text{mm}^{-3}$ and $\mathbf{e}_r$ describing the radial direction of the current position $\mathbf{x}$ onto the surface of the tube. In essence, this quasi-contact condition penalises the radial dilation of the outer surface of the tube. Using equation (5.10), the weak constraint can be re-expressed in terms of the equivalent referential traction as

$$\mathbf{t}_0^{\text{mech,ext}}\,\text{dA} \equiv \boldsymbol{\sigma} \cdot \mathbf{n}\,\text{da} = \left[\mathbf{t}_t^{\text{mech,ext}} \otimes \mathbf{n}\right] \cdot \text{cof}\,(\mathbf{F}) \cdot \mathbf{N}\,\text{dA} \tag{11.5}$$

Returning to the variational formulation of the coupled problem, as stated in Section 5.3.7.2, in contrast to equation (5.184) the now solution-dependent external mechanical energy is

$$\delta\Pi^{\mathrm{ext}} = -\int_{\partial\mathscr{B}_0^{\mathrm{t}}} \delta\boldsymbol{\varphi}\cdot\mathbf{t}_0^{\mathrm{mech,ext}}(\boldsymbol{\varphi})\,\mathrm{dA} \quad . \tag{11.6}$$

Noting that it is necessary to consider the dependence of both the penalty term and the normal vector on the solution, the resulting linearisation of the external energy is

$$\Delta\delta\Pi^{\mathrm{ext}} = -\int_{\partial\mathscr{B}_0^{\mathrm{t}}} \delta\boldsymbol{\varphi}\cdot\frac{\mathrm{d}\mathbf{t}_0^{\mathrm{mech,ext}}}{\mathrm{d}\boldsymbol{\varphi}}\cdot\Delta\boldsymbol{\varphi}\,\mathrm{dA} = -\int_{\partial\mathscr{B}_0^{\mathrm{t}}} \delta\boldsymbol{\varphi}\cdot\left[\frac{\partial\mathbf{t}_0^{\mathrm{mech,ext}}}{\partial\boldsymbol{\varphi}}\cdot\Delta\boldsymbol{\varphi} + \frac{\partial\mathbf{t}_0^{\mathrm{mech}}}{\partial\mathbf{F}}:\Delta\mathbf{F}\right]\mathrm{dA} \quad . \tag{11.7}$$

The constitutive law for the solid medium is chosen to be similar in nature to that presented previously in equation (7.80). Ignoring the free field energy within the body ($M_0 = 0$), the phenomenological model representing the volumetric, matrix and chain components of the total strain energy (respectively a polyconvex function, a coupled Neo–Hookean type model, and a Fung-type constitutive law) are

$$\begin{aligned} W_0^{\mathrm{vol}} &= \frac{\kappa_{\mathrm{e}}}{4}\left[J^2 - 1 - 2\ln(J)\right] \quad , \\ W_0^{\mathrm{m}} &= \frac{\mu_{\mathrm{e}}}{2}\left[I_1\left(\overline{\mathbf{C}}\right) - \mathbf{I}:\mathbf{I}\right] + n_2 I_4(\mathbb{H}) + n_3 I_7(\mathbf{C},\mathbb{H}) \quad , \\ W_0^{\mathrm{c}} &= \frac{\mu_{\mathrm{c}}}{2}\left[\exp\left(\left[I_1(\mathbf{C}^{\mathrm{c}}) - \mathbf{G}:\mathbf{I}\right]^2\right) - 1\right] + n_5 I_4(\mathbb{H}^{\mathrm{c}}) + n_6 I_7(\mathbf{C},\mathbb{H}^{\mathrm{c}}) \quad . \end{aligned} \tag{11.8}$$

The parameter κ_{e} is the elastic bulk modulus, while μ_{c} denotes the effective shear modulus of the particle chains. The pertinent fixed material constants are listed in Table 11.3. Lastly, the mesh update was performed by choosing $\alpha = 0$ in conjunction with the power law stiffness function (equation $(11.3)_2$ with the exponent $p = 1$) and the metric $d(\mathbf{X})$ represented the FE cell diameter.

Table 11.3: Constitutive parameters chosen for the magnetisable tube immersed in free space.

Parameter	Value	Unit	Parameter	Value	Unit
μ_{e}	30	kPa	n_2, n_5	$0.5\mu_0$	$\mathrm{H\,m^{-1}}$
κ_{e}	1490	kPa	n_3, n_6	$-\mu_0$	$\mathrm{H\,m^{-1}}$
θ	45	deg			

Figure 11.7 presents the magnetic fields and solid body displacement generated under the maximal magnetic loading when there exists no material anisotropy. As expected, the magnetic induction within the magnetoelastic body is predominantly aligned with the magnetic field. Although the magnetic field gradient developed at the axial extent of the elastic body promotes extension of the elastomer, the complex mechanical and

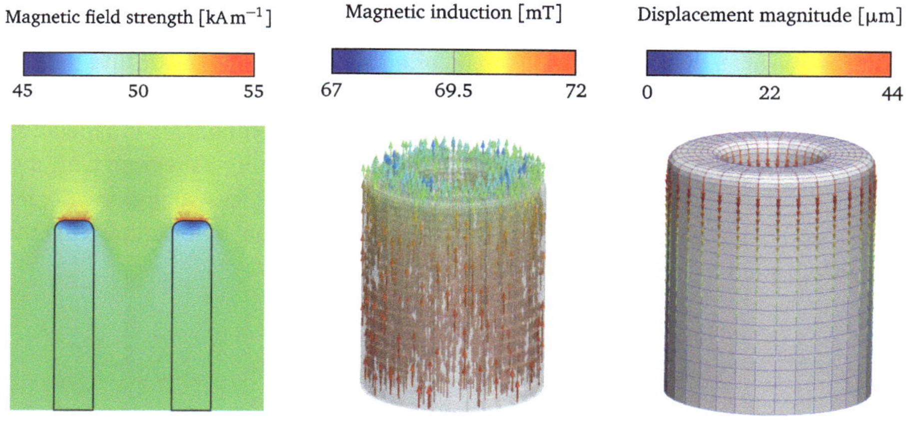

(a) Magnetic field strength. **(b)** Magnetic induction. **(c)** Displacement.

Figure 11.7: Solution fields for a dispersion parameter of $\kappa = \frac{1}{3}$ (isotropic). The average chain orientation angle $\theta = 45°$ and the chains are assumed to be mechanically weak ($\frac{\mu_c}{2\mu_e} = \frac{1}{10000}$). [472, figs. 18a,18b,19a (reproduced with permission)]

magnetic loading conditions in conjunction with the chosen constitutive law results in an overall shortening of the material and inwards radial expansion.

When introducing some preferred directionality into the material, the magnetic induction developed within $\mathscr{B}_0$ aligns with the average orientation $\overline{\mathbf{M}}$ of the underlying chain-like structures and its strength is reinforced. This in turn induces a component of material contraction in the circumferential direction, thus introducing a torsion effect into the material. Both of these phenomena are observable in Figure 11.8, and suggest that this device may be used as a torsional actuator.

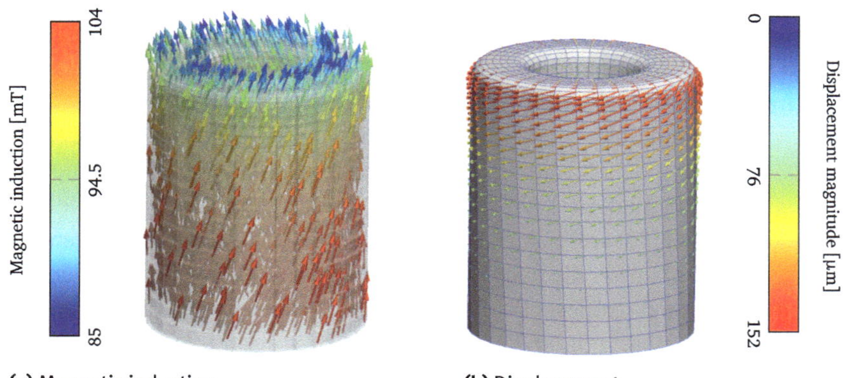

(a) Magnetic induction. **(b)** Displacement.

Figure 11.8: Solution fields for a dispersion parameter of $\kappa = \frac{1}{3000}$ (almost completely transversely isotropic). The average chain orientation angle $\theta = 45°$ and the chains are assumed to be mechanically weak ($\frac{\mu_c}{2\mu_e} = \frac{1}{10000}$). [472, figs. 18c, 19b (reproduced with permission)]

The overall behaviour exhibited by this MAP is influenced by several factors, including the degree of chain dispersion, the chain orientation, and the constitutive parameters selected for both the bulk material and the underlying microstructure (in particular, their mechanical, magnetic and magneto-mechanical components independent of one another). As is observed in Figure 11.9, when considering a medium with mechanically weak particle chains (i. e. one with large interparticle spacing) the degree of dispersion has a greater influence on the degree of rotational deformation that the material will exhibit than its axial contraction.

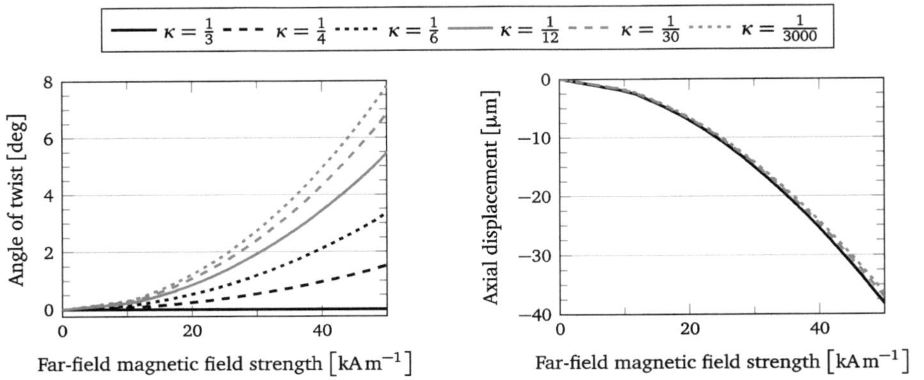

(a) Angle of twist. **(b)** Axial displacement.

Figure 11.9: Influence of dispersion parameter on deformation at measurement point **P** when the chains are mechanically weak ($\frac{\mu_c}{2\mu_e} = \frac{1}{10000}$). The deformation field shown in Figure 11.8b is considered to have undergone positive angle of twist. [472, fig. 20]

However, as can be seen in Figure 11.10, for which mechanically strong particle chains have been modelled, the dispersion of the microstructure has a much greater role for both the axial and rotational motion of the material. It is immediately observable that the torsion direction has switched between this case and that presented in Figure 11.9. This is because the mechanical reinforcement along the particle orientation axis leads to the preferred direction of deformation being in the plane of transverse isotropy as the material is more compliant in this direction. The total twist and overall shortening of the tube decreases significantly. Interestingly, when $\frac{\mu_c}{\mu_e} \approx 10$ (not shown) then the magnetically-generated contractile force in the azimuthal direction is balanced almost precisely by the additional stiffness provided by the particle chains in this direction.

By modifying the orientation angle of the microstructure, it is possible to maximise the magnetostrictive behaviour of the tube. Shown in Figure 11.11 is the twist angle and amount of contraction exhibited by the MAP for a range of microstructural orientations. For a non-arbitrary orientation (in this case, 60° for the examined parameters), the rotation induced within the short sample as well as its contraction is signifi-

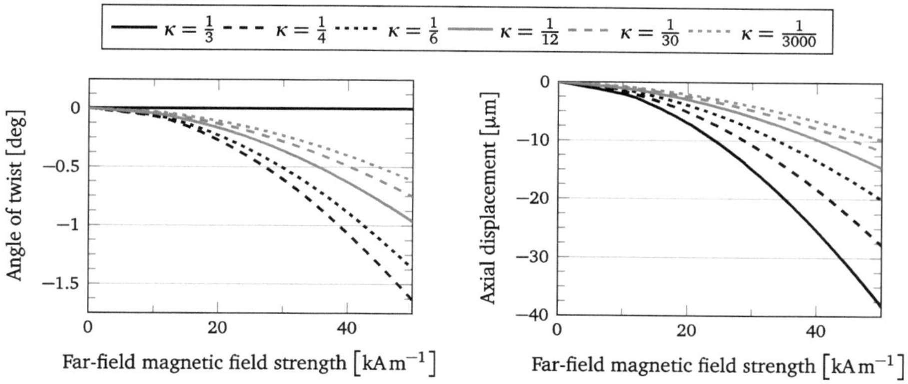

(a) Angle of twist. **(b)** Axial displacement.

Figure 11.10: Influence of dispersion parameter on deformation at measurement point **P** when the chains are mechanically stiff ($\frac{\mu_c}{2\mu_e} = 100$). [472, fig. 23]

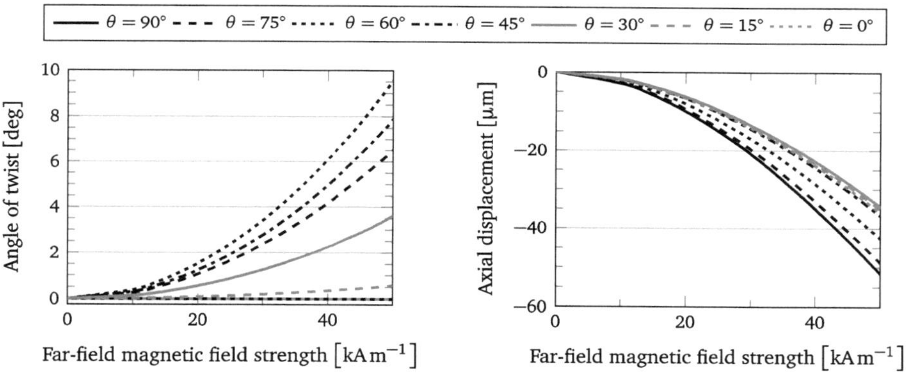

(a) Angle of twist. **(b)** Axial displacement.

Figure 11.11: Influence of chain orientation on deformation at measurement point **P**. When $\theta = 90°$ the chains are, on average, axially aligned while $\theta = 0°$ indicated azimuthal alignment. The dispersion parameter $\kappa = \frac{1}{3000}$ (almost completely transversely isotropic) and the chains are assumed to be mechanically weak ($\frac{\mu_c}{2\mu_e} = \frac{1}{10000}$). [472, fig. 21]

cantly increased when compared to the baseline orientation. When the microstructure is aligned with the magnetic field then the contractile properties are maximised, while a tangential orientation causes the embedded particle chains to have no magnetically induced influence on the macroscopic body's response to the magnetic field in which it is immersed.

11.3 Mixed variational approach for quasi-incompressible media

As it is a rubber-like material, the matrix component of the MAP described in Chapter 2 exhibits near incompressible behaviour. Due to the inherent stiffness of the embedded CIP particles, the composite also exhibits quasi-incompressible behaviour at the macro-scale. When simulating this media using the FEM it is therefore necessary to take this into consideration. Low-order finite elements tend to exhibit volumetric locking in quasi-incompressible solids (as well as shear locking in bending dominated problems). However, due to their added computational expense, in practical applications the use of higher-order elements has historically often not been preferred. Numerical tools, such as the mixed finite element method, offer a compromise between these two approaches by enabling the use of low-order finite elements through the addition of, in some cases, only a few additional DoFs.

The Veubeke–Hu–Washizu principle [150] is considered the foundational component in the development of mixed methods applied to the FEM to prevent the occurrence of locking in elastic problems. Simo et al. [495] presented a formulation for non-linear quasi-incompressible elasticity, based on the seminal work by Nagtegaal et al. [394], that utilises two additional scalar fields; further detail is given in [494, 358, 61, 204, 559], and is expanded on in [489] in the context of viscoelasticity. This particular formulation, although not being without some drawbacks, offers a balance between computational expense and implementational simplicity [61] when compared to other mixed methods [482, 479, 169] and alternatives approaches, such as that of enhanced strains [490, 578, 559]. As is expressed by Vogel et al. [531], the application of these approaches in coupled problems is not only limited to the elastic fields. Hu–Washizu mixed formulations have been successfully applied to the formulation of coupled problems by Ask et al. [18, 19, 20], Jabareen [229], and Ortigosa and Gil [409], albeit for conditions that exclude the surrounding free space.

In the following sections, we present the formulation detailed by Pelteret et al. [417] (which in turn was inspired by Simo et al. [495] and Ask et al. [20]) that considers the decomposition of the domain into the elastic and free space regions. In this way, we provide a more explicit expression of the energies and resulting system of equations than was previously presented in Section 5.3.7 and Chapter 6. Furthermore, it allows us to present the rationale behind certain key aspects carefully taken into consideration during the computational modelling of an industrially relevant problem.

Variational formulation

In this section, we present a mixed variational approach from which we derive the weak form of the governing equations. Using Section 5.3.7.2 as a starting point, we now define the total potential energy functional as

$$
\begin{aligned}
\Pi &= \Pi^{\text{int}} + \Pi^{\text{ext}} \\
&= \int_{\mathscr{B}_0} \Bigg[\underbrace{W_0^{\text{vol}}(\tilde{J}) + \overline{W}_0\left(\overline{\mathbf{C}}, \mathbb{H}\right) + M_0\left(\tilde{J}, \mathbf{C}, \mathbb{H}\right)}_{\Omega_0(\tilde{J},\mathbf{C},\mathbb{H})} + \tilde{p}\,[J - \tilde{J}] \Bigg] \,\mathrm{dV} + \int_{\mathscr{S}_0} M_0\left(J, \mathbf{C}, \mathbb{H}\right) \,\mathrm{dV} \\
&\quad - \int_{\mathscr{B}_0} \boldsymbol{\varphi} \cdot \mathbf{b}_0^{\text{mech}} \,\mathrm{dV} - \int_{\partial\mathscr{B}_0^{\text{t}}} \boldsymbol{\varphi} \cdot \mathbf{t}_0^{\text{mech,ext}} \,\mathrm{dA} - \int_{\partial\mathscr{S}_0^{\mathbb{B}}} \Phi\,[\mathbb{B}_\infty \cdot \mathbf{N}_\infty] \,\mathrm{dA} \quad .
\end{aligned}
\tag{11.9}
$$

It is evident from equation (11.9) that there are three unknown elastic field variables and one related to the magnetic problem. We recognise the Lagrange multiplier $\tilde{p}$ as the pressure response, which is related to the hydrostatic pressure by $\tilde{p} = -p^{\text{hyd}}$. In equation (11.9), the fourth term in the integral over the solid body serves to penalise the constraint stating that there should exist no difference between the pointwise volumetric Jacobian $J = J(\boldsymbol{\varphi}) = \det(\mathbf{F})$ and the dilatation $\tilde{J}$. The additive split of $\Omega_0\left(\tilde{J}, \mathbf{C}, \mathbb{H}\right)$, the total strain energy function within the magnetisable body, is related to that described in equations (7.11) and (7.14). It should be noted that the coupled material isochoric strain energy function is to be expressed in terms of the dilatation, an isochoric deformation tensor and the referential magnetic field vector; in contrast, the free space energy within the body is parameterised by the canonical deformation tensor and, now, the dilatation (due to the mixed formulation). Furthermore, unlike Simo and Taylor [494], there is no direct imposition of the incompressibility condition via a penalty or Lagrange multiplier method. Instead, a bulk modulus that is representative of a near-incompressible material will be prescribed.

First variation

The stationary (saddle-)point [42] $\min_{\boldsymbol{\varphi},\tilde{J}} \max_{\tilde{p},\Phi} \Pi \;\Rightarrow\; \delta\Pi = 0$ defines the equilibrium solution for the boundary value problem. Using the Gâteaux derivative

$$
\delta\Pi = \left[\delta_{\boldsymbol{\varphi}}\Pi^{\text{int}} + \delta_{\tilde{p}}\Pi^{\text{int}} + \delta_{\tilde{J}}\Pi^{\text{int}} + \delta_{\Phi}\Pi^{\text{int}}\right] + \left[\delta_{\boldsymbol{\varphi}}\Pi^{\text{ext}} + \delta_{\Phi}\Pi^{\text{ext}}\right] \quad ,
\tag{11.10}
$$

the components of the first variation are

$$
\delta_{\boldsymbol{\varphi}}\Pi^{\text{int}} = \int_{\mathscr{B}_0} \delta\mathbf{E} : \underbrace{\left[\overline{\mathbf{S}} + \mathbf{S}^{\max} + \tilde{p}J\mathbf{C}^{-1}\right]}_{\mathbf{S}^{\text{tot}}} \,\mathrm{dV} + \int_{\mathscr{S}_0} \delta\mathbf{E} : \mathbf{S}^{\max} \,\mathrm{dV} \quad ,
\tag{11.11a}
$$

$$
\delta_{\tilde{p}}\Pi^{\text{int}} = \int_{\mathscr{B}_0} \delta\tilde{p}\,[J - \tilde{J}] \,\mathrm{dV} \quad ,
\tag{11.11b}
$$

$$
\delta_{\tilde{J}}\Pi^{\text{int}} = \int_{\mathscr{B}_0} \delta\tilde{J} \left[\frac{\partial W_0^{\text{vol}}}{\partial\tilde{J}} + \frac{\partial M_0}{\partial\tilde{J}} - \tilde{p}\right] \mathrm{dV} \quad ,
\tag{11.11c}
$$

$$
\delta_{\Phi}\Pi^{\text{int}} = \int_{\mathscr{B}_0 \cup \mathscr{S}_0} \delta\mathbb{H} \cdot \mathbb{B} \,\mathrm{dV} \quad ,
\tag{11.11d}
$$

where the variation for the Green–Lagrange strain tensor is given by

$$\delta\mathbf{E} = \text{sym}\left[\mathbf{F}^T \cdot \delta\mathbf{F}\right] \quad , \tag{11.12}$$

and that of the deformation gradient and magnetic field are respectively stated in equation $(5.170)_1$ and equation (5.184). The first variation of the external energy is stated in equation (5.185). As derived from equation (7.10), the total stress and magnetic induction within the elastic body $\mathscr{B}_0$ are

$$\mathbf{S}^{\text{tot}} = \overline{\mathbf{S}} + \mathbf{S}^{\text{max}} + \tilde{p}J\mathbf{C}^{-1} \quad , \quad \overline{\mathbf{S}} = 2\frac{\partial \overline{W}_0}{\partial \mathbf{C}} = \left[2\frac{\partial \overline{W}_0}{\partial \overline{\mathbf{C}}}\right] : \widehat{\mathscr{P}} \quad , \quad \mathbb{B} = -\frac{\partial \overline{W}_0}{\partial \mathbb{H}} + \mathbb{B}^{\text{max}} \tag{11.13}$$

with the referential isochoric projection tensor $\widehat{\mathscr{P}}$ defined in equation (7.20). Note that due to the addition of the pressure response field, the definition of the total stress differs to that given in equation (7.16). As it is a non-magnetisable medium, we recall from equation (5.90) that the total stress in the free space reduces to the Maxwell contribution. Therefore, we may express the Maxwell stress and magnetic induction in the free space as

$$\mathbf{S}^{\text{max}} = 2\frac{\partial M_0}{\partial \mathbf{C}} \quad , \quad \mathbb{B}^{\text{max}} = -\frac{\partial M_0}{\partial \mathbb{H}} \quad . \tag{11.14}$$

Note that we have expressed the Maxwell contributions in generic form as the parameterisations are different within the body $\mathscr{B}_0$ (where $\mathbf{S}^{\text{max}} = \mathbf{S}^{\text{max}}(\tilde{J}, \mathbf{C}, \mathbb{H})$ and $\mathbb{B}^{\text{max}} = \mathbb{B}^{\text{max}}(\tilde{J}, \mathbf{C}, \mathbb{H})$) and the surrounding free space $\mathscr{S}_0$ (in which $\mathbf{S}^{\text{max}} = \mathbf{S}^{\text{max}}(J, \mathbf{C}, \mathbb{H})$ and $\mathbb{B}^{\text{max}} = \mathbb{B}^{\text{max}}(J, \mathbf{C}, \mathbb{H})$).

The relationship between the developed weak form and the strong form is collectively detailed in [73, 69] (for a more simplified form of the electro- and magneto-static problems) and [204] (for the three-field elastostatic problem). The Euler–Lagrange equations associated with the residual are equation (5.135) and equation $(5.114)_2$, along with

$$\tilde{p} = \frac{\partial \Omega_0}{\partial \tilde{J}} = \frac{\partial W_0^{\text{vol}}}{\partial \tilde{J}} + \frac{\partial M_0}{\partial \tilde{J}} \quad \text{and} \quad J = \tilde{J} \quad . \tag{11.15}$$

From equations (11.11), it may be observed that the Sobolev space in which the variations must lie are

$$\delta\Phi \in \mathscr{H}^1\left(\overline{\mathscr{B}_0 \cup \mathscr{S}_0}\right) \quad , \quad \delta\boldsymbol{\varphi} \in \mathscr{H}^1\left(\overline{\mathscr{B}_0 \cup \mathscr{S}_0}\right) \quad , \quad \delta\tilde{p}, \delta\tilde{J} \in \mathscr{L}^2\left(\overline{\mathscr{B}_0}\right) \tag{11.16}$$

subject to the constraints

$$\delta\Phi = 0 \quad \text{on} \quad \partial\mathscr{S}_0^{\Phi} \quad , \quad \delta\boldsymbol{\varphi} = \mathbf{0} \quad \text{on} \quad \partial\mathscr{B}_0^{\boldsymbol{\varphi}} \cup \left[\overline{\mathscr{S}_0} \backslash \Gamma_0^{\mathscr{B}\mathscr{S}}\right] \tag{11.17}$$

when considering the argumentation preceding equation (6.10). This aligns with what was stated previously in Chapter 6 for the two-field problem derived directly from the strong form.

Linearisation

The resulting linearisation of equations (11.11) using a first-order Taylor expansion is

$$0 \doteq \delta\Pi\big|_{\boldsymbol{\varphi},\tilde{p},\tilde{J},\Phi} + \left[\Delta\delta_{\boldsymbol{\varphi}}\Pi^{\text{int}} + \Delta\delta_{\tilde{p}}\Pi^{\text{int}} + \Delta\delta_{\tilde{J}}\Pi^{\text{int}} + \Delta\delta_{\Phi}\Pi^{\text{int}}\right] \quad . \tag{11.18}$$

Assuming a dead load, the details of the direct terms in the linearisation are

$$\begin{aligned} \Delta_{\boldsymbol{\varphi}}\delta_{\boldsymbol{\varphi}}\Pi^{\text{int}} &= \int_{\mathscr{S}_0} \Delta\delta\mathbf{E} : \mathbf{S}^{\max}\,\mathrm{dV} + \int_{\mathscr{S}_0} \delta\mathbf{E} : \mathscr{H}^{\max} : \Delta\mathbf{E}\,\mathrm{dV} + \int_{\mathscr{B}_0} \Delta\delta\mathbf{E} : \mathbf{S}^{\text{tot}}\,\mathrm{dV} \\ &+ \int_{\mathscr{B}_0} \delta\mathbf{E} : \underbrace{\left[\overline{\mathscr{H}} + \mathscr{H}^{\max} + \tilde{p}J\left[\mathbf{C}^{-1}\otimes\mathbf{C}^{-1} - 2\mathbf{C}^{-1}\overline{\otimes}\mathbf{C}^{-1}\right]\right]}_{\mathscr{H}^{\text{tot}}} : \Delta\mathbf{E}\,\mathrm{dV} \quad , \end{aligned} \tag{11.19a}$$

$$\Delta_{\tilde{J}}\delta_{\tilde{J}}\Pi^{\text{int}} = \int_{\mathscr{B}_0} \delta\tilde{J}\left[\frac{\partial^2 W_0^{\text{vol}}}{\partial\tilde{J}^2} + \frac{\partial^2 M_0}{\partial\tilde{J}^2}\right]\Delta\tilde{J}\,\mathrm{dV} \quad , \tag{11.19b}$$

$$\Delta_{\Phi}\delta_{\Phi}\Pi^{\text{int}} = -\int_{\mathscr{B}_0\cup\mathscr{S}_0} \delta\mathbb{H}\cdot\mathbf{D}\cdot\Delta\mathbb{H}\,\mathrm{dV} \quad , \tag{11.19c}$$

where the linearisation of the Green–Lagrange strain tensor and its variation are

$$\Delta\mathbf{E} = \text{sym}\left[\mathbf{F}^T\cdot\Delta\mathbf{F}\right] \quad , \quad \Delta\delta\mathbf{E} = \text{sym}\left[\Delta\mathbf{F}^T\cdot\delta\mathbf{F}\right] \quad . \tag{11.20}$$

Furthermore, the coupling terms arising from the linearisation are

$$\Delta_{\tilde{p}}\delta_{\boldsymbol{\varphi}}\Pi^{\text{int}} = \int_{\mathscr{B}_0} \delta\mathbf{E} : J\mathbf{C}^{-1}\,\Delta\tilde{p}\,\mathrm{dV} \quad , \quad \Delta_{\boldsymbol{\varphi}}\delta_{\tilde{p}}\Pi^{\text{int}} = \int_{\mathscr{B}_0} \delta\tilde{p}\,J\mathbf{C}^{-1} : \Delta\mathbf{E}\,\mathrm{dV} \quad , \tag{11.21a}$$

$$\Delta_{\tilde{J}}\delta_{\boldsymbol{\varphi}}\Pi^{\text{int}} = \int_{\mathscr{B}_0} \delta\mathbf{E} : \mathbf{R}\,\Delta\tilde{J}\,\mathrm{dV} \quad , \quad \Delta_{\boldsymbol{\varphi}}\delta_{\tilde{J}}\Pi^{\text{int}} = \int_{\mathscr{B}_0} \delta\tilde{J} : \mathbf{R}\,\Delta\mathbf{E}\,\mathrm{dV} \quad , \tag{11.21b}$$

$$\begin{aligned} \Delta_{\Phi}\delta_{\boldsymbol{\varphi}}\Pi^{\text{int}} &= -\int_{\mathscr{B}_0\cup\mathscr{S}_0} \delta\mathbf{E} : \mathfrak{P}\cdot\Delta\mathbb{H}\,\mathrm{dV} \quad , \\ \Delta_{\boldsymbol{\varphi}}\delta_{\Phi}\Pi^{\text{int}} &= -\int_{\mathscr{B}_0\cup\mathscr{S}_0} \delta\mathbb{H}\cdot\mathfrak{P}^T : \Delta\mathbf{E}\,\mathrm{dV} \quad , \end{aligned} \tag{11.21c}$$

$$\Delta_{\tilde{J}}\delta_{\tilde{p}}\Pi^{\text{int}} = -\int_{\mathscr{B}_0} \delta\tilde{p}\,\Delta\tilde{J}\,\mathrm{dV} \quad , \quad \Delta_{\tilde{p}}\delta_{\tilde{J}}\Pi^{\text{int}} = -\int_{\mathscr{B}_0} \delta\tilde{J}\,\Delta\tilde{p}\,\mathrm{dV} \quad , \tag{11.21d}$$

$$\Delta_{\Phi}\delta_{\tilde{J}}\Pi^{\text{int}} = -\int_{\mathscr{B}_0} \delta\tilde{J}\,\mathbb{Q}\cdot\Delta\mathbb{H}\,\mathrm{dV} \quad , \quad \Delta_{\tilde{J}}\delta_{\Phi}\Pi^{\text{int}} = -\int_{\mathscr{B}_0} \delta\mathbb{H}\cdot\mathbb{Q}\,\Delta\tilde{J}\,\mathrm{dV} \quad . \tag{11.21e}$$

The elastic tangent $\mathscr{H}$, the magnetoelasticity tensor $\mathfrak{P}$, the magnetostatic tensor $\mathbf{D}$, and the auxiliary coupling tensor $\mathbf{R}$ and vector $\mathbb{Q}$ in the elastic body $\mathscr{B}_0$ are defined as

$$\begin{aligned} \mathscr{H}^{\text{tot}} &= \overline{\mathscr{H}} + \mathscr{H}^{\max} + \tilde{p}J\left[\mathbf{C}^{-1}\otimes\mathbf{C}^{-1} - 2\mathbf{C}^{-1}\overline{\otimes}\mathbf{C}^{-1}\right] \quad , \\ \overline{\mathscr{H}} &= 2\frac{\mathrm{d}\overline{\mathbf{S}}}{\mathrm{d}\mathbf{C}} = 2\frac{\mathrm{d}\widehat{\mathscr{P}}}{\mathrm{d}\mathbf{C}} : \left[2\frac{\partial\overline{W}_0}{\partial\overline{\mathbf{C}}}\right] + \widehat{\mathscr{P}}^T : \left[4\frac{\partial^2\overline{W}_0}{\partial\overline{\mathbf{C}}\otimes\mathrm{d}\overline{\mathbf{C}}}\right] : \widehat{\mathscr{P}} \quad , \end{aligned} \tag{11.22a}$$

$$\mathbf{D} = \frac{\mathrm{d}\mathbb{B}}{\mathrm{d}\mathbb{H}} = -\frac{\partial^2 \overline{W}_0}{\partial \mathbb{H} \otimes \mathrm{d}\mathbb{H}} + \mathbf{D}^{\max} \quad , \tag{11.22b}$$

$$\mathbf{R} = \frac{\mathrm{d}\mathbf{S}^{\mathrm{tot}}}{\mathrm{d}\widetilde{J}} = 2\frac{\partial^2 M_0}{\mathrm{d}\widetilde{J}\partial \mathbf{C}} \quad , \quad \mathbb{Q} = \frac{\mathrm{d}\mathbb{B}}{\mathrm{d}\widetilde{J}} = -\frac{\partial^2 M_0}{\mathrm{d}\widetilde{J}\partial \mathbb{H}} \quad , \tag{11.22c}$$

$$\mathfrak{P} = -\frac{\mathrm{d}\mathbf{S}^{\mathrm{tot}}}{\mathrm{d}\mathbb{H}} = \widehat{\mathscr{P}}^{T} : \left[-2\frac{\partial^2 \overline{W}_0}{\partial \overline{\mathbf{C}} \otimes \mathrm{d}\mathbb{H}}\right] + \mathfrak{P}^{\max} \quad ,$$

$$\mathfrak{P}^{T} = 2\frac{\mathrm{d}\mathbb{B}^{\mathrm{tot}}}{\mathrm{d}\mathbf{C}} = \left[-2\frac{\partial^2 \overline{W}_0}{\partial \mathbb{H} \otimes \mathrm{d}\overline{\mathbf{C}}}\right] : \widehat{\mathscr{P}} + [\mathfrak{P}^{\max}]^{T} \quad , \tag{11.22d}$$

while their Maxwell counterparts in the free space $\mathscr{S}_0$ are simply

$$\mathscr{H}^{\max} = 2\frac{\mathrm{d}\mathbf{S}^{\max}}{\mathrm{d}\mathbf{C}} = 4\frac{\partial^2 M_0}{\partial \mathbf{C} \otimes \mathrm{d}\mathbf{C}} \quad , \quad \mathfrak{P}^{\max} = -\frac{\mathrm{d}\mathbf{S}^{\max}}{\mathrm{d}\mathbb{H}} = -2\frac{\partial^2 M_0}{\partial \mathbf{C} \otimes \mathrm{d}\mathbb{H}} \quad , \tag{11.23a}$$

$$[\mathfrak{P}^{\max}]^{T} = 2\frac{\mathrm{d}\mathbb{B}^{\max}}{\mathrm{d}\mathbf{C}} = -2\frac{\partial^2 M_0}{\partial \mathbb{H} \otimes \mathrm{d}\mathbf{C}} \quad , \quad \mathbf{D}^{\max} = \frac{\mathrm{d}\mathbb{B}^{\max}}{\mathrm{d}\mathbb{H}} = -\frac{\partial^2 M_0}{\partial \mathbb{H} \otimes \mathrm{d}\mathbb{H}} \quad . \tag{11.23b}$$

The first term of the elastic tangent stated in equation $(11.22a)_2$ can be expanded using the identities shown in Appendix C.9.5. For completeness, the exposition of the isochoric contribution to the total material elastic tangent is

$$\begin{aligned}\overline{\mathscr{H}}^{\mathrm{tot}} &= \frac{2}{\mathrm{d}}J^{\frac{-2}{\mathrm{d}}}\left[\left[2\frac{\partial W_0}{\partial \overline{\mathbf{C}}} : \mathbf{C}\right]\left[\frac{1}{\mathrm{d}}\mathbf{C}^{-1} \otimes \mathbf{C}^{-1} - \frac{\mathrm{d}\mathbf{C}^{-1}}{\mathrm{d}\mathbf{C}}\right] - \left[\left[2\frac{\partial W_0}{\partial \overline{\mathbf{C}}}\right] \otimes \mathbf{C}^{-1} + \mathbf{C}^{-1} \otimes \left[2\frac{\partial W_0}{\partial \overline{\mathbf{C}}}\right]\right]\right] \\ &\quad + \widehat{\mathscr{P}}^{T} : \left[4\frac{\partial^2 W_0}{\partial \overline{\mathbf{C}} \otimes \mathrm{d}\overline{\mathbf{C}}}\right] : \widehat{\mathscr{P}} \quad ;\end{aligned} \tag{11.24}$$

this differs from the traditional description [559, 204] due to its presentation in dimension independent form.

Finite element discretisation

We describe the ansatz for the displacement and magnetic scalar potential fields and their variations, as well as the corresponding gradients, in the same manner as was listed in Sections 6.1.1 and 6.1.3. Furthermore, the discretisations of the unknown pressure response and dilatation fields are expressed as

$$\widetilde{p}\,(\mathbf{X}) \approx \sum_{I} \widetilde{p}^{I} \Psi^{I}_{\widetilde{p}}\,(\mathbf{X}) \quad , \quad \widetilde{J}\,(\mathbf{X}) \approx \sum_{I} \widetilde{J}^{I} \Psi^{I}_{\widetilde{J}}\,(\mathbf{X}) \quad , \tag{11.25}$$

while those of their variations and increments follow similarly. As was discussed previously in Chapter 6, through the application of the discretisation for the various fields we obtain the contributions to the residual for each DoF. In order for the discrete saddle point problem to be well posed, it is necessary for the FE discretisation to satisfy the Ladyzenskaja–Babuška–Brezzi (LBB) conditions. It is therefore sufficient to select identical discontinuous constant elements to approximate the pressure response and dilatation fields. This discretisation is based on the well known $Q^n - P^{n-1} - P^{n-1}$ solid element that is known, for the lowest possible order $n = 1$, not to strictly satisfy the LBB conditions [222]. However, it has been shown in applications that this choice is in general robust [495, 489, 494, 358]. The ansatz to be applied to each subdomain of the primary mixed variational magnetoelastic formulation is detailed in Table 11.4.

Table 11.4: Summary of finite element basis applied to the mixed variational formulation for magnetoelastic coupled problem using the magnetic scalar potential. The element types FE_Q and FE_DGM are respectively the continuous Lagrange FE and discontinuous monomial FE. The polynomial order is denoted by n (with equal-complexity element pairs tabulated) and vector elements are highlighted by **bold** font.

Relevant (continuous) equations		Subdomain	Field						
Magnetic scalar potential			$\boldsymbol{\varphi}$		$\widetilde{p}$		$\widetilde{J}$		Φ
Residual	11.11a–11.11d	$\mathscr{B}_0$	**FE_Q** (n)	×	FE_DGM ($n-1$)	×	FE_DGM ($n-1$)	×	FE_Q (n)
Linearisation	11.19a–11.19c,	$\mathscr{S}_0^{\mathscr{B}}$	**FE_Q** (n)	×	—	×	—	×	FE_Q (n)
	11.21a–11.21e	$\mathscr{S}_0^{\mathrm{h}} \backslash \mathscr{S}_0^{\mathscr{B}}$	—	×	—	×	—	×	FE_Q (n)

Defining the discrete residual for the mixed problem by

$$r^I = \delta\Pi^I_{\boldsymbol{\varphi}} + \delta\Pi^I_{\widetilde{p}} + \delta\Pi^I_{\widetilde{J}} + \delta\Pi^I_{\Phi} \tag{11.26}$$

the internal elastic contributions to the total residual at each DoF I, as computed from equations (11.11a) to (11.11c), are

$$\delta_{\boldsymbol{\varphi}}\Pi^{\text{int}} \approx \sum_I^{n_{\Psi_{\boldsymbol{\varphi}}}} \delta\varphi^I \underbrace{\left[\int_{\mathscr{B}_0} \text{sym}\left[\mathbf{F}^T \cdot \nabla_0 \boldsymbol{\Psi}^I_{\boldsymbol{\varphi}}\right] : \left[\overline{\mathbf{S}} + \mathbf{S}^{\max} + \widetilde{p} J \mathbf{C}^{-1}\right] \, \mathrm{dV} + \int_{\mathscr{S}_0} \text{sym}\left[\mathbf{F}^T \cdot \nabla_0 \boldsymbol{\Psi}^I_{\boldsymbol{\varphi}}\right] : \mathbf{S}^{\max} \, \mathrm{dV} \right]}_{\left[\delta\Pi^{\text{int}}_{\boldsymbol{\varphi}}\right]^I} \tag{11.27a}$$

$$\delta_{\widetilde{p}}\Pi^{\text{int}} \approx \sum_I^{n_{\Psi_{\widetilde{p}}}} \delta\widetilde{p}^I \underbrace{\int_{\mathscr{B}_0} \Psi^I_{\widetilde{p}} \left[J - \widetilde{J}\right] \, \mathrm{dV}}_{\left[\delta\Pi_{\widetilde{p}}\right]^I} \quad , \tag{11.27b}$$

$$\delta_{\widetilde{J}}\Pi^{\text{int}} \approx \sum_I^{n_{\Psi_{\widetilde{J}}}} \delta\widetilde{J}^I \underbrace{\int_{\mathscr{B}_0} \Psi^I_{\widetilde{J}} \left[\frac{\partial W_0^{\text{vol}}}{\partial \widetilde{J}} + \frac{\partial M_0}{\partial \widetilde{J}} - \widetilde{p} \right] \mathrm{dV}}_{\left[\delta\Pi_{\widetilde{J}}\right]^I} \tag{11.27c}$$

while that of the magnetic scalar potential is given by equation (6.24). The discretised expression for the linearisation follows similarly; its components have a form similar to that presented in equations (11.19a) to (11.19c) and equations (11.21a) to (11.21e).

Solution of linear iteration step

The sparse linear system that arises from the linearisation of the residual is

$$\begin{bmatrix} \delta\boldsymbol{\varphi} \\ \delta\tilde{p} \\ \delta\tilde{J} \\ \delta\Phi \end{bmatrix} \cdot \underbrace{\begin{bmatrix} \boldsymbol{K}_{\varphi\varphi} & \boldsymbol{K}_{\varphi\tilde{p}} & \boldsymbol{K}_{\varphi\tilde{J}} & \boldsymbol{K}_{\varphi\Phi} \\ \boldsymbol{K}_{\tilde{p}\varphi} & \mathbf{0} & \boldsymbol{K}_{\tilde{p}\tilde{J}} & \mathbf{0} \\ \boldsymbol{K}_{\tilde{J}\varphi} & \boldsymbol{K}_{\tilde{J}\tilde{p}} & \boldsymbol{K}_{\tilde{J}\tilde{J}} & \boldsymbol{K}_{\tilde{J}\Phi} \\ \boldsymbol{K}_{\Phi\varphi} & \mathbf{0} & \boldsymbol{K}_{\Phi\tilde{J}} & \boldsymbol{K}_{\Phi\Phi} \end{bmatrix}}_{\boldsymbol{K}} \cdot \underbrace{\begin{bmatrix} \Delta\boldsymbol{d}_{\varphi} \\ \Delta\boldsymbol{d}_{\tilde{p}} \\ \Delta\boldsymbol{d}_{\tilde{J}} \\ \Delta\boldsymbol{d}_{\Phi} \end{bmatrix}}_{\Delta\boldsymbol{d}} = \begin{bmatrix} \delta\boldsymbol{\varphi} \\ \delta\tilde{p} \\ \delta\tilde{J} \\ \delta\Phi \end{bmatrix} \cdot \underbrace{\begin{bmatrix} \boldsymbol{f}_{\varphi} \\ \boldsymbol{f}_{\tilde{p}} \\ \boldsymbol{f}_{\tilde{J}} \\ \boldsymbol{f}_{\Phi} \end{bmatrix}}_{\boldsymbol{f}=-\boldsymbol{r}} \quad . \tag{11.28}$$

When used in conjunction with a non-linear solution scheme (such as the Newton–Raphson method outlined in Section 6.3.2), it can be used to incrementally solve for the solution field at a given time step.

In addition to the numerical issues mentioned in Section 6.3.3, this system has a further challenge that must be overcome. Since $\tilde{p}$ is a Lagrange multiplier, there is no pressure-pressure coupling ($\boldsymbol{K}_{\tilde{p}\tilde{p}} = \mathbf{0}$); this renders the global system singular [461]. Furthermore, the coupling of the global system makes the implementation of an iterative solution scheme using block-elimination difficult. As a simplification, one may restrict the material strain-energy function applied in equation (11.9) to be of the form

$$W_0 + M_0 := \overline{W}_0\left(\overline{\mathbf{C}}, \mathbb{H}\right) + \overline{M}_0\left(\overline{\mathbf{C}}, \mathbb{H}\right) \quad \text{in} \quad \mathscr{B}_0 \quad ,$$

thus dismissing any coupling of the magnetic field with the dilatation field. This implies that the stored energy due to material dilation as a response to the applied magnetic field is negligible. Contributions of the form given by equation (5.178) can be modified to

$$M_0 := \overline{M}_0\left(\overline{\mathbf{C}}, \mathbb{H}\right) = -\frac{\mu_0}{2}\left[\overline{\mathbf{C}}^{-1} : \mathbb{H} \otimes \mathbb{H}\right] \quad \text{in} \quad \mathscr{B}_0 \quad , \tag{11.29}$$

which is a reasonable approximation particularly for the special case of a near incompressible body. Under this simplification, the full magneto-mechanical energy function that is now considered in equation (11.9) for the magnetisable body is

$$\Omega_0\left(\tilde{J}, \mathbf{C}, \mathbb{H}\right) := W_0^{\text{vol}}\left(\tilde{J}\right) + \overline{W}_0\left(\overline{\mathbf{C}}, \mathbb{H}\right) + \overline{M}_0\left(\overline{\mathbf{C}}, \mathbb{H}\right) \quad \text{in} \quad \mathscr{B}_0 \quad . \tag{11.30}$$

As a result, the modified linear system is

$$\begin{bmatrix} \boldsymbol{K}_{\varphi\varphi} & \boldsymbol{K}_{\varphi\tilde{p}} & \mathbf{0} & \boldsymbol{K}_{\varphi\Phi} \\ \boldsymbol{K}_{\tilde{p}\varphi} & \mathbf{0} & \boldsymbol{K}_{\tilde{p}\tilde{J}} & \mathbf{0} \\ \mathbf{0} & \boldsymbol{K}_{\tilde{J}\tilde{p}} & \boldsymbol{K}_{\tilde{J}\tilde{J}} & \mathbf{0} \\ \boldsymbol{K}_{\Phi\varphi} & \mathbf{0} & \mathbf{0} & \boldsymbol{K}_{\Phi\Phi} \end{bmatrix} \cdot \begin{bmatrix} \Delta\boldsymbol{d}_{\varphi} \\ \Delta\boldsymbol{d}_{\tilde{p}} \\ \Delta\boldsymbol{d}_{\tilde{J}} \\ \Delta\boldsymbol{d}_{\Phi} \end{bmatrix} = \begin{bmatrix} \boldsymbol{f}_{\varphi} \\ \boldsymbol{f}_{\tilde{p}} \\ \boldsymbol{f}_{\tilde{J}} \\ \boldsymbol{f}_{\Phi} \end{bmatrix} \tag{11.31}$$

since the contributions $\mathbf{R}$ and $\mathbb{Q}$ are null (as $\overline{M}_0$ has no dependence on $\tilde{J}$) and, consequently, the coupling terms $\boldsymbol{K}_{\varphi\tilde{J}}, \boldsymbol{K}_{\tilde{J}\varphi}, \boldsymbol{K}_{\tilde{J}\Phi}$ and $\boldsymbol{K}_{\Phi\tilde{J}}$ vanish a priori. Observe as well that the definitions for contributions within the body $\mathscr{B}_0$ stated in equations (11.14) and (11.23) must now be expanded to accommodate the isochoric parametrisation of the “pseudo-Maxwell energy” defined in equation (11.29).

From this augmented system of linear equations, several iterative solution approaches may be formulated using Gaussian elimination of the full block system, two of which are detailed below. Firstly, the displacement update may be expressed as

$$\underbrace{\left[\boldsymbol{K}_{\varphi\varphi} + \boldsymbol{K}_{\varphi\tilde{p}}\boldsymbol{K}_{\tilde{J}\tilde{p}}^{-1}\boldsymbol{K}_{\tilde{J}\tilde{J}}\boldsymbol{K}_{\tilde{p}\tilde{J}}^{-1}\boldsymbol{K}_{\tilde{p}\varphi} - \boldsymbol{K}_{\varphi\Phi}\,\boldsymbol{K}_{\Phi\Phi}^{-1}\,\boldsymbol{K}_{\Phi\varphi}\right]}_{\widehat{\boldsymbol{K}}}\Delta\boldsymbol{d}_{\varphi}$$
$$= \underbrace{\boldsymbol{f}_{\varphi} - \boldsymbol{K}_{\varphi\tilde{p}}\boldsymbol{K}_{\tilde{J}\tilde{p}}^{-1}\left[\boldsymbol{f}_{\tilde{J}} - \boldsymbol{K}_{\tilde{J}\tilde{J}}\boldsymbol{K}_{\tilde{p}\tilde{J}}^{-1}\boldsymbol{f}_{\tilde{p}}\right] - \boldsymbol{K}_{\varphi\Phi}\boldsymbol{K}_{\Phi\Phi}^{-1}\boldsymbol{f}_{\Phi}}_{\widehat{\boldsymbol{f}}} \tag{11.32}$$

where $\widehat{\boldsymbol{K}}$ is analogous to the Schur matrix given in equation (6.41). The incremental updates for the scalar potential, dilatation and pressure response fields are given by

$$\Delta\boldsymbol{d}_{\Phi} = \boldsymbol{K}_{\Phi\Phi}^{-1}\left[\boldsymbol{f}_{\Phi} - \boldsymbol{K}_{\Phi\varphi}\,\Delta\boldsymbol{d}_{\varphi}\right] \quad , \tag{11.33a}$$
$$\Delta\boldsymbol{d}_{\tilde{J}} = \boldsymbol{K}_{\tilde{p}\tilde{J}}^{-1}\left[\boldsymbol{f}_{\tilde{p}} - \boldsymbol{K}_{\tilde{p}\varphi}\Delta\boldsymbol{d}_{\varphi}\right] \quad , \tag{11.33b}$$
$$\Delta\boldsymbol{d}_{\tilde{p}} = \boldsymbol{K}_{\tilde{J}\tilde{p}}^{-1}\left[\boldsymbol{f}_{\tilde{J}} - \boldsymbol{K}_{\tilde{J}\tilde{J}}\Delta\boldsymbol{d}_{\tilde{J}}\right] \tag{11.33c}$$

respectively.

As an alternative approach, one may also exploit the discontinuous nature of the pressure response and dilatation fields; in this way one may perform static condensation to remove these DoFs from the global system. The condensed form of the linear problem is

$$\underbrace{\left[\widetilde{\boldsymbol{K}}_{\varphi\varphi} - \boldsymbol{K}_{\varphi\Phi}\,\boldsymbol{K}_{\Phi\Phi}^{-1}\,\boldsymbol{K}_{\Phi\varphi}\right]}_{\widetilde{\boldsymbol{K}}}\Delta\boldsymbol{d}_{\varphi} = \underbrace{\widetilde{\boldsymbol{f}}_{\varphi} - \boldsymbol{K}_{\varphi\Phi}\boldsymbol{K}_{\Phi\Phi}^{-1}\boldsymbol{f}_{\Phi}}_{\widetilde{\boldsymbol{f}}} \quad , \tag{11.34}$$

where the augmented stiffness matrix and right-hand side vector contributions are

$$\widetilde{\boldsymbol{K}}_{\varphi\varphi} = \boldsymbol{K}_{\varphi\varphi} + \boldsymbol{K}_{\varphi\tilde{p}}\overline{\boldsymbol{K}}_{\tilde{p}\tilde{p}}\boldsymbol{K}_{\tilde{p}\varphi} \quad , \quad \widetilde{\boldsymbol{f}}_{\varphi} = \boldsymbol{f}_{\varphi} - \boldsymbol{K}_{\varphi\tilde{p}}\left[\boldsymbol{K}_{\tilde{J}\tilde{p}}^{-1}\boldsymbol{f}_{\tilde{J}} - \overline{\boldsymbol{K}}_{\tilde{p}\tilde{p}}\boldsymbol{f}_{\tilde{p}}\right] \quad . \tag{11.35a}$$

Here we have defined the auxiliary matrix, which retains a block-sparse structure, as

$$\overline{\boldsymbol{K}}_{\tilde{p}\tilde{p}} := \boldsymbol{K}_{\tilde{J}\tilde{p}}^{-1}\boldsymbol{K}_{\tilde{J}\tilde{J}}\boldsymbol{K}_{\tilde{p}\tilde{J}}^{-1} \quad . \tag{11.35b}$$

Note that the solutions to the two strategies

$$\Delta\boldsymbol{d}_{\varphi} = \widetilde{\boldsymbol{K}}^{-1}\widetilde{\boldsymbol{f}} \equiv \widehat{\boldsymbol{K}}^{-1}\widehat{\boldsymbol{f}} \tag{11.36}$$

are equivalent but their implementation differs. For example, if the matrix $\boldsymbol{K}_{\varphi\varphi}$ is modified in-place to produce $\widetilde{\boldsymbol{K}}_{\varphi\varphi}$, then the numerical qualities of this matrix (such as its condition number) differs from that of its original form. This primarily influences the effectiveness of preconditioners for $\widetilde{\boldsymbol{K}}$ that are applied to solve the linear system.

Numerical example: Quasi-incompressible magneto-active valve

Pelteret et al. [417] examined two arrangements for a simplified magneto-active valve, which were idealised versions the coil-activated designs presented by Böse et al. [55, figure 2]. The truncated geometry of the valve is depicted in Figure 11.12; the MAP is denoted as $\mathscr{B}_0$, while the iron inner yoke and casing that are indicated by $\mathscr{S}_0^{I,i}$ and $\mathscr{S}_0^{I,o}$, respectively. Although the very stiff iron components were considered to be fully immobile (and, therefore, not represented as an elastic solid), the magnetisable polymer was modelled as a deformable solid. The height and radius of the truncated far-field $\mathscr{S}_0$ were 24 mm and 33 mm, respectively. The height of the inner yoke was set at 9 mm, while its radius was 3 mm. The casing had a height of 7 mm, and an inner and outer radius of 14 mm and 18 mm, respectively. The outer radius of the polymer was 12 mm and its overall thickness was 5 mm. The initial gap between the polymer and casing was therefore 2 mm. The free corners of the polymer were rounded with a fillet of radius 0.25 mm to prevent the generation of magnetic singularities along these edges. However, sharp corners of the iron components were retained. The problem, based on the geometry, is axisymmetric; however, to demonstrate the mixed formulation it was simulated using one quarter of the full three-dimensional model. Within a radius of 0.25 mm of the three sharp corners marked **R** in Figure 11.12, local h-refinement was employed to capture the magnetic singularity that was expected to develop in these locations.

As for the boundary conditions, the elastic body was constrained at its interface to the inner yoke. The central displacement DoFs were prescribed as fixed in space,

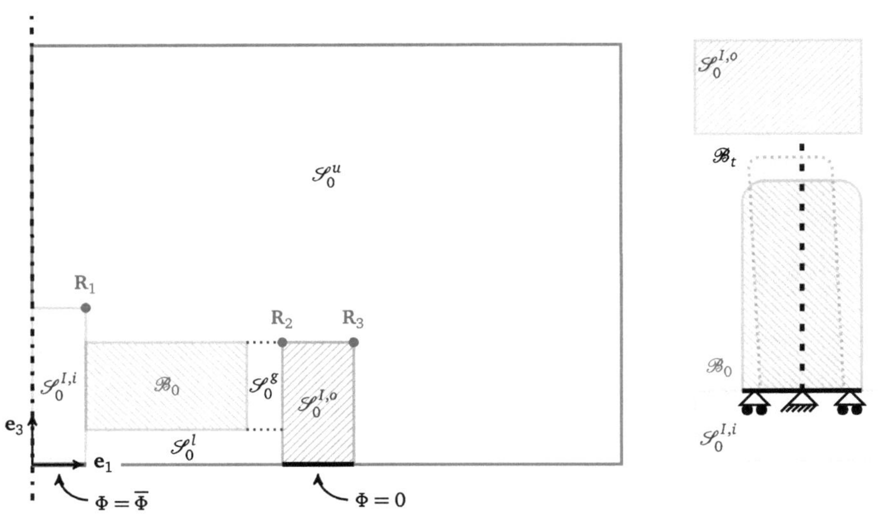

Figure 11.12: A scale representation of the cross-section of MAP valve geometry and boundary conditions. The original position of the polymer $\mathscr{B}_0$ is shown in red and its expected displaced configuration $\mathscr{B}_t$ shown in grey. [417, fig. 14]

while the rest of the surface could slide along the interface. The energised coil could not be directly represented using the magnetic scalar potential formulation. However, through the appropriate choice of potential boundary conditions (namely prescribing the magnetic potential at the surface of truncation of the casing and inner yoke) an approximation of its influence was captured. The casing was set to have zero magnetic potential and that at the inner yoke was defined as

$$\overline{\Phi} = \sqrt{t}\Phi^{\max} \tag{11.37}$$

with $t \in [0,1]$ and $\Phi^{\max} = 310$ A. With respect to the mesh update problem, the mesh in $\mathscr{S}_0$ was allowed to slide along "solid" geometries (the outer casing and iron yoke), as well as along the boundary of the free space at the edge of the computational domain that is represented by $\partial\mathscr{S}_0$. With respect to the mesh update algorithm, the following parameters rendered the most favourable results: the blend parameter $\alpha = -1$, the stiffness function was given by equation $(11.3)_2$ in conjunction with the exponent $p = 1.25$ and the metric $d(\mathbf{X})$ represented the distance to the nearest solid or "pseudo-solid" vertex.

The volumetric and isochoric components of the additively decomposed polymer strain energy density function was chosen as

$$W_0^{\text{vol}}\left(\widetilde{J}\right) = \frac{\kappa}{4}\left[\widetilde{J}^2 - 1 - 2\ln\left(\widetilde{J}\right)\right] \quad , \tag{11.38}$$

$$\begin{aligned} \overline{W}_0\left(\overline{\mathbf{C}}, \mathbb{H}\right) + \overline{M}_0\left(\overline{\mathbf{C}}, \mathbb{H}\right) &= \frac{\mu}{2}\left[I_1\left(\overline{\mathbf{C}}\right) - \mathrm{d}\right] + \alpha\mu_0 I_4\left(\mathbb{H}\right) \\ &\quad + \beta\mu_0 I_5\left(\overline{\mathbf{C}}, \mathbb{H}\right) + \eta\mu_0 I_7\left(\overline{\mathbf{C}}, \mathbb{H}\right) \quad , \end{aligned} \tag{11.39}$$

implying that the isochoric response is governed by a coupled Neo–Hookean type model while the volumetric response is determined by a polyconvex function. The total stored energy function for the (inelastic) iron components was

$$W_0\left(J, \mathbf{C}, \mathbb{H}\right) = \mu_r M_0\left(J, \mathbf{C}, \mathbb{H}\right) \tag{11.40}$$

with the relative magnetic permeability of the iron $\mu_r = 5000$. The constitutive parameters for the polymer are given in Table 11.5, and the bulk modulus $\kappa = \frac{2\mu[1+\nu]}{3[1-2\nu]}$.

Table 11.5: Constitutive parameters chosen for the quasi-incompressible MAP valve constituting the valve diaphragm.

Parameter	Value	Unit	Parameter	Value	Unit
μ	30	kPa	β	−4	—
ν	0.4999	—	η	−0.5	—
α	−0.5	—			

Figure 11.13a illustrates the magnetic fields predicted in the first design, for which the polymeric diaphragm was attached to the inner core, and a gap existed between it and the outer casing. It was shown in [417] that, for the fully static case, the solution for the magnetic field was both qualitatively and quantitatively similar to the equivalent geometry evaluated using the magnetic vector potential formulation. The plot of the radial displacement along the polymer mid-plane, shown in Figure 11.13b, indicated that at the maximum loading conditions the gap radius was reduced by 25 %. Although the chosen energy function is quadratic in $|\mathbb{H}|$, the radial displacement was not found to be directly proportional to $\sqrt{t}$. This observation was attributed to the influence of the ponderomotive force (driven by the jump in the magnetic field at the interface) and,

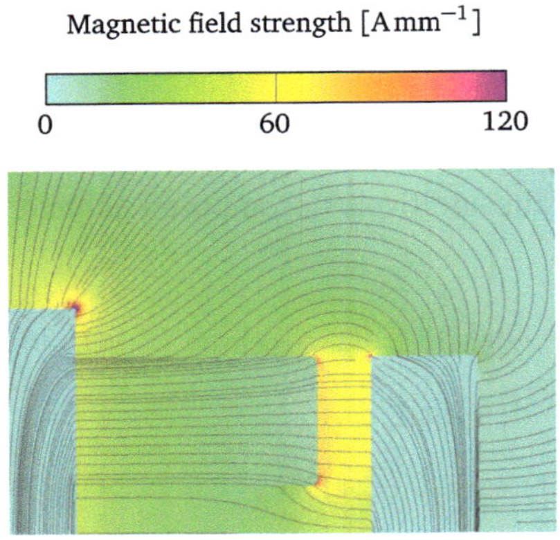

(a) Magnetic field and flux lines for magneto-static case with $\overline{\Phi} = 310\,\mathrm{A}$.

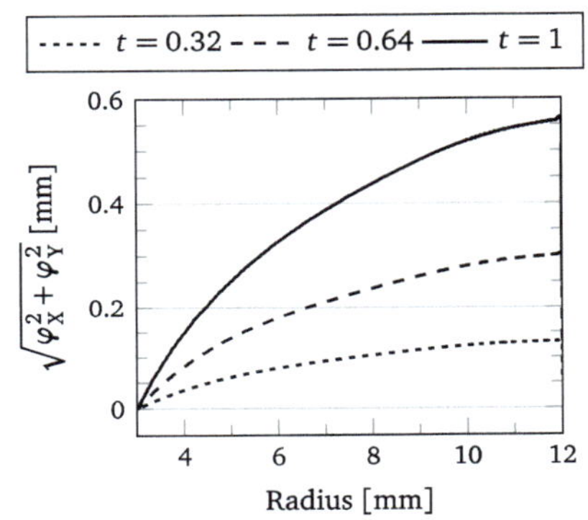

(b) Radial displacement.

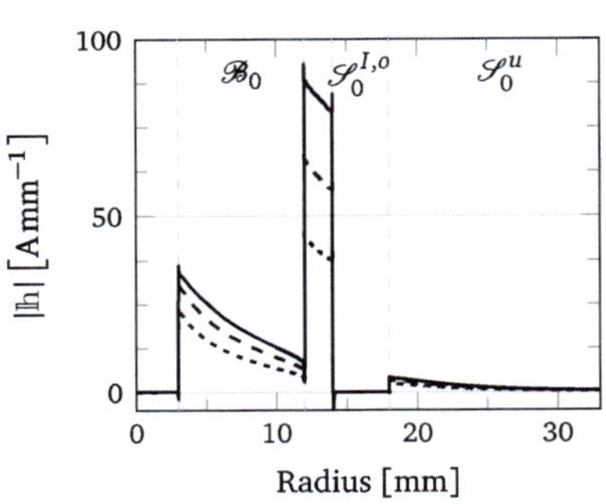

(c) Spatial magnetic field strength.

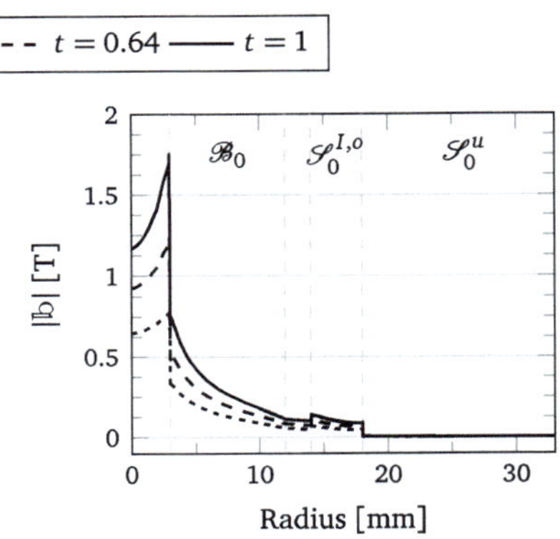

(d) Spatial magnetic induction.

Figure 11.13: Evaluation of an MAP valve, for which the polymeric diaphragm was attached to the inner yoke. Measurements were taken radially (along reference coordinates) at the height $Z = 4.5\,\mathrm{mm}$. [417, fig. 15b (reproduced with permission)], [417, fig. 17a,16] to a lesser extent, the non-linearity of the material itself. The large radial gradient in the magnetic field at $\overline{\mathscr{B}_0} \cap \overline{\mathscr{S}_0^g}$ produced a large ponderomotive traction at this surface. Also observed, but not shown here, was a measurable axial deflection of the polymer primarily due the axially asymmetric magnetic loading.

The spatial magnetic field and induction measured along the mid-plane of the polymer, found to be qualitatively similar to that shown by Böse et al. [55], are presented in Figures 11.13c and 11.13d. The magnetic field strength was weak in the iron yoke and casing, and increased in strength in the less permeable materials. As a result of the non-linear material constitutive law and material deformation, the field strength in the gap and polymer decreased non-linearly as the radius increased. It was also observed that the jump in the magnetic field strength at the polymer-air interface $\overline{\mathscr{B}_0} \cap \overline{\mathscr{S}_0^g}$ became disproportionally larger with increasing load step. As for the magnetic induction field, the proximity of the inner yoke to the source of the potential resulted in the induction in this region being very large. In response to the non-linear magnetic field the induction measured within the polymer decayed non-linearly from the inner to the outer radius. As a point of comparison, the induction in the air-gap appeared almost constant for a given magnetic load.

A second valve configuration, whose performance and functioning is illustrated in Figure 11.14, that was simulated had the polymer attached to the outer casing. The operation of this design relied on the expansion of the MAP inwards under the application of the magnetic field. Two modifications to the imposed boundary conditions were necessary when modelling this new geometry. Firstly, the pinned boundary condition was moved from the centreline of the polymer to the upper edge coinciding with $\mathbf{R}_2$. Secondly, as a result of issues related to numerical stability, the maximum applied magnetic potential was lowered to $\Phi^{max} = 243$ A.

Comparing Figure 11.13a with Figure 11.14a, it was observed that the magnetic field strength in the air-gap was significantly greater for the same applied magnetic potential. When considering the magnetoelastic case for this second geometry, the applied potential at $t = 1$ was nearly equivalent to that for the first scenario at $t = 0.7$. Under equivalent conditions for deformable conditions, it was determined through comparison of Figures 11.14c and 11.14d to Figures 11.13c and 11.13d that the material magnetisation decreased considerably in the second design. However, the jump in the magnetic field at the radial polymer-free space interface was significantly larger than before, thus indicating that the ponderomotive traction was stronger for this particular geometry when compared to the first. Ultimately, when magnetic potentials of similar magnitudes were applied to both designs, the maximum radial displacement was comparable for the two geometries.

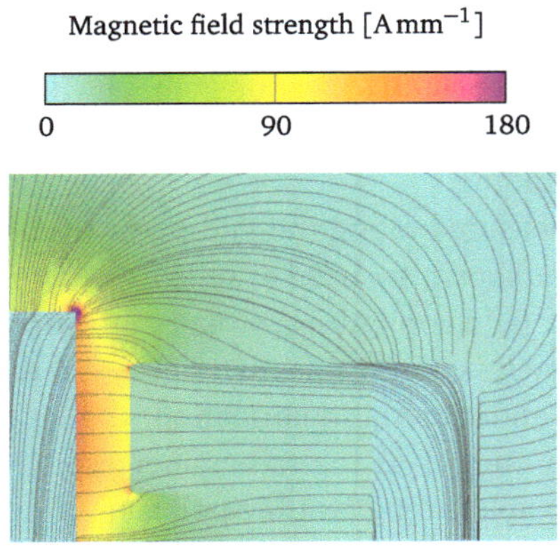

(a) Magnetic field and flux lines for magnetostatic case with $\overline{\Phi}$ = 310 A.

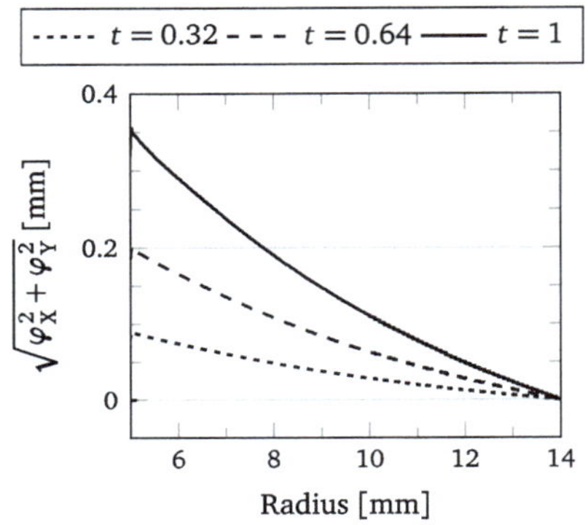

(b) Radial displacement.

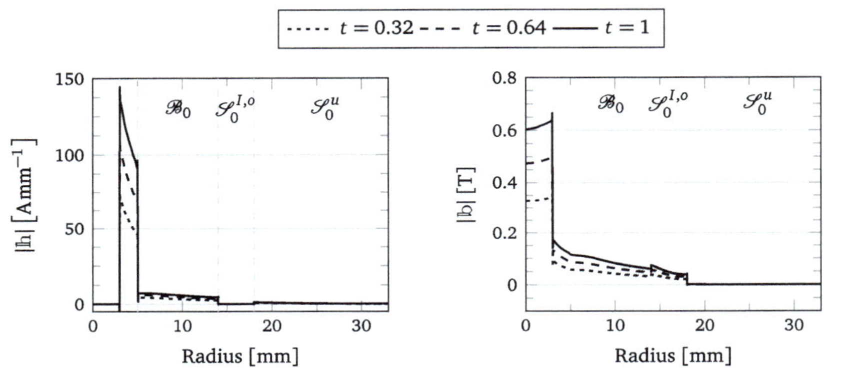

(c) Spatial magnetic field strength.

(d) Spatial magnetic induction.

Figure 11.14: Evaluation of an alternative design for the valve, for which the polymeric diaphragm was attached to the outer casing. Measurements were taken radially (along reference coordinates) at the height Z = 4.5 mm. [417, fig. 20a (reproduced with permission)], [417, fig. 20b,20c,20d]

12 Further reading

In much of this book, we have restricted the discussion on thermodynamics, the continuum mechanics framework, constitutive modelling and numerical simulation to those aspects applicable to examining magneto-coupled, (rubber-like) viscoelastic composites at both the macro- and micro-scale, and within a quasi-static setting. However, in many engineering applications these materials cannot be considered in isolation as they form part of a more sophisticated construction. Therefore, the influences of further physical phenomena may also need to be taken into consideration. In anticipation of the need to extend the content that has been presented, we conclude with a concise summary of select further topics of interest.

Transient behaviour

In many common and exotic [524] applications, the transient nature of electromagnetic phenomena cannot be neglected as it is core to the functioning of many classes of devices. Thus, in order to model electromagnets that employ alternating currents (such as those used in transformers, inductors, and a variety of motors), the time-dependent terms in Maxwell's equations cannot be ignored. Similar holds for machinery driven by changing magnetic fields, such as those that act as electromagnetic brakes and clutches [153, 547], and electrodynamic suspension or magnetic levitation devices [284]. Associated with a switching applied current, or movement of an electronic device through a magnetic field, is the generation of Eddy currents as governed by Lenz's law [301, 343]. The thermal energy generated by devices that either employ strong currents, or those that have large Eddy currents induced in them, may therefore also require consideration of heat dissipation within the components.

From a solid-mechanics perspective, inertial effects may no longer be ignored in applications involving high strain-rates or the acceleration of large masses. Examples of technologies involving the rapid actuation of MAPs includes magneto-active valves [55] and vibration damping [76, 309] devices. For elastodynamics problems, particular attention needs to be paid to the applied time integration schemes in order to ensure that (i) they are conservative in terms of both linear and angular momentum, and (ii) they exhibit sufficiently good stability characteristics and numerical qualities. For further details on this broad topic see, for instance, [493, 222, 177, 14, 45, 36, 559].

Dissipative material response

Many of the materials used in device construction exhibit dissipative properties, or are subject to loading that requires the consideration of energy dissipation. For the latter, these conditions include that of frictional contact [558] and damage [489, 505, 353, 90, 94]. Temperature-induced effects become significant when the problem cannot be treated as isothermal. Energy may also be dissipated in materials due to the

https://doi.org/10.1515/9783110418576-012

internal friction mechanisms of rubber-like materials, potentially also leading to the development of strong thermal gradients, and permanent deformation. The combination of admissible dissipation mechanisms leads to topics related to phenomenological modelling at a macroscopic level that are inclusive of dissipative magnetostriction [370], elasto-plasticity (e. g. modelled using a hypoelastic formulation [495, 491], as single crystals [367, 491] or using gradient-based approaches [491, 346, 362, 445, 270, 363, 375, 373, 78]), thermo-elasticity [204], thermo-viscoelasticity [207, 447], thermo-plasticity [75, 33], and thermo-viscoplasticity [500].

Micromagnetism

Micromagnetics encompasses the theory and modelling of the response of magnetic materials at a length scale on which magnetic structures, such as domain walls and vortices, are relevant. It describes the fundamental mechanisms that, for example, govern the formation of the magnetic structures in permanent magnets. Dependent on the spontaneous arrangement of magnetic dipoles within the domains, the homogenised material property of materials that have domain microstructures may be that reflecting a ferromagnetic, ferrimagnetic or anti-ferromagnetic material. All of these materials have the ability to exhibit remanence properties, and thus retain their local and overall magnetisation in the absence of an externally generated field. In contrast, the overall response of MAPs is aligned with that of paramagnetic or diamagnetic materials, namely one in which the magnetic dipoles formed by the CIPs align in response to an externally applied magnetic field.

To resolve the details of the magnetic ordering, the length scales that must be considered are thus smaller than the micrometre scale that define the characteristic size of a CIP, but are still larger than the atomistic scale at which the continuum hypothesis is no longer valid. As such, the domain theory of magnetisation proposed for rigid bodies by Landau and Lifshitz [285], and extended into the realm of magnetoelastic bodies by Brown [65], accommodates the magneto-mechanical coupling due to the formation and evolution (and consequent property change) of magnetic microstructures. More recently, the Landau–Lifshitz–Gilbert (LLG) equation [170] was proposed to describe the dynamic development of magnetic domains, effectively capturing the time-dependent evolution of magnetisation as influenced by magnetic torques.

Domain wall and micromagnetic theory, and aspects of materials with magnetic ordering, is more carefully elucidated in the works of Prohl [437], Kronmüller and Fähnle [280], Cullity and Graham [103], and Abert [2], among others. Applications simulated numerically include those pertaining to the domain structures themselves [336], shape memory alloys [407], thin magnetic films [509, 106] and tubes [231], and nanowires [194] and chains of nanospheres [106]. The LLG equation may also be used in combination with, for example, the magnetostatic Maxwell equations to consider multi-domain problems for which there exist two distinct length scales. Coupling between the two macro- and micro-scopic problems may be achieved by applying the appropriate discretisation to each domain and, for example, using BEM–FEM techniques [66] or by computational homogenisation such as through the phase field approach [365, 137, 499]. As a reference to its implementation, an open-source implementation of micromagnetism using finite elements is documented in [1].

Surface effects

As motivated in Section 2.1, surface coatings may be applied to fillers used in MAPs in order to change their interaction with the surrounding polymeric matrix, or to enhance their magnetic properties. As the chemical interactions typically remain localised to a particles' surface, the thickness of the "coating" that forms on the filler is orders of magnitude smaller than both the particle radius and average interparticle distance. Numerical simulation of the outcome of such processes, particularly within the context of a microstructural modelling framework, therefore poses a great challenge since direct simulation of this additional material layer is typically not feasible. However, as the size of the particles decreases the surface-to-volume ratio, the influence of the surface on the mechanics of the composite increases and its influence cannot be neglected.

There exist two established methods to investigate the mechanics and thermodynamics of surfaces, namely (i) the zero-thickness approach attributed to Gibbs [168], and (ii) finite-thickness layer method described by van der Waals [525]. Guggenheim [181] provides a detailed comparison of these two approaches. Following the zero-thickness method, Gurtin and Murdoch [185] and [384] offer different view-points from which they develop the balance laws that govern the behaviour of the surface. Additional key contributions to the theory of thermodynamically responsive surfaces include [392, 104, 113, 503] among many others. A thorough examination of the history of development of surface theory is provided within the context of thermo-elasticity by Javili [233]. To supplement the above, McBride et al. [345] have documented and provided a numerical implementation of surface elasticity at finite strains, and Esmaeili et al. [136] present extensions of the theory for surface plasticity.

The explicit derivation of the continuum mechanics that forms the basis of surface magnetoelasticity theory involves revisiting each component of Sections 5.3 and 5.4. As this is beyond the scope of this manuscript, we instead refer the interested reader to the detailed formulations given by Chatzigeorgiou et al. [83]. Due to some of the assumptions made therein, we note that the surface magnetoelastic theory presented in [83] is restricted to a parameterisation in terms of the magnetic induction.

Electroelasticity

The literature on topics related to both the theory and numerical modelling of electroelastic media is extensive. However, due to the similarity in the governing equations for (stationary) magnetoelasticity and electroelasticity, much of the seminal literature on the topic, their subsequent formulation and numerical treatment have a significant overlap. As a result of their polymeric nature, both electro-active polymers (EAPs) and MAPs often exhibit both incompressible and viscoelastic behaviour. Consequently, like for MAPs the analytical examination of the constitutive response of these materials [555, 124, 72, 212, 18, 471] commonly (but not always [70]) assumes that the material is incompressible. Material instabilities may also be exhibited by EAPs; for example, in dielectric actuators, application of strong electric fields as needed to obtain large deformations (driven by Coulombic forces) may generate an unstable response [390, 546, 458].

Focussing on FE formulations and numerical aspects of electroelastic media, there exists a plethora of examples of finite-strain coupled numerical models of EAPs including [563, 541, 581, 19, 496, 67, 347, 532, 351], among many others. Furthermore, mixed variational frameworks have been formalised by [564, 456, 73], and Hu–Washizu mixed formulations have been successfully adopted for the coupled problem within a numerical context by [20, 229, 581, 409]. Extensions on the coupled response of these materials have been made, with both electro-thermoelastic [351], and magneto-electroelastic [371, 322, 260] composites having been considered.

In all the aforementioned works for EAPs, we note that only their material bodies were accounted for and/or discretised using the FEM. This is a reasonable approach for considering condensator-like structures, as the influence of the external fields (and therefore the Maxwell stress) is eliminated due to the geometry of the system. Similar holds in the simulation of piezoelectric materials [434] under electric stimulation except during special cases such as fracture [306, 352]. However, in dealing with EAPs the contribution of the free space can in some cases become significant [272, 216] and must be taken into account [504, 530]. How this may be accomplished has been discussed in the context of BEM–FEM [540, 536], and mixed variational approaches [169, 417].

A Identities

As a shorthand to indicate where previously determined equations have been used, a left arrow ($\leftarrow$) indicates that the stated results are applied to the current line, while a right arrow ($\rightarrow$) indicates that the stated results are necessary to achieve what follows on the next line.

To assist in the presentation of some tensor identities, we denote s as an arbitrary scalar, $\mathbf{a}, \mathbf{b}, \mathbf{c}$ as arbitrary vectors, and $\mathbf{T}$ as an arbitrary rank-2 tensor. We also denote any constant vector field by $(\bullet)^*$.

A.1 Operation identities

Cross product

Vector field

$$\mathbf{a} \times \mathbf{b} \equiv -\mathbf{b} \times \mathbf{a} \Rightarrow \epsilon_{ijk} b_k a_j \mathbf{e}_i \tag{A.1}$$

Tensor field

$$\mathbf{a} \times [\mathbf{T} \cdot \mathbf{b}] \Rightarrow \epsilon_{ijk} T_{kl} b_l a_j \mathbf{e}_i$$

$$[\mathbf{T} \cdot \mathbf{a}] \times \mathbf{b} \Rightarrow \epsilon_{ijk} b_k T_{jl} a_l \mathbf{e}_i \tag{A.2}$$

$$\mathbf{a} \times \mathbf{T} \Rightarrow \epsilon_{ijk} T_{kl} a_j \mathbf{e}_i \otimes \mathbf{e}_l \tag{A.3}$$

$$\mathbf{T} \times \mathbf{a} \Rightarrow \epsilon_{ijk} T_{jl} a_k \mathbf{e}_i \otimes \mathbf{e}_l \tag{A.4}$$

Cofactor (of a non-singular tensor)

$$\operatorname{cof}(\mathbf{T}) := \det(\mathbf{T})\, \mathbf{T}^{-T} \tag{A.5}$$

Permutation operators

Cyclic and anti-cyclic permutations

$$\epsilon_{ijk} = \epsilon_{jki} = \epsilon_{kij} \quad , \quad \epsilon_{ijk} = -\epsilon_{ikj} = -\epsilon_{jik} = -\epsilon_{kji} \tag{A.6}$$

Single contraction of two permutation tensors

$$\epsilon_{ijk}\epsilon_{imn} = \delta_{jm}\delta_{kn} - \delta_{jn}\delta_{km} \tag{A.7}$$

Double contraction of a vector outer product

$$\boldsymbol{\epsilon} : [\mathbf{a} \otimes \mathbf{b}] = \mathbf{a} \times \mathbf{b} \tag{A.8}$$

Double contraction of a symmetric tensor

$$\boldsymbol{\epsilon} : \operatorname{sym}[\mathbf{T}] = \mathbf{0} \tag{A.9}$$

https://doi.org/10.1515/9783110418576-013

Cross product with a scalar factor

$$s\,[\mathbf{a}\times\mathbf{b}] = [s\mathbf{a}]\times\mathbf{b} = \mathbf{a}\times[s\mathbf{b}] \tag{A.10}$$

Cross product with a common operator

$$\begin{aligned}[\mathbf{T}\cdot\mathbf{a}]\times[\mathbf{T}\cdot\mathbf{b}] &= \det(\mathbf{T})\,\mathbf{T}^{-T}\cdot[\mathbf{a}\times\mathbf{b}] = \det(\mathbf{T})\,[\mathbf{a}\times\mathbf{b}]\cdot\mathbf{T}^{-1} && \text{(A.11)}\\ &= \operatorname{cof}(\mathbf{T})\cdot[\mathbf{a}\times\mathbf{b}] = [\mathbf{a}\times\mathbf{b}]\cdot\operatorname{cof}\left(\mathbf{T}^{T}\right) && \leftarrow \text{eq. (A.5)}\end{aligned}$$

Scalar triple product

$$\mathbf{a}\cdot[\mathbf{b}\times\mathbf{c}] = \mathbf{b}\cdot[\mathbf{c}\times\mathbf{a}] = \mathbf{c}\cdot[\mathbf{a}\times\mathbf{b}] \tag{A.12}$$

Scalar triple product with common operator

$$\begin{aligned}[\mathbf{T}\cdot\mathbf{a}]\cdot[[\mathbf{T}\cdot\mathbf{b}]\times[\mathbf{T}\cdot\mathbf{c}]] &= [\mathbf{T}\cdot\mathbf{a}]\cdot\det(\mathbf{T})\,\mathbf{T}^{-T}\cdot[\mathbf{b}\times\mathbf{c}] && \leftarrow \text{eq. (A.11)}\\ &= \det(\mathbf{T})\,\mathbf{a}\cdot\mathbf{T}^{T}\cdot\mathbf{T}^{-T}\cdot[\mathbf{b}\times\mathbf{c}] &&\\ &= \det(\mathbf{T})\,\mathbf{a}\cdot[\mathbf{b}\times\mathbf{c}] && \text{(A.13)}\end{aligned}$$

Vector triple product

$$\begin{aligned}\varepsilon_{mlk}c_l\,[\varepsilon_{kij}a_ib_j] &= \varepsilon_{kij}\varepsilon_{kml}a_ib_jc_l && \rightarrow \text{eq. (A.7)}\\ &= [\delta_{im}\delta_{jl}-\delta_{il}\delta_{jm}]\,a_ib_jc_l &&\\ &= \delta_{im}\delta_{jl}a_ib_jc_l-\delta_{il}\delta_{jm}a_ib_jc_l &&\\ &= a_mb_lc_l-a_lb_mc_l &&\\ \mathbf{c}\times[\mathbf{a}\times\mathbf{b}] &= \mathbf{a}\,[\mathbf{b}\cdot\mathbf{c}]-\mathbf{b}\,[\mathbf{a}\cdot\mathbf{c}] && \rightarrow \text{eq. (A.7)}\\ &\equiv \mathbf{a}\,[\mathbf{b}\cdot\mathbf{c}]-[\mathbf{b}\otimes\mathbf{c}]\cdot\mathbf{a} = [[\mathbf{b}\cdot\mathbf{c}]\,\mathbf{I}-[\mathbf{b}\otimes\mathbf{c}]]\cdot\mathbf{a} && \text{(A.14)}\end{aligned}$$

Note also that

$$\begin{aligned}\varepsilon_{mkl}\,[\varepsilon_{kij}a_ib_j]\,c_l &= \varepsilon_{kij}\varepsilon_{klm}a_ib_jc_l && \rightarrow \text{eq. (A.7)}\\ &= [\delta_{il}\delta_{jm}-\delta_{im}\delta_{jl}]\,a_ib_jc_l &&\\ &= \delta_{il}\delta_{jm}a_ib_jc_l-\delta_{im}\delta_{jl}a_ib_jc_l &&\\ &= a_ib_mc_i-a_mb_jc_j &&\\ [\mathbf{a}\times\mathbf{b}]\times\mathbf{c} &= \mathbf{b}\,[\mathbf{a}\cdot\mathbf{c}]-[\mathbf{a}\otimes\mathbf{c}]\cdot\mathbf{b} && \text{(A.15)}\\ &\equiv -\mathbf{c}\times[\mathbf{a}\times\mathbf{b}] = -\,[[\mathbf{b}\cdot\mathbf{c}]\,\mathbf{I}-[\mathbf{b}\otimes\mathbf{c}]]\cdot\mathbf{a} && \leftarrow \text{eq. (A.7)}\end{aligned}$$

Jacobi identity

$$[\mathbf{a}\times\mathbf{b}]\times\mathbf{c} = \mathbf{a}\times[\mathbf{b}\times\mathbf{c}] - \mathbf{b}\times[\mathbf{a}\times\mathbf{c}] \tag{A.16}$$

$$[\mathbf{a}\times\mathbf{b}]\times\mathbf{c} + [\mathbf{b}\times\mathbf{c}]\times\mathbf{a} + [\mathbf{c}\times\mathbf{a}]\times\mathbf{b} = 0 \qquad \leftarrow \text{eq. (A.1)}$$

Jumps and averages

Defining a general vector product operator $\star$ (which denotes a vector contraction, cross product, or any other valid operator acting on two arbitrary order quantities), then

$$\begin{array}{rcl} \underbrace{\tfrac{1}{2}[\mathbf{a}^+ + \mathbf{a}^-]}_{\{\mathbf{a}\}} \star \underbrace{[\mathbf{b}^+ - \mathbf{b}^-]}_{[\mathbf{b}]} & = & \tfrac{1}{2}[\mathbf{a}^+\star\mathbf{b}^+ + \mathbf{a}^-\star\mathbf{b}^+ - \mathbf{a}^+\star\mathbf{b}^- - \mathbf{a}^-\star\mathbf{b}^-] \\ + \underbrace{[\mathbf{a}^+ - \mathbf{a}^-]}_{[\mathbf{a}]} \star \underbrace{\tfrac{1}{2}[\mathbf{b}^+ + \mathbf{b}^-]}_{\{\mathbf{b}\}} & = & \tfrac{1}{2}[\mathbf{a}^+\star\mathbf{b}^+ - \mathbf{a}^-\star\mathbf{b}^+ + \mathbf{a}^+\star\mathbf{b}^- - \mathbf{a}^-\star\mathbf{b}^-] \\ \hline \{\!\{\mathbf{a}\}\!\}\star[\![\mathbf{b}]\!] + [\![\mathbf{a}]\!]\star\{\!\{\mathbf{b}\}\!\} & = & [\![\mathbf{a}\star\mathbf{b}]\!] \end{array} \tag{A.17}$$

while noting that, in general, the operation $\star$ does not permit the permutation of its operands.

For the case where $\mathbf{n} = \{\!\{\mathbf{n}\}\!\}$, then

$$\begin{aligned} [\![[\mathbf{a}\times\mathbf{n}]\times\mathbf{b}]\!] &= [\![[\mathbf{a}\cdot\mathbf{b}]\,\mathbf{n} - [\mathbf{a}\otimes\mathbf{b}]\cdot\mathbf{n}]\!] && \rightarrow \text{eq. (A.15)} \\ &= [\![[\mathbf{a}\cdot\mathbf{b}]\,\mathbf{i} - \mathbf{a}\otimes\mathbf{b}]\!]\cdot\mathbf{n} \\ &= [[\![\mathbf{a}\cdot\mathbf{b}]\!]\,\mathbf{i} - [\![\mathbf{a}\otimes\mathbf{b}]\!]]\cdot\mathbf{n} \\ &= [[\mathbf{a}^+\cdot\mathbf{b}^+]\,\mathbf{i} - \mathbf{a}^+\otimes\mathbf{b}^+]\cdot\mathbf{n} - [[\mathbf{a}^-\cdot\mathbf{b}^-]\,\mathbf{i} - \mathbf{a}^-\otimes\mathbf{b}^-]\cdot\mathbf{n} && \rightarrow \text{eq. (A.15)} \\ &= [\mathbf{a}^+\times\mathbf{n}]\times\mathbf{b}^+ - [\mathbf{a}^-\times\mathbf{n}]\times\mathbf{b}^- && \rightarrow \text{eq. (A.17)} \\ &= \{\!\{\mathbf{a}\times\mathbf{n}\}\!\}\times[\![\mathbf{b}]\!] + [\![\mathbf{a}\times\mathbf{n}]\!]\times\{\!\{\mathbf{b}\}\!\} \\ &= [\{\!\{\mathbf{a}\}\!\}\times\mathbf{n}]\times[\![\mathbf{b}]\!] + [[\![\mathbf{a}]\!]\times\mathbf{n}]\times\{\!\{\mathbf{b}\}\!\} && \text{(A.18)} \end{aligned}$$

and, from some of the intermediate calculations in the above,

$$\begin{aligned} [\![[\mathbf{a}\cdot\mathbf{b}]\,\mathbf{n} - [\mathbf{a}\otimes\mathbf{b}]\cdot\mathbf{n}]\!] &\equiv [[\![\mathbf{a}\cdot\mathbf{b}]\!]\,\mathbf{i} - [\![\mathbf{a}\otimes\mathbf{b}]\!]]\cdot\mathbf{n} && \rightarrow \text{eq. (A.17)} \\ &= [[\{\!\{\mathbf{a}\}\!\}\cdot[\![\mathbf{b}]\!] + [\![\mathbf{a}]\!]\cdot\{\!\{\mathbf{b}\}\!\}]\,\mathbf{i} - [[\![\mathbf{a}]\!]\otimes\{\!\{\mathbf{b}\}\!\} + \{\!\{\mathbf{a}\}\!\}\otimes[\![\mathbf{b}]\!]]]\cdot\mathbf{n} \\ &= \mathbf{n}\,[\{\!\{\mathbf{a}\}\!\}\cdot[\![\mathbf{b}]\!] + [\![\mathbf{a}]\!]\cdot\{\!\{\mathbf{b}\}\!\}] - [\![\mathbf{a}]\!]\,[\{\!\{\mathbf{b}\}\!\}\cdot\mathbf{n}] + \{\!\{\mathbf{a}\}\!\}\,[[\![\mathbf{b}]\!]\cdot\mathbf{n}] \quad . \end{aligned} \tag{A.19}$$

A.2 Generic differential identities

Although in the subsequent section we define the following differential identities in terms of the spatial differential operator ∇ valid on $\mathscr{B}_t$, it should be noted that the structure remains the same when expressing the equivalent operations on $\mathscr{B}_0$ in terms of $\boldsymbol{\nabla}_0$.

Divergence

Vector field

$$\boldsymbol{\nabla}\cdot\mathbf{a} \Rightarrow \frac{\mathrm{d}a_i}{\mathrm{d}x_i} \tag{A.20}$$

Tensor field

$$\boldsymbol{\nabla}\cdot\mathbf{T} = [\boldsymbol{\nabla}\mathbf{T}] : \mathbf{I} \Rightarrow \frac{\mathrm{d}T_{ij}}{\mathrm{d}x_j} = \frac{\mathrm{d}T_{ij}}{\mathrm{d}x_k}\delta_{kj} \tag{A.21}$$

Curl

Vector field

$$\varepsilon_{ijk}\frac{\mathrm{d}a_k}{\mathrm{d}x_j} = \frac{\mathrm{d}}{\mathrm{d}x_j}\left[\varepsilon_{ijk}a_k\right]$$

$$\boldsymbol{\nabla}\times\mathbf{a} := \boldsymbol{\nabla}\cdot[\boldsymbol{\epsilon}\cdot\mathbf{a}] \tag{A.22}$$

Tensor field

$$\varepsilon_{kji}\frac{\mathrm{d}T_{li}}{\mathrm{d}x_j} = \frac{\mathrm{d}}{\mathrm{d}x_j}\left[\varepsilon_{kji}T_{li}\right] = \frac{\mathrm{d}}{\mathrm{d}x_j}\left[T_{li}\varepsilon_{ikj}\right] \qquad \leftarrow \text{eq. (A.21)}$$

$$\boldsymbol{\nabla}\times\mathbf{T} := \left[\boldsymbol{\nabla}\cdot[\mathbf{T}\cdot\boldsymbol{\epsilon}]\right]^T \tag{A.23}$$

Curl of a gradient

$$\varepsilon_{kji}\frac{\mathrm{d}}{\mathrm{d}x_j}\left[\frac{\mathrm{d}\left[a_l + a_l^*\right]}{\mathrm{d}x_i}\right] = \varepsilon_{kji}\left[\frac{\mathrm{d}^2[a_l + a_l^*]}{\mathrm{d}x_j\mathrm{d}x_i}\right] = 0 \qquad \leftarrow \text{eqs. (A.23,A.9)}$$

$$\boldsymbol{\nabla}\times\left[\boldsymbol{\nabla}\left[\mathbf{a}+\mathbf{a}^*\right]\right] = \mathbf{0} \tag{A.24}$$

Divergence of a curl

$$\frac{\mathrm{d}}{\mathrm{d}x_i}\left[\varepsilon_{ijk}\frac{\mathrm{d}\left[a_k + a_k^*\right]}{\mathrm{d}x_j}\right] = \varepsilon_{ijk}\frac{\mathrm{d}^2[a_k + a_k^*]}{\mathrm{d}x_i\mathrm{d}x_j} = 0 \qquad \leftarrow \text{eqs. (A.22,A.9)}$$

$$\boldsymbol{\nabla}\cdot\left[\boldsymbol{\nabla}\times\left[\mathbf{a}+\mathbf{a}^*\right]\right] = 0 \tag{A.25}$$

Curl of a curl

$$\varepsilon_{ijk}\frac{\mathrm{d}}{\mathrm{d}x_j}\left[\varepsilon_{klm}\frac{\mathrm{d}a_m}{\mathrm{d}x_l}\right] = \varepsilon_{kij}\varepsilon_{klm}\frac{\mathrm{d}^2a_m}{\mathrm{d}x_j\mathrm{d}x_l} \qquad \leftarrow \text{eq. (A.22)}\rightarrow \text{eq. (A.7)}$$

$$= [\delta_{il}\delta_{jm} - \delta_{im}\delta_{jl}] \frac{\mathrm{d}^2 a_m}{\mathrm{d}x_j \mathrm{d}x_l}$$

$$= \frac{\mathrm{d}^2 a_j}{\mathrm{d}x_j \mathrm{d}x_i} - \frac{\mathrm{d}^2 a_i}{\mathrm{d}x_j \mathrm{d}x_j}$$

$$= \frac{\mathrm{d}}{\mathrm{d}x_i}\left[\frac{\mathrm{d}a_j}{\mathrm{d}x_j}\right] - \frac{\mathrm{d}}{\mathrm{d}x_j}\left[\frac{\mathrm{d}a_i}{\mathrm{d}x_j}\right]$$

$$\nabla \times [\nabla \times \mathbf{a}] = \nabla [\nabla \cdot \mathbf{a}] - \nabla \cdot [\nabla \mathbf{a}] \tag{A.26}$$

Gradient of a scalar

$$\nabla s = \nabla \cdot [s\mathbf{I}] \Rightarrow \frac{\mathrm{d}s}{\mathrm{d}x_i} = \frac{\mathrm{d}}{\mathrm{d}x_j}[s\delta_{ij}] \tag{A.27}$$

Divergence of a scaled vector

$$\frac{\mathrm{d}}{\mathrm{d}x_i}[sa_i] = \frac{\mathrm{d}s}{\mathrm{d}x_i} a_i + s\frac{\mathrm{d}a_i}{\mathrm{d}x_i}$$

$$\nabla \cdot [s\mathbf{a}] = [\nabla s] \cdot \mathbf{a} + s[\nabla \cdot \mathbf{a}] \tag{A.28}$$

Gradient of a scaled vector

$$\frac{\mathrm{d}}{\mathrm{d}x_j}[sa_i] = \frac{\mathrm{d}s}{\mathrm{d}x_j} a_i + s\frac{\mathrm{d}a_i}{\mathrm{d}x_j}$$

$$\nabla [s\mathbf{a}] = \mathbf{a} \otimes [\nabla s] + s[\nabla \mathbf{a}] \tag{A.29}$$

Curl of a scaled vector

$$\varepsilon_{ijk}\frac{\mathrm{d}[sa_k]}{\mathrm{d}x_j} = s\varepsilon_{ijk}\frac{\mathrm{d}a_k}{\mathrm{d}x_j} + \varepsilon_{ijk}a_k\frac{\mathrm{d}s}{\mathrm{d}x_j} \qquad \leftarrow \text{eq. (A.22)} \rightarrow \text{eq. (A.1)}$$

$$\nabla \times [s\mathbf{a}] = s[\nabla \times \mathbf{a}] + [\nabla s] \times \mathbf{a} \tag{A.30}$$

Divergence of a vector cross product

$$\frac{\mathrm{d}}{\mathrm{d}x_i}[\varepsilon_{ijk}b_k a_j] = \varepsilon_{ijk}\frac{\mathrm{d}b_k}{\mathrm{d}x_i}a_j + \varepsilon_{ijk}b_k\frac{\mathrm{d}a_j}{\mathrm{d}x_i} \qquad \leftarrow \text{eq. (A.3)}$$

$$= b_k\varepsilon_{kij}\frac{\mathrm{d}a_j}{\mathrm{d}x_i} - a_j\varepsilon_{jik}\frac{\mathrm{d}b_k}{\mathrm{d}x_i} \qquad \rightarrow \text{eq. (A.22)}$$

$$\nabla \cdot [\mathbf{a} \times \mathbf{b}] = \mathbf{b} \cdot [\nabla \times \mathbf{a}] - \mathbf{a} \cdot [\nabla \times \mathbf{b}] \tag{A.31}$$

Gradient of a vector cross product

$$\frac{\mathrm{d}}{\mathrm{d}x_l}\left[\varepsilon_{ijk}b_k a_j\right] = \varepsilon_{ijk}\frac{\mathrm{d}b_k}{\mathrm{d}x_l}a_j + \varepsilon_{ijk}b_k\frac{\mathrm{d}a_j}{\mathrm{d}x_l} \qquad \leftarrow \text{eqs. (A.3,A.4)}$$

$$\nabla\left[\mathbf{a}\times\mathbf{b}\right] = \left[\nabla\mathbf{a}\right]\times\mathbf{b} + \mathbf{a}\times\left[\nabla\mathbf{b}\right] \tag{A.32}$$

Divergence of the tensor product of two vectors

$$\frac{\mathrm{d}}{\mathrm{d}x_j}\left[a_i b_j\right] = \frac{\mathrm{d}a_i}{\mathrm{d}x_j}b_j + a_i\frac{\mathrm{d}b_j}{\mathrm{d}x_j} \qquad \leftarrow \text{eq. (A.21)}$$

$$\nabla\cdot\left[\mathbf{a}\otimes\mathbf{b}\right] = \left[\nabla\mathbf{a}\right]\cdot\mathbf{b} + \mathbf{a}\left[\nabla\cdot\mathbf{b}\right] \tag{A.33}$$

Divergence of a scaled tensor

$$\frac{\mathrm{d}}{\mathrm{d}x_j}\left[sT_{ij}\right] = T_{ij}\frac{\mathrm{d}s}{\mathrm{d}x_j} + s\frac{\mathrm{d}T_{ij}}{\mathrm{d}x_j}$$

$$\nabla\cdot\left[s\mathbf{T}\right] = \mathbf{T}\cdot\left[\nabla s\right] + s\left[\nabla\cdot\mathbf{T}\right] \tag{A.34}$$

Divergence of a vector-tensor inner product

$$\frac{\mathrm{d}}{\mathrm{d}x_j}\left[a_i T_{ij}\right] = \frac{\mathrm{d}a_i}{\mathrm{d}x_j}T_{ij} + a_i\frac{\mathrm{d}T_{ij}}{\mathrm{d}x_j} \qquad \leftarrow \text{eq. (A.20)} \rightarrow \text{eq. (A.21)}$$

$$\nabla\cdot\left[\mathbf{a}\cdot\mathbf{T}\right] = \left[\nabla\mathbf{a}\right] : \mathbf{T} + \mathbf{a}\cdot\left[\nabla\cdot\mathbf{T}\right] \tag{A.35}$$

Divergence of a tensor-tensor inner product

$$\frac{\mathrm{d}}{\mathrm{d}x_j}\left[S_{ik}T_{kj}\right] = \frac{\mathrm{d}S_{ik}}{\mathrm{d}x_j}T_{kj} + S_{ik}\frac{\mathrm{d}T_{kj}}{\mathrm{d}x_j} \qquad \leftarrow \text{eq. (A.21)} \rightarrow \text{eq. (A.21)}$$

$$\nabla\cdot\left[\mathbf{S}\cdot\mathbf{T}\right] = \left[\nabla\mathbf{S}\right] : \mathbf{T} + \mathbf{S}\cdot\left[\nabla\cdot\mathbf{T}\right] \tag{A.36}$$

Divergence of a vector-tensor outer product

$$\frac{\mathrm{d}}{\mathrm{d}x_k}\left[a_i T_{jk}\right] = \frac{\mathrm{d}a_i}{\mathrm{d}x_k}T_{jk} + a_i\frac{\mathrm{d}T_{jk}}{\mathrm{d}x_k} \qquad \leftarrow \text{eq. (A.21)}$$

$$\nabla\cdot\left[\mathbf{a}\otimes\mathbf{T}\right] = \left[\nabla\mathbf{a}\right]\cdot\mathbf{T}^T + \mathbf{a}\otimes\left[\nabla\cdot\mathbf{T}\right] \tag{A.37}$$

Divergence of a vector-tensor cross product

$$\frac{\mathrm{d}}{\mathrm{d}x_l}\left[\varepsilon_{ijk}T_{kl}b_j\right] = \varepsilon_{ijk}\frac{\mathrm{d}b_j}{\mathrm{d}x_l}T_{kl} + \varepsilon_{ijk}b_j\frac{\mathrm{d}T_{kl}}{\mathrm{d}x_l} \qquad \leftarrow \text{eqs. (A.21,A.3)}$$

$$\nabla\cdot\left[\mathbf{b}\times\mathbf{T}\right] = \boldsymbol{\epsilon} : \left[\left[\nabla\mathbf{b}\right]\cdot\mathbf{T}^T\right] + \mathbf{b}\times\left[\nabla\cdot\mathbf{T}\right] \tag{A.38}$$

Curl of a cross product

$$
\begin{aligned}
\varepsilon_{mlk}\frac{\mathrm{d}}{\mathrm{d}x_l}\left[\varepsilon_{ijk}a_ib_j\right] &= \varepsilon_{kij}\varepsilon_{kml}\frac{\mathrm{d}}{\mathrm{d}x_l}\left[a_ib_j\right] && \to \text{eq. (A.7)}\\
&= \left[\delta_{im}\delta_{jl}-\delta_{il}\delta_{jm}\right]\left[\frac{\mathrm{d}a_i}{\mathrm{d}x_l}b_j + a_i\frac{\mathrm{d}b_j}{\mathrm{d}x_l}\right]\\
&= \delta_{im}\delta_{jl}\left[\frac{\mathrm{d}a_i}{\mathrm{d}x_l}b_j + a_i\frac{\mathrm{d}b_j}{\mathrm{d}x_l}\right] - \delta_{il}\delta_{jm}\left[\frac{\mathrm{d}a_i}{\mathrm{d}x_l}b_j + a_i\frac{\mathrm{d}b_j}{\mathrm{d}x_l}\right]\\
&= \frac{\mathrm{d}a_m}{\mathrm{d}x_l}b_l + a_m\frac{\mathrm{d}b_l}{\mathrm{d}x_l} - \frac{\mathrm{d}a_l}{\mathrm{d}x_l}b_m - a_l\frac{\mathrm{d}b_m}{\mathrm{d}x_l}\\
\nabla\times\left[\mathbf{a}\times\mathbf{b}\right] &= \mathbf{a}\left[\nabla\cdot\mathbf{b}\right] - \mathbf{b}\left[\nabla\cdot\mathbf{a}\right] + \left[\nabla\mathbf{a}\right]\cdot\mathbf{b} - \left[\nabla\mathbf{b}\right]\cdot\mathbf{a} && \text{(A.39)}
\end{aligned}
$$

Cross product of a curl

$$
\begin{aligned}
\varepsilon_{mlk}b_l\left[\varepsilon_{kji}\frac{\mathrm{d}a_i}{\mathrm{d}x_j}\right] &= -\varepsilon_{kij}\varepsilon_{kml}b_l\frac{\mathrm{d}a_i}{\mathrm{d}x_j} && \to \text{eq. (A.7)}\\
&= \left[\delta_{il}\delta_{jm}-\delta_{im}\delta_{jl}\right]b_l\frac{\mathrm{d}a_i}{\mathrm{d}x_j}\\
&= \delta_{il}\delta_{jm}b_l\frac{\mathrm{d}a_i}{\mathrm{d}x_j} - \delta_{im}\delta_{jl}b_l\frac{\mathrm{d}a_i}{\mathrm{d}x_j}\\
&= b_l\frac{\mathrm{d}a_l}{\mathrm{d}x_m} - b_l\frac{\mathrm{d}a_m}{\mathrm{d}x_l}\\
\mathbf{b}\times\left[\nabla\times\mathbf{a}\right] &= \mathbf{b}\cdot\left[\left[\nabla\mathbf{a}\right]-\left[\nabla\mathbf{a}\right]^T\right] = \mathbf{b}\cdot\left[\nabla\mathbf{a}\right] - \left[\nabla\mathbf{a}\right]\cdot\mathbf{b} && \text{(A.40)}
\end{aligned}
$$

$$
\begin{aligned}
\varepsilon_{mkl}\left[\varepsilon_{kji}\frac{\mathrm{d}a_i}{\mathrm{d}x_j}\right]b_l &= \varepsilon_{klm}\varepsilon_{kji}\frac{\mathrm{d}a_i}{\mathrm{d}x_j}b_l && \to \text{eq. (A.7)}\\
&= \left[\delta_{lj}\delta_{mi}-\delta_{li}\delta_{mj}\right]\frac{\mathrm{d}a_i}{\mathrm{d}x_j}b_l\\
&= \frac{\mathrm{d}a_m}{\mathrm{d}x_l}b_l - \frac{\mathrm{d}a_l}{\mathrm{d}x_m}b_l\\
\left[\nabla\times\mathbf{a}\right]\times\mathbf{b} &= \left[\nabla\mathbf{a}\right]\cdot\mathbf{b} - \mathbf{b}\cdot\left[\nabla\mathbf{a}\right] && \text{(A.41)}
\end{aligned}
$$

Gradient of a vector-vector inner product

$$
\begin{aligned}
\frac{\mathrm{d}}{\mathrm{d}x_j}\left[a_ib_i\right] &\equiv \frac{\mathrm{d}}{\mathrm{d}x_k}\left[\left[a_ib_i\right]\delta_{jk}\right]\\
&= b_i\frac{\mathrm{d}a_i}{\mathrm{d}x_j} + a_i\frac{\mathrm{d}b_i}{\mathrm{d}x_j}\\
\nabla\left[\mathbf{a}\cdot\mathbf{b}\right] &\equiv \nabla\cdot\left[\left[\mathbf{a}\cdot\mathbf{b}\right]\mathbf{I}\right]\\
&= \mathbf{b}\cdot\left[\nabla\mathbf{a}\right] + \mathbf{a}\cdot\left[\nabla\mathbf{b}\right] && \to \text{eqs. (A.40,A.41)}\\
&\equiv \mathbf{b}\times\left[\nabla\times\mathbf{a}\right] + \mathbf{a}\times\left[\nabla\times\mathbf{b}\right] + \left[\nabla\mathbf{a}\right]\cdot\mathbf{b} + \left[\nabla\mathbf{b}\right]\cdot\mathbf{a} && \text{(A.42)}
\end{aligned}
$$

A.3 Differential and rate identities: Continuum mechanics

Derivatives with respect to the deformation gradient tensor [559]

Volumetric Jacobian

$$\frac{\partial J}{\partial \mathbf{F}} = J\mathbf{F}^{-T} \tag{A.43}$$

Right Cauchy–Green deformation tensor

$$\begin{aligned}
\frac{\partial C_{AB}}{\partial F_{iJ}} &= \frac{\partial \left[F_{kA}F_{kB}\right]}{\partial F_{iJ}} \\
&= \frac{\partial F_{kA}}{\partial F_{iJ}} F_{kB} + F_{kA}\frac{\partial F_{kB}}{\partial F_{iJ}} \\
&= \delta_{ki}\delta_{AJ}F_{kB} + F_{kA}\delta_{ki}\delta_{BJ} \\
&= \delta_{AJ}F_{iB} + \delta_{BJ}F_{iA} \\
\frac{\partial \mathbf{C}}{\partial \mathbf{F}} &= \mathbf{I}\underline{\otimes}\mathbf{F}^{T} + \mathbf{F}^{T}\overline{\otimes}\mathbf{I}
\end{aligned} \tag{A.44}$$

Product of the volumetric Jacobian and inverse of the right Cauchy–Green deformation tensor

$$\begin{aligned}
\frac{\partial \left[JC^{-1}_{AB}\right]}{\partial F_{iJ}} &= \frac{\partial \left[JC^{-1}_{AB}\right]}{\partial C_{CD}} : \frac{\partial C_{CD}}{\partial F_{iJ}} = \left[\frac{\partial J}{\partial C_{CD}} C^{-1}_{AB} + J\frac{\partial C^{-1}_{AB}}{\partial C_{CD}}\right]\frac{\partial C_{CD}}{\partial F_{iJ}} \qquad \rightarrow \text{eqs. (A.47a,A.46,A.44)} \\
&= \left[\frac{1}{2}JC^{-1}_{CD}C^{-1}_{AB} - \frac{1}{2}J\left[C^{-1}_{AC}C^{-1}_{BD} + C^{-1}_{AD}C^{-1}_{BC}\right]\right]\left[\delta_{CJ}F_{iD} + \delta_{DJ}F_{iC}\right] \\
&= \frac{1}{2}J\left[C^{-1}_{CD}C^{-1}_{AB} - C^{-1}_{AC}C^{-1}_{BD} - C^{-1}_{AD}C^{-1}_{BC}\right]\left[\delta_{CJ}F_{iD} + \delta_{DJ}F_{iC}\right] \\
&= \frac{1}{2}J\left[C^{-1}_{CD}C^{-1}_{AB}\delta_{CJ}F_{iD} - C^{-1}_{AC}C^{-1}_{BD}\delta_{CJ}F_{iD} - C^{-1}_{AD}C^{-1}_{BC}\delta_{CJ}F_{iD}\right] \\
&+ \frac{1}{2}J\left[C^{-1}_{CD}C^{-1}_{AB}\delta_{DJ}F_{iC} - C^{-1}_{AC}C^{-1}_{BD}\delta_{DJ}F_{iC} - C^{-1}_{AD}C^{-1}_{BC}\delta_{DJ}F_{iC}\right] \\
&= \frac{1}{2}J\left[C^{-1}_{AB}F^{-1}_{Ji} - C^{-1}_{AJ}F^{-1}_{Bi} - C^{-1}_{BJ}F^{-1}_{Ai}\right] + \frac{1}{2}J\left[C^{-1}_{AB}F^{-1}_{Ji} - C^{-1}_{BJ}F^{-1}_{Ai} - C^{-1}_{AJ}F^{-1}_{Bi}\right] \\
&= J\left[C^{-1}_{AB}F^{-1}_{Ji} - C^{-1}_{AJ}F^{-1}_{Bi} - C^{-1}_{BJ}F^{-1}_{Ai}\right] \\
\frac{\partial \left[J\mathbf{C}^{-1}\right]}{\partial \mathbf{F}} &= \frac{\partial \left[J\mathbf{C}^{-1}\right]}{\partial \mathbf{C}} : \frac{\partial \mathbf{C}}{\partial \mathbf{F}} = \left[\frac{\partial J}{\partial \mathbf{C}} \otimes \mathbf{C}^{-1} + J\frac{\partial \mathbf{C}^{-1}}{\partial \mathbf{C}}\right] : \frac{\partial \mathbf{C}}{\partial \mathbf{F}} \qquad \rightarrow \text{eqs. (A.47a,A.46,A.44)} \\
&= J\left[\mathbf{C}^{-1} \otimes \mathbf{F}^{-T} - \mathbf{C}^{-1}\underline{\otimes}\mathbf{F}^{-1} - \mathbf{F}^{-1}\overline{\otimes}\mathbf{C}^{-1}\right]
\end{aligned} \tag{A.45}$$

Derivatives with respect to the right Cauchy–Green deformation tensor [559]

Inverse of the right Cauchy–Green deformation tensor

$$\frac{\partial C^{-1}_{AB}}{\partial C_{CD}} = -\frac{1}{2}\left[C^{-1}_{AC}C^{-1}_{BD} + C^{-1}_{AD}C^{-1}_{BC}\right]$$

$$\frac{\partial \mathbf{C}^{-1}}{\partial \mathbf{C}} = -\frac{1}{2}\left[\mathbf{C}^{-1}\overline{\otimes}\mathbf{C}^{-1} + \mathbf{C}^{-1}\underline{\otimes}\mathbf{C}^{-1}\right] \qquad \text{(A.46)}$$

Volumetric Jacobian

$$\frac{\partial J}{\partial \mathbf{C}} = \frac{1}{2}J\mathbf{C}^{-1} \qquad \text{(A.47a)}$$

$$\begin{aligned}
\frac{\partial^2 J}{\partial \mathbf{C}\otimes\partial \mathbf{C}} &= \frac{\partial}{\partial \mathbf{C}}\left[\frac{1}{2}J\mathbf{C}^{-1}\right] && \leftarrow \text{eq. (A.47a)}\\
&= \frac{1}{2}\left[\mathbf{C}^{-1}\otimes\frac{\partial J}{\partial \mathbf{C}} + J\frac{\partial \mathbf{C}^{-1}}{\partial \mathbf{C}}\right] && \rightarrow \text{eqs. (A.47a,A.46)}\\
&= \frac{1}{2}\left[\frac{1}{2}J\mathbf{C}^{-1}\otimes\mathbf{C}^{-1} - \frac{1}{2}J\left[\mathbf{C}^{-1}\overline{\otimes}\mathbf{C}^{-1} + \mathbf{C}^{-1}\underline{\otimes}\mathbf{C}^{-1}\right]\right]\\
&= \frac{1}{4}J\left[\mathbf{C}^{-1}\otimes\mathbf{C}^{-1} - \mathbf{C}^{-1}\overline{\otimes}\mathbf{C}^{-1} - \mathbf{C}^{-1}\underline{\otimes}\mathbf{C}^{-1}\right] && \text{(A.47b)}
\end{aligned}$$

Volumetric Jacobian with an exponent of the negative reciprocal of an arbitrary spatial dimension

$$\begin{aligned}
\frac{\partial[J^{-\frac{2}{\mathrm{d}}}]}{\partial \mathbf{C}} &= -\frac{2}{\mathrm{d}}J^{[-\frac{2}{\mathrm{d}}-1]}\frac{\partial J}{\partial \mathbf{C}} && \rightarrow \text{eq. (A.47a)}\\
&= -\frac{2}{\mathrm{d}}J^{[-\frac{2}{\mathrm{d}}-1]}\frac{1}{2}J\mathbf{C}^{-1}\\
&= -\frac{1}{\mathrm{d}}J^{-\frac{2}{\mathrm{d}}}\mathbf{C}^{-1} && \text{(A.48)}
\end{aligned}$$

Material time derivatives [51, 559]

Deformation gradient

$$\begin{aligned}
\dot{\mathbf{F}} = \nabla_0\left[\dot{\boldsymbol{\varphi}}\right] &= \frac{\partial \mathbf{v}}{\partial \mathbf{x}}\cdot\frac{\partial \mathbf{x}}{\partial \mathbf{X}} && \rightarrow \text{eq. (5.3)}\\
&= \nabla\mathbf{v}\cdot\mathbf{F} && \text{(A.49)}
\end{aligned}$$

Inverse of the deformation gradient

$$\begin{aligned}
\mathrm{D}_t\left[\mathbf{F}\cdot\mathbf{F}^{-1}\right] = \mathrm{D}_t[\mathbf{i}] = \mathbf{0} \quad &\Rightarrow \quad \mathbf{0} = \dot{\mathbf{F}}\cdot\mathbf{F}^{-1} + \mathbf{F}\cdot\mathrm{D}_t\left[\mathbf{F}^{-1}\right] && \rightarrow \text{eq. (A.49)}\\
&\Rightarrow \quad \mathrm{D}_t\left[\mathbf{F}^{-1}\right] = -\mathbf{F}^{-1}\cdot\nabla\mathbf{v} && \text{(A.50)}
\end{aligned}$$

Volumetric Jacobian

$$\begin{aligned}
\dot{J} &= \frac{\partial J}{\partial \mathbf{F}} : \dot{\mathbf{F}} && \rightarrow \text{eqs. (A.43,A.49,A.21)}\\
&= J\nabla\mathbf{v} : \mathbf{i} = J\nabla\cdot\mathbf{v} && \text{(A.51)}
\end{aligned}$$

Cofactor of the deformation gradient

$$\begin{aligned}
\mathrm{D}_t[\mathrm{cof}(\mathbf{F})] &= \mathrm{D}_t\left[\det(\mathbf{F})\,\mathbf{F}^{-T}\right]\\
&= \mathrm{D}_t\left[J\mathbf{F}^{-T}\right] = \dot{J}\mathbf{F}^{-T} + J\mathrm{D}_t\left[\mathbf{F}^{-T}\right] && \rightarrow \text{eqs. (A.51,A.50)}
\end{aligned}$$

$$
\begin{aligned}
&= [J\nabla\cdot\mathbf{v}]\,\mathbf{F}^{-T} + J\left[-\mathbf{F}^{-1}\cdot\nabla\mathbf{v}\right]^{T} \\
&= [\nabla\cdot\mathbf{v}]\,J\mathbf{F}^{-T} - [\nabla\mathbf{v}]^{T}\cdot\left[J\mathbf{F}^{-T}\right] && \rightarrow \text{eq. (A.5)} \\
&= [\nabla\cdot\mathbf{v}]\,\mathrm{cof}\,(\mathbf{F}) - [\nabla\mathbf{v}]^{T}\cdot\mathrm{cof}\,(\mathbf{F}) && \text{(A.52)}
\end{aligned}
$$

B Calculus

The notation used to define general control volumes and surfaces, and their associated director fields, is presented in Section 5.2. Like in Appendix A.2, where possible we will make generic statements for integral theorems in terms of the differential operator ∇. However, when the domain of integration is not arbitrary this will be reflected by the limits of the integral and the subsequent choice of notation for the differential operators and fields.

Furthermore, to facilitate the statement of general continuum theorems, we denote s as an arbitrary scalar, $\mathbf{a}, \mathbf{b}$ as arbitrary vectors, and $\mathbf{T}$ as an arbitrary rank-2 tensor.

B.1 Microscopic theorems

Dirac delta function [133]

We define $\delta(\mathbf{x})$ as the Dirac delta function that has the property

$$\int_{\mathscr{V}} \delta(\mathbf{x})\ \mathrm{dv} = 1 \quad ; \tag{B.1}$$

then for a continuous function $f(\mathbf{x})$

$$\int_{\mathscr{V}} \delta(\mathbf{x} - \mathbf{x}') f(\mathbf{x})\ \mathrm{dv} = f(\mathbf{x}') \quad . \tag{B.2}$$

Gradient operator [414]

$$\nabla \left[\frac{1}{|\mathbf{x} - \mathbf{x}'|} \right] = - \frac{\mathbf{x} - \mathbf{x}'}{|\mathbf{x} - \mathbf{x}'|^3} \tag{B.3}$$

Divergence operator [414, 133, 230]

$$\nabla \cdot \left[\frac{\mathbf{x} - \mathbf{x}'}{|\mathbf{x} - \mathbf{x}'|^3} \right] = 4\pi\delta(\mathbf{x} - \mathbf{x}') \tag{B.4}$$

Laplace operator [133, 230]

$$\nabla^2 \left[\frac{1}{|\mathbf{x} - \mathbf{x}'|} \right] = \nabla \cdot \left[\nabla \left[\frac{1}{|\mathbf{x} - \mathbf{x}'|} \right] \right] \qquad \rightarrow \text{eqs. (B.3,B.4)}$$

$$= -4\pi\delta(\mathbf{x} - \mathbf{x}') \tag{B.5}$$

https://doi.org/10.1515/9783110418576-014

Curl operator [414, 133]

$$\nabla \times \left[\frac{\mathbf{x} - \mathbf{x}'}{|\mathbf{x} - \mathbf{x}'|^3} \right] = \mathbf{0} \tag{B.6}$$

B.2 Continuum theorems

Volume transformation

Considering the reference volume element,

$$\begin{aligned} \mathrm{dV} &= \mathrm{d}\mathbf{A} \cdot \mathrm{d}\mathbf{L} = [\mathrm{d}\mathbf{X}_1 \times \mathrm{d}\mathbf{X}_2] \cdot \mathrm{d}\mathbf{X}_3 & \leftarrow \text{eq. (5.9)} \\ &= \left[\left[\mathbf{F}^{-1} \cdot \mathrm{d}\mathbf{x}_1\right] \times \left[\mathbf{F}^{-1} \cdot \mathrm{d}\mathbf{x}_2\right]\right] \cdot \mathrm{d}\mathbf{X}_3 & \rightarrow \text{eq. (A.11)} \\ &= \det\left(\mathbf{F}^{-1}\right) [\mathrm{d}\mathbf{x}_1 \times \mathrm{d}\mathbf{x}_2] \cdot \mathbf{F} \cdot \mathrm{d}\mathbf{X}_3 \\ &= \det\left(\mathbf{F}^{-1}\right) \mathrm{d}\mathbf{a} \cdot \mathrm{d}\mathbf{l} \\ &\equiv J^{-1}\,\mathrm{dv} \quad , & \text{(B.7)} \end{aligned}$$

where the spatial cross-sectional area vector $\mathrm{d}\mathbf{a} = \mathrm{d}\mathbf{x}_1 \times \mathrm{d}\mathbf{x}_2$ and $\mathrm{d}\mathbf{l} = \mathbf{F} \cdot \mathrm{d}\mathbf{X}_3$ is the spatial length vector.

Area transformation by Nanson's formula

Assuming that the referential volume element is on the boundary, such that $\mathrm{dA} \in \partial\mathscr{B}_0$ and, therefore, $\mathrm{d}\mathbf{L} \equiv \mathbf{N}$, then

$$\begin{aligned} \mathrm{d}\mathbf{A} \cdot \mathrm{d}\mathbf{L} &= \left[\det\left(\mathbf{F}^{-1}\right) [\mathrm{d}\mathbf{x}_1 \times \mathrm{d}\mathbf{x}_2] \cdot \mathbf{F}\right] \cdot \mathbf{N} & \leftarrow \text{eqs. (5.9,A.11,A.5)} \\ &= \left[\mathrm{d}\mathbf{a} \cdot \mathrm{cof}\left(\mathbf{F}^{-T}\right)\right] \cdot \mathbf{N} & \text{(B.8)} \end{aligned}$$

which, since **N** is arbitrary, leads to the definition given in equation (5.10).

Integration by parts

Scaled vector

$$\int_{\mathscr{V}} s\,[\nabla \cdot \mathbf{a}]\,\mathrm{dv} = \int_{\mathscr{V}} \nabla \cdot [s\mathbf{a}]\,\mathrm{dv} - \int_{\mathscr{V}} [\nabla s] \cdot \mathbf{a}\,\mathrm{dv} \tag{B.9}$$

Vector inner product

$$\int_{\mathscr{V}} \mathbf{a} \cdot [\nabla \mathbf{b}]\,\mathrm{dv} = \int_{\mathscr{V}} \nabla\,[\mathbf{a} \cdot \mathbf{b}]\,\mathrm{dv} - \int_{\mathscr{V}} \mathbf{b} \cdot [\nabla \mathbf{a}]\,\mathrm{dv} \tag{B.10}$$

Vector outer product

$$\int_{\mathscr{V}} \mathbf{a}\,[\nabla \cdot \mathbf{b}]\,\mathrm{dv} = \int_{\mathscr{V}} \nabla \cdot [\mathbf{a} \otimes \mathbf{b}]\,\mathrm{dv} - \int_{\mathscr{V}} [\nabla \mathbf{a}] \cdot \mathbf{b}\,\mathrm{dv} \tag{B.11}$$

Vector cross product

$$\int_{\mathscr{V}} a_i\,[\epsilon_{ijk}\partial_j b_k] = \int_{\mathscr{V}} \epsilon_{ijk} a_i\,[\partial_j b_k] \qquad \rightarrow \text{eq. (B.10)}$$

$$= \int_{\mathscr{V}} \epsilon_{ijk}\partial_j\,[a_i b_k] - \int_{\mathscr{V}} \epsilon_{ijk} b_k\,[\partial_j a_i] \qquad \rightarrow \text{eq. (A.6)}$$

$$= \int_{\mathscr{V}} \partial_j\,[\epsilon_{jki} b_k a_i] + \int_{\mathscr{V}} b_k \epsilon_{kji}\,[\partial_j a_i] \qquad \rightarrow \text{eq. (A.8)}$$

$$\int_{\mathscr{V}} \mathbf{a}\cdot[\nabla\times\mathbf{b}]\ \mathrm{dv} = \int_{\mathscr{V}} \nabla\cdot[\mathbf{b}\times\mathbf{a}]\ \mathrm{dv} + \int_{\mathscr{V}} \mathbf{b}\cdot[\nabla\times\mathbf{a}]\ \mathrm{dv} \tag{B.12}$$

Vector-tensor inner product

$$\int_{\mathscr{V}} \mathbf{a}\cdot[\nabla\cdot\mathbf{T}]\ \mathrm{dv} = \int_{\mathscr{V}} \nabla\cdot[\mathbf{a}\cdot\mathbf{T}]\ \mathrm{dv} - \int_{\mathscr{V}} [\nabla\mathbf{a}] : \mathbf{T}\ \mathrm{dv} \tag{B.13}$$

Kelvin–Stokes theorem

$$\int_{\mathscr{A}} [\nabla\times\mathbf{b}]\cdot\mathrm{d}\mathbf{a} = \int_{\mathscr{A}} [\nabla\times\mathbf{b}]\cdot\mathbf{n}\ \mathrm{da} = \oint_{\partial\mathscr{A}} \mathbf{b}\cdot\ \mathrm{d}\mathbf{l} \tag{B.14}$$

$$-\int_{\mathscr{A}} \boldsymbol{\epsilon} : [\nabla s\otimes\ \mathrm{d}\mathbf{a}] = -\int_{\mathscr{A}} \nabla s\times\mathrm{d}\mathbf{a} = \int_{\mathscr{A}} \mathbf{n}\times\nabla s\ \mathrm{da} = \oint_{\partial\mathscr{A}} s\ \mathrm{d}\mathbf{l} \tag{B.15}$$

Divergence theorem

The divergence (or Gauss) theorem states that

$$\int_{\mathscr{V}} \nabla\cdot\mathbf{b}\ \mathrm{dv} = \int_{\partial\mathscr{V}} \mathbf{b}\cdot\mathbf{n}\ \mathrm{da} = \int_{\partial\mathscr{V}} \mathbf{b}\cdot\ \mathrm{d}\mathbf{a} \quad \text{and}$$

$$\int_{\mathscr{V}} \nabla\cdot\mathbf{T}\ \mathrm{dv} = \int_{\partial\mathscr{V}} \mathbf{T}\cdot\mathbf{n}\ \mathrm{da} = \int_{\partial\mathscr{V}} \mathbf{T}\cdot\ \mathrm{d}\mathbf{a}\quad . \tag{B.16}$$

Gradient theorem

It can be shown that for the gradient of an arbitrary scalar field

$$\int_{\mathscr{V}} \nabla s\ \mathrm{dv} = \int_{\partial\mathscr{V}} s\mathbf{n}\ \mathrm{da} = \int_{\partial\mathscr{V}} s\ \mathrm{d}\mathbf{a} \tag{B.17}$$

by using equation (B.16) and choosing $\mathbf{b} = s\mathbf{g}^*$, where $\mathbf{g}^*$ is an arbitrary constant vector field. The gradient of an arbitrary vector field is

$$\int_{\mathscr{V}} \nabla\mathbf{b}\ \mathrm{dv} = \int_{\mathscr{V}} \nabla\cdot[\mathbf{i}\otimes\mathbf{b}]\ \mathrm{dv} \qquad \leftarrow \text{eq. (A.21)} \rightarrow \text{eq. (B.16)}$$

$$= \int_{\partial \mathscr{V}} [\mathbf{b} \otimes \mathbf{i}] \cdot \mathbf{n} \, \mathrm{da} = \int_{\partial \mathscr{V}} \mathbf{b} \otimes \mathbf{n} \, \mathrm{da} \quad . \tag{B.18}$$

Piola identity [204]

$$\int_{\mathscr{V}_0} \nabla_0 \cdot \left[J \mathbf{F}^{-T}\right] \, \mathrm{dV} = \int_{\partial \mathscr{V}_0} \left[J \mathbf{F}^{-T}\right] \cdot \mathbf{N} \, \mathrm{dA} \qquad \leftarrow \text{eq. (B.16)} \rightarrow \text{eq. (5.10)}$$

$$= \int_{\partial \mathscr{V}_t} \mathbf{n} \, \mathrm{da} = \int_{\partial \mathscr{V}_t} \mathbf{i} \cdot \mathbf{n} \, \mathrm{da} \qquad \rightarrow \text{eq. (B.16)}$$

$$= \int_{\mathscr{V}_t} \underbrace{\nabla \cdot \mathbf{i}}_{=\mathbf{0}} \, \mathrm{dv}$$

$$\Longrightarrow \quad \nabla_0 \cdot \left[J \mathbf{F}^{-T}\right] = \mathbf{0} \tag{B.19}$$

Leibniz integral rule

Derived as a specialisation of equation (B.21), the (one-dimensional) Leibniz integral rule is stated as

$$\frac{\mathrm{d}}{\mathrm{d}t} \left[\int_{s_0(t)}^{s_1(t)} f(x,t) \, \mathrm{d}x \right] = \int_{s_0(t)}^{s_1(t)} \partial_t f(x,t) \Big|_x \mathrm{d}x + \left[f(s_1(t),t) \frac{\mathrm{d}s_1(t)}{\mathrm{d}t} - f(s_0(t),t) \frac{\mathrm{d}s_0(t)}{\mathrm{d}t} \right] \quad . \tag{B.20}$$

B.2.1 Materials without discontinuities

Reynolds transport theorem: Control volume

In order to derive the transport equation for control volumes, we first define a generic scalar field $\Phi = \Phi(\mathbf{x}, t)$ and vector field $\boldsymbol{\Phi} = \boldsymbol{\Phi}(\mathbf{x}, t)$ that represent volumetric densities.

For a spatial scalar field, the Reynolds transport theorem for control volumes states that [133, 204]

$$\mathrm{D}_t \int_{\mathscr{V}_t} \Phi_t \, \mathrm{dv} = \mathrm{D}_t \int_{\mathscr{V}_0} \Phi_t J \, \mathrm{dV} \qquad \leftarrow \text{eq. (5.11)}$$

$$= \int_{\mathscr{V}_0} \mathrm{D}_t [\Phi_t J] \, \mathrm{dV}$$

$$= \int_{\mathscr{V}_0} [\mathrm{D}_t [\Phi_t] J + \Phi_t \mathrm{D}_t [J]] \, \mathrm{dV} \qquad \rightarrow \text{eqs. (5.15,A.51)}$$

$$= \int_{\mathscr{V}_0} [J [\mathrm{d}_t \Phi_t + \nabla \Phi_t \cdot \mathbf{v}] + \Phi_t J \nabla \cdot \mathbf{v}] \, \mathrm{dV}$$

$$= \int_{\mathcal{V}_0} [\mathrm{d}_t\Phi_t + [\nabla\Phi_t \cdot \mathbf{v} + \Phi_t\nabla \cdot \mathbf{v}]]\,J\,\mathrm{dV} \qquad \to \text{eqs. (B.9,5.11)}$$

$$= \int_{\mathcal{V}_t} [\mathrm{d}_t\Phi_t + \nabla \cdot [\Phi_t\mathbf{v}]]\ \mathrm{dv} \tag{B.21}$$

while for a spatial vector quantity

$$\mathrm{D}_t \int_{\mathcal{V}_t} \boldsymbol{\Phi}_t\ \mathrm{dv} = \mathrm{D}_t \int_{\mathcal{V}_0} \boldsymbol{\Phi}_t J\ \mathrm{dV} \qquad \leftarrow \text{eq. (5.11)}$$

$$= \int_{\mathcal{V}_0} \mathrm{D}_t[\boldsymbol{\Phi}_t J]\ \mathrm{dV}$$

$$= \int_{\mathcal{V}_0} [\mathrm{D}_t[\boldsymbol{\Phi}_t]J + \boldsymbol{\Phi}_t\mathrm{D}_t[J]]\ \mathrm{dV} \qquad \to \text{eqs. (5.15,A.51)}$$

$$= \int_{\mathcal{V}_0} [J\,[\mathrm{d}_t\boldsymbol{\Phi}_t + \nabla\boldsymbol{\Phi}_t \cdot \mathbf{v}] + \boldsymbol{\Phi}_t J\nabla \cdot \mathbf{v}]\ \mathrm{dV}$$

$$= \int_{\mathcal{V}_0} [\mathrm{d}_t\boldsymbol{\Phi}_t + [\nabla\boldsymbol{\Phi}_t \cdot \mathbf{v} + \boldsymbol{\Phi}_t\nabla \cdot \mathbf{v}]]\,J\,\mathrm{dV} \qquad \to \text{eqs. (B.11,5.11)}$$

$$= \int_{\mathcal{V}_t} [\mathrm{d}_t\boldsymbol{\Phi}_t + \nabla \cdot [\boldsymbol{\Phi}_t \otimes \mathbf{v}]]\ \mathrm{dv} \quad . \tag{B.22}$$

Considering the localised form of the above, this defines the (spatial) nominal time derivatives of scalar and vectorial densities per unit spatial volume as

$$\mathrm{n}_t\Phi_t \coloneqq \mathrm{d}_t\Phi_t + \nabla \cdot [\Phi_t\mathbf{v}] = J^{-1}\mathrm{D}_t[J\Phi_t] \quad \text{and} \tag{B.23a}$$

$$\mathrm{n}_t\boldsymbol{\Phi}_t \coloneqq \mathrm{d}_t\boldsymbol{\Phi}_t + \nabla \cdot [\boldsymbol{\Phi}_t \otimes \mathbf{v}] = J^{-1}\mathrm{D}_t[J\boldsymbol{\Phi}_t] \tag{B.23b}$$

for scalar quantities Φ_t and vectorial quantities $\boldsymbol{\Phi}_t$ in a volume. We may also similarly define the (material) nominal time derivatives of scalar and vectorial densities per unit reference volume as

$$\mathrm{N}_t\Phi_0 \coloneqq \mathrm{D}_t\Phi_0 + \nabla_0 \cdot [\Phi_0\mathbf{V}] = J\mathrm{d}_t\left[J^{-1}\Phi_0\right] \quad \text{and} \tag{B.24a}$$

$$\mathrm{N}_t\boldsymbol{\Phi}_0 \coloneqq \mathrm{D}_t\boldsymbol{\Phi}_0 + \nabla_0 \cdot [\boldsymbol{\Phi}_0 \otimes \mathbf{V}] = J\mathrm{d}_t\left[J^{-1}\boldsymbol{\Phi}_0\right] \quad ; \tag{B.24b}$$

these definitions are verified later in Appendix C.5.4 with the assumption that the relationship between the densities in the reference and current configurations are $\Phi_t = J^{-1}\Phi_0$ and $\boldsymbol{\Phi}_t = J^{-1}\boldsymbol{\Phi}_0$ (that is, equations (5.21a) and (5.21b)).

Observe that, through the application of equation (B.16), the above can be transformed into the commonly presented from of Reynolds transport equation, namely

$$\mathrm{D}_t \int_{\mathcal{V}_t} \Phi_t\ \mathrm{dv} = \int_{\mathcal{V}_t} \mathrm{d}_t\Phi_t\ \mathrm{dv} + \int_{\partial\mathcal{V}_t} [\Phi_t\mathbf{v}] \cdot \mathbf{n}\ \mathrm{da} \quad \text{and}$$

$$\mathrm{D}_t \int\limits_{\mathscr{V}_t} \boldsymbol{\Phi}_t \,\mathrm{dv} = \int\limits_{\mathscr{V}_t} \mathrm{d}_t \boldsymbol{\Phi}_t \,\mathrm{dv} + \int\limits_{\partial\mathscr{V}_t} [\boldsymbol{\Phi}_t \otimes \mathbf{v}] \cdot \mathbf{n} \,\mathrm{da} \quad .$$

In this case, it is important to note that, due to the use of a closed control volume, the surface integrals are evaluated over a closed set of surfaces bounding the volume.

Reynolds transport theorem: Control surface

For a spatial vector field $\boldsymbol{\psi} = \boldsymbol{\psi}(\mathbf{x}, t)$ that represents a flux, the Reynolds transport theorem for control areas relates to the time rate of change of the normal flux of the vector field across a (possibly open) surface. It is given by

$$\begin{aligned}
\mathrm{D}_t \int\limits_{\mathscr{A}_t} \boldsymbol{\psi} \cdot \mathbf{m} \,\mathrm{da} &= \mathrm{D}_t \int\limits_{\mathscr{A}_0} [\boldsymbol{\psi} \cdot \operatorname{cof}(\mathbf{F})] \cdot \mathbf{M} \,\mathrm{dA} && \leftarrow \text{eq. (5.10)} \\
&= \int\limits_{\mathscr{A}_0} [\mathrm{D}_t[\boldsymbol{\psi}] \cdot \operatorname{cof}(\mathbf{F}) + \boldsymbol{\psi} \cdot \mathrm{D}_t[\operatorname{cof}(\mathbf{F})]] \cdot \mathbf{M} \,\mathrm{dA} && \rightarrow \text{eq. (A.52)} \\
&= \int\limits_{\mathscr{A}_0} \left[\mathrm{D}_t[\boldsymbol{\psi}] \cdot \operatorname{cof}(\mathbf{F}) + \boldsymbol{\psi} \cdot \left[[\nabla \cdot \mathbf{v}] \operatorname{cof}(\mathbf{F}) - [\nabla \mathbf{v}]^T \cdot \operatorname{cof}(\mathbf{F})\right]\right] \cdot \mathbf{M} \,\mathrm{dA} \\
& && \rightarrow \text{eq. (5.10)} \\
&= \int\limits_{\mathscr{A}_t} \left[\mathrm{D}_t[\boldsymbol{\psi}] + [\nabla \cdot \mathbf{v}] \boldsymbol{\psi} - \boldsymbol{\psi} \cdot [\nabla \mathbf{v}]^T\right] \cdot \mathbf{m} \,\mathrm{da} && \rightarrow \text{eq. (5.15)} \\
&= \int\limits_{\mathscr{A}_t} [\mathrm{d}_t \boldsymbol{\psi} + [\nabla \boldsymbol{\psi}] \cdot \mathbf{v} + [\nabla \cdot \mathbf{v}] \boldsymbol{\psi} - [\nabla \mathbf{v}] \cdot \boldsymbol{\psi}] \cdot \mathbf{m} \,\mathrm{da} && \rightarrow \text{eq. (A.39)} \\
&= \int\limits_{\mathscr{A}_t} [\mathrm{d}_t \boldsymbol{\psi} + \nabla \times [\boldsymbol{\psi} \times \mathbf{v}] + [\nabla \cdot \boldsymbol{\psi}] \mathbf{v}] \cdot \mathbf{m} \,\mathrm{da} \quad . && \text{(B.25)}
\end{aligned}$$

Considering the localised form of the above, this defines the (spatial) nominal time derivative

$$\mathrm{n}_t \boldsymbol{\psi} := \mathrm{d}_t \boldsymbol{\psi} + \nabla \times [\boldsymbol{\psi} \times \mathbf{v}] + [\nabla \cdot \boldsymbol{\psi}] \mathbf{v} = \mathrm{D}_t[\boldsymbol{\psi} \cdot \operatorname{cof}(\mathbf{F})] \cdot \operatorname{cof}\left(\mathbf{F}^{-1}\right) \tag{B.26}$$

for spatial flux quantities $\boldsymbol{\psi}$ on a surface. We may also similarly define the (material) nominal time derivative

$$\mathrm{N}_t \boldsymbol{\Psi} := \mathrm{D}_t \boldsymbol{\Psi} + \nabla_0 \times [\boldsymbol{\Psi} \times \mathbf{V}] + [\nabla_0 \cdot \boldsymbol{\Psi}] \mathbf{V} = \mathrm{d}_t\left[\boldsymbol{\Psi} \cdot \operatorname{cof}\left(\mathbf{F}^{-1}\right)\right] \cdot \operatorname{cof}(\mathbf{F}) \tag{B.27}$$

for referential flux quantities $\boldsymbol{\Psi}$ on a surface; this definition is verified later in Appendix C.5.4 with the assumption that the relationship between the flux in the reference and current configurations is $\boldsymbol{\psi} = \boldsymbol{\Psi} \cdot \operatorname{cof}(\mathbf{F}^{-1}) = J^{-1}\mathbf{F} \cdot \boldsymbol{\Psi}$ (i. e. equation (5.21c)).

Global balance law: Control volume

The rate of change of any density field is equal to the combined increase due to volume sources for the field and the normal component of its flux over the control volume. This global balance law is expressed as [133, 530]

$$\mathrm{D}_t \int_{\mathscr{V}_t} \Phi_t \,\mathrm{dv} = \int_{\mathscr{V}_t} \mathrm{d}_t \Phi_t \,\mathrm{dv} + \int_{\partial \mathscr{V}_t} [\Phi_t \mathbf{v}] \cdot \mathbf{n} \,\mathrm{da} \qquad \leftarrow \text{eq. (B.21)}$$

$$\equiv \int_{\mathscr{V}_t} s_\Phi (\Phi_t) \,\mathrm{dv} + \int_{\partial \mathscr{V}_t} \mathbf{s}_\Phi (\Phi_t) \cdot \mathbf{n} \,\mathrm{da} \tag{B.28}$$

for a scalar field, with s_Φ representing the scalar volume source and $\mathbf{s}_\Phi$ the vectorial flux, and

$$\mathrm{D}_t \int_{\mathscr{V}_t} \boldsymbol{\Phi}_t \,\mathrm{dv} = \int_{\mathscr{V}_t} \mathrm{d}_t \boldsymbol{\Phi}_t \,\mathrm{dv} + \int_{\partial \mathscr{V}_t} [\boldsymbol{\Phi}_t \otimes \mathbf{v}] \cdot \mathbf{n} \,\mathrm{da} \qquad \leftarrow \text{eq. (B.22)}$$

$$\equiv \int_{\mathscr{V}_t} \mathbf{s}_{\boldsymbol{\Phi}} (\boldsymbol{\Phi}_t) \,\mathrm{dv} + \int_{\partial \mathscr{V}_t} \mathbf{S}_{\boldsymbol{\Phi}} (\boldsymbol{\Phi}_t) \cdot \mathbf{n} \,\mathrm{da} \tag{B.29}$$

for a vector field for which $\mathbf{s}_{\boldsymbol{\Phi}}$ denotes the volume source vector and $\mathbf{S}_{\boldsymbol{\Phi}}$ the flux tensor.

Global balance law: Control surface

The rate of change of any flux field moving orthogonally to a surface is equal to the combined increase due to sources for the field within the area and the flux circulating around the control area. This global balance law is expressed as [133, 530]

$$\mathrm{D}_t \int_{\mathscr{A}_t} \boldsymbol{\psi} \cdot \mathbf{M} \,\mathrm{da} = \int_{\mathscr{A}_t} [\mathrm{d}_t \boldsymbol{\psi} + \nabla \times [\boldsymbol{\psi} \times \mathbf{v}] + [\nabla \cdot \boldsymbol{\psi}] \mathbf{v}] \cdot \mathbf{M} \,\mathrm{da} \qquad \leftarrow \text{eq. (B.25)}$$

$$\equiv \int_{\mathscr{A}_t} \hat{\mathbf{s}}_\psi (\boldsymbol{\psi}) \cdot \mathbf{M} \,\mathrm{da} + \oint_{\partial \mathscr{A}_t} \tilde{\mathbf{s}}_\psi (\boldsymbol{\psi}) \cdot \mathbf{l} \,\mathrm{dl} \quad . \tag{B.30}$$

B.2.2 Materials with discontinuities

To motivate the starting point for the derivations of Reynolds transport theorem accounting for material discontinuities we must recognise that, due to its migratory nature and independence with respect to the continuum body, the material interface as viewed from the continuum reference configuration $\mathscr{B}_0$ is transient. That is why, as stated in Section 5.2.1, the material position of the interface $\mathbf{X}(t) \in \overline{\mathscr{B}_0^+ \cap \mathscr{B}_0^-}$ has associated with it the material velocity $\mathbf{W}$. The material velocity effectively describes the velocity of the interface in this configuration as taken with respect to a truly fixed configurations $\mathscr{B}_\square$. This must, therefore, be taken into account through the application of equation (5.15) with $\mathscr{B}_\square$ taken as the stationary frame. We may then define an "absolute" time derivative by

$$\begin{aligned}\mathscr{D}_t(\bullet)_0 &\coloneqq \partial_t(\bullet)_0\big|_{\mathbf{X}} + \nabla_0\,(\bullet)_0\,\big|_t \cdot \partial_t\mathbf{X}\big|_{\mathbf{X}_\square} \\ &= \mathrm{D}_t(\bullet) + \nabla_0(\bullet)\cdot\mathbf{W} \quad . \end{aligned} \tag{B.31}$$

Furthermore, with reference to Figures 5.4 and 5.5, the moving interface has no normal component at the boundary of the control volume

$$\mathbf{W}\cdot\mathbf{N} = 0 \quad \text{on} \quad \partial\mathscr{V}_0 \tag{B.32}$$

and moves tangentially to the control area, which is to say that

$$\mathbf{W}\cdot\mathbf{M} = 0 \quad \text{on} \quad \mathscr{A}_0 \quad \text{and} \tag{B.33a}$$

$$\mathbf{W}\times\mathbf{L} = \mathbf{0} \quad \text{on} \quad \partial\mathscr{A}_0 \quad . \tag{B.33b}$$

Reynolds transport theorem: Control volume

With the consideration of the contribution from the moving interface, the amended form of the Reynolds transport theorem for scalar densities (related to equation (B.21)) is

$$\begin{aligned}
\mathscr{D}_t \int_{\mathscr{V}_\square} \Phi_\square \,\mathrm{dV}_\square &= \int_{\mathscr{V}_0} \mathrm{D}_t\Phi_0 \,\mathrm{dV} + \int_{\mathscr{V}_0} \nabla_0\cdot[\Phi_0\mathbf{W}]\,\mathrm{dV} && \leftarrow \text{eqs. (B.31,B.21)} \rightarrow \text{eq. (5.44)} \\
&= \int_{\mathscr{V}_0} \mathrm{D}_t\Phi_0 \,\mathrm{dV} + \int_{\partial\mathscr{V}_0} \cancelto{0}{[\Phi_0\mathbf{W}]\cdot\mathbf{N}}\,\mathrm{dA} - \int_{\mathscr{I}_0} [\![\Phi_0\mathbf{W}]\!]\cdot\mathbf{N}_{\mathscr{I}}\,\mathrm{dA} && \rightarrow \text{eq. (B.32)} \\
&= \int_{\mathscr{V}_0} \mathrm{D}_t\Phi_0 \,\mathrm{dV} - \int_{\mathscr{I}_0} [\![\Phi_0\mathbf{W}]\!]\cdot\mathbf{N}_{\mathscr{I}}\,\mathrm{dA} && \rightarrow \text{eqs. (5.22a,5.10)} \\
&= \int_{\mathscr{V}_t} \mathrm{n}_t\Phi_t \,\mathrm{dv} - \int_{\mathscr{I}_t} [\![\Phi_0\mathbf{W}]\!]\cdot\mathrm{cof}\left(\mathbf{F}^{-1}\right)\cdot\mathbf{n}_{\mathscr{I}}\,\mathrm{da} && \rightarrow \text{eq. (5.48)} \\
&= \int_{\mathscr{V}_t} \mathrm{n}_t\Phi_t \,\mathrm{dv} - \int_{\mathscr{I}_t} \left[\!\left[\Phi_0\mathbf{W}\cdot\mathrm{cof}\left(\mathbf{F}^{-1}\right)\right]\!\right]\cdot\mathbf{n}_{\mathscr{I}}\,\mathrm{da} && \rightarrow \text{eq. (A.5)} \\
&= \int_{\mathscr{V}_t} \mathrm{n}_t\Phi_t \,\mathrm{dv} - \int_{\mathscr{I}_t} \left[\!\left[J^{-1}\Phi_0\mathbf{W}\cdot\mathbf{F}^T\right]\!\right]\cdot\mathbf{n}_{\mathscr{I}}\,\mathrm{da} && \rightarrow \text{eqs. (5.21a,5.43)} \\
&= \int_{\mathscr{V}_t} \mathrm{n}_t\Phi_t \,\mathrm{dv} - \int_{\mathscr{I}_t} [\![\Phi_t\,[\mathbf{w}-\mathbf{v}]]\!]\cdot\mathbf{n}_{\mathscr{I}}\,\mathrm{da} && \text{(B.34)} \\
&= \int_{\mathscr{V}_t} \mathrm{d}_t\Phi_t \,\mathrm{dv} + \int_{\mathscr{V}_t} \nabla\cdot[\Phi_t\mathbf{v}]\,\mathrm{dv} - \int_{\mathscr{I}_t} [\![\Phi_t\,[\mathbf{w}-\mathbf{v}]]\!]\cdot\mathbf{n}_{\mathscr{I}}\,\mathrm{da} \\
& && \leftarrow \text{eq. (5.19a)} \rightarrow \text{eq. (5.44)}
\end{aligned}$$

$$
\begin{aligned}
&= \int_{\mathscr{V}_t} \mathrm{d}_t\Phi_t \,\mathrm{dv} + \int_{\partial\mathscr{V}_t} [\Phi_t\mathbf{v}]\cdot\mathbf{n}\,\mathrm{da} - \int_{\mathscr{I}_t} [\![\Phi_t\mathbf{v}]\!]\cdot\mathbf{n}_{\mathscr{I}}\,\mathrm{da} - \int_{\mathscr{I}_t} [\![\Phi_t\,[\mathbf{w}-\mathbf{v}]]\!]\cdot\mathbf{n}_{\mathscr{I}}\,\mathrm{da}\\
&= \int_{\mathscr{V}_t} \mathrm{d}_t\Phi_t \,\mathrm{dv} + \int_{\partial\mathscr{V}_t} [\Phi_t\mathbf{v}]\cdot\mathbf{n}\,\mathrm{da} - \int_{\mathscr{I}_t} [\![\Phi_t\mathbf{w}]\!]\cdot\mathbf{n}_{\mathscr{I}}\,\mathrm{da}\\
&= \mathscr{D}_t \int_{\mathscr{V}_t} \Phi_t\,\mathrm{dv} \quad .
\end{aligned} \tag{B.35}
$$

To physically interpret the term $-\int_{\mathscr{I}_0} [\![\Phi_0\mathbf{W}]\!]\cdot\mathbf{N}_{\mathscr{I}}\,\mathrm{dA}$ representing the contribution across the discontinuity, observe the following: If the jump in the scalar field $[\![\Phi_0]\!]$ increases, then there is a net increase in the scalar field from $\mathscr{V}_0^-$ to $\mathscr{V}_0^+$. If the interface velocity $-\mathbf{W}$ increases, then there is also a net increase in the scalar field from $\mathscr{V}_0^-$ to $\mathscr{V}_0^+$. This corresponds to the interface retreating into $\mathscr{V}_0^-$.

Similarly, when taking into account the moving interface, the adjusted expression for Reynolds transport theorem for vector densities (related to equation (B.22)) is

$$
\begin{aligned}
\mathscr{D}_t \int_{\mathscr{V}_\square} \boldsymbol{\Phi}_\square\,\mathrm{dV}_\square &= \int_{\mathscr{V}_0} \mathrm{D}_t\boldsymbol{\Phi}_0\,\mathrm{dV} + \int_{\mathscr{V}_0} \nabla_0\cdot[\boldsymbol{\Phi}_0\otimes\mathbf{W}]\,\mathrm{dV} && \leftarrow \text{eqs. (B.31,B.22)} \rightarrow \text{eq. (5.44)}\\
&= \int_{\mathscr{V}_0} \mathrm{D}_t\boldsymbol{\Phi}_0\,\mathrm{dV} + \underbrace{\int_{\partial\mathscr{V}_0} [\boldsymbol{\Phi}_0\otimes\mathbf{W}]\cdot\mathbf{N}\,\mathrm{dA}}_{\mathbf{0}} - \int_{\mathscr{I}_0} [\![\boldsymbol{\Phi}_0\otimes\mathbf{W}]\!]\cdot\mathbf{N}_{\mathscr{I}}\,\mathrm{dA}\\
& && \rightarrow \text{eq. (B.32)}\\
&= \int_{\mathscr{V}_0} \mathrm{D}_t\boldsymbol{\Phi}_0\,\mathrm{dV} - \int_{\mathscr{I}_0} [\![\boldsymbol{\Phi}_0\otimes\mathbf{W}]\!]\cdot\mathbf{N}_{\mathscr{I}}\,\mathrm{dA} && \rightarrow \text{eqs. (5.22b,5.10)}\\
&= \int_{\mathscr{V}_t} \mathrm{n}_t\boldsymbol{\Phi}_t\,\mathrm{dv} - \int_{\mathscr{I}_t} [\![\boldsymbol{\Phi}_0\otimes\mathbf{W}]\!]\cdot\operatorname{cof}\left(\mathbf{F}^{-1}\right)\cdot\mathbf{n}_{\mathscr{I}}\,\mathrm{da} && \rightarrow \text{eq. (5.48)}\\
&= \int_{\mathscr{V}_t} \mathrm{n}_t\boldsymbol{\Phi}_t\,\mathrm{dv} - \int_{\mathscr{I}_t} \left[\!\!\left[[\boldsymbol{\Phi}_0\otimes\mathbf{W}]\cdot\operatorname{cof}\left(\mathbf{F}^{-1}\right)\right]\!\!\right]\cdot\mathbf{n}_{\mathscr{I}}\,\mathrm{da} && \rightarrow \text{eq. (A.5)}\\
&= \int_{\mathscr{V}_t} \mathrm{n}_t\boldsymbol{\Phi}_t\,\mathrm{dv} - \int_{\mathscr{I}_t} \left[\!\!\left[J^{-1}\boldsymbol{\Phi}_0\otimes\left[\mathbf{W}\cdot\mathbf{F}^T\right]\right]\!\!\right]\cdot\mathbf{n}_{\mathscr{I}}\,\mathrm{da} && \rightarrow \text{eqs. (5.21b,5.43)}\\
&= \int_{\mathscr{V}_t} \mathrm{n}_t\boldsymbol{\Phi}_t\,\mathrm{dv} - \int_{\mathscr{I}_t} [\![\boldsymbol{\Phi}_t\otimes[\mathbf{w}-\mathbf{v}]]\!]\cdot\mathbf{n}_{\mathscr{I}}\,\mathrm{da} && \text{(B.36)}\\
&= \int_{\mathscr{V}_t} \mathrm{d}_t\boldsymbol{\Phi}_t\,\mathrm{dv} + \int_{\mathscr{V}_t} \nabla\cdot[\boldsymbol{\Phi}_t\otimes\mathbf{v}]\,\mathrm{dv} - \int_{\mathscr{I}_t} [\![\boldsymbol{\Phi}_t\otimes[\mathbf{w}-\mathbf{v}]]\!]\cdot\mathbf{n}_{\mathscr{I}}\,\mathrm{da}\\
& && \leftarrow \text{eq. (5.19b)} \rightarrow \text{eq. (5.44)}\\
&= \int_{\mathscr{V}_t} \mathrm{d}_t\boldsymbol{\Phi}_t\,\mathrm{dv} + \int_{\partial\mathscr{V}_t} [\boldsymbol{\Phi}_t\otimes\mathbf{v}]\cdot\mathbf{n}\,\mathrm{da} - \int_{\mathscr{I}_t} [\![\boldsymbol{\Phi}_t\otimes\mathbf{v}]\!]\cdot\mathbf{n}_{\mathscr{I}}\,\mathrm{da}
\end{aligned}
$$

$$
\begin{aligned}
&- \int_{\mathscr{I}_t} [\![\boldsymbol{\Phi}_t \otimes [\mathbf{w} - \mathbf{v}]]\!] \cdot \mathbf{n}_{\mathscr{I}} \, \mathrm{da} \\
&= \int_{\mathscr{V}_t} \mathrm{d}_t \boldsymbol{\Phi}_t \, \mathrm{dv} + \int_{\partial \mathscr{V}_t} [\boldsymbol{\Phi}_t \otimes \mathbf{v}] \cdot \mathbf{n} \, \mathrm{da} - \int_{\mathscr{I}_t} [\![\boldsymbol{\Phi}_t \otimes \mathbf{w}]\!] \cdot \mathbf{n}_{\mathscr{I}} \, \mathrm{da} \\
&= \mathscr{D}_t \int_{\mathscr{V}_t} \boldsymbol{\Phi}_t \, \mathrm{dv} \quad .
\end{aligned} \tag{B.37}
$$

Reynolds transport theorem: Control area

With the consideration of the contribution from the moving interface, the amended form of the Reynolds transport theorem for vector fluxes (related to equation (B.25)) is

$$
\begin{aligned}
\mathscr{D}_t \int_{\mathscr{A}_\square} \boldsymbol{\Psi}_\square \cdot \mathbf{M}_\square \, \mathrm{dA}_\square &= \int_{\mathscr{A}_0} [\mathrm{D}_t \boldsymbol{\psi} + \nabla_0 \times [\boldsymbol{\psi} \times \mathbf{W}] + [\nabla_0 \cdot \boldsymbol{\psi}] \, \mathbf{W}] \cdot \mathbf{M} \, \mathrm{dA} && \leftarrow \text{eqs. (B.31,B.25)} \\
&= \int_{\mathscr{A}_0} [\mathrm{D}_t \boldsymbol{\psi}] \cdot \mathbf{M} \, \mathrm{dA} + \int_{\mathscr{A}_0} \underbrace{[\nabla_0 \cdot \boldsymbol{\psi}] \, \mathbf{W} \cdot \mathbf{M}}_{\to 0} \, \mathrm{dA} \\
&\quad + \int_{\mathscr{A}_0} [\nabla_0 \times [\boldsymbol{\psi} \times \mathbf{W}]] \cdot \mathbf{M} \, \mathrm{dA} && \to \text{eq. (B.33a)} \\
&= \int_{\mathscr{A}_0} \mathrm{D}_t \boldsymbol{\psi} \cdot \mathbf{M} \, \mathrm{dA} + \int_{\mathscr{A}_0} [\nabla_0 \times [\boldsymbol{\psi} \times \mathbf{W}]] \cdot \mathbf{M} \, \mathrm{dA} && \to \text{eq. (5.57)} \\
&= \int_{\mathscr{A}_0} \mathrm{D}_t \boldsymbol{\psi} \cdot \mathbf{M} \, \mathrm{dA} + \oint_{\mathscr{A}_0} [\boldsymbol{\psi} \times \mathbf{W}] \cdot \mathbf{L} \, \mathrm{dL} - \int_{\mathscr{I}_0} [\![\boldsymbol{\psi} \times \mathbf{W}]\!] \cdot \mathbf{L}_{\mathscr{I}} \, \mathrm{dL} \\
& && \to \text{eq. (A.12)} \\
&= \int_{\mathscr{A}_0} \mathrm{D}_t \boldsymbol{\psi} \cdot \mathbf{M} \, \mathrm{dA} + \oint_{\mathscr{A}_0} \underbrace{[\mathbf{W} \times \mathbf{L}]}_{\to \mathbf{0}} \cdot \boldsymbol{\psi} \, \mathrm{dL} - \int_{\mathscr{I}_0} [\![\boldsymbol{\psi} \times \mathbf{W}]\!] \cdot \mathbf{L}_{\mathscr{I}} \, \mathrm{dL} \\
& && \to \text{eqs. (5.22c,B.33b)} \\
&= \int_{\mathscr{A}_0} [\mathrm{D}_t \boldsymbol{\Psi}] \cdot \mathbf{M} \, \mathrm{dA} - \int_{\mathscr{I}_0} [\![\boldsymbol{\Psi} \times \mathbf{W}]\!] \cdot \underbrace{\mathbf{L}_{\mathscr{I}} \, \mathrm{dL}}_{\mathrm{d}\mathbf{X}} && \to \text{eqs. (5.22c,5.8)} \\
&= \int_{\mathscr{A}_t} [\mathrm{n}_t \boldsymbol{\psi}] \cdot \mathbf{m} \, \mathrm{da} - \int_{\mathscr{I}_t} [\![\boldsymbol{\Psi} \times \mathbf{W}]\!] \cdot \mathbf{F}^{-1} \cdot \mathbf{l}_{\mathscr{I}} \, \mathrm{dl} && \to \text{eq. (5.48)} \\
&= \int_{\mathscr{A}_t} [\mathrm{n}_t \boldsymbol{\psi}] \cdot \mathbf{m} \, \mathrm{da} - \int_{\mathscr{I}_t} \left[\!\left[[\boldsymbol{\Psi} \times \mathbf{W}] \cdot \mathbf{F}^{-1} \right]\!\right] \cdot \mathbf{l}_{\mathscr{I}} \, \mathrm{dl} \\
&= \int_{\mathscr{A}_t} [\mathrm{n}_t \boldsymbol{\psi}] \cdot \mathbf{m} \, \mathrm{da} - \int_{\mathscr{I}_t} \left[\!\left[J^{-1} \, [\boldsymbol{\Psi} \times \mathbf{W}] \cdot \left[J \mathbf{F}^{-1} \right] \right]\!\right] \cdot \mathbf{l}_{\mathscr{I}} \, \mathrm{dl} \\
& && \to \text{eqs. (A.10,A.5,A.11)}
\end{aligned}
$$

$$= \int_{\mathscr{A}_t} [\mathrm{n}_t\boldsymbol{\psi}] \cdot \mathbf{m}\, \mathrm{da} - \int_{\mathscr{I}_t} \left[\!\left[\left[J^{-1}\mathbf{F} \cdot \boldsymbol{\Psi}\right] \times [\mathbf{F} \cdot \mathbf{W}] \right]\!\right] \cdot \mathbf{l}_{\mathscr{I}}\, \mathrm{dl}$$

$\rightarrow$ eqs. (5.21c,5.43)

$$= \int_{\mathscr{A}_t} [\mathrm{n}_t\boldsymbol{\psi}] \cdot \mathbf{m}\, \mathrm{da} - \int_{\mathscr{I}_t} [\![\boldsymbol{\psi} \times [\mathbf{w} - \mathbf{v}]]\!] \cdot \mathbf{l}_{\mathscr{I}}\, \mathrm{dl} \tag{B.38}$$

$$= \int_{\mathscr{A}_t} [\mathrm{d}_t\boldsymbol{\psi} + [\nabla \cdot \boldsymbol{\psi}]\, \mathbf{v}] \cdot \mathbf{m}\, \mathrm{da} + \int_{\mathscr{A}_t} [\nabla \times [\boldsymbol{\psi} \times \mathbf{v}]] \cdot \mathbf{m}\, \mathrm{da}$$

$$- \int_{\mathscr{I}_t} [\![\boldsymbol{\psi} \times [\mathbf{w} - \mathbf{v}]]\!] \cdot \mathbf{l}_{\mathscr{I}}\, \mathrm{dl}$$

$\leftarrow$ eq. (5.22c) $\rightarrow$ eq. (5.57)

$$= \int_{\mathscr{A}_t} [\mathrm{d}_t\boldsymbol{\psi} + [\nabla \cdot \boldsymbol{\psi}]\, \mathbf{v}] \cdot \mathbf{m}\, \mathrm{da} + \oint_{\partial\mathscr{A}_t} [\boldsymbol{\psi} \times \mathbf{v}] \cdot \mathbf{l}\, \mathrm{dl}$$

$$- \int_{\mathscr{I}_t} [\![\boldsymbol{\psi} \times \mathbf{v}]\!] \cdot \mathbf{l}_{\mathscr{I}}\, \mathrm{dl} - \int_{\mathscr{I}_t} [\![\boldsymbol{\psi} \times [\mathbf{w} - \mathbf{v}]]\!] \cdot \mathbf{l}_{\mathscr{I}}\, \mathrm{dl}$$

$$= \int_{\mathscr{A}_t} [\mathrm{d}_t\boldsymbol{\psi} + [\nabla \cdot \boldsymbol{\psi}]\, \mathbf{v}] \cdot \mathbf{m}\, \mathrm{da} + \oint_{\partial\mathscr{A}_t} [\boldsymbol{\psi} \times \mathbf{v}] \cdot \mathbf{l}\, \mathrm{dl} - \int_{\mathscr{I}_t} [\![\boldsymbol{\psi} \times \mathbf{w}]\!] \cdot \mathbf{l}_{\mathscr{I}}\, \mathrm{dl}$$

$$= \mathscr{D}_t \int_{\mathscr{A}_t} \boldsymbol{\psi} \cdot \mathbf{m}\, \mathrm{da} \tag{B.39}$$

B.3 Continuum identities

Volume integral of a curl

$$\int_{\mathscr{V}} \nabla \times \mathbf{a}\, \mathrm{dv} = \int_{\mathscr{V}} \nabla \cdot [\boldsymbol{\epsilon} \cdot \mathbf{a}]\, \mathrm{dv}$$

$\leftarrow$ eq. (A.22)

$$= \int_{\partial\mathscr{V}} [\boldsymbol{\epsilon} \cdot \mathbf{a}] \cdot \mathbf{n}\, \mathrm{da} = \int_{\partial\mathscr{V}} \boldsymbol{\epsilon} : [\mathbf{n} \otimes \mathbf{a}]\, \mathrm{da}$$

$\leftarrow$ eqs. (B.16,A.8)

$$= \int_{\partial\mathscr{V}} \mathbf{n} \times \mathbf{a}\, \mathrm{da} \tag{B.40}$$

Volume integral of the divergence of a contravariant quantity

As contravariant vectors and tensors get pushed forward using the Piola transform, we can define the relationship between a general spatial vectorial or tensorial quantity $\mathbf{t}$ and its referential counterpart $\mathbf{T}$ by $\mathbf{t} = \mathbf{T} \cdot \mathrm{cof}\,(\mathbf{F}^{-1}) = J^{-1}\mathbf{T} \cdot \mathbf{F}^T$. Therefore,

$$\int_{\mathscr{V}_0} \nabla_0 \cdot \mathbf{T}\, \mathrm{dV} = \int_{\partial\mathscr{V}_0} \mathbf{T} \cdot\, \mathrm{d}\mathbf{A} = \int_{\partial\mathscr{V}_t} \mathbf{T} \cdot \mathrm{cof}\left(\mathbf{F}^{-1}\right) \cdot\, \mathbf{da}$$

$\leftarrow$ eqs. (B.16,5.10)

$$=: \int_{\partial \mathscr{V}_t} \mathbf{t} \cdot \,\mathrm{d}\mathbf{a} = \int_{\mathscr{V}_t} \nabla \cdot \mathbf{t} \,\mathrm{dv} \quad , \tag{B.41}$$

from which we can also identify that

$$\nabla_0 \cdot \mathbf{T} = \nabla_0 \cdot \left[J\mathbf{t} \cdot \mathbf{F}^T \right] = J \nabla \cdot \mathbf{t} \quad . \tag{B.42}$$

In the case of the vectorial quantities **B** and **b**, this can also be expressed as

$$\nabla_0 \cdot \mathbf{B} = \nabla_0 \cdot \left[J\mathbf{F}^{-1} \cdot \mathbf{b} \right] = J \nabla \cdot \mathbf{b} \quad . \tag{B.43}$$

These can also be proved directly using the Piola identity stated in equation (B.19), with the assistance of equations (A.35), (A.36) and (5.7), as

$$\nabla_0 \cdot \mathbf{T} = \nabla_0 \cdot \left[\mathbf{t} \cdot \left[J\mathbf{F}^{-T} \right] \right] = J \left[\nabla_0 \mathbf{t} \right] : \mathbf{F}^{-T} + \mathbf{t} \cdot \underbrace{\nabla_0 \cdot \left[J\mathbf{F}^{-T} \right]}_{=\mathbf{0}} = J \nabla \cdot \mathbf{t} \quad , \tag{B.44a}$$

$$\nabla_0 \cdot \mathbf{B} = \nabla_0 \cdot \left[\mathbf{b} \cdot \left[J\mathbf{F}^{-T} \right] \right] = J \left[\nabla_0 \mathbf{b} \right] : \mathbf{F}^{-T} + \mathbf{b} \cdot \underbrace{\nabla_0 \cdot \left[J\mathbf{F}^{-T} \right]}_{=\mathbf{0}} = J \nabla \cdot \mathbf{b} \quad . \tag{B.44b}$$

Area integral of the curl of a covariant vector

A general spatial vector **h** is transformed from its material description **H** by $\mathbf{h} = \mathbf{F}^{-T} \cdot \mathbf{H}$. Thus

$$\int_{\mathscr{A}_0} \left[\nabla_0 \times \mathbf{H} \right] \cdot \,\mathrm{d}\mathbf{A} = \int_{\partial \mathscr{A}_0} \mathbf{H} \cdot \,\mathrm{d}\mathbf{L} = \int_{\partial \mathscr{A}_t} \mathbf{H} \cdot \mathbf{F}^{-1} \cdot \,\mathrm{d}\mathbf{l} \qquad \leftarrow \text{eqs. (B.14,5.8)}$$

$$=: \int_{\partial \mathscr{A}_t} \mathbf{h} \cdot \,\mathrm{d}\mathbf{l} = \int_{\mathscr{A}_t} \left[\nabla \times \mathbf{h} \right] \cdot \,\mathrm{d}\mathbf{a} \quad , \tag{B.45}$$

from which, in conjunction with equation (5.10), we can also identify that

$$\nabla_0 \times \mathbf{H} = \nabla_0 \times \left[\mathbf{h} \cdot \mathbf{F} \right] = \left[\nabla \times \mathbf{h} \right] \cdot \left[J\mathbf{F}^{-T} \right] \quad . \tag{B.46}$$

C Derivations and proofs

This appendix serves as to provide derivations and proofs for the theory presented throughout the main chapters of this book. Herein, we heavily leverage the identities previously determined in Appendices A and B. Many of the proofs derived here are also nicely expressed, in summary, part or full, by Brown [64], Ogden [406], Chatzigeorgiou et al. [82], and Vogel [530].

C.1 Fundamentals of electromagnetism

It is important to observe in the following equations that the quantities $\varrho_t(\mathbf{x}')$ and $\mathbb{j}(\mathbf{x}')$ are evaluated at a fixed position $\mathbf{x}'$, and are therefore considered constant on the domain $\mathscr{D}_t$. Additionally, all externally generated fields are considered spatially constant. As a consequence of this, any differential operations performed on these quantities over the integrated interval $\mathscr{D}_t$ equate to zero.

C.1.1 Electrostatics

Divergence of electric field vector

$$\begin{aligned}
\nabla \cdot \mathbb{e}(\mathbf{x}) &= \nabla \cdot \left[\frac{1}{4\pi\varepsilon_0} \int_{\mathscr{D}_t} \frac{\varrho_t(\mathbf{x}')}{|\mathbf{x}-\mathbf{x}'|^2} \frac{\mathbf{x}-\mathbf{x}'}{|\mathbf{x}-\mathbf{x}'|} \, \mathrm{dv} \right] && \leftarrow \text{eq. (5.26)} \\
&= \frac{1}{4\pi\varepsilon_0} \int_{\mathscr{D}_t} \nabla \cdot \left[\varrho_t(\mathbf{x}') \frac{\mathbf{x}-\mathbf{x}'}{|\mathbf{x}-\mathbf{x}'|^3} \right] \mathrm{dv} \\
&= \frac{1}{4\pi\varepsilon_0} \int_{\mathscr{D}_t} \varrho_t(\mathbf{x}') \nabla \cdot \left[\frac{\mathbf{x}-\mathbf{x}'}{|\mathbf{x}-\mathbf{x}'|^3} \right] \mathrm{dv} && \rightarrow \text{eq. (B.4)} \\
&= \frac{1}{4\pi\varepsilon_0} \int_{\mathscr{D}_t} \left[4\pi \varrho_t(\mathbf{x}') \delta(\mathbf{x}-\mathbf{x}') \right] \mathrm{dv} && \rightarrow \text{eq. (B.2)} \\
&= \frac{\varrho_t(\mathbf{x})}{\varepsilon_0} && \text{(C.1)}
\end{aligned}$$

Negative gradient of electric potential field

$$\begin{aligned}
-\nabla \Phi(\mathbf{x}) &= \nabla \left[-\Phi^{\text{ext}} \right] - \nabla \frac{1}{4\pi\varepsilon_0} \int_{\mathscr{D}_t} \frac{\varrho_t(\mathbf{x}')}{|\mathbf{x}-\mathbf{x}'|} \, \mathrm{dv} && \leftarrow \text{eq. (5.29)} \\
&= -\frac{1}{4\pi\varepsilon_0} \int_{\mathscr{D}_t} \nabla \left[\frac{\varrho_t(\mathbf{x}')}{|\mathbf{x}-\mathbf{x}'|} \right] \mathrm{dv}
\end{aligned}$$

https://doi.org/10.1515/9783110418576-015

$$= -\frac{1}{4\pi\varepsilon_0} \int\limits_{\mathscr{D}_t} \varrho_t(\mathbf{x}') \nabla \left[\frac{1}{|\mathbf{x} - \mathbf{x}'|} \right] \mathrm{dv} \qquad \rightarrow \text{eq. (B.3)}$$

$$= \frac{1}{4\pi\varepsilon_0} \int\limits_{\mathscr{D}_t} \varrho_t(\mathbf{x}') \frac{\mathbf{x} - \mathbf{x}'}{|\mathbf{x} - \mathbf{x}'|^3} \mathrm{dv} \qquad \rightarrow \text{eq. (5.26)}$$

$$= \mathbb{e}(\mathbf{x}) \tag{C.2}$$

C.1.2 Magnetostatics

Perfect path integral over a closed circuit [230]

Provided that the circuit path is closed or extends to infinity, then

$$\oint\limits_{\mathscr{C}_i} \frac{\mathbf{r}_{ji}}{|\mathbf{r}_{ji}|^3} \cdot \mathrm{d}\mathbb{i}_i = \oint\limits_{\mathscr{C}_i} \left[\nabla \frac{1}{\mathbf{r}_{ji}} \right] \cdot \mathrm{d}\mathbb{i}_i \qquad \leftarrow \text{eq. (B.3)}$$

$$\propto \oint\limits_{\mathscr{C}_i} \left[\nabla \frac{1}{\mathbf{r}_{ji}} \right] \cdot \mathrm{d}\mathbf{r}_i \qquad \rightarrow \text{eq. (B.14)}$$

$$= \int\limits_{\mathscr{A}_i} \left[\nabla \times \left[\nabla \frac{1}{\mathbf{r}_{ji}} \right] \right] \cdot \mathrm{d}\mathbf{a}_i \qquad \rightarrow \text{eq. (A.24)}$$

$$= 0 \quad . \tag{C.3}$$

Note that here we have also made use of the relation $\mathrm{d}\mathbb{i} = q\mathbb{v} = i\mathrm{d}\mathbf{r} \propto \mathrm{d}\mathbf{r}$; see [230] for a detailed discussion on the definition of the electric current element.

Force acting on a circuit

$$\mathbf{f}_{ij} = \oint\limits_{\mathscr{C}_i} \mathrm{d}\mathbb{i}_i \times \frac{\mu_0}{4\pi} \oint\limits_{\mathscr{C}_j} \frac{\mathrm{d}\mathbb{i}_j}{|\mathbf{r}_{ji}|^2} \times \frac{\mathbf{r}_{ji}}{|\mathbf{r}_{ji}|}$$

$$= \frac{\mu_0}{4\pi} \oint\limits_{\mathscr{C}_i} \oint\limits_{\mathscr{C}_j} [\mathrm{d}\mathbb{i}_i \times [\mathrm{d}\mathbb{i}_j \times \mathbf{r}_{ji}]] \frac{1}{|\mathbf{r}_{ji}|^3} \qquad \rightarrow \text{eq. (A.14)}$$

$$= \frac{\mu_0}{4\pi} \oint\limits_{\mathscr{C}_i} \oint\limits_{\mathscr{C}_j} \left[\mathrm{d}\mathbb{i}_j \frac{\mathbf{r}_{ji} \cdot \mathrm{d}\mathbb{i}_i}{|\mathbf{r}_{ji}|^3} - \frac{\mathbf{r}_{ji}}{|\mathbf{r}_{ji}|^3} [\mathrm{d}\mathbb{i}_i \cdot \mathrm{d}\mathbb{i}_j] \right]$$

$$= \frac{\mu_0}{4\pi} \left[\oint\limits_{\mathscr{C}_j} \mathrm{d}\mathbb{i}_j \oint\limits_{\mathscr{C}_i} \frac{\mathbf{r}_{ji}}{|\mathbf{r}_{ji}|^3} \cdot \mathrm{d}\mathbb{i}_i - \oint\limits_{\mathscr{C}_i} \oint\limits_{\mathscr{C}_j} \frac{\mathbf{r}_{ji}}{|\mathbf{r}_{ji}|^3} [\mathrm{d}\mathbb{i}_i \cdot \mathrm{d}\mathbb{i}_j] \right] \qquad \rightarrow \text{eq. (C.3)}$$

$$= -\frac{\mu_0}{4\pi} \oint\limits_{\mathscr{C}_i} \oint\limits_{\mathscr{C}_j} \left[\frac{\mathbf{r}_{ji}}{|\mathbf{r}_{ji}|^3} [\mathrm{d}\mathbb{i}_i \cdot \mathrm{d}\mathbb{i}_j] \right] \tag{C.4}$$

Divergence of magnetic induction vector

$$\nabla \cdot \mathbb{b}(\mathbf{x}) = \nabla \cdot \left[\frac{\mu_0}{4\pi} \int_{\mathscr{D}_t} \frac{\mathbb{j}(\mathbf{x}')}{|\mathbf{x}-\mathbf{x}'|^2} \times \frac{\mathbf{x}-\mathbf{x}'}{|\mathbf{x}-\mathbf{x}'|} \, \mathrm{dv} \right] \qquad \leftarrow \text{eq. (5.35)}$$

$$= \frac{\mu_0}{4\pi} \int_{\mathscr{D}_t} \nabla \cdot \left[\mathbb{j}(\mathbf{x}') \times \frac{\mathbf{x}-\mathbf{x}'}{|\mathbf{x}-\mathbf{x}'|^3} \right] \mathrm{dv} \qquad \rightarrow \text{eq. (B.12)}$$

$$= \frac{\mu_0}{4\pi} \int_{\mathscr{D}_t} \left[\mathbb{j}(\mathbf{x}') \cdot \underbrace{\left[-\nabla \times \frac{\mathbf{x}-\mathbf{x}'}{|\mathbf{x}-\mathbf{x}'|^3} \right]}_{\mathbf{0}} + \frac{\mathbf{x}-\mathbf{x}'}{|\mathbf{x}-\mathbf{x}'|^3} \cdot \underbrace{[\nabla \times \mathbb{j}(\mathbf{x}')]}_{\mathbf{0}} \right] \mathrm{dv} \qquad \rightarrow \text{eq. (B.6)}$$

$$= 0 \tag{C.5}$$

Curl of magnetic vector potential field

$$\nabla \times \mathbb{a}(\mathbf{x}) = \nabla \times \mathbb{a}^{\text{ext}} + \nabla \times \left[\frac{\mu_0}{4\pi} \int_{\mathscr{D}_t} \frac{\mathbb{j}(\mathbf{x}')}{|\mathbf{x}-\mathbf{x}'|} \, \mathrm{dv} \right] \qquad \leftarrow \text{eq. (5.38)}$$

$$= \frac{\mu_0}{4\pi} \int_{\mathscr{D}_t} \nabla \times \left[\frac{\mathbb{j}(\mathbf{x}')}{|\mathbf{x}-\mathbf{x}'|} \right] \mathrm{dv} \qquad \rightarrow \text{eq. (A.30)}$$

$$= \frac{\mu_0}{4\pi} \int_{\mathscr{D}_t} \left[\frac{1}{|\mathbf{x}-\mathbf{x}'|} \underbrace{[\nabla \times \mathbb{j}(\mathbf{x}')]}_{\mathbf{0}} + \nabla \left[\frac{1}{|\mathbf{x}-\mathbf{x}'|} \right] \times \mathbb{j}(\mathbf{x}') \right] \mathrm{dv} \qquad \rightarrow \text{eqs. (B.3,A.1)}$$

$$= \frac{\mu_0}{4\pi} \int_{\mathscr{D}_t} \mathbb{j}(\mathbf{x}') \times \frac{\mathbf{x}-\mathbf{x}'}{|\mathbf{x}-\mathbf{x}'|^3} \, \mathrm{dv} \qquad \rightarrow \text{eq. (5.35)}$$

$$= \mathbb{b}(\mathbf{x}) \tag{C.6}$$

C.2 Continuum mechanics for magnetoelasticity

Push forward of referential magnetic field, magnetisation and electric current
Due to the balance law given in equation $(5.105)_1$, the push forward of the magnetic field and magnetisation vectors

$$\mathbb{h} = \mathbb{H} \cdot \mathbf{F}^{-1} \quad , \quad \mathbb{m} = \mathbb{M} \cdot \mathbf{F}^{-1} \tag{C.7}$$

can be directly inferred from equation (B.45). Additionally, using equation (5.10), the proof of this identity can be manipulated to identify

$$[\nabla \times \mathbb{m}] = \nabla_0 \times \mathbb{M} \cdot \mathrm{cof}\left(\mathbf{F}^{-1}\right) \qquad \rightarrow \text{eq. (5.71)}$$

$$\Rightarrow \quad \mathbb{j} = \mathbb{J} \cdot \mathrm{cof}\left(\mathbf{F}^{-1}\right) \quad . \tag{C.8}$$

Push forward of referential magnetic induction

Due to the balance law given in equation $(5.105)_2$, the transformation of the magnetic induction from the referential to the current configuration

$$\mathbb{b} = \mathbb{B} \cdot \operatorname{cof}\left(\mathbf{F}^{-1}\right) \tag{C.9}$$

can be inferred directly from equation (B.41).

Pull back of magnetic fundamental constitutive relation

From equation (5.118),

$$\begin{aligned}
\mathbb{b} &= \mu_0 \left[\mathbb{h} + \mathbb{m}\right] && \rightarrow \text{eqs. } (5.102_{1,2,3}) \\
J^{-1}\mathbf{F} \cdot \mathbb{B} &= \mu_0 \left[\mathbf{F}^{-T} \cdot \mathbb{H} + \mathbf{F}^{-T} \cdot \mathbb{M}\right] \\
J^{-1}\mathbf{F}^T \cdot \mathbf{F} \cdot \mathbb{B} &= \mu_0 \left[\mathbb{H} + \mathbb{M}\right] && \rightarrow \text{eq. } (5.5) \\
J^{-1}\mathbf{C} \cdot \mathbb{B} &= \mu_0 \left[\mathbb{H} + \mathbb{M}\right] \quad . && \text{(C.10)}
\end{aligned}$$

Referential surface electric charge density

Noting that from equation (5.10) the definition of the surface Jacobian, that is the ratio of spatial and referential area elements, is

$$\begin{aligned}
\mathbf{n} \cdot [\mathbf{n}\, \mathrm{da}] &= \mathbf{n} \cdot \left[J\mathbf{F}^{-T} \cdot \mathbf{N}\, \mathrm{dA}\right] && \leftarrow \text{eq. } (5.10) \\
\Rightarrow \quad \widehat{J} &:= \frac{\mathrm{da}}{\mathrm{dA}} = J\mathbf{n} \cdot \mathbf{F}^{-T} \cdot \mathbf{N} \quad . && \text{(C.11)}
\end{aligned}$$

Analogous to equation (5.101), the relationship between the spatial and material surface electric charge densities is

$$\widehat{\varrho}_t = \widehat{J}^{-1} \widehat{\varrho}_0 \quad . \tag{C.12}$$

Combining these renders the result that

$$\begin{aligned}
\int_{\partial \mathcal{V}_t} \widehat{\varrho}_t \, \mathrm{da} &= \int_{\partial \mathcal{V}_t} \widehat{\varrho}_t \mathbf{n} \cdot \, \mathbf{da} && \rightarrow \text{eq. } (5.10) \\
&= \int_{\partial \mathcal{V}_0} \widehat{\varrho}_t \mathbf{n} \cdot J\mathbf{F}^{-T} \cdot \mathbf{N}\, \mathrm{dA} && \rightarrow \text{eq. } (C.11) \\
&= \int_{\partial \mathcal{V}_0} \widehat{\varrho}_t \widehat{J}\, \mathrm{dA} && \rightarrow \text{eq. } (C.12) \\
&= \int_{\partial \mathcal{V}_0} \widehat{\varrho}_0 \, \mathrm{dA} \quad . && \text{(C.13)}
\end{aligned}$$

Referential total surface current vector

For reasons motivated later in the derivation of equation (C.107), we define the pull back of the surface electric current vector to be

$$\widehat{\mathbb{J}} := \widehat{J}\mathbf{F}^{-1} \cdot \widehat{\mathbb{j}} \quad . \tag{C.14}$$

Jump of spatial magnetic field and induction vectors for decoupled magnetic fields

Assuming quasi-static electric conditions, then from equation $(5.110)_3$

$$\begin{aligned}
\widehat{\mathbb{j}}^{\mathrm{f}} \times \mathbf{n} &= [\mathbf{n} \times [\![\mathbb{h}]\!]] \times \mathbf{n} && \rightarrow \text{eq. (A.15)} \\
&= [\![\mathbb{h}]\!] \, [\mathbf{n} \cdot \mathbf{n}] - \mathbf{n} \, [[\![\mathbb{h}]\!] \cdot \mathbf{n}] \\
\Rightarrow \quad [\![\mathbb{h}]\!] &= \widehat{\mathbb{j}}^{\mathrm{f}} \times \mathbf{n} + \mathbf{n} \, [[\![\mathbb{h}]\!] \cdot \mathbf{n}] && \leftarrow \text{eq. (5.118)} \\
&= \widehat{\mathbb{j}}^{\mathrm{f}} \times \mathbf{n} + \mathbf{n} \left[\frac{1}{\mu_0} [\![\mathbb{b}]\!] \cdot \mathbf{n} \right] - \mathbf{n} \, [[\![\mathbb{m}]\!] \cdot \mathbf{n}] && \rightarrow \text{eq. } (5.110_4) \\
&= \widehat{\mathbb{j}}^{\mathrm{f}} \times \mathbf{n} - \mathbf{n} \, [[\![\mathbb{m}]\!] \cdot \mathbf{n}] \quad . && \text{(C.15)}
\end{aligned}$$

Similarly, starting from equation $(5.117)_2$,

$$\begin{aligned}
\widehat{\mathbb{j}}^{\mathrm{b}} \times \mathbf{n} &= [\mathbf{n} \times [\![\mathbb{m}]\!]] \times \mathbf{n} && \rightarrow \text{eq. (A.15)} \\
&= [\![\mathbb{m}]\!] \, [\mathbf{n} \cdot \mathbf{n}] - \mathbf{n} \, [[\![\mathbb{m}]\!] \cdot \mathbf{n}] \\
\Rightarrow \quad \mathbf{n} \, [[\![\mathbb{m}]\!] \cdot \mathbf{n}] &= [\![\mathbb{m}]\!] - \widehat{\mathbb{j}}^{\mathrm{b}} \times \mathbf{n} \quad . && \text{(C.16)}
\end{aligned}$$

Therefore, from the above,

$$\begin{aligned}
[\![\mathbb{h}]\!] &= \widehat{\mathbb{j}}^{\mathrm{f}} \times \mathbf{n} + \widehat{\mathbb{j}}^{\mathrm{b}} \times \mathbf{n} - [\![\mathbb{m}]\!] && \leftarrow \text{eqs. (C.15,C.16)} \\
&= \widehat{\mathbb{j}} \times \mathbf{n} - [\![\mathbb{m}]\!] \quad , && \text{(C.17)}
\end{aligned}$$

with the total spatial surface current defined (in analogy with equation $(5.70)_2$) as

$$\widehat{\mathbb{j}} := \widehat{\mathbb{j}}^{\mathrm{f}} + \widehat{\mathbb{j}}^{\mathrm{b}} \quad , \tag{C.18}$$

and from equation (5.118)

$$\begin{aligned}
[\![\mathbb{b}]\!] &= \mu_0 \, [[\![\mathbb{h}]\!] + [\![\mathbb{m}]\!]] && \rightarrow \text{eq. (C.17)} \\
&= \mu_0 \widehat{\mathbb{j}} \times \mathbf{n} \quad . && \text{(C.19)}
\end{aligned}$$

Observe as well that

$$[\![\mathbb{b}]\!] \cdot \mathbf{n} = [\mu_0 \widehat{\mathbb{j}} \times \mathbf{n}] \cdot \mathbf{n} = 0 \tag{C.20}$$

as the jump in the magnetic induction is orthogonal to the normal.

Cross product of the spatial surface normal and the jump of the spatial magnetic field and induction vectors for decoupled magnetic fields
We note that from equation (C.15)

$$\begin{aligned} \mathbf{n} \times [\![\mathbb{h}]\!] &= \mathbf{n} \times \left[\widehat{\mathbb{j}}^{\mathrm{f}} \times \mathbf{n} - \mathbf{n}\left[[\![\mathbb{m}]\!] \cdot \mathbf{n}\right]\right] && \rightarrow \text{eq. (C.17)} \\ &= \mathbf{n} \times \left[\widehat{\mathbb{j}}^{\mathrm{f}} \times \mathbf{n}\right] - \mathbf{n} \times \mathbf{n}\left[[\![\mathbb{m}]\!] \cdot \mathbf{n}\right] \\ &= \mathbf{n} \times \left[\widehat{\mathbb{j}}^{\mathrm{f}} \times \mathbf{n}\right] && \rightarrow \text{eq. (A.14)} \\ &= \widehat{\mathbb{j}}^{\mathrm{f}}\left[\mathbf{n} \cdot \mathbf{n}\right] - \mathbf{n}\left[\mathbf{n} \cdot \widehat{\mathbb{j}}^{\mathrm{f}}\right] \\ &= \widehat{\mathbb{j}}^{\mathrm{f}} \quad , && \text{(C.21)} \end{aligned}$$

which aligns with the original statement extracted from equation $(5.110)_3$, and from equation (C.19) that

$$\begin{aligned} \mathbf{n} \times [\![\mathbb{b}]\!] &= \mathbf{n} \times \mu_0 \left[\widehat{\mathbb{j}} \times \mathbf{n}\right] && \rightarrow \text{eq. (A.14)} \\ &= \mu_0 \left[\widehat{\mathbb{j}}\left[\mathbf{n} \cdot \mathbf{n}\right] - \mathbf{n}\left[\mathbf{n} \cdot \widehat{\mathbb{j}}\right]\right] \\ &= \mu_0 \widehat{\mathbb{j}} && \text{(C.22)} \end{aligned}$$

since, by definition, the surface current is tangential to any material interface

$$\mathbf{n} \cdot \widehat{\mathbb{j}} = 0 \quad . \tag{C.23}$$

Cross product of the reference surface normal and the jump of the reference magnetic field and induction vectors for decoupled magnetic fields
From equation (C.21),

$$\begin{aligned} \widehat{\mathbb{j}}^{\mathrm{f}} &= \mathbf{n} \times [\![\mathbb{h}]\!] && \rightarrow \text{eqs. } (5.10, 5.102_1) \\ &= \left[\widehat{J}^{-1} J \mathbf{F}^{-T} \cdot \mathbf{N}\right] \times \left[\!\left[\mathbf{F}^{-T} \cdot \mathbb{H}\right]\!\right] && \rightarrow \text{eq. (A.10)} \\ &= \widehat{J}^{-1} J \left[\mathbf{F}^{-T} \cdot \mathbf{N}\right] \times \left[\!\left[\mathbf{F}^{-T} \cdot \mathbb{H}\right]\!\right] && \rightarrow \text{eq. (A.11)} \\ &= \widehat{J}^{-1} J \det\left(\mathbf{F}^{-T}\right) \mathbf{F} \cdot \left[\mathbf{N} \times [\![\mathbb{H}]\!]\right] \\ \mathbf{N} \times [\![\mathbb{H}]\!] &= \widehat{J} \mathbf{F}^{-1} \cdot \widehat{\mathbb{j}}^{\mathrm{f}} && \rightarrow \text{eq. (C.14)} \\ &= \widehat{\mathbb{J}}^{\mathrm{f}} \quad . && \text{(C.24)} \end{aligned}$$

This aligns with what is proven later in equation (C.107) when assuming a static electric field.

From equation (5.74),

$$\begin{aligned} \mathbf{N} \times \left[\!\left[J^{-1} \mathbf{C} \cdot \mathbb{B}\right]\!\right] &= \mathbf{N} \times [\![\mu_0 \left[\mathbb{H} + \mathbb{M}\right]]\!] && \rightarrow \text{eq. (5.74)} \\ &= \mu_0 \mathbf{N} \times \left[[\![\mathbb{H}]\!] + [\![\mathbb{M}]\!]\right] && \rightarrow \text{eqs. (C.24, 5.79, 5.83)} \end{aligned}$$

$$= \mu_0 \left[\mathbb{J}^{\mathrm{f}} + \mathbb{J}^{\mathrm{b}} \right] \qquad \rightarrow \text{eq. (C.26)}$$
$$= \mu_0 \widehat{\mathbb{J}} \tag{C.25}$$

where, motivated by the relationships given in equations (C.14) and (C.18), the total referential surface current has been defined as

$$\widehat{\mathbb{J}} := \widehat{\mathbb{J}}^{\mathrm{f}} + \widehat{\mathbb{J}}^{\mathrm{b}} \quad . \tag{C.26}$$

C.3 Stress tensors

From this juncture forward, we always assume quasi-static conditions thereby permitting that the electric and magnetic problems be decoupled. Assuming that the electric fields and polarisation are small, all ponderomotive stresses then depend only on magnetic fields and the free current.

C.3.1 Definitions

Spatial ponderomotive stress tensor
To begin, we assume that the definition of the spatial ponderomotive stress tensor, as given in equation (5.131a), is correct. It will be shown in Appendix C.3.2 that this definition satisfies the definition of the Lorentz body force in a magnetisable solid and the divergence-free condition in the free space.

Two-point ponderomotive stress tensor
From the definition given in equation (5.131a),

$$\begin{aligned}
\mathbf{P}^{*\mathrm{pon}} &= \boldsymbol{\sigma}^{*\mathrm{pon}} \cdot \mathrm{cof}\,(\mathbf{F}) = J\boldsymbol{\sigma}^{*\mathrm{pon}} \cdot \mathbf{F}^{-T} && \leftarrow \text{eqs. (A.5,5.99)} \rightarrow \text{eq. (5.131a)} \\
&= J\left[\frac{1}{2\mu_0}\left[\mathbb{b} \cdot \mathbb{b}\right]\mathbf{i} - \left[\mathbb{h} \cdot \mathbb{b}\right]\mathbf{i} + \mathbb{h} \otimes \mathbb{b}\right] \cdot \mathbf{F}^{-T} && \rightarrow \text{eq. (5.102)} \\
&= \frac{J}{2\mu_0}\left[\left[J^{-1}\mathbf{F} \cdot \mathbb{B}\right] \cdot \left[J^{-1}\mathbf{F} \cdot \mathbb{B}\right]\right]\mathbf{F}^{-T} - J\left[\left[\mathbf{F}^{-T} \cdot \mathbb{H}\right] \cdot \left[J^{-1}\mathbf{F} \cdot \mathbb{B}\right]\right]\mathbf{F}^{-T} \\
&\quad + J\left[\left[\mathbf{F}^{-T} \cdot \mathbb{H}\right] \otimes \left[J^{-1}\mathbf{F} \cdot \mathbb{B}\right]\right] \cdot \mathbf{F}^{-T} \\
&= \frac{J^{-1}}{2\mu_0}\left[\left[\mathbf{F}^{T} \cdot \mathbf{F} \cdot \mathbb{B}\right] \cdot \mathbb{B}\right]\mathbf{F}^{-T} - \left[\left[\mathbf{F}^{T} \cdot \mathbf{F}^{-T} \cdot \mathbb{H}\right] \cdot \mathbb{B}\right]\mathbf{F}^{-T} + \left[\left[\mathbf{F}^{-T} \cdot \mathbb{H}\right] \otimes \left[\mathbb{B} \cdot \mathbf{F}^{T} \cdot \mathbf{F}^{-T}\right]\right] \\
&&& \rightarrow \text{eq. (5.5)} \\
&= \left[\frac{J^{-1}}{2\mu_0}\mathbf{C} : \left[\mathbb{B} \otimes \mathbb{B}\right]\right]\mathbf{F}^{-T} - \left[\mathbb{H} \cdot \mathbb{B}\right]\mathbf{F}^{-T} + \left[\mathbf{F}^{-T} \cdot \mathbb{H}\right] \otimes \mathbb{B} \quad . && \text{(C.27)}
\end{aligned}$$

Spatial Maxwell stress tensor
From the definition given in equation (5.131a) with $\mathbb{m} = \mathbf{0}$ (as there is no magnetisation of the free space),

$$
\begin{aligned}
\boldsymbol{\sigma}^{*\max} &= \boldsymbol{\sigma}^{*\mathrm{pon}}\big|_{\mathbb{m}=\mathbf{0}} && \to \text{eq. (5.131a)}\\
&= \left[\frac{1}{2\mu_0}\left[\mathbb{b}\cdot\mathbb{b}\right]\mathbf{i} - \left[\mathbb{h}\cdot\mathbb{b}\right]\mathbf{i} + \mathbb{h}\otimes\mathbb{b}\right]\Bigg|_{\mathbb{m}=\mathbf{0}} && \to \text{eq. (5.118)}\\
&= \left[\frac{1}{2\mu_0}\left[\mathbb{b}\cdot\mathbb{b}\right]\mathbf{i} - \left[\left[\frac{1}{\mu_0}\mathbb{b}-\mathbb{m}\right]\cdot\mathbb{b}\right]\mathbf{i} + \left[\frac{1}{\mu_0}\mathbb{b}-\mathbb{m}\right]\otimes\mathbb{b}\right]\Bigg|_{\mathbb{m}=\mathbf{0}}\\
&= \frac{1}{\mu_0}\mathbb{b}\otimes\mathbb{b} - \frac{1}{2\mu_0}\left[\mathbb{b}\cdot\mathbb{b}\right]\mathbf{i} \quad . && \text{(C.28)}
\end{aligned}
$$

Two-point Maxwell stress tensor
From the definition given in equation (5.131c),

$$
\begin{aligned}
\mathbf{P}^{*\max} &= \boldsymbol{\sigma}^{*\max}\cdot\operatorname{cof}(\mathbf{F}) = J\boldsymbol{\sigma}^{*\max}\cdot\mathbf{F}^{-T} && \leftarrow \text{eqs. (A.5,5.99)}\\
&= J\left[\frac{1}{\mu_0}\mathbb{b}\otimes\mathbb{b} - \frac{1}{2\mu_0}\left[\mathbb{b}\cdot\mathbb{b}\right]\mathbf{i}\right]\cdot\mathbf{F}^{-T} && \to \text{eq. (5.102)}\\
&= \frac{J}{\mu_0}\left[J^{-1}\mathbf{F}\cdot\mathbb{B}\right]\otimes\left[J^{-1}\mathbf{F}\cdot\mathbb{B}\right]\cdot\mathbf{F}^{-T} - \frac{J}{2\mu_0}\left[\left[J^{-1}\mathbf{F}\cdot\mathbb{B}\right]\cdot\left[J^{-1}\mathbf{F}\cdot\mathbb{B}\right]\right]\mathbf{F}^{-T}\\
&= \frac{J^{-1}}{\mu_0}\left[\mathbf{F}\cdot\mathbb{B}\right]\otimes\left[\mathbb{B}\cdot\mathbf{F}^{T}\cdot\mathbf{F}^{-T}\right] - \frac{J^{-1}}{2\mu_0}\left[\left[\mathbf{F}^{T}\cdot\mathbf{F}\cdot\mathbb{B}\right]\cdot\mathbb{B}\right]\mathbf{F}^{-T}\\
&= \left[\frac{J^{-1}}{\mu_0}\mathbf{F}\cdot\mathbb{B}\right]\otimes\mathbb{B} - \left[\frac{J^{-1}}{2\mu_0}\mathbf{C}:\left[\mathbb{B}\otimes\mathbb{B}\right]\right]\mathbf{F}^{-T} \quad . && \text{(C.29)}
\end{aligned}
$$

Spatial magnetisation stress tensor
From the definitions given in equations (5.131a) and (5.131c),

$$
\begin{aligned}
\boldsymbol{\sigma}^{*\mathrm{mag}} &= \boldsymbol{\sigma}^{*\mathrm{pon}} - \boldsymbol{\sigma}^{*\max} && \to \text{eqs. (5.131a,5.131c)}\\
&= \left[\frac{1}{2\mu_0}\left[\mathbb{b}\cdot\mathbb{b}\right]\mathbf{i} - \left[\mathbb{h}\cdot\mathbb{b}\right]\mathbf{i} + \mathbb{h}\otimes\mathbb{b}\right] - \left[\frac{1}{\mu_0}\mathbb{b}\otimes\mathbb{b} - \frac{1}{2\mu_0}\left[\mathbb{b}\cdot\mathbb{b}\right]\mathbf{i}\right]\\
&= \frac{1}{\mu_0}\left[\mathbb{b}\cdot\mathbb{b}\right]\mathbf{i} - \left[\mathbb{h}\cdot\mathbb{b}\right]\mathbf{i} + \mathbb{h}\otimes\mathbb{b} - \frac{1}{\mu_0}\mathbb{b}\otimes\mathbb{b}\\
&= \left[\left[\frac{1}{\mu_0}\mathbb{b}-\mathbb{h}\right]\cdot\mathbb{b}\right]\mathbf{i} - \left[\frac{1}{\mu_0}\mathbb{b}-\mathbb{h}\right]\otimes\mathbb{b} && \to \text{eq. (5.118)}\\
&= \left[\mathbb{m}\cdot\mathbb{b}\right]\mathbf{i} - \mathbb{m}\otimes\mathbb{b} \quad . && \text{(C.30)}
\end{aligned}
$$

Two-point magnetisation stress tensor
From the definition given in equation (5.131b),

$$
\mathbf{P}^{*\mathrm{mag}} = \boldsymbol{\sigma}^{*\mathrm{mag}}\cdot\operatorname{cof}(\mathbf{F}) = J\boldsymbol{\sigma}^{*\mathrm{mag}}\cdot\mathbf{F}^{-T} \qquad \leftarrow \text{eqs. (A.5,5.99)} \to \text{eq. (5.131b)}
$$

$$
\begin{aligned}
&= J\left[[\mathbb{m}\cdot\mathbb{b}]\,\mathbf{i} - \mathbb{m}\otimes\mathbb{b}\right]\cdot\mathbf{F}^{-T} && \rightarrow \text{eq. (5.102)}\\
&= J\left[\left[\mathbf{F}^{-T}\cdot\mathbb{M}\right]\cdot\left[J^{-1}\mathbf{F}\cdot\mathbb{B}\right]\right]\mathbf{F}^{-T} - J\left[\mathbf{F}^{-T}\cdot\mathbb{M}\right]\otimes\left[J^{-1}\mathbf{F}\cdot\mathbb{B}\right]\cdot\mathbf{F}^{-T}\\
&= \left[\left[\mathbf{F}^{T}\cdot\mathbf{F}^{-T}\cdot\mathbb{M}\right]\cdot\mathbb{B}\right]\mathbf{F}^{-T} - \left[\mathbf{F}^{-T}\cdot\mathbb{M}\right]\otimes\left[\mathbb{B}\cdot\mathbf{F}^{T}\cdot\mathbf{F}^{-T}\right]\\
&= [\mathbb{M}\cdot\mathbb{B}]\,\mathbf{F}^{-T} - \left[\mathbf{F}^{-T}\cdot\mathbb{M}\right]\otimes\mathbb{B} \quad . && \text{(C.31)}
\end{aligned}
$$

C.3.2 Divergences

Following the proposition made by Pao [414], the ponderomotive body force $\mathbf{b}_t^{\text{pon}}$ can be expressed as the divergence of a stress tensor. Due to the additive decomposition of the ponderomotive stress, equivalent expressions for the body force arising from material magnetisation and the Maxwell contribution result from their counterpart stress tensors.

Ponderomotive stress tensor

Starting with the definition of the magnetic contribution to the ponderomotive stress, listed in equation (5.131a), then

$$
\begin{aligned}
\nabla\cdot\boldsymbol{\sigma}^{*\text{pon}} &= \nabla\cdot\left[\frac{1}{2\mu_0}[\mathbb{b}\cdot\mathbb{b}]\,\mathbf{i} - [\mathbb{h}\cdot\mathbb{b}]\,\mathbf{i} + \mathbb{h}\otimes\mathbb{b}\right] && \leftarrow \text{eq. (5.131a)}\\
&= \left[\frac{1}{\mu_0}\mathbb{b}\cdot[\nabla\mathbb{b}] - \mathbb{b}\cdot[\nabla\mathbb{h}] - \mathbb{h}\cdot[\nabla\mathbb{b}] + [\nabla\mathbb{h}]\cdot\mathbb{b} + \mathbb{h}\cancelto{0}{[\nabla\cdot\mathbb{b}]}\right] && \rightarrow \text{eq. } (5.105_2)\\
&= \left[[\nabla\mathbb{h}]\cdot\mathbb{b} - \mathbb{b}\cdot[\nabla\mathbb{h}] + \left[\frac{1}{\mu_0}\mathbb{b} - \mathbb{h}\right]\cdot[\nabla\mathbb{b}]\right] && \rightarrow \text{eq. (5.118,A.41)}\\
&= [\nabla\times\mathbb{h}]\times\mathbb{b} + \mathbb{m}\cdot[\nabla\mathbb{b}] && \rightarrow \text{eq. } (5.105_1)\\
&= \mathbb{j}^{\text{f}}\times\mathbb{b} + \mathbb{m}\cdot[\nabla\mathbb{b}] && \text{(C.32)}\\
&\equiv J^{-1}\nabla_0\cdot\mathbf{P}^{*\text{pon}} \quad . && \leftarrow \text{eq. (B.42)}
\end{aligned}
$$

This result can be verified by considering the decomposition of the ponderomotive stress as given in equation (5.130) in conjunction with equations (C.33) and (C.34), both of which will be derived next. It also shows that the chosen definition for the ponderomotive stress satisfies the fundamental definition of the Lorentz body force given in equation $(5.127)_2$.

Magnetisation stress tensor

$$
\begin{aligned}
\nabla\cdot\boldsymbol{\sigma}^{*\text{mag}} &= \nabla\cdot\left[[\mathbb{m}\cdot\mathbb{b}]\,\mathbf{i} - \mathbb{m}\otimes\mathbb{b}\right] && \leftarrow \text{eq. (5.131b)}\\
&= \mathbb{m}\cdot[\nabla\mathbb{b}] + \mathbb{b}\cdot[\nabla\mathbb{m}] - [\nabla\mathbb{m}]\cdot\mathbb{b} - \mathbb{m}\cancelto{0}{[\nabla\cdot\mathbb{b}]} && \rightarrow \text{eqs. } (\text{A.40},5.114_2)\\
&= \mathbb{m}\cdot[\nabla\mathbb{b}] + \mathbb{b}\times[\nabla\times\mathbb{m}] && \rightarrow \text{eqs. } (5.117_1\text{A.1})
\end{aligned}
$$

$$= \mathbb{m} \cdot [\nabla \mathbb{b}] - \mathbb{j}^{\mathrm{b}} \times \mathbb{b} \tag{C.33}$$

$$\equiv J^{-1} \nabla_0 \cdot \mathbf{P}^{*\mathrm{mag}} \qquad \leftarrow \text{eq. (B.42)}$$

Maxwell stress tensor

$$\begin{aligned}
\nabla \cdot \boldsymbol{\sigma}^{*\mathrm{max}} &= \nabla \cdot \left[\frac{1}{\mu_0} \left[\mathbb{b} \otimes \mathbb{b} - \frac{1}{2} [\mathbb{b} \cdot \mathbb{b}] \, \mathbf{i} \right] \right] && \leftarrow \text{eq. (5.131c)} \\
&= \frac{1}{\mu_0} \left[[\nabla \mathbb{b}] \cdot \mathbb{b} + \mathbb{b} \cancelto{0}{[\nabla \cdot \mathbb{b}]} - \mathbb{b} \cdot [\nabla \mathbb{b}] \right] && \rightarrow \text{eqs. (A.41,5.114}_2\text{)} \\
&= \frac{1}{\mu_0} [\nabla \times \mathbb{b}] \times \mathbb{b} && \rightarrow \text{eq. (5.118)} \\
&= [\nabla \times [\mathbb{m} + \mathbb{h}]] \times \mathbb{b} && \rightarrow \text{eqs. (5.114}_1\text{,5.117}_1\text{)} \\
&= \mathbb{j} \times \mathbb{b} && \text{(C.34)} \\
&\equiv J^{-1} \nabla_0 \cdot \mathbf{P}^{*\mathrm{max}} && \leftarrow \text{eq. (B.42)}
\end{aligned}$$

C.3.3 Jumps

As is stated in equations (5.98) and (5.126), the traction continuity condition is expressed in terms of the jump of the stress across a surface. Below we derive the magnetic contribution to the jump in the total Cauchy stress across the interface. Note that the referential counterparts of these expressions can be derived with the consideration of equation (B.41).

Magnetisation stress tensor

$$\begin{aligned}
[\![\boldsymbol{\sigma}^{*\mathrm{mag}}]\!] \cdot \mathbf{n} &= [\![[\mathbb{m} \cdot \mathbb{b}] \, \mathbf{i} - \mathbb{m} \otimes \mathbb{b}]\!] \cdot \mathbf{n} && \leftarrow \text{eq. (5.131b)} \rightarrow \text{eq. (A.19)} \\
&= [\{\!\{\mathbb{m}\}\!\} \cdot [\![\mathbb{b}]\!] + [\![\mathbb{m}]\!] \cdot \{\!\{\mathbb{b}\}\!\}] \, \mathbf{n} - [\![\mathbb{m}]\!] \, [\{\!\{\mathbb{b}\}\!\} \cdot \mathbf{n}] + \{\!\{\mathbb{m}\}\!\} \cancelto{0}{[[\![\mathbb{b}]\!] \cdot \mathbf{n}]} \\
& && \rightarrow \text{eqs. (A.15,C.20)} \\
&= [\{\!\{\mathbb{m}\}\!\} \cdot [\![\mathbb{b}]\!]] \, \mathbf{n} + [\mathbf{n} \times [\![\mathbb{m}]\!]] \times \{\!\{\mathbb{b}\}\!\} && \rightarrow \text{eq. (5.117}_2\text{)} \\
&= [\{\!\{\mathbb{m}\}\!\} \cdot [\![\mathbb{b}]\!]] \, \mathbf{n} - \hat{\mathbb{j}}^{\mathrm{b}} \times \{\!\{\mathbb{b}\}\!\} && \text{(C.35)}
\end{aligned}$$

Maxwell stress tensor

$$\begin{aligned}
[\![\boldsymbol{\sigma}^{*\mathrm{max}}]\!] \cdot \mathbf{n} &= \left[\!\!\left[\frac{1}{\mu_0} \left[\mathbb{b} \otimes \mathbb{b} - \frac{1}{2} [\mathbb{b} \cdot \mathbb{b}] \, \mathbf{i} \right] \right]\!\!\right] \cdot \mathbf{n} && \leftarrow \text{eq. (5.131c)} \\
&= \left[\!\!\left[-\frac{1}{\mu_0} \left[\frac{1}{2} [\mathbb{b} \cdot \mathbb{b}] \, \mathbf{i} - \mathbb{b} \otimes \mathbb{b} \right] \right]\!\!\right] \cdot \mathbf{n} \\
&= \left[\!\!\left[-\frac{1}{\mu_0} \left[[[\mathbb{b} \cdot \mathbb{b}] \, \mathbf{i} - \mathbb{b} \otimes \mathbb{b}] - \frac{1}{2} [\mathbb{b} \cdot \mathbb{b}] \, \mathbf{i} \right] \right]\!\!\right] \cdot \mathbf{n}
\end{aligned}$$

$$
\begin{aligned}
&= -\frac{1}{\mu_0} \llbracket [\mathbb{b}\cdot\mathbb{b}]\,\mathbf{i} - \mathbb{b}\otimes\mathbb{b} \rrbracket \cdot \mathbf{n} + \frac{1}{2\mu_0} \llbracket \mathbb{b}\cdot\mathbb{b} \rrbracket\,\mathbf{n} && \rightarrow \text{eqs. (A.18,A.17)}\\
&= -\frac{1}{\mu_0} \left[[\{\!\{\mathbb{b}\}\!\} \times \mathbf{n}] \times \llbracket \mathbb{b} \rrbracket + [\llbracket \mathbb{b} \rrbracket \times \mathbf{n}] \times \{\!\{\mathbb{b}\}\!\} \right] + \frac{1}{\mu_0} \left[\{\!\{\mathbb{b}\}\!\} \cdot \llbracket \mathbb{b} \rrbracket \right] \mathbf{n} \\
& && \rightarrow \text{eq. (A.15)}\\
&= -\frac{1}{\mu_0} \left[[\{\!\{\mathbb{b}\}\!\}\cdot\llbracket \mathbb{b} \rrbracket]\,\mathbf{n} - [\{\!\{\mathbb{b}\}\!\} \otimes \llbracket \mathbb{b} \rrbracket]\cdot\mathbf{n} + [\llbracket \mathbb{b} \rrbracket \times \mathbf{n}] \times \{\!\{\mathbb{b}\}\!\} \right] + \frac{1}{\mu_0} \left[\{\!\{\mathbb{b}\}\!\}\cdot\llbracket \mathbb{b} \rrbracket \right] \mathbf{n}\\
&= \frac{1}{\mu_0} \left[[\{\!\{\mathbb{b}\}\!\} \otimes \llbracket \mathbb{b} \rrbracket] \cdot \mathbf{n} - [\llbracket \mathbb{b} \rrbracket \times \mathbf{n}] \times \{\!\{\mathbb{b}\}\!\} \right]\\
&= \{\!\{\mathbb{b}\}\!\} \cancelto{0}{[\llbracket \mathbb{b} \rrbracket \cdot \mathbf{n}]} + \left[\mathbf{n} \times \frac{1}{\mu_0} \llbracket \mathbb{b} \rrbracket \right] \times \{\!\{\mathbb{b}\}\!\} && \rightarrow \text{eq. (5.117}_2\text{,C.20)}\\
&= \left[\mathbf{n} \times [\hat{\mathbb{j}} \times \mathbf{n}] \right] \times \{\!\{\mathbb{b}\}\!\}\\
&= \left[[\mathbf{n} \times \hat{\mathbb{j}}] \times \mathbf{n} \right] \times \{\!\{\mathbb{b}\}\!\} && \rightarrow \text{eq. (A.15)}\\
&= \left[\hat{\mathbb{j}} \underbrace{[\mathbf{n}\cdot\mathbf{n}]}_{=1} - [\mathbf{n} \otimes \mathbf{n}] \cdot \hat{\mathbb{j}} \right] \times \{\!\{\mathbb{b}\}\!\}\\
&= \left[\hat{\mathbb{j}} - \mathbf{n} \cancelto{0}{[\mathbf{n} \cdot \hat{\mathbb{j}}]} \right] \times \{\!\{\mathbb{b}\}\!\} && \rightarrow \text{eq. (C.23)}\\
&= \hat{\mathbb{j}} \times \{\!\{\mathbb{b}\}\!\} && \text{(C.36)}
\end{aligned}
$$

Ponderomotive stress tensor

$$
\begin{aligned}
\llbracket \boldsymbol{\sigma}^{*\mathrm{pon}} \rrbracket \cdot \mathbf{n} &= \llbracket \boldsymbol{\sigma}^{*\mathrm{mag}} + \boldsymbol{\sigma}^{*\mathrm{max}} \rrbracket \cdot \mathbf{n} && \leftarrow \text{eq. (5.130)}\\
&= \left[\llbracket \boldsymbol{\sigma}^{*\mathrm{mag}} \rrbracket + \llbracket \boldsymbol{\sigma}^{*\mathrm{max}} \rrbracket \right] \cdot \mathbf{n} && \rightarrow \text{eqs. (C.35,C.36)}\\
&= \left[\{\!\{\mathbb{m}\}\!\} \cdot \llbracket \mathbb{b} \rrbracket \right] \mathbf{n} - \hat{\mathbb{j}}^{\mathrm{b}} \times \{\!\{\mathbb{b}\}\!\} + \hat{\mathbb{j}} \times \{\!\{\mathbb{b}\}\!\}\\
&= \left[\{\!\{\mathbb{m}\}\!\} \cdot \llbracket \mathbb{b} \rrbracket \right] \mathbf{n} + \left[\hat{\mathbb{j}} - \hat{\mathbb{j}}^{\mathrm{b}} \right] \times \{\!\{\mathbb{b}\}\!\} && \rightarrow \text{eq. (C.18)}\\
&= \left[\{\!\{\mathbb{m}\}\!\} \cdot \llbracket \mathbb{b} \rrbracket \right] \mathbf{n} + \hat{\mathbb{j}}^{\mathrm{f}} \times \{\!\{\mathbb{b}\}\!\} && \text{(C.37)}\\
&= \left[\{\!\{\mathbb{m}\}\!\} \times \mathbf{n} \right] \times \llbracket \mathbb{b} \rrbracket + \left[\{\!\{\mathbb{m}\}\!\} \otimes \llbracket \mathbb{b} \rrbracket \right] \cdot \mathbf{n} + \hat{\mathbb{j}}^{\mathrm{f}} \times \{\!\{\mathbb{b}\}\!\} && \leftarrow \text{eq. (A.15)}\\
&= \left[\{\!\{\mathbb{m}\}\!\} \times \mathbf{n} \right] \times \llbracket \mathbb{b} \rrbracket + \{\!\{\mathbb{m}\}\!\} \left[\llbracket \mathbb{b} \rrbracket \cdot \mathbf{n} \right] + \hat{\mathbb{j}}^{\mathrm{f}} \times \{\!\{\mathbb{b}\}\!\} && \rightarrow \text{eq. (5.110}_2\text{)}\\
&= \left[\{\!\{\mathbb{m}\}\!\} \times \mathbf{n} \right] \times \llbracket \mathbb{b} \rrbracket + \hat{\mathbb{j}}^{\mathrm{f}} \times \{\!\{\mathbb{b}\}\!\} && \text{(C.38)}
\end{aligned}
$$

C.4 Ponderomotive forces and tractions

C.4.1 Definitions of the Lorentz forces

Spatial total ponderomotive force vector

Adopting the notation given in Figure 5.1, we define a domain $\mathscr{D}_t = \mathscr{B}_t \cup \mathscr{S}_t$ which is a control volume completely enclosing the solid $\mathscr{B}_t$ with $\overline{\partial\mathscr{B}_t} = \overline{\mathscr{B}_t} \cap \overline{\mathscr{S}_t}$. Then, in general, the total ponderomotive (Lorentz) force is acting over the entire quasi-static domain is [414, 133, 530]

$$\mathbf{f}_t^{\mathrm{pon}} := \int_{\mathscr{D}_t} [\varrho_t \mathbb{e} + \mathbb{j} \times \mathbb{b}] \; \mathrm{dv} + \int_{\partial\mathscr{B}_t} [\widehat{\varrho}_t \, \{\!\{\mathbb{e}\}\!\} + \widehat{\mathbb{j}} \times \{\!\{\mathbb{b}\}\!\}] \; \mathrm{da} \quad , \tag{C.39}$$

which is a generalisation and extension of equation (5.40) using equation (5.41). When neglecting the contributions due to electric fields (i. e. assuming that all considered materials are non-ferrous and decoupled [133, 530]), this reduces to

$$\mathbf{f}_t^{\mathrm{pon}} = \int_{\mathscr{D}_t} \mathbb{j} \times \mathbb{b} \; \mathrm{dv} + \int_{\partial\mathscr{B}_t} \widehat{\mathbb{j}} \times \{\!\{\mathbb{b}\}\!\} \; \mathrm{da} \tag{C.40}$$

that consists only of the magnetic Lorentz force, due to a free current running through a portion of the domain, and the force acting on magnetic dipoles. With some manipulation, this can be reformulated as

$$\begin{aligned} \mathbf{f}_t^{\mathrm{pon}} &= \int_{\mathscr{D}_t} \left[\mathbb{j}^{\mathrm{f}} + \mathbb{j}^{\mathrm{b}}\right] \times \mathbb{b} \; \mathrm{dv} + \int_{\partial\mathscr{B}_t} \widehat{\mathbb{j}} \times \{\!\{\mathbb{b}\}\!\} \; \mathrm{da} && \leftarrow \text{eq. } (5.111_2) \rightarrow \text{eq. } (5.117) \\ &= \int_{\mathscr{D}_t} \left[\mathbb{j}^{\mathrm{f}} + \boldsymbol{\nabla} \times \mathbb{m}\right] \times \mathbb{b} \; \mathrm{dv} + \int_{\partial\mathscr{B}_t} \widehat{\mathbb{j}} \times \{\!\{\mathbb{b}\}\!\} \; \mathrm{da} \quad . \end{aligned} \tag{C.41}$$

As an aside, we derive the identity

$$[\boldsymbol{\nabla} \times \mathbb{m}] \times \mathbb{b} = -\boldsymbol{\nabla} \cdot [[\mathbb{m} \cdot \mathbb{b}] \, \mathbf{i}] + \mathbb{m} \times [\boldsymbol{\nabla} \times \mathbb{b}] + [\boldsymbol{\nabla}\mathbb{m}] \cdot \mathbb{b} + [\boldsymbol{\nabla}\mathbb{b}] \cdot \mathbb{m} \tag{C.42}$$

using a rearrangement of equation (A.42). Noting that with equation $(5.133)_2$ we can repose

$$\begin{aligned} [\boldsymbol{\nabla}\mathbb{m}] \cdot \mathbb{b} &\equiv [\boldsymbol{\nabla}\mathbb{m}] \cdot \mathbb{b} + \mathbb{m} \, [\boldsymbol{\nabla} \cdot \mathbb{b}] && \rightarrow \text{eq. } (5.133_2) \\ &= \boldsymbol{\nabla} \cdot [\mathbb{m} \otimes \mathbb{b}] \quad ; \end{aligned} \tag{C.43}$$

equation (C.41) then becomes

$$\begin{aligned} \mathbf{f}_t^{\mathrm{pon}} &= \int_{\mathscr{D}_t} \left[\mathbb{j}^{\mathrm{f}} \times \mathbb{b} - \boldsymbol{\nabla} \cdot [[\mathbb{m} \cdot \mathbb{b}] \, \mathbf{i}] + \mathbb{m} \times [\boldsymbol{\nabla} \times \mathbb{b}] + \boldsymbol{\nabla} \cdot [\mathbb{m} \otimes \mathbb{b}] + [\boldsymbol{\nabla}\mathbb{b}] \cdot \mathbb{m}\right] \mathrm{dv} \\ &\quad + \int_{\partial\mathscr{B}_t} \widehat{\mathbb{j}} \times \{\!\{\mathbb{b}\}\!\} \; \mathrm{da} && \rightarrow \text{eqs. (C.42,C.43)} \\ &= \int_{\mathscr{D}_t} \mathbb{j}^{\mathrm{f}} \times \mathbb{b} - \boldsymbol{\nabla} \cdot [[\mathbb{m} \cdot \mathbb{b}] \, \mathbf{i}] + \boldsymbol{\nabla} \cdot [\mathbb{m} \otimes \mathbb{b}] + [\mathbb{m} \times [\boldsymbol{\nabla} \times \mathbb{b}] + [\boldsymbol{\nabla}\mathbb{b}] \cdot \mathbb{m}] \; \mathrm{dv} \\ &\quad + \int_{\partial\mathscr{B}_t} \widehat{\mathbb{j}} \times \{\!\{\mathbb{b}\}\!\} \; \mathrm{da} && \rightarrow \text{eq. (A.40)} \end{aligned}$$

$$= \int_{\mathscr{D}_t} \left[\mathbb{j}^\mathrm{f} \times \mathbb{b} - \boldsymbol{\nabla} \cdot [[\mathbb{m} \cdot \mathbb{b}]\, \mathbf{i}] + \boldsymbol{\nabla} \cdot [\mathbb{m} \otimes \mathbb{b}] + \mathbb{m} \cdot [\boldsymbol{\nabla} \mathbb{b}] \right] \mathrm{dv} + \int_{\partial \mathscr{B}_t} \hat{\mathbb{j}} \times \{\!\{\mathbb{b}\}\!\}\ \mathrm{da} \quad \text{(C.44)}$$

with equation (A.40). Using equation (B.16), the divergence terms in the volume integral are transferred to a surface integral with the result that

$$\mathbf{f}_t^\mathrm{pon} = \int_{\mathscr{D}_t} \left[\mathbb{j}^\mathrm{f} \times \mathbb{b} + \mathbb{m} \cdot [\boldsymbol{\nabla} \mathbb{b}] \right] \mathrm{dv} + \int_{\partial \mathscr{B}_t} \hat{\mathbb{j}} \times \{\!\{\mathbb{b}\}\!\}\ \mathrm{da} - \int_{\partial \mathscr{D}_t} [[\mathbb{m} \cdot \mathbb{b}]\, \mathbf{i} - \mathbb{m} \otimes \mathbb{b}] \cdot \mathbf{n}\ \mathrm{da}$$

$\leftarrow$ eq. (B.16)

$$= \int_{\mathscr{D}_t} \left[\mathbb{j}^\mathrm{f} \times \mathbb{b} + \mathbb{m} \cdot [\boldsymbol{\nabla} \mathbb{b}] \right] \mathrm{dv} + \int_{\partial \mathscr{B}_t} \hat{\mathbb{j}} \times \{\!\{\mathbb{b}\}\!\}\ \mathrm{da} - \int_{\partial \mathscr{D}_t} \boldsymbol{\sigma}^{*\mathrm{mag}} \cdot \mathbf{n}\ \mathrm{da} \qquad \leftarrow \text{eq. (5.131b)}$$

$$= \int_{\mathscr{D}_t} \left[\mathbb{j}^\mathrm{f} \times \mathbb{b} + \mathbb{m} \cdot [\boldsymbol{\nabla} \mathbb{b}] \right] \mathrm{dv} + \int_{\partial \mathscr{B}_t} \left[\hat{\mathbb{j}} \times \{\!\{\mathbb{b}\}\!\} + [\![\boldsymbol{\sigma}^{*\mathrm{mag}}]\!] \cdot \mathbf{n} \right] \mathrm{da} - \int_{\partial \mathscr{S}_t} \boldsymbol{\sigma}^{*\mathrm{mag}} \cdot \mathbf{n}\ \mathrm{da}$$

(C.45)

with the definition given in equation (5.131b). Given that in the free space $\mathscr{S}_t$ the free currents and magnetisation vanish ($\mathbb{j}^\mathrm{f} = \mathbf{0}, \mathbb{m} = \mathbf{0} \Rightarrow \mathbb{j}^\mathrm{b} = \mathbf{0}$), the third integral equals zero. Thus, with the result given in equation (C.35),

$$\mathbf{f}_t^\mathrm{pon} = \int_{\mathscr{B}_t} \underbrace{\left[\mathbb{j}^\mathrm{f} \times \mathbb{b} + \mathbb{m} \cdot [\boldsymbol{\nabla} \mathbb{b}] \right]}_{\mathbf{b}_t^\mathrm{pon}} \mathrm{dv} + \int_{\partial \mathscr{B}_t} \underbrace{\left[\hat{\mathbb{j}}^\mathrm{f} \times \{\!\{\mathbb{b}\}\!\} + [\{\!\{\mathbb{m}\}\!\} \cdot [\![\mathbb{b}]\!]]\, \mathbf{n} \right]}_{\mathbf{t}_t^\mathrm{pon}} \mathrm{da} \quad \text{(C.46)}$$

$$= \int_{\mathscr{B}_t} \boldsymbol{\nabla} \cdot \boldsymbol{\sigma}^{*\mathrm{pon}}\ \mathrm{dv} + \int_{\partial \mathscr{B}_t} [\![\boldsymbol{\sigma}^{*\mathrm{pon}}]\!] \cdot \mathbf{n}\ \mathrm{da} \quad . \qquad \leftarrow \text{eqs. (C.32,C.37)}$$

Referential ponderomotive body force density vector

$$\int_{\mathscr{D}_0} \mathbf{b}_0^\mathrm{pon}\ \mathrm{dV} = \int_{\mathscr{D}_0} \boldsymbol{\nabla}_0 \cdot \mathbf{P}^{*\mathrm{pon}}\ \mathrm{dV} = \int_{\mathscr{D}_0} J \boldsymbol{\nabla} \cdot \boldsymbol{\sigma}^{*\mathrm{pon}}\ \mathrm{dV} = \int_{\mathscr{D}_t} \mathbf{b}_t^\mathrm{pon}\ \mathrm{dv}$$

$\leftarrow$ eq. $(5.127_1) \rightarrow$ eq. (C.32)

$$= \int_{\mathscr{D}_t} \left[\mathbb{j}^\mathrm{f} \times \mathbb{b} + \mathbb{m} \cdot [\boldsymbol{\nabla} \mathbb{b}] \right] \mathrm{dv} \qquad \rightarrow \text{eqs. (5.7,5.102)}$$

$$= \int_{\mathscr{D}_t} \left[\left[J^{-1} \mathbf{F} \cdot \mathbb{J}^\mathrm{f} \right] \times \left[J^{-1} \mathbf{F} \cdot \mathbb{B} \right] + \left[\mathbf{F}^{-T} \cdot \mathbb{M} \right] \cdot \left[[\boldsymbol{\nabla}_0 \mathbb{b}] \cdot \mathbf{F}^{-1} \right] \right] \mathrm{dv}$$

$\rightarrow$ eqs. (A.11,A.10)

$$= \int_{\mathscr{D}_t} \left[J^{-2} \det(\mathbf{F})\, \mathbf{F}^{-T} \cdot \left[\mathbb{J}^\mathrm{f} \times \mathbb{B} \right] + \left[\mathbf{F}^{-T} \cdot \mathbb{M} \right] \cdot \left[\boldsymbol{\nabla}_0 \left[J^{-1} \mathbf{F} \cdot \mathbb{B} \right] \cdot \mathbf{F}^{-1} \right] \right] \mathrm{dv}$$

$$= \int_{\mathscr{D}_t} \mathbf{F}^{-T} \cdot \left[J^{-1} \left[\mathbb{J}^\mathrm{f} \times \mathbb{B} \right] + \mathbb{M} \cdot \left[\boldsymbol{\nabla}_0 \left[J^{-1} \mathbf{F} \cdot \mathbb{B} \right] \cdot \mathbf{F}^{-1} \right] \right] \mathrm{dv} \qquad \rightarrow \text{eq. (5.11)}$$

$$= \int_{\mathscr{D}_0} \mathbf{F}^{-T} \cdot \left[J^{-1} \left[\mathbb{J}^{\mathrm{f}} \times \mathbb{B} \right] + \mathbb{M} \cdot \left[\boldsymbol{\nabla}_0 \left[J^{-1} \mathbf{F} \cdot \mathbb{B} \right] \cdot \mathbf{F}^{-1} \right] \right] J \, \mathrm{dV}$$

$$= \int_{\mathscr{D}_0} \mathbf{F}^{-T} \cdot \left[\left[\mathbb{J}^{\mathrm{f}} \times \mathbb{B} \right] + J \mathbb{M} \cdot \left[\boldsymbol{\nabla}_0 \left[J^{-1} \mathbf{F} \cdot \mathbb{B} \right] \cdot \mathbf{F}^{-1} \right] \right] \mathrm{dV} \tag{C.47}$$

Referential ponderomotive traction vector

$$\int_{\partial \mathscr{B}_0} \mathbf{t}_0^{\mathrm{pon}} \, \mathrm{dA} = \int_{\partial \mathscr{B}_0} [\![\mathbf{P}^{*\mathrm{pon}}]\!] \cdot \mathbf{N} \, \mathrm{dA} = \int_{\partial \mathscr{B}_t} [\![\boldsymbol{\sigma}^{*\mathrm{pon}}]\!] \cdot \mathbf{n} \, \mathrm{da} = \int_{\partial \mathscr{B}_t} \mathbf{t}_t^{\mathrm{pon}} \, \mathrm{da}$$

$\leftarrow$ eqs. (5.128,B.41) $\rightarrow$ eq. (C.37)

$$= \int_{\partial \mathscr{B}_t} \left[\hat{\mathbb{j}}^{\mathrm{f}} \times \{\!\{ \mathbb{b} \}\!\} + [\{\!\{ \mathbb{m} \}\!\} \cdot [\![\mathbb{b}]\!]] \, \mathbf{n} \right] \mathrm{da} \qquad \rightarrow \text{eqs. (C.14,5.102)}$$

$$= \int_{\partial \mathscr{B}_t} \left[\left[\hat{J}^{-1} \mathbf{F} \cdot \hat{\mathbb{J}}^{\mathrm{f}} \right] \times \left\{\!\!\left\{ J^{-1} \mathbf{F} \cdot \mathbb{B} \right\}\!\!\right\} + \left[\left\{\!\!\left\{ \mathbf{F}^{-T} \cdot \mathbb{M} \right\}\!\!\right\} \cdot \left[\!\!\left[J^{-1} \mathbf{F} \cdot \mathbb{B} \right]\!\!\right] \right] \mathbf{n} \right] \mathrm{da}$$

$\rightarrow$ eqs. (C.11,5.10)

$$= \int_{\partial \mathscr{B}_t} \left[\left[\mathbf{F} \cdot \hat{\mathbb{J}}^{\mathrm{f}} \right] \times \left\{\!\!\left\{ J^{-1} \mathbf{F} \cdot \mathbb{B} \right\}\!\!\right\} + \left[\left\{\!\!\left\{ \mathbf{F}^{-T} \cdot \mathbb{M} \right\}\!\!\right\} \cdot \left[\!\!\left[J^{-1} \mathbf{F} \cdot \mathbb{B} \right]\!\!\right] \right] J \mathbf{F}^{-T} \cdot \mathbf{N} \right] \mathrm{dA}$$

$\rightarrow$ eqs. (A.10,A.11)

$$= \int_{\partial \mathscr{B}_t} \left[J^{-1} J \mathbf{F}^{-T} \cdot \left[\hat{\mathbb{J}}^{\mathrm{f}} \times \{\!\{ \mathbb{B} \}\!\} \right] + \left[\{\!\{ \mathbb{M} \}\!\} \cdot \left[J^{-1} \mathbf{F}^{-1} \cdot \mathbf{F} \cdot [\![\mathbb{B}]\!] \right] \right] J \mathbf{F}^{-T} \cdot \mathbf{N} \right] \mathrm{dA}$$

$$= \int_{\partial \mathscr{B}_t} \left[\mathbf{F}^{-T} \cdot \left[\hat{\mathbb{J}}^{\mathrm{f}} \times \{\!\{ \mathbb{B} \}\!\} + [\{\!\{ \mathbb{M} \}\!\} \cdot [\![\mathbb{B}]\!]] \, \mathbf{N} \right] \right] \mathrm{dA} \tag{C.48}$$

C.5 Thermomechanical and electromagnetic balance laws

C.5.1 Derivation of spatial statement of Maxwell's equations in a non-relativistic Eulerian reference frame

A note on the setting for Maxwell's equations

The Maxwell equations are derived most generally in a relativistic setting accounting for the non-Euclidean geometry of space-time. The discussion of the governing laws in this general setting is well beyond the scope of this manuscript, and we refer the reader to [325, 523, 133, 230, 277] to offer the relevant background on this topic.

Hereafter, we consider the framework in which Maxwell's equations for electromagnetic fields are presented to have a Euclidean geometry and to be non-relativistic (one wherein the velocity magnitude $|\mathbf{v}| \ll c$, the speed of light). Therein, to relate a formulation in a truly Eulerian framework (with terms indicated as $(\bullet)'$) to a formulation in a Lagrangian framework (in which we follow the motion of individual physical "particles") we introduce the following transformations:

$$\mathbb{d}' := \mathbb{d} \tag{C.49a}$$

$$\mathbb{e}' := \mathbb{e} + \mathbb{b} \times \mathbf{v} \tag{C.49b}$$

$$\mathbb{j}' := \mathbb{j} + \varrho_t \mathbf{v} \tag{C.49c}$$

$$\mathbb{b}' := \mathbb{b} \tag{C.49d}$$

$$\mathbb{h}' := \mathbb{h} - \mathbb{d} \times \mathbf{v} \quad . \tag{C.49e}$$

It is important to note that all quantities are expressed in SI units. Note as well that these transformations shall not be confused with those used to transform quantities in the *aether frame* (in which the description of the governing laws is universal) into the *Euclidean frame*.

Maxwell's equations: Gauss' law

Gauss' flux theorem states that the relationship between the distribution of free charge and the resulting electric displacement field is

$$0 = \int_{\mathcal{V}_t} \varrho_t^{\mathrm{f}}(\mathbf{x}, t)\ \mathrm{dv} - \int_{\partial \mathcal{V}_t} \mathbb{d}(\mathbf{x}, t) \cdot \mathbf{n}\ \mathrm{da} \quad . \tag{C.50}$$

The sources listed in Table 5.1 are inferred by comparison of the weak form of the conservation law to equation (5.52). Inserting these definitions into equations (5.53) and (5.56a) renders, respectively, the local form of the governing equation

$$0 = \varrho_t^{\mathrm{f}} - \nabla \cdot \mathbb{d} \quad \forall \mathbf{x} \quad \text{in} \quad \mathcal{V}_t \tag{C.51}$$

and the jump term

$$[\![\mathbb{d}]\!] \cdot \mathbf{n} = \widehat{\varrho}_t^{\mathrm{f}} \quad \forall \mathbf{x} \quad \text{on} \quad \mathscr{A}_t \quad . \tag{C.52}$$

These correspond to what is described in equation $(5.104)_2$ and equation $(5.109)_2$ when additional surface contributions (due to the presence of surface charges) are considered. Alternative, however entirely equivalent, statements of Gauss' law in free space and in matter are assembled in Table 5.5.

Maxwell's equations: Gauss' magnetism law

Gauss' magnetism law, or the statement of no magnetic monopoles, is that

$$0 = \int_{\partial \mathcal{V}_t} \mathbb{b}(\mathbf{x}, t) \cdot \mathbf{n}\ \mathrm{da} \quad . \tag{C.53}$$

The sources listed in Table 5.1 are inferred by comparison of the weak form of the conservation law to equation (5.52). Inserting these definitions into equations (5.53) and (5.56a) renders, respectively, the local form of the governing equation within the volume

$$0 = \nabla \cdot \mathbb{b} \quad \forall \mathbf{x} \quad \text{in} \quad \mathscr{V}_t \tag{C.54}$$

and the jump term

$$[\![\mathbb{b}]\!] \cdot \mathbf{n} = 0 \quad \forall \mathbf{x} \quad \text{on} \quad \mathscr{A}_t \quad . \tag{C.55}$$

These correspond to what is described in equation $(5.105)_2$ and equation $(5.110)_2$. As is impressed in Table 5.5, Gauss' magnetism law remains unchanged in free space and in matter.

Maxwell's equations: Faraday's law

The statement of Faraday's law of induction is

$$\oint_{\partial\mathscr{A}_t} \mathbb{e}(\mathbf{x}, t) \cdot \mathbf{l} \, \mathrm{dl} = -\mathrm{D}_t \int_{\mathscr{A}_t} \mathbb{b}(\mathbf{x}, t) \cdot \mathbf{n} \, \mathrm{da} \quad . \tag{C.56}$$

Comparison to equation (5.61) leads to the definition of the sources given in Table 5.3. From equation (5.63), localisation of Faraday's law thus leads to the result that

$$\begin{aligned} \mathrm{d}_t \mathbb{b} + [\nabla \cdot \mathbb{b}]\, \mathbf{v} + \nabla \times [\mathbb{b} \times \mathbf{v}] &= -\nabla \times \mathbb{e} && \rightarrow \text{eqs. } (5.105_2, \text{C.49a}) \\ \mathrm{d}_t \mathbb{b} &= -\nabla \times \mathbb{e}' \quad \forall \mathbf{x} \quad \text{in} \quad \mathscr{V}_t && \end{aligned} \tag{C.57}$$

which equates to the law presented in equation (5.104). The associated jump condition, extracted from equation (5.64), is

$$\mathbf{n} \times [\![\mathbb{e}]\!] = \mathbf{0} \quad \forall \mathbf{x} \quad \text{on} \quad \mathscr{A}_t \tag{C.58}$$

and can be identified as that listed in equation $(5.109)_1$. Alternative, however entirely equivalent, statements of Faraday's law in terms of material, spatial and nominal time derivatives are assembled in Table 5.5.

Maxwell's equations: Ampére's law

The statement of Ampére's law is

$$\oint_{\partial\mathscr{A}_t} \mathbb{h}(\mathbf{x}, t) \cdot \mathbf{l} \, \mathrm{dl} = \mathrm{D}_t \int_{\mathscr{A}_t} \mathbb{d}(\mathbf{x}, t) \cdot \mathbf{n} \, \mathrm{da} + \int_{\mathscr{A}_t} \mathbb{j}^{\mathrm{f}}(\mathbf{x}, t) \cdot \mathbf{n} \, \mathrm{da} \quad . \tag{C.59}$$

The sources given in Table 5.3 are again deduced by observing similarities between the above and equation (5.61). The local expression for Ampére's law is then derived through the source definitions applied to equation (5.63), which renders the strong form

$$
\begin{aligned}
\mathrm{d}_t \mathbb{d} + [\boldsymbol{\nabla} \cdot \mathbb{d}]\, \mathbf{v} + \boldsymbol{\nabla} \times [\mathbb{d} \times \mathbf{v}] &= -\mathbb{j}^{\mathrm{f}} + \boldsymbol{\nabla} \times \mathbb{h} && \rightarrow \text{eqs. } (5.104_2) \\
\mathrm{d}_t \mathbb{d} + \varrho_t^{\mathrm{f}} \mathbf{v} + \boldsymbol{\nabla} \times [\mathbb{d} \times \mathbf{v}] &= -\mathbb{j}^{\mathrm{f}} + \boldsymbol{\nabla} \times \mathbb{h} && \rightarrow \text{eqs. (C.49c,C.49e)} \\
\mathrm{d}_t \mathbb{d} &= -\mathbb{j}^{\mathrm{f}\prime} + \boldsymbol{\nabla} \times \mathbb{h}' \quad \forall\, \mathbf{x} \quad \text{in} \quad \mathscr{V}_t \quad , && \text{(C.60)}
\end{aligned}
$$

which corresponds with equation $(5.105)_1$. The associated jump condition is

$$
\mathbf{n} \times [\![\mathbb{h}]\!] = \hat{\mathbb{j}}^{\mathrm{f}} \quad \forall\, \mathbf{x} \quad \text{on} \quad \mathscr{A}_t \quad ; \qquad \text{(C.61)}
$$

when ignoring contributions that may arise due to the presence of surface currents and flow of surface charges, the associated continuity expression is

$$
\mathbf{n} \times [\![\mathbb{h}]\!] = \mathbf{0} \quad \forall\, \mathbf{x} \quad \text{on} \quad \mathscr{A}_t \qquad \text{(C.62)}
$$

which relates to equation $(5.110)_1$. Alternative, however entirely equivalent, statements of Ampére's law in terms of material, spatial and nominal time derivatives are assembled in Table 5.5.

Conservation of electric charge

Starting from equation $(5.105)_2$, and taking the divergence of both sides renders

$$
\begin{aligned}
\boldsymbol{\nabla} \cdot [\boldsymbol{\nabla} \times \mathbb{h}'] &= \boldsymbol{\nabla} \cdot \mathbb{j}^{\mathrm{f}\prime} + \boldsymbol{\nabla} \cdot [\mathrm{d}_t \mathbb{d}] && \rightarrow \text{eq. (A.25)} \\
0 &= \boldsymbol{\nabla} \cdot \mathbb{j}^{\mathrm{f}\prime} + \mathrm{d}_t [\boldsymbol{\nabla} \cdot \mathbb{d}] + \boldsymbol{\nabla} \cdot [\boldsymbol{\nabla} \times \mathbb{m}] && \rightarrow \text{eqs. } (5.104_2, 5.117_1) \\
0 &= \boldsymbol{\nabla} \cdot \mathbb{j}^{\mathrm{f}\prime} + \mathrm{d}_t \varrho_t^{\mathrm{f}} + \boldsymbol{\nabla} \cdot \mathbb{j}^{\mathrm{b}\prime} + \mathrm{d}_t \varrho_t^{\mathrm{b}} && \rightarrow \text{eq. } (5.111_2) \\
0 &= \mathrm{d}_t \varrho_t + \boldsymbol{\nabla} \cdot \mathbb{j}' \quad . && \text{(C.63)}
\end{aligned}
$$

For deforming bodies, it is implied that

$$
\int_{\mathscr{V}_t} \varrho_t\,(\mathbf{x}, t)\ \mathrm{dv} = \int_{\mathscr{V}_0} \varrho_0\,(\mathbf{X}, t)\ \mathrm{dV} \quad , \qquad \text{(C.64)}
$$

under the consideration of equation (5.101), the rate of which is balanced by its corresponding flux. This is stated by

$$
\mathrm{D}_t \int_{\mathscr{V}_t} \varrho_t\,(\mathbf{x}, t)\ \mathrm{dv} = -\int_{\partial \mathscr{V}_t} \mathbb{j} \cdot \mathbf{n}\ \mathrm{da} \quad . \qquad \text{(C.65)}
$$

When comparing this to equation (5.52), the sources presented in Table 5.1 can be inferred. Localisation of the charge conservation law is derived through application of these sources to equation (5.53). This returns the strong form statement that

$$
0 = \mathrm{d}_t \varrho_t + \boldsymbol{\nabla} \cdot [\varrho_t \mathbf{v}] + \boldsymbol{\nabla} \cdot \mathbb{j} \quad \forall\, \mathbf{x} \quad \text{in} \quad \mathscr{V}_t \qquad \text{(C.66)}
$$

which, through application of equation (C.49c), can be shown to be the equivalent of equation (C.63).

C.5.2 Derivation of spatial statement of mechanical conservation laws (for magnetostatic systems)

Conservation of mass

The conservation of mass for deforming bodies states that, in conjunction with equation (5.100),

$$\int_{\mathscr{V}_t} \rho_t(\mathbf{x}, t)\ \mathrm{dv} = \int_{\mathscr{V}_0} \rho_0(\mathbf{X}, t)\ \mathrm{dV} = \text{constant} \quad , \tag{C.67}$$

which, when stated in rate form, implies

$$0 = \mathrm{D}_t \int_{\mathscr{V}_t} \rho_t(\mathbf{x}, t)\ \mathrm{dv} = \int_{\mathscr{V}_t} \dot{\rho_0}(\mathbf{X}, t)\ \mathrm{dV} \quad . \tag{C.68}$$

Consideration of the above to equation (5.52) leads to the sources given in Table 5.1. Applying these sources to equation (5.53) then renders the localisation given in equation (5.119), namely that

$$\begin{aligned} 0 &= \mathrm{d}_t\rho_t + \nabla \cdot [\rho_t \mathbf{v}] && \rightarrow \text{eq. (A.28)} \\ &= \mathrm{d}_t\rho_t + [\nabla \rho_t] \cdot \mathbf{v} + \rho_t [\nabla \cdot \mathbf{v}] && \rightarrow \text{eq. (5.15)} \\ &\equiv \mathrm{D}_t\rho_t + \rho_t [\nabla \cdot \mathbf{v}] \quad \forall\, \mathbf{x} \quad \text{in} \quad \mathscr{V}_t \quad . && \text{(C.69)} \end{aligned}$$

Traction continuity

On the surface $\partial\mathscr{B}_t$, as stated by Cauchy's stress theorem there is a balance of the internally generated mechanical forces with the externally applied tractions. We have already established that, in the presence of a magnetic field, there exists a ponderomotive traction that acts on the body's surface in addition to any applied mechanical tractions. These two sets of forces are in equilibrium and, therefore, the balance of tractions can be expressed as

$$\begin{aligned} \mathbf{t}_t^{\text{int}} &= \mathbf{t}_t^{\text{ext}} && \rightarrow \text{eq. (5.124), fig. (5.1)} \\ \Longrightarrow \quad \boldsymbol{\sigma}^{\text{mech}} \cdot \mathbf{n}^+ &= \mathbf{t}_t^{\text{mech,ext}} + \mathbf{t}_t^{\text{pon}} && \text{(C.70)} \\ -\left[\!\left[\boldsymbol{\sigma}^{\text{mech}}\right]\!\right] \cdot \mathbf{n}^+ &= \mathbf{t}_t^{\text{mech,ext}} + \left[\!\left[\boldsymbol{\sigma}^{*\text{pon}}\right]\!\right] \cdot \mathbf{n}^+ && \leftarrow \text{eqs. (5.45a,5.128)} \\ -\left[\!\left[\boldsymbol{\sigma}^{\text{mech}} + \boldsymbol{\sigma}^{*\text{pon}}\right]\!\right] \cdot \mathbf{n}^+ &= \mathbf{t}_t^{\text{mech,ext}} && \rightarrow \text{eq. (5.130)} \\ -\left[\!\left[\boldsymbol{\sigma}^{\text{tot}}\right]\!\right] \cdot \mathbf{n}^+ &= \mathbf{t}_t^{\text{mech,ext}} \quad \forall\, \mathbf{x} \quad \text{on} \quad \partial\mathscr{B}_t && \text{(C.71)} \end{aligned}$$

noting that $\boldsymbol{\sigma}^{\text{mech}} = \mathbf{0}$ in $\mathscr{S}_t$.

Balance of linear momentum for polar media [523, 133, 204]

The rate of change of linear momentum of any subregion $\mathscr{V}_t \in \mathscr{B}_t$ of a deforming body

$$\mathbf{f}_t^{\text{tot}} = \mathrm{D}_t \int_{\mathscr{V}_t} \rho_t \mathbf{v}(\mathbf{X},t)\ \mathrm{dv} \tag{C.72}$$

is, in general, equilibrated by

$$\begin{aligned}
\mathbf{f}_t^{\text{mech}} + \mathbf{f}_t^{\text{pon}} &= \int_{\mathscr{V}_t} \left[\mathbf{b}_t^{\text{mech}} + \mathbf{b}_t^{\text{pon}}\right] \mathrm{dv} + \int_{\partial\mathscr{V}_t} \left[\mathbf{t}_t^{\text{mech,ext}} + \mathbf{t}_t^{\text{pon}}\right] \mathrm{da} && \rightarrow \text{eqs. } (5.127_2, \text{C}.70)\\
&= \int_{\mathscr{V}_t} \left[\mathbf{b}_t^{\text{mech}} + \nabla\cdot\boldsymbol{\sigma}^{*\text{pon}}\right] \mathrm{dv} + \int_{\partial\mathscr{V}_t} \boldsymbol{\sigma}^{\text{mech}}\cdot\mathbf{n}^+\ \mathrm{da} && \rightarrow \text{eq. (B.16)}\\
&= \int_{\mathscr{V}_t} \mathbf{b}_t^{\text{mech}}\ \mathrm{dv} + \int_{\partial\mathscr{V}_t} \left[\boldsymbol{\sigma}^{\text{mech}} + \boldsymbol{\sigma}^{*\text{pon}}\right]\cdot\mathbf{n}^+\ \mathrm{da} \quad , && \text{(C.73)}
\end{aligned}$$

which represents the total external force acting on the control volume while accounting for a material discontinuity at $\partial\mathscr{V}_t$. For a localisation within the volume we can thus state that

$$\mathrm{D}_t \int_{\mathscr{V}_t} \rho_t \mathbf{v}(\mathbf{X},t)\ \mathrm{dv} = \int_{\mathscr{V}_t} \mathbf{b}_t^{\text{mech}}\ \mathrm{dv} + \int_{\partial\mathscr{V}_t} \left[\boldsymbol{\sigma}^{\text{mech}} + \boldsymbol{\sigma}^{*\text{pon}}\right]\cdot\mathbf{n}\ \mathrm{da} \tag{C.74}$$

and thereby identify the sources listed in Table 5.2 by comparison of the above to equation (5.54). Through application of the sources to equation (5.55), in association with the identity

$$\begin{aligned}
&\mathrm{d}_t[\rho_t\mathbf{v}] + \nabla\cdot[[\rho_t\mathbf{v}]\otimes\mathbf{v}] && \rightarrow \text{eq. (A.33)}\\
&= [\mathrm{d}_t\rho_t]\,\mathbf{v} + \rho_t\,[\mathrm{d}_t\mathbf{v}] + [\nabla\,[\rho_t\mathbf{v}]]\cdot\mathbf{v} + [\rho_t\mathbf{v}]\,[\nabla\cdot\mathbf{v}] && \rightarrow \text{eq. (A.29)}\\
&= [\mathrm{d}_t\rho_t]\,\mathbf{v} + \rho_t\,[\mathrm{d}_t\mathbf{v}] + [[\mathbf{v}\otimes\mathbf{v}]\cdot[\nabla\rho_t] + \rho_t\,[\nabla\mathbf{v}]\cdot\mathbf{v}] + [\rho_t\mathbf{v}]\,[\nabla\cdot\mathbf{v}]\\
&= [\mathrm{d}_t\rho_t]\,\mathbf{v} + [[\nabla\rho_t]\cdot\mathbf{v} + \rho_t\,[\nabla\cdot\mathbf{v}]]\,\mathbf{v} + \rho_t\,[[\mathrm{d}_t\mathbf{v}] + [\nabla\mathbf{v}]\cdot\mathbf{v}] && \rightarrow \text{eqs. (A.28,5.15)}\\
&= [\underbrace{\mathrm{d}_t\rho_t + \nabla\cdot[\rho_t\mathbf{v}]}_{0}]\mathbf{v} + \rho_t\,[\mathrm{D}_t\mathbf{v}] && \rightarrow \text{eqs. (5.119,5.13b)}\\
&= \rho_t\mathbf{a} \quad , && \text{(C.75)}
\end{aligned}$$

the localised form for the balance of linear momentum is, therefore, expressed as

$$\rho_t\mathbf{a} = \mathbf{b}_t^{\text{mech}} + \nabla\cdot\left[\boldsymbol{\sigma}^{\text{mech}} + \boldsymbol{\sigma}^{*\text{pon}}\right] \equiv \left[\mathbf{b}_t^{\text{mech}} + \mathbf{b}_t^{\text{pon}}\right] + \nabla\cdot\boldsymbol{\sigma}^{\text{mech}} \quad \forall\,\mathbf{x} \ \text{ in } \ \mathscr{V}_t \quad . \tag{C.76}$$

This corresponds to what is given in equation (5.120). The associated jump term, as derived from equation (5.56b) for an unloaded surface is

$$[\![\boldsymbol{\sigma}^{\text{mech}} + \boldsymbol{\sigma}^{*\text{pon}}]\!]\cdot\mathbf{n} = \mathbf{0} \quad \forall\,\mathbf{x} \ \text{ on } \ \mathscr{A}_t \quad . \tag{C.77}$$

Further consideration of the influence of external mechanical tractions renders the expression provided in equation (5.126) for the for the solid body-free space interface.

Balance of angular momentum for polar media [523, 133, 204]

The principle of conservation of angular momentum states that the time rate of change of rotational momentum of a control volume $\mathscr{V}_t \in \mathscr{B}_t$

$$\mathbf{m}_t^{\text{tot}} = \mathrm{D}_t \int_{\mathscr{V}_t} \mathbf{r} \times \rho_t \mathbf{v}(\mathbf{X}, t) \ \mathrm{dv} \tag{C.78}$$

is due to the influence of the external sources

$$\begin{aligned}
\mathbf{m}_t^{\text{mech}} + \mathbf{m}_t^{\text{pon}} &= \int_{\mathscr{V}_t} \left[\mathbf{r} \times \left[\mathbf{b}_t^{\text{mech}} + \mathbf{b}_t^{\text{pon}}\right] + \mathbf{m}_t^{\text{mag}}\right] \mathrm{dv} \\
&+ \int_{\partial\mathscr{V}_t} \left[\mathbf{r} \times \left[\mathbf{t}_t^{\text{mech,ext}} + \mathbf{t}_t^{\text{pon}}\right] + \widehat{\mathbf{m}}_t^{\text{mag}}\right] \mathrm{da} && \to \text{eqs. } (5.127_2, \text{C.70}) \\
&= \int_{\mathscr{V}_t} \left[\mathbf{r} \times \left[\mathbf{b}_t^{\text{mech}} + \nabla \cdot \boldsymbol{\sigma}^{*\text{pon}}\right] + \mathbf{m}_t^{\text{mag}}\right] \mathrm{dv} \\
&+ \int_{\partial\mathscr{V}_t} \left[\mathbf{r} \times \left[\boldsymbol{\sigma}^{\text{mech}} \cdot \mathbf{n}^+\right] + \widehat{\mathbf{m}}_t^{\text{mag}}\right] \mathrm{da} && \to \text{eq. (A.38)} \\
&= \int_{\mathscr{V}_t} \left[\mathbf{r} \times \mathbf{b}_t^{\text{mech}} + \nabla \cdot \left[\mathbf{r} \times \boldsymbol{\sigma}^{*\text{pon}}\right] - \boldsymbol{\epsilon} : \left[[\nabla \mathbf{r}] \cdot \left[\boldsymbol{\sigma}^{*\text{pon}}\right]^T\right] + \mathbf{m}_t^{\text{mag}}\right] \mathrm{dv} \\
&+ \int_{\partial\mathscr{V}_t} \left[\mathbf{r} \times \left[\boldsymbol{\sigma}^{\text{mech}} \cdot \mathbf{n}^+\right] + \widehat{\mathbf{m}}_t^{\text{mag}}\right] \mathrm{da} && \to \text{eq. (B.16)} \\
&= \int_{\mathscr{V}_t} \left[\mathbf{r} \times \mathbf{b}_t^{\text{mech}} - \boldsymbol{\epsilon} : \left[\boldsymbol{\sigma}^{*\text{pon}}\right]^T + \mathbf{m}_t^{\text{mag}}\right] \mathrm{dv} \\
&+ \int_{\partial\mathscr{V}_t} \left[\mathbf{r} \times \left[\left[\boldsymbol{\sigma}^{\text{mech}} + \boldsymbol{\sigma}^{*\text{pon}}\right] \cdot \mathbf{n}^+\right] + \widehat{\mathbf{m}}_t^{\text{mag}}\right] \mathrm{da} \quad .
\end{aligned} \tag{C.79}$$

Here, we have denoted the arbitrary position vector $\mathbf{r}(\mathbf{x}, t) = \mathbf{x}(\mathbf{X}, t) - \mathbf{x}_0$ where $\mathbf{x}_0$ is the origin around which the moment is measured; it may therefore be observed that $\nabla \mathbf{r} = \nabla \mathbf{x} = \mathbf{i}$. The vectorial quantities $\mathbf{m}_t^{\text{mag}}$ and $\widehat{\mathbf{m}}_t^{\text{mag}}$ are, respectively, the extra body and traction couples that arise due to magnetisation.

For a localisation inside the body, away from surface discontinuities and, therefore, where $\widehat{\mathbf{m}}_t^{\text{mag}} = \mathbf{0}$, we can then state that

$$\begin{aligned}
\mathrm{D}_t \int_{\mathscr{V}_t} \mathbf{r} \times \rho_t \mathbf{v}(\mathbf{X}, t) \ \mathrm{dv} = &\int_{\mathscr{V}_t} \left[\mathbf{r} \times \mathbf{b}_t^{\text{mech}} - \boldsymbol{\epsilon} : \left[\boldsymbol{\sigma}^{*\text{pon}}\right]^T + \mathbf{m}_t^{\text{mag}}\right] \mathrm{dv} \\
&+ \int_{\partial\mathscr{V}_t} \left[\mathbf{r} \times \left[\boldsymbol{\sigma}^{\text{mech}} + \boldsymbol{\sigma}^{*\text{pon}}\right]\right] \cdot \mathbf{n} \ \mathrm{da} \quad .
\end{aligned} \tag{C.80}$$

This leads to the identification of the sources for equation (5.54) as those listed in Table 5.2. For expansion of equation (5.55), we first observe that

$$
\begin{aligned}
&\mathrm{d}_t[\mathbf{r}\times\rho_t\mathbf{v}] + \nabla\cdot[[\mathbf{r}\times\rho_t\mathbf{v}]\otimes\mathbf{v}] && \rightarrow \text{eq. (A.10)}\\
&= \mathrm{d}_t[\rho_t\,[\mathbf{r}\times\mathbf{v}]] + \nabla\cdot[\rho_t\,[\mathbf{r}\times\mathbf{v}]\otimes\mathbf{v}] && \rightarrow \text{eq. (A.34)}\\
&= [\mathrm{d}_t\rho_t]\,[\mathbf{r}\times\mathbf{v}] + \rho_t\,[\mathrm{d}_t[\mathbf{r}\times\mathbf{v}]] + [[\mathbf{r}\times\mathbf{v}]\otimes\mathbf{v}]\cdot[\nabla\rho_t] + \rho_t\nabla\cdot[[\mathbf{r}\times\mathbf{v}]\otimes\mathbf{v}] && \rightarrow \text{eqs. (A.33)}\\
&= [\mathrm{d}_t\rho_t]\,[\mathbf{r}\times\mathbf{v}] + \rho_t\,[\mathrm{d}_t[\mathbf{r}\times\mathbf{v}]] + [\mathbf{r}\times\mathbf{v}]\,[[\nabla\rho_t]\cdot\mathbf{v}]\\
&+\rho_t\,[\nabla\,[\mathbf{r}\times\mathbf{v}]]\cdot\mathbf{v} + \rho_t\,[\mathbf{r}\times\mathbf{v}]\,[\nabla\cdot\mathbf{v}] && \rightarrow \text{eq. (A.32)}\\
&= [\mathbf{r}\times\mathbf{v}]\,\underbrace{\Big[\mathrm{d}_t\rho_t + [\nabla\rho_t]\cdot\mathbf{v} + \rho_t\,[\nabla\cdot\mathbf{v}]\Big]}_{0} && \leftarrow \text{eq. (5.119)}\\
&+\rho_t\,[\mathrm{d}_t[\mathbf{r}\times\mathbf{v}] + [\nabla\,[\mathbf{r}\times\mathbf{v}]]\cdot\mathbf{v}] && \rightarrow \text{eq. (5.15)}\\
&= \rho_t\mathrm{D}_t[\mathbf{r}\times\mathbf{v}] \equiv \mathbf{r}\times\rho_t\mathbf{a} && \text{(C.81)}
\end{aligned}
$$

noting that $\mathrm{D}_t\mathbf{r} = \mathrm{D}_t\mathbf{x} = \mathbf{v}$ and $\mathbf{v}\times\mathbf{v} = \mathbf{0}$. After substitution of the above and the source definitions, equation (5.55) becomes

$$
\begin{aligned}
\mathbf{r}\times\rho_t\mathbf{a} &= \mathbf{r}\times\mathbf{b}_t^{\mathrm{mech}} - \boldsymbol{\epsilon} : \left[\boldsymbol{\sigma}^{*\mathrm{pon}}\right]^T + \mathbf{m}_t^{\mathrm{mag}} + \nabla\cdot\left[\mathbf{r}\times\left[\boldsymbol{\sigma}^{\mathrm{mech}} + \boldsymbol{\sigma}^{*\mathrm{pon}}\right]\right] && \rightarrow \text{eq. (A.38)}\\
&= \mathbf{r}\times\mathbf{b}_t^{\mathrm{mech}} - \boldsymbol{\epsilon} : \left[[\nabla\mathbf{r}]\cdot\boldsymbol{\sigma}^{*\mathrm{pon}}\right]^T + \mathbf{m}_t^{\mathrm{mag}}\\
&+\boldsymbol{\epsilon} : \left[[\nabla\mathbf{r}]\cdot\left[\boldsymbol{\sigma}^{\mathrm{mech}} + \boldsymbol{\sigma}^{*\mathrm{pon}}\right]^T\right] + \mathbf{r}\times\left[\nabla\cdot\left[\boldsymbol{\sigma}^{\mathrm{mech}} + \boldsymbol{\sigma}^{*\mathrm{pon}}\right]\right]\\
&= \mathbf{r}\times\mathbf{b}_t^{\mathrm{mech}} + \mathbf{m}_t^{\mathrm{mag}} + \boldsymbol{\epsilon} : \left[\mathbf{i}\cdot\left[\boldsymbol{\sigma}^{\mathrm{mech}}\right]^T\right] + \mathbf{r}\times\left[\nabla\cdot\left[\boldsymbol{\sigma}^{\mathrm{mech}} + \boldsymbol{\sigma}^{*\mathrm{pon}}\right]\right]
\end{aligned}
$$

that, with some rearrangement of terms, renders

$$
\mathbf{r}\times\underbrace{\left[\rho_t\mathbf{a} - \mathbf{b}_t^{\mathrm{mech}} - \nabla\cdot\left[\boldsymbol{\sigma}^{\mathrm{mech}} + \boldsymbol{\sigma}^{*\mathrm{pon}}\right]\right]}_{\mathbf{0}} = \boldsymbol{\epsilon} : \left[\boldsymbol{\sigma}^{\mathrm{mech}}\right]^T + \mathbf{m}_t^{\mathrm{mag}}
$$

$$\leftarrow \text{eqs. (5.120,5.130)} \rightarrow \text{eq. (A.12)}$$

$$\mathbf{0} =: \boldsymbol{\epsilon} : \left[\boldsymbol{\sigma}^{\mathrm{tot}}\right]^T \quad \forall\,\mathbf{x}\ \text{ in }\ \mathscr{V}_t\ . \qquad \text{(C.82)}$$

As a consequence of this, then from equation (5.121),

$$
\begin{aligned}
\mathbf{0} &= \boldsymbol{\epsilon} : \left[\boldsymbol{\sigma}^{\mathrm{tot}}\right]^T && \leftarrow \text{eq. (5.121)} \rightarrow \text{eq. (5.130)}\\
&= \boldsymbol{\epsilon} : \left[\boldsymbol{\sigma}^{\mathrm{mech}} + \boldsymbol{\sigma}^{*\mathrm{pon}}\right]^T && \rightarrow \text{eq. (5.131a)}\\
&= \boldsymbol{\epsilon} : \left[\boldsymbol{\sigma}^{\mathrm{mech}} + \frac{1}{2\mu_0}\,[\mathbb{b}\cdot\mathbb{b}]\,\mathbf{i} - [\mathbb{h}\cdot\mathbb{b}]\,\mathbf{i} + \mathbb{h}\otimes\mathbb{b}\right]^T && \rightarrow \text{eq. (A.9)}\\
&= \boldsymbol{\epsilon} : \left[\boldsymbol{\sigma}^{\mathrm{mech}} + \mathbb{h}\otimes\mathbb{b}\right]^T && \rightarrow \text{eq. (5.118)}\\
&= \boldsymbol{\epsilon} : \left[\boldsymbol{\sigma}^{\mathrm{mech}} + \left[\frac{1}{\mu_0}\mathbb{b} - \mathbb{m}\right]\otimes\mathbb{b}\right]^T && \rightarrow \text{eq. (A.9)}
\end{aligned}
$$

$$
\begin{aligned}
&= \boldsymbol{\epsilon} : \left[\boldsymbol{\sigma}^{\text{mech}} - \mathbb{m} \otimes \mathbb{b}\right]^T \\
&= \boldsymbol{\epsilon} : \left[\boldsymbol{\sigma}^{\text{mech}}\right]^T - \boldsymbol{\epsilon} : [\mathbb{b} \otimes \mathbb{m}] && \rightarrow \text{eq. (A.8)} \\
&= \boldsymbol{\epsilon} : \left[\boldsymbol{\sigma}^{\text{mech}}\right]^T + \underbrace{\mathbb{m} \times \mathbb{b}}_{\mathbf{m}_t^{\text{mag}}} && \text{(C.83)}
\end{aligned}
$$

from which the definition of the magnetic moment as given in equation (5.129) is identified. This fortified relationship, namely $\boldsymbol{\epsilon} : \left[\boldsymbol{\sigma}^{*\text{pon}}\right]^T = \mathbf{m}_t^{\text{mag}}$, can then be used to simplify equation (C.80) and, therefore, the sources listed in Table 5.2.

C.5.3 Derivation of additional balance and jump identities for potentials

Magnetostatics

For the magnetostatic case, we recognise the similarities between equation (5.94) and the static form of Faraday's law given in equation (5.84). We can therefore adopt the same approach as was used to derive the jump condition for the compatibility equation and can then immediately deduce that

$$
\mathbf{0} = -\int_{\mathscr{I}_0} \mathbf{N} \times [\![\mathbb{H}]\!] \, dA \quad . \tag{C.84}
$$

This is the equivalent of the result given in equation $(5.85)_1$. Equation (5.82), the case where free currents are not negligible, requires the further consideration of equation (C.24).

We can also apply the same approach to Gauss' magnetism law. Using equation (B.41), equation $(5.67)_2$ as viewed across a material interface becomes

$$
\begin{aligned}
\int_{\mathscr{V}_0^+} \nabla_0 \cdot \mathbb{B} \, dV &= \int_{\partial\mathscr{V}_0^+ \setminus \mathscr{I}_t} \mathbf{N}^+ \cdot \mathbb{B} \, dA - \int_{\mathscr{I}_0} \mathbf{N}_{\mathscr{I}} \cdot \mathbb{B}^+ \, dA \quad \text{and} \\
\int_{\mathscr{V}_0^-} \nabla_0 \cdot \mathbb{B} \, dV &= \int_{\partial\mathscr{V}_0^- \setminus \mathscr{I}_t} \mathbf{N}^- \cdot \mathbb{B} \, dA + \int_{\mathscr{I}_0} \mathbf{N}_{\mathscr{I}} \cdot \mathbb{B}^- \, dA \quad .
\end{aligned}
$$

Combining the above returns the result that

$$
\int_{\mathscr{V}_0} \nabla_0 \cdot \mathbb{B} \, dV = \int_{\partial\mathscr{V}_0} \mathbf{N} \cdot \mathbb{B} \, dA - \int_{\mathscr{I}_0} \mathbf{N}_{\mathscr{I}} \cdot [\![\mathbb{B}]\!] \, dA \quad , \tag{C.85}
$$

and subsequent localisation onto a point on the interface renders the jump condition specified in equation $(5.85)_2$,

$$
0 = -\int_{\mathscr{I}_0} \mathbf{N} \cdot [\![\mathbb{B}]\!] \, dA \quad . \tag{C.86}
$$

Magnetic vector potential

The continuity condition for the magnetic vector potential is derived from equation (C.86) using equation (5.141). We conclude the result stated in equation (5.142) from

$$\begin{aligned} 0 &= -\int_{\mathscr{I}_0} \mathbf{N} \cdot [\![\nabla_0 \times \mathbb{A}]\!] \; \mathrm{dA} && \leftarrow \text{eqs. (C.86,5.141)} \\ &= -\int_{\mathscr{I}_0} \mathbf{N} \cdot [\nabla_0 \times [\![\mathbb{A}]\!]] \; \mathrm{dA} && \rightarrow \text{eq. (A.31)} \\ &= \int_{\mathscr{I}_0} \nabla_0 \cdot \underbrace{[\mathbf{N} \times [\![\mathbb{A}]\!]]}_{\doteq \mathbf{0}} \; \mathrm{dA} \quad , && \text{(C.87)} \end{aligned}$$

while noting that $\nabla_0 \times \mathbf{N} = \mathbf{0}$.

Magnetic scalar potential

Derivation of the continuity condition for the magnetic scalar potential requires only the substitution of equation (5.146) into equation (C.84). From this, we observe that

$$\begin{aligned} \mathbf{0} &= -\int_{\mathscr{I}_0} \mathbf{N} \times [\![-\nabla_0 \Phi]\!] \; \mathrm{dA} && \leftarrow \text{eq. (C.84)} \rightarrow \text{eq. (5.146)} \\ &= \int_{\mathscr{I}_0} \mathbf{N} \times \underbrace{[\nabla_0 [\![\Phi]\!]]}_{\doteq \mathbf{0}} \; \mathrm{dA} && \text{(C.88)} \end{aligned}$$

which is the equivalent statement to that given in equation $(5.147)_1$.

C.5.4 Transformation of conservation laws to their referential description

As the governing equations are primarily derived in a fully Eulerian setting, what remains is to transform them into a Lagrangian setting using referential variables; this is done in the sections below for the nominal time derivatives, as well as the balance laws and continuity equations governing electro-magnetomechanics.

C.5.4.1 Time derivatives

In the following sections, we validate the expressions for the material nominal time derivatives deduced in Appendix B.2 from Reynolds transport theorem in control volumes and on control surfaces.

Material nominal time derivative: Scalar densities

We define the general spatial scalar quantity Φ_t which is related to its material counterpart through the relation $\Phi_t = J^{-1}\Phi_0$. The material time derivative of the material scalar quantity is then

$$\begin{aligned}\dot{\Phi}_0 &= \mathrm{D}_t[J\Phi_t] = \dot{J}\Phi_t + J\dot{\Phi}_t && \rightarrow \text{eqs. (A.51,5.15)} \\ &= J\Phi_t\,[\nabla \cdot \mathbf{v}] + J\,[\mathrm{d}_t\Phi_t + [\nabla\Phi_t] \cdot \mathbf{v}] \quad . && \text{(C.89)}\end{aligned}$$

Rearranging equation (C.89), we now have

$$\begin{aligned}J\mathrm{d}_t[\Phi_t] &= \dot{\Phi}_0 - J\,[\Phi_t\,[\nabla \cdot \mathbf{v}] + [\nabla\Phi_t] \cdot \mathbf{v}] && \rightarrow \text{eq. (A.28)} \\ &= \dot{\Phi}_0 - J\nabla \cdot [\Phi_t \mathbf{v}] && \rightarrow \text{eq. (B.43)} \\ &= \dot{\Phi}_0 - \nabla_0 \cdot \left[J\Phi_t \mathbf{F}^{-1} \cdot \mathbf{v}\right] && \rightarrow \text{eq. (5.17)}\end{aligned}$$

rendering the definition

$$\begin{aligned}\mathrm{N}_t\Phi_0 &= J\mathrm{d}_t[\Phi_t] = J\mathrm{d}_t\left[J^{-1}\Phi_t\right] \\ &\equiv \dot{\Phi}_0 + \nabla_0 \cdot [\Phi_0 \mathbf{V}] \quad . && \text{(C.90)}\end{aligned}$$

If we identify Φ_0 as a scalar density per unit reference volume, then the above furnishes the identical result to that given in equation (B.24a).

Material nominal time derivative: Vector densities

Similar to the case for scalar densities, we define the general spatial vector quantity $\boldsymbol{\Phi}_t$ that is transforms to its corresponding material description through the relation $\boldsymbol{\Phi}_t = J^{-1}\boldsymbol{\Phi}_0$. The material time derivative of the material vector quantity is then

$$\begin{aligned}\dot{\boldsymbol{\Phi}}_0 &= \mathrm{D}_t[J\boldsymbol{\Phi}_t] = \dot{J}\boldsymbol{\Phi}_t + J\dot{\boldsymbol{\Phi}}_t && \rightarrow \text{eqs. (A.51,5.15)} \\ &= J\boldsymbol{\Phi}_t\,[\nabla \cdot \mathbf{v}] + J\,[\mathrm{d}_t\boldsymbol{\Phi}_t + [\nabla\boldsymbol{\Phi}_t] \cdot \mathbf{v}] \quad , && \text{(C.91)}\end{aligned}$$

which may be rearranged to

$$\begin{aligned}J\left[\mathrm{d}_t\left[J^{-1}\boldsymbol{\Phi}_0\right]\right] &= \dot{\boldsymbol{\Phi}}_0 - J\,[\boldsymbol{\Phi}_t\,[\nabla \cdot \mathbf{v}] + [\nabla[\boldsymbol{\Phi}_t]] \cdot \mathbf{v}] && \rightarrow \text{eq. (A.33)} \\ &= \dot{\boldsymbol{\Phi}}_0 - J\nabla \cdot [\boldsymbol{\Phi}_t \otimes \mathbf{v}] && \rightarrow \text{eq. (B.42)} \\ &= \dot{\boldsymbol{\Phi}}_0 - \nabla \cdot \left[\left[J\mathbf{F}^{-1} \cdot \boldsymbol{\Phi}_t\right] \otimes \mathbf{v}\right] \\ &= \dot{\boldsymbol{\Phi}}_0 - \nabla \cdot \left[[J\boldsymbol{\Phi}_t] \otimes \left[\mathbf{v} \cdot \mathbf{F}^{-T}\right]\right] \quad . && \rightarrow \text{eq. (5.17)}\end{aligned}$$

The final definition of the nominal time derivative of the material vector field is, therefore,

$$\begin{aligned}\mathrm{N}_t\boldsymbol{\Phi}_0 &:= J\mathrm{d}_t\left[J^{-1}\boldsymbol{\Phi}_0\right] \\ &\equiv \dot{\boldsymbol{\Phi}}_0 + \nabla_0 \cdot [\boldsymbol{\Phi}_0 \otimes \mathbf{V}] \quad . && \text{(C.92)}\end{aligned}$$

We can observe that this is the identical result to that given in equation (B.24b) when we identify $\boldsymbol{\Phi}_0$ as a vector density per unit reference volume.

Material nominal time derivative: Flux vectors [406, 133]

To expand the material nominal time derivative for flux vectors, we first define the general contravariant spatial vector $\boldsymbol{\psi}$ which is related to its material counterpart through the Piola transformation $\boldsymbol{\psi} = \boldsymbol{\Psi} \cdot \operatorname{cof}(\mathbf{F}^{-1}) = J^{-1}\mathbf{F} \cdot \boldsymbol{\Psi}$. The material time derivative of the material vector is then

$$
\begin{aligned}
\dot{\boldsymbol{\Psi}} &= \mathrm{D}_t[\boldsymbol{\psi} \cdot \operatorname{cof}(\mathbf{F})] = \mathrm{D}_t\left[J\mathbf{F}^{-1} \cdot \boldsymbol{\psi}\right] \\
&= \dot{J}\mathbf{F}^{-1} \cdot \boldsymbol{\psi} + J\mathrm{D}_t\left[\mathbf{F}^{-1}\right] \cdot \boldsymbol{\psi} + J\mathbf{F}^{-1} \cdot \mathrm{D}_t[\boldsymbol{\psi}] && \rightarrow \text{eqs. (A.49, A.50, A.51)} \\
&= J\,[\nabla \cdot \mathbf{v}]\,\mathbf{F}^{-1} \cdot \boldsymbol{\psi} - J\mathbf{F}^{-1} \cdot \nabla\mathbf{v} \cdot \boldsymbol{\psi} + J\mathbf{F}^{-1} \cdot \mathrm{D}_t[\boldsymbol{\psi}] \\
&= J\mathbf{F}^{-1}\,[\mathrm{D}_t[\boldsymbol{\psi}] - \nabla\mathbf{v} \cdot \boldsymbol{\psi} + [\nabla \cdot \mathbf{v}]\,\boldsymbol{\psi}] \quad . && \text{(C.93)}
\end{aligned}
$$

The bracketed terms can be manipulated such that

$$
\begin{aligned}
\mathrm{D}_t[\boldsymbol{\psi}] - \nabla\mathbf{v} \cdot \boldsymbol{\psi} + [\nabla \cdot \mathbf{v}]\,\boldsymbol{\psi} &= \mathrm{d}_t\boldsymbol{\psi} + [\nabla\boldsymbol{\psi}] \cdot \mathbf{v} - \nabla\mathbf{v} \cdot \boldsymbol{\psi} + [\nabla \cdot \mathbf{v}]\,\boldsymbol{\psi} \\
& \qquad \leftarrow \text{eq. (5.15)} \rightarrow \text{eq. (A.39)} \\
&= \mathrm{d}_t\boldsymbol{\psi} - \nabla \times [\mathbf{v} \times \boldsymbol{\psi}] + [\nabla \cdot \boldsymbol{\psi}]\,\mathbf{v} \quad , && \text{(C.94)}
\end{aligned}
$$

with which we can rearrange equation (C.93) as

$$
\dot{\boldsymbol{\Psi}} = J\mathbf{F}^{-1} \cdot [\mathrm{d}_t\boldsymbol{\psi} - \nabla \times [\mathbf{v} \times \boldsymbol{\psi}] + [\nabla \cdot \boldsymbol{\psi}]\,\mathbf{v}] \qquad \leftarrow \text{eq. (C.94)}
$$

which itself can be rearranged and further manipulated

$$
\begin{aligned}
J\mathbf{F}^{-1} \cdot [\mathrm{d}_t\boldsymbol{\psi}] &= \dot{\boldsymbol{\Psi}} + J\mathbf{F}^{-1} \cdot [\nabla \times [\mathbf{v} \times \boldsymbol{\psi}] - [\nabla \cdot \boldsymbol{\psi}]\,\mathbf{v}] && \rightarrow \text{eq. (B.45)} \\
&= \dot{\boldsymbol{\Psi}} + \nabla_0 \times \left[\mathbf{F}^T \cdot [\mathbf{v} \times \boldsymbol{\psi}]\right] - [J\nabla \cdot \boldsymbol{\psi}]\,\mathbf{F}^{-1} \cdot \mathbf{v} && \rightarrow \text{eqs. (A.10,A.11,B.43)} \\
&= \dot{\boldsymbol{\Psi}} + \nabla_0 \times \left[\left[\mathbf{F}^{-1} \cdot \mathbf{v}\right] \times \left[J\mathbf{F}^{-1} \cdot \boldsymbol{\psi}\right]\right] - [\nabla_0 \cdot \boldsymbol{\Psi}]\,\mathbf{F}^{-1} \cdot \mathbf{v} && \rightarrow \text{eqs. (A.5,5.17)}
\end{aligned}
$$

finally rendering the definition

$$
\begin{aligned}
\mathrm{N}_t\boldsymbol{\Psi} &:= [\mathrm{d}_t\boldsymbol{\psi}] \cdot \operatorname{cof}(\mathbf{F}) = \left[\mathrm{d}_t\left[\boldsymbol{\Psi} \cdot \operatorname{cof}\left(\mathbf{F}^{-1}\right)\right]\right] \cdot \operatorname{cof}(\mathbf{F}) \\
&\equiv \dot{\boldsymbol{\Psi}} - \nabla_0 \times [\mathbf{V} \times \boldsymbol{\Psi}] + [\nabla_0 \cdot \boldsymbol{\Psi}]\,\mathbf{V} \quad . && \text{(C.95)}
\end{aligned}
$$

This is the identical result to that given in equation (B.27) when $\boldsymbol{\Psi}$ is interpreted as a referential flux vector.

C.5.4.2 Balance laws and fundamental relations in a volume

Pull back of electric charge conservation law

$$
\begin{aligned}
&\int_{\mathscr{V}_t} [\mathrm{d}_t\varrho_t + \nabla \cdot \mathbb{j}'] \;\mathrm{d}v && \leftarrow \text{eq. (5.103)} \rightarrow \text{eq. (C.49c)} \\
&= \int_{\mathscr{V}_t} [\mathrm{d}_t\varrho_t + \nabla \cdot [\mathbb{j} + \varrho_t\mathbf{v}]] \;\mathrm{d}v
\end{aligned}
$$

$$= \int_{\mathscr{V}_t} \Big[\underbrace{[\mathrm{d}_t \varrho_t + \boldsymbol{\nabla} \cdot [\varrho_t \mathbf{v}]]}_{=\mathrm{n}_t \varrho_t} + \boldsymbol{\nabla} \cdot \mathbb{j} \Big] \; \mathrm{dv} \qquad \rightarrow \text{eqs. (5.19a,B.43)}$$

$$= \int_{\mathscr{V}_t} \big[J^{-1} \mathrm{D}_t [J \varrho_t] + J^{-1} \boldsymbol{\nabla}_0 \cdot \mathbb{J} \big] \; \mathrm{dv} \qquad \rightarrow \text{eqs. (5.101,5.11)}$$

$$= \int_{\mathscr{V}_0} [\dot{\varrho}_0 + \boldsymbol{\nabla}_0 \cdot \mathbb{J}] \; \mathrm{dV} \tag{C.96}$$

Pull back of Maxwell's equations: Gauss' law

$$0 = \int_{\mathscr{V}_t} \big[\boldsymbol{\nabla} \cdot \mathbb{d} - \varrho_t^{\mathrm{f}} \big] \; \mathrm{dv} \qquad \leftarrow \text{eq. } (5.104_2) \rightarrow \text{eqs. (B.41,5.11)}$$

$$= \int_{\mathscr{V}_0} \big[\boldsymbol{\nabla}_0 \cdot \mathbb{D} - J \varrho_t^{\mathrm{f}} \big] \; \mathrm{dV} \qquad \rightarrow \text{eq. (5.101)}$$

$$= \int_{\mathscr{V}_0} \big[\boldsymbol{\nabla}_0 \cdot \mathbb{D} - \varrho_0^{\mathrm{f}} \big] \; \mathrm{dV} \tag{C.97}$$

Pull back of Maxwell's equations: Gauss' magnetism law

$$0 = \int_{\mathscr{V}_t} \boldsymbol{\nabla} \cdot \mathbb{b} \; \mathrm{dv} \qquad \leftarrow \text{eq. } (5.104_4) \rightarrow \text{eq. (B.41)}$$

$$= \int_{\mathscr{V}_0} \boldsymbol{\nabla}_0 \cdot \mathbb{B} \; \mathrm{dV} \tag{C.98}$$

Pull back of transient Maxwell's equations: Faraday's law

From equation $(5.104)_1$, considering both sides of the equation individually but simultaneously,

$$\int_{\mathscr{A}_t} [\boldsymbol{\nabla} \times \mathbb{e}'] \cdot \; \mathrm{d}\mathbf{a} = - \int_{\mathscr{A}_t} \mathrm{d}_t \mathbb{b} \cdot \; \mathrm{d}\mathbf{a} \qquad \rightarrow \text{eqs. (5.10,B.45)}$$

$$\int_{\mathscr{A}_0} [\boldsymbol{\nabla}_0 \times \mathbb{E}'] \cdot \; \mathrm{d}\mathbf{A} = - \int_{\mathscr{A}_0} \underbrace{[\mathrm{d}_t \mathbb{b} \cdot \mathrm{cof}\,(\mathbf{F})]}_{=\mathrm{N}_t \mathbb{B}} \cdot \mathrm{d}\mathbf{A} \qquad \rightarrow \text{tbl. (5.5), eqs. } (5.102_3, 5.20\text{a})$$

$$\int_{\mathscr{A}_0} [\boldsymbol{\nabla}_0 \times [\mathbb{E} - \mathbb{B} \times \mathbf{V}]] \cdot \; \mathrm{d}\mathbf{A} = - \int_{\mathscr{A}_0} \Big[\mathrm{D}_t \mathbb{B} + \boldsymbol{\nabla}_0 \times [\mathbb{B} \times \mathbf{V}] + \underbrace{[\boldsymbol{\nabla}_0 \cdot \mathbb{B}]}_{0} \mathbf{V} \Big] \cdot \; \mathrm{d}\mathbf{A}$$

$$\rightarrow \text{eqs. } (5.67_2, \text{A.1})$$

$$\Longrightarrow \quad \int_{\mathscr{A}_0} [\boldsymbol{\nabla}_0 \times \mathbb{E}] \cdot \; \mathrm{d}\mathbf{A} = - \int_{\mathscr{A}_0} \dot{\mathbb{B}} \cdot \; \mathrm{d}\mathbf{A} \quad . \tag{C.99}$$

From the nature of quantities for which equation (B.45) holds, it may be identified from the left-hand side of the above equations that

$$\mathbb{e} = \mathbf{F}^{-T} \cdot \mathbb{E} \quad . \tag{C.100}$$

Pull back of transient Maxwell's equations: Ampére's law

Similar to the above, from equation $(5.105)_1$

$$\int_{\mathscr{A}_t} [\nabla \times \mathbb{h}'] \cdot \, \mathbf{da} = \int_{\mathscr{A}_t} \left[\mathbb{j}^{f\prime} + \mathrm{d}_t \mathbb{d}\right] \cdot \, \mathbf{da} \qquad \rightarrow \text{eqs. (5.10,B.45)}$$

$$\int_{\mathscr{A}_0} [\nabla_0 \times \mathbb{H}'] \cdot \, \mathbf{dA} = \int_{\mathscr{A}_0} \left[\mathbb{j}^{f\prime} \cdot \operatorname{cof}(\mathbf{F}) + \underbrace{\mathrm{d}_t \mathbb{d} \cdot \operatorname{cof}(\mathbf{F})}_{=\mathrm{N}_t \mathbb{D}} \right] \cdot \, \mathbf{dA}$$

$$\rightarrow \text{tbl. (5.5), eqs. } (5.102_4, 5.20\text{a})$$

$$\int_{\mathscr{A}_0} [\nabla_0 \times [\mathbb{H} + \mathbb{D} \times \mathbf{V}]] \cdot \, \mathbf{dA} = \int_{\mathscr{A}_0} \left[\mathbb{J}^{f\prime} + \mathrm{D}_t \mathbb{D} + [\nabla_0 \cdot \mathbb{D}] \, \mathbf{V} + \nabla_0 \times [\mathbb{D} \times \mathbf{V}]\right] \cdot \, \mathbf{dA}$$

$$\int_{\mathscr{A}_0} [\nabla_0 \times \mathbb{H}] \cdot \, \mathbf{dA} = \int_{\mathscr{A}_0} \left[\left[\mathbb{J}^{f} - \varrho_0^{f} \mathbf{V}\right] + \mathrm{D}_t \mathbb{D} + \underbrace{[\nabla_0 \cdot \mathbb{D}]}_{=\varrho_0^{f}} \mathbf{V} \right] \cdot \, \mathbf{dA}$$

$$\rightarrow \text{tbl. (5.5), eq. } (5.75_1)$$

$$\int_{\mathscr{A}_0} [\nabla_0 \times \mathbb{H}] \cdot \, \mathbf{dA} = \int_{\mathscr{A}_0} \left[\mathbb{J}^{f} + \dot{\mathbb{D}}\right] \cdot \, \mathbf{dA} \quad . \tag{C.101}$$

As we have identified the relationship

$$\int_{\mathscr{A}_t} \mathrm{d}_t \mathbb{d} \cdot \, \mathbf{da} := \int_{\mathscr{A}_0} \mathrm{N}_t \mathbb{D} \, \mathbf{dA} \quad , \tag{C.102}$$

it follows from the nature of quantities for which equation (C.95) holds that

$$\mathbb{d} = \mathbb{D} \cdot \operatorname{cof}\left(\mathbf{F}^{-1}\right) = J^{-1} \mathbf{F} \cdot \mathbb{D} \quad . \tag{C.103}$$

This is in agreement with the assumptions leading to equation (C.97).

Referential balance of angular momentum

$$\mathbf{0} = \boldsymbol{\epsilon} : \left[J^{-1} \mathbf{P}^{\text{tot}} \cdot \mathbf{F}^T\right]^T \qquad \leftarrow \text{eq. (5.93)} \rightarrow \text{eq. (5.90)}$$

$$= \boldsymbol{\epsilon} : \left[J^{-1} \mathbf{P}^{\text{mech}} \cdot \mathbf{F}^T + J^{-1} \mathbf{P}^{*\text{pon}} \cdot \mathbf{F}^T\right]^T \qquad \rightarrow \text{eq. (5.91a)}$$

$$= \boldsymbol{\epsilon} : \left[J^{-1} \mathbf{P}^{\text{mech}} \cdot \mathbf{F}^T + J^{-1} \left[\left[\frac{J^{-1}}{2\mu_0} \mathbf{C} : [\mathbb{B} \otimes \mathbb{B}]\right] \mathbf{F}^{-T} - [\mathbb{H} \cdot \mathbb{B}] \, \mathbf{F}^{-T} + \left[\mathbf{F}^{-T} \cdot \mathbb{H}\right] \otimes \mathbb{B}\right] \cdot \mathbf{F}^T\right]^T$$

$$= \boldsymbol{\epsilon} : \left[J^{-1}\mathbf{P}^{\text{mech}} \cdot \mathbf{F}^T + J^{-1} \left[\left[\frac{J^{-1}}{2\mu_0}\mathbf{C} : [\mathbb{B} \otimes \mathbb{B}] \right] \mathbf{I} - [\mathbb{H} \cdot \mathbb{B}]\, \mathbf{I} + \left[\mathbf{F}^{-T} \cdot \mathbb{H}\right] \otimes [\mathbf{F} \cdot \mathbb{B}] \right] \right]^T \quad \rightarrow \text{eq. (A.9)}$$

$$= \boldsymbol{\epsilon} : \left[J^{-1}\mathbf{P}^{\text{mech}} \cdot \mathbf{F}^T + \left[\mathbf{F}^{-T} \cdot \mathbb{H}\right] \otimes \left[J^{-1}\mathbf{F} \cdot \mathbb{B}\right] \right]^T \quad \rightarrow \text{eq. (5.74)}$$

$$= \boldsymbol{\epsilon} : \left[J^{-1}\mathbf{P}^{\text{mech}} \cdot \mathbf{F}^T + \left[\mathbf{F}^{-T} \cdot \left[\frac{1}{\mu_0 J}\mathbf{C} \cdot \mathbb{B} - \mathbb{M} \right] \right] \otimes \left[J^{-1}\mathbf{F} \cdot \mathbb{B}\right] \right]^T$$

$$= \boldsymbol{\epsilon} : \left[J^{-1}\mathbf{P}^{\text{mech}} \cdot \mathbf{F}^T + \left[\frac{1}{\mu_0 J}\mathbf{F} \cdot \mathbb{B} \right] \otimes \left[J^{-1}\mathbf{F} \cdot \mathbb{B}\right] - \left[\mathbf{F}^{-T} \cdot \mathbb{M}\right] \otimes \left[J^{-1}\mathbf{F} \cdot \mathbb{B}\right] \right]^T \quad \rightarrow \text{eq. (A.9)}$$

$$= \boldsymbol{\epsilon} : \left[J^{-1}\mathbf{P}^{\text{mech}} \cdot \mathbf{F}^T - \left[\mathbf{F}^{-T} \cdot \mathbb{M}\right] \otimes \left[J^{-1}\mathbf{F} \cdot \mathbb{B}\right] \right]^T$$

$$= \boldsymbol{\epsilon} : \left[J^{-1}\mathbf{P}^{\text{mech}} \cdot \mathbf{F}^T \right]^T - \boldsymbol{\epsilon} : \left[\left[J^{-1}\mathbf{F} \cdot \mathbb{B}\right] \otimes \left[\mathbf{F}^{-T} \cdot \mathbb{M}\right] \right] \quad \rightarrow \text{eq. (A.8)}$$

$$= \boldsymbol{\epsilon} : \left[J^{-1}\mathbf{P}^{\text{mech}} \cdot \mathbf{F}^T \right]^T + \left[\mathbf{F}^{-T} \cdot \mathbb{M}\right] \times \left[J^{-1}\mathbf{F} \cdot \mathbb{B}\right] \tag{C.104}$$

C.5.4.3 Jump conditions and fundamental relations on a surface

Jump condition associated with Faraday's law

$$\mathbf{0} = \int_{\mathscr{A}_t} [\mathbf{n} \times [\![\mathbb{e}']\!] - [\![\mathbb{b}]\!]\,[\mathbf{v} \cdot \mathbf{n}]] \;\mathrm{da} \quad \leftarrow \text{eq. } (5.109_1) \rightarrow \text{tbl. (5.5)}$$

$$= \int_{\mathscr{A}_t} [\mathbf{n} \times [\![\mathbb{e} + \mathbb{b} \times \mathbf{v}]\!] - [\![\mathbb{b}]\!]\,[\mathbf{v} \cdot \mathbf{n}]] \;\mathrm{da}$$

$$= \int_{\mathscr{A}_t} [\mathbf{n} \times [\![\mathbb{e}]\!] + \mathbf{n} \times [\![\mathbb{b}]\!] \times \mathbf{v} - [\![\mathbb{b}]\!]\,[\mathbf{v} \cdot \mathbf{n}]] \;\mathrm{da} \quad \rightarrow \text{eq. (A.14)}$$

$$= \int_{\mathscr{A}_t} \left[\mathbf{n} \times [\![\mathbb{e}]\!] + [\![\mathbb{b}]\!]\,[\mathbf{n} \cdot \mathbf{v}] - \mathbf{v}\underbrace{[[\![\mathbb{b}]\!] \cdot \mathbf{n}]}_{\to 0} - [\![\mathbb{b}]\!]\,[\mathbf{v} \cdot \mathbf{n}]\right] \;\mathrm{da}$$

$$\leftarrow \text{eq. (C.20)} \rightarrow \text{eqs. (C.11,C.100)}$$

$$= \int_{\mathscr{A}_t} \mathbf{n} \times [\![\mathbb{e}]\!] \;\mathrm{da} = \int_{\mathscr{A}_t} \left[J\widehat{J}^{-1}\mathbf{F}^{-T} \cdot \mathbf{N}\right] \times \left[\!\left[\mathbf{F}^{-T} \cdot \mathbb{E}\right]\!\right] \;\mathrm{da} \quad \rightarrow \text{eq. (A.11)}$$

$$= \int_{\mathscr{A}_t} J \det\left(\mathbf{F}^{-1}\right) \mathbf{F} \cdot [\mathbf{N} \times [\![\mathbb{E}]\!]]\,\widehat{J}^{-1} \;\mathrm{da} \quad \rightarrow \text{eq. (C.11)}$$

$$= \int_{\mathscr{A}_0} \mathbf{F} \cdot \underbrace{[\mathbf{N} \times [\![\mathbb{E}]\!]]}_{\doteq \mathbf{0}} \;\mathrm{dA} \tag{C.105}$$

Jump condition associated with Gauss' law

$$0 = \int_{\mathscr{A}_t} \left[\mathbf{n} \cdot [\![\mathbb{d}]\!] - \widehat{\varrho}^{\mathrm{f}}_t\right] \;\mathrm{da} \quad \leftarrow \text{eq. } (5.109_2)$$

$$= \int_{\mathcal{A}_t} [\![\mathbb{d}]\!] \cdot \, \mathrm{d}\mathbf{a} - \int_{\mathcal{A}_t} \hat{\varrho}_t^{\mathrm{f}} \cdot \, \mathrm{da} \qquad \rightarrow \text{eqs. (5.10,C.13)}$$

$$= \int_{\mathcal{A}_0} [\![\mathbb{d}]\!] \cdot \mathrm{cof}(\mathbf{F}) \cdot \, \mathrm{d}\mathbf{A} - \int_{\mathcal{A}_0} \hat{\varrho}_0^{\mathrm{f}} \cdot \, \mathrm{dA} \qquad \rightarrow \text{eq. (C.103)}$$

$$= \int_{\mathcal{A}_0} \left[\mathbf{N} \cdot [\![\mathbb{D}]\!] - \hat{\varrho}_0^{\mathrm{f}}\right] \mathrm{dA} \qquad \text{(C.106)}$$

Jump condition associated with Ampére's law

$$\mathbf{0} = \int_{\mathcal{A}_t} \left[\mathbf{n} \times [\![\mathbb{h}']\!] - \hat{\mathbb{j}}^{\mathrm{f}\prime} + [\![\mathbb{d}]\!]\,[\mathbf{v} \cdot \mathbf{n}]\right] \mathrm{da} \qquad \leftarrow \text{eq. } (5.110_1) \rightarrow \text{tbl. (5.6)}$$

$$= \int_{\mathcal{A}_t} \left[\mathbf{n} \times [\![\mathbb{h} - \mathbb{d} \times \mathbf{v}]\!] - \left[\hat{\mathbb{j}}^{\mathrm{f}} + \hat{\varrho}_t^{\mathrm{f}}\mathbf{v}\right] + [\![\mathbb{d}]\!]\,[\mathbf{v} \cdot \mathbf{n}]\right] \mathrm{da}$$

$$= \int_{\mathcal{A}_t} \left[\mathbf{n} \times [\![\mathbb{h}]\!] - \mathbf{n} \times [\![\mathbb{d}]\!] \times \mathbf{v} - \left[\hat{\mathbb{j}}^{\mathrm{f}} + \hat{\varrho}_t^{\mathrm{f}}\mathbf{v}\right] + [\![\mathbb{d}]\!]\,[\mathbf{v} \cdot \mathbf{n}]\right] \mathrm{da} \qquad \rightarrow \text{eqs. (A.14,5.109)}$$

$$= \int_{\mathcal{A}_t} \left[\mathbf{n} \times [\![\mathbb{h}]\!] - [\![\mathbb{d}]\!]\,[\mathbf{n} \cdot \mathbf{v}] + \mathbf{v}\,[[\![\mathbb{d}]\!] \cdot \mathbf{n}] - \left[\hat{\mathbb{j}}^{\mathrm{f}} + [\mathbf{n} \cdot [\![\mathbb{d}]\!]]\,\mathbf{v}\right] + [\![\mathbb{d}]\!]\,[\mathbf{v} \cdot \mathbf{n}]\right] \mathrm{da}$$

$$= \int_{\mathcal{A}_t} \left[\mathbf{n} \times [\![\mathbb{h}]\!] - \hat{\mathbb{j}}^{\mathrm{f}}\right] \mathrm{da} \qquad \rightarrow \text{eqs. } (\text{C.11}, 5.102_1, \text{C.14})$$

$$= \int_{\mathcal{A}_t} \left[\left[J\hat{J}^{-1}\mathbf{F}^{-T} \cdot \mathbf{N}\right] \times \left[\!\left[\mathbf{F}^{-T} \cdot \mathbb{H}\right]\!\right] - \hat{J}^{-1}\mathbf{F} \cdot \hat{\mathbb{J}}^{\mathrm{f}}\right] \mathrm{da} \qquad \rightarrow \text{eq. (A.11)}$$

$$= \int_{\mathcal{A}_t} \mathbf{F} \cdot \left[J \det\left(\mathbf{F}^{-1}\right)[\mathbf{N} \times [\![\mathbb{H}]\!]] - \hat{\mathbb{J}}^{\mathrm{f}}\right] \hat{J}^{-1} \, \mathrm{da} \qquad \rightarrow \text{eq. (C.11)}$$

$$= \int_{\mathcal{A}_0} \mathbf{F} \cdot \underbrace{\left[\mathbf{N} \times [\![\mathbb{H}]\!] - \hat{\mathbb{J}}^{\mathrm{f}}\right]}_{\doteq \mathbf{0}} \mathrm{dA} \qquad \text{(C.107)}$$

Jump condition associated with Gauss' magnetism law

$$0 = \int_{\mathcal{A}_t} \mathbf{n} \cdot [\![\mathbb{b}]\!] \, \mathrm{da} \qquad \leftarrow \text{eq. } (5.110_2) \rightarrow \text{eq. (5.10)}$$

$$= \int_{\mathcal{A}_0} [\![\mathbb{b}]\!] \cdot \mathrm{cof}(\mathbf{F}) \, \mathrm{d}\mathbf{A} = \int_{\mathcal{A}_0} [\![\mathbb{b} \cdot \mathrm{cof}(\mathbf{F})]\!] \, \mathrm{d}\mathbf{A} \qquad \rightarrow \text{eq. (5.102)}$$

$$= \int_{\mathcal{A}_0} \mathbf{N} \cdot [\![\mathbb{B}]\!] \, \mathrm{dA} \qquad \text{(C.108)}$$

Cauchy stress theorem

The referential counterpart to the Cauchy stress theorem, as given in equation (5.97), arises immediately from equation (B.41) as

$$\int_{\partial\mathscr{V}_t} \mathbf{t}_t \,\mathrm{da} = \int_{\partial\mathscr{V}_t} \boldsymbol{\sigma}\cdot\mathbf{n}\,\mathrm{da} \qquad \leftarrow \text{eq. (5.124)} \rightarrow \text{eqs. (B.41,5.97)}$$

$$= \int_{\partial\mathscr{V}_0} \mathbf{P}\cdot\mathbf{N}\,\mathrm{dA} =: \int_{\partial\mathscr{V}_0} \mathbf{t}_0\,\mathrm{dA} \quad . \qquad \text{(C.109)}$$

Pull back of traction continuity

$$\int_{\partial\mathscr{B}_t} \mathbf{t}_t^{\text{mech,ext}}\,\mathrm{da} = -\int_{\partial\mathscr{B}_t} [\![\boldsymbol{\sigma}^{\text{tot}}]\!]\cdot\mathbf{n}\,\mathrm{da} \qquad \leftarrow \text{eq. (5.126)} \rightarrow \text{eq. (B.41)}$$

$$\equiv -\int_{\partial\mathscr{B}_0} [\![\mathbf{P}^{\text{tot}}]\!]\cdot\mathbf{N}\,\mathrm{dA} = \int_{\partial\mathscr{B}_0} \mathbf{t}_0^{\text{mech,ext}}\,\mathrm{dA} \qquad \rightarrow \text{fig. (5.1)}$$

$$\Longrightarrow \quad -[\![\mathbf{P}^{\text{tot}}]\!]\cdot\mathbf{N}^+ = \mathbf{t}_0^{\text{mech,ext}} \quad \text{on} \quad \partial\mathscr{B}_0 \qquad \text{(C.110)}$$

C.5.5 Weak formulation of quasi-static balance of linear momentum

$$0 = \int_{\mathscr{B}_0\cup\mathscr{S}_0} \delta\boldsymbol{\varphi}\cdot\left[\nabla_0\cdot\mathbf{P}^{\text{tot}}\right]\mathrm{dV} + \int_{\mathscr{B}_0} \delta\boldsymbol{\varphi}\cdot\mathbf{b}_0^{\text{mech}}\,\mathrm{dV} \qquad \leftarrow \text{eq. (5.87)} \rightarrow \text{eq. (5.90)}$$

$$= \int_{\mathscr{B}_0} \delta\boldsymbol{\varphi}\cdot\left[\nabla_0\cdot\mathbf{P}^{\text{tot}}\right]\mathrm{dV} + \int_{\mathscr{S}_0} \delta\boldsymbol{\varphi}\cdot\left[\nabla_0\cdot\mathbf{P}^{*\max}\right]\mathrm{dV} + \int_{\mathscr{B}_0} \delta\boldsymbol{\varphi}\cdot\mathbf{b}_0^{\text{mech}}\,\mathrm{dV}$$

$$\rightarrow \text{eq. (B.13)}$$

$$= -\int_{\mathscr{B}_0} \nabla_0\delta\boldsymbol{\varphi} : \mathbf{P}^{\text{tot}}\,\mathrm{dV} + \int_{\mathscr{B}_0} \nabla_0\cdot\left[\delta\boldsymbol{\varphi}\cdot\mathbf{P}^{\text{tot}}\right]\mathrm{dV}$$

$$- \int_{\mathscr{S}_0} \nabla_0\delta\boldsymbol{\varphi} : \mathbf{P}^{*\max}\,\mathrm{dV} + \int_{\mathscr{S}_0} \nabla_0\cdot\left[\delta\boldsymbol{\varphi}\cdot\mathbf{P}^{*\max}\right]\mathrm{dV} + \int_{\mathscr{B}_0} \delta\boldsymbol{\varphi}\cdot\mathbf{b}_0^{\text{mech}}\,\mathrm{dV}$$

$$\Longrightarrow \quad \int_{\mathscr{B}_0} \nabla_0\delta\boldsymbol{\varphi} : \mathbf{P}^{\text{tot}}\,\mathrm{dV} + \int_{\mathscr{S}_0} \nabla_0\delta\boldsymbol{\varphi} : \mathbf{P}^{*\max}\,\mathrm{dV}$$

$$= \int_{\mathscr{B}_0} \delta\boldsymbol{\varphi}\cdot\mathbf{b}_0^{\text{mech}}\,\mathrm{dV} + \int_{\mathscr{B}_0} \nabla_0\cdot\left[\delta\boldsymbol{\varphi}\cdot\mathbf{P}^{\text{tot}}\right]\mathrm{dV} + \int_{\mathscr{S}_0} \nabla_0\cdot\left[\delta\boldsymbol{\varphi}\cdot\mathbf{P}^{*\max}\right]\mathrm{dV}$$

$$\rightarrow \text{eq. (B.16)}$$

$$= \int_{\mathscr{B}_0} \delta\boldsymbol{\varphi}\cdot\mathbf{b}_0^{\text{mech}}\,\mathrm{dV} + \int_{\partial\mathscr{B}_0} \delta\boldsymbol{\varphi}\cdot\left[\mathbf{P}^{\text{tot}}\cdot\mathbf{N}\right]\mathrm{dA} + \int_{\partial\mathscr{B}_0\cup\partial\mathscr{S}_0} \delta\boldsymbol{\varphi}\cdot\left[\mathbf{P}^{*\max}\cdot\mathbf{N}\right]\mathrm{dA}$$

$$\rightarrow \text{fig. (5.1)}$$

$$= \int_{\mathscr{B}_0} \delta\boldsymbol{\varphi} \cdot \mathbf{b}_0^{\text{mech}} \, dV + \int_{\partial\mathscr{B}_0} \delta\boldsymbol{\varphi} \cdot \left[\left[\mathbf{P}^{\text{tot}}\right]^- \cdot \mathbf{N}^+\right] dA$$
$$+ \int_{\partial\mathscr{S}_0} \delta\boldsymbol{\varphi} \cdot [\mathbf{P}^{*\max} \cdot \mathbf{N}_\infty] \, dA + \int_{\partial\mathscr{B}_0} \delta\boldsymbol{\varphi} \cdot \left[[\mathbf{P}^{*\max}]^+ \cdot \mathbf{N}^-\right] dA$$
$$= \int_{\mathscr{B}_0} \delta\boldsymbol{\varphi} \cdot \mathbf{b}_0^{\text{mech}} \, dV + \int_{\partial\mathscr{B}_0} \delta\boldsymbol{\varphi} \cdot \left[\left[\mathbf{P}^{\text{tot}}\right]^- \cdot \mathbf{N}^+\right] dA$$
$$+ \int_{\partial\mathscr{S}_0} \delta\boldsymbol{\varphi} \cdot [\mathbf{P}^{*\max} \cdot \mathbf{N}_\infty] \, dA - \int_{\partial\mathscr{B}_0} \delta\boldsymbol{\varphi} \cdot \left[[\mathbf{P}^{*\max}]^+ \cdot \mathbf{N}^+\right] dA$$

→ eqs. (5.45a,5.90)

$$= \int_{\mathscr{B}_0} \delta\boldsymbol{\varphi} \cdot \mathbf{b}_0^{\text{mech}} \, dV - \int_{\partial\mathscr{B}_0^t} \delta\boldsymbol{\varphi} \cdot [\![\mathbf{P}^{\text{tot}}]\!] \cdot \mathbf{N} \, dA + \int_{\partial\mathscr{S}_0} \delta\boldsymbol{\varphi} \cdot [\mathbf{P}^{*\max} \cdot \mathbf{N}_\infty] \, dA$$

→ eqs. (5.98,5.97)

$$= \int_{\mathscr{B}_0} \delta\boldsymbol{\varphi} \cdot \mathbf{b}_0^{\text{mech}} \, dV + \int_{\partial\mathscr{B}_0^t} \delta\boldsymbol{\varphi} \cdot \mathbf{t}_0^{\text{mech,ext}} \, dA + \int_{\partial\mathscr{S}_0} \delta\boldsymbol{\varphi} \cdot \mathbf{t}_0^{\text{mag,ext}} \, dA \qquad (C.111)$$

C.6 Legendre transformations

As was done in Chapter 7, and analogous to equation (5.163), we can in general describe a constitutive law governing the behaviour of a magneto-sensitive material (and having undergone no Legendre transformations or re-parameterisation) by

$$\Psi_0^*(\mathbf{F}, \mathbb{B}) = U_0^*(\mathbf{F}, \mathbb{B}) + M_0^*(\mathbf{F}, \mathbb{B}) \quad . \qquad (C.112)$$

Therein it is assumed that the total free energy can be expressed as an additive decomposition of energy stored in the magnetic field (M_0^*, as given by equation (5.165)) and U_0^* which represents the combined recoverable and dissipated energy stored in the material. This fundamental decomposition was stated in Table 5.9, and forms the basis on which the following derivations of the Legendre transformed energies is constructed. From this (and complementary to equation (5.90)), we can describe the relationship between the free energy function U_0^* and the stresses arising from mechanical deformation and material magnetisation, namely

$$\mathbf{P}^{\text{mech}} + \mathbf{P}^{*\text{mag}} = \frac{\partial U_0^*}{\partial \mathbf{F}} \quad , \quad \mathbf{P}^{*\max} = \frac{\partial M_0^*}{\partial \mathbf{F}} \quad . \qquad (C.113)$$

Furthermore, from thermodynamic arguments the magnetisation vector can also be related to the magnetic induction through the material free energy by [120]

$$\mathbb{M} = -\frac{\partial U_0^*}{\partial \mathbb{B}} \quad , \quad \mathbb{H} + \mathbb{M} = \frac{\partial M_0^*}{\partial \mathbb{B}} \quad . \qquad (C.114)$$

Total free energy

Starting from the total free energy, and applying a Legendre transformation in the form given by equation (C.112),

$$\begin{aligned}
\Psi_0(\mathbf{F},\mathbb{H}) &= \Psi_0^*(\mathbf{F},\mathbb{B}(\mathbb{H})) - \mathbb{H}\cdot\mathbb{B}(\mathbb{H}) && \leftarrow \text{eq. (7.2)}\rightarrow \text{eq. (C.112)}\\
&= U_0^*(\mathbf{F},\mathbb{B}(\mathbb{H})) + M_0^*(\mathbf{F},\mathbb{B}(\mathbb{H})) - \mathbb{H}\cdot\mathbb{B}(\mathbb{H}) && \rightarrow \text{eq. (5.165)}\\
&= U_0^*(\mathbf{F},\mathbb{B}(\mathbb{H})) + \frac{1}{2\mu_0}\left[\mathbb{B}(\mathbb{H})\otimes\mathbb{B}(\mathbb{H})\right]:J^{-1}\mathbf{C} - \mathbb{H}\cdot\mathbb{B}(\mathbb{H}) && \rightarrow \text{eq. (5.74)}\\
&= U_0^*(\mathbf{F},\mathbb{B}(\mathbb{H})) - \mathbb{H}\cdot\left[\mu_0 J\mathbf{C}^{-1}\cdot[\mathbb{H}+\mathbb{M}(\mathbb{H})]\right]\\
&\quad + \frac{1}{2\mu_0}\left[\left[\mu_0 J\mathbf{C}^{-1}\cdot[\mathbb{H}+\mathbb{M}(\mathbb{H})]\right]\otimes\left[\mu_0 J\mathbf{C}^{-1}\cdot[\mathbb{H}+\mathbb{M}(\mathbb{H})]\right]\right]:J^{-1}\mathbf{C}\\
&= U_0^*(\mathbf{F},\mathbb{B}(\mathbb{H})) - \mu_0\mathbb{H}\cdot\left[J\mathbf{C}^{-1}\cdot[\mathbb{H}+\mathbb{M}(\mathbb{H})]\right]\\
&\quad + \frac{\mu_0}{2}\left[[\mathbb{H}+\mathbb{M}(\mathbb{H})]\otimes\left[J\mathbf{C}^{-1}\cdot[\mathbb{H}+\mathbb{M}(\mathbb{H})]\right]\right]:\left[\mathbf{C}\cdot\mathbf{C}^{-T}\right]\\
&= U_0^*(\mathbf{F},\mathbb{B}(\mathbb{H})) - \mu_0\left[\mathbb{H}\otimes[\mathbb{H}+\mathbb{M}(\mathbb{H})]\right]:J\mathbf{C}^{-1}\\
&\quad + \frac{\mu_0}{2}\left[[\mathbb{H}+\mathbb{M}(\mathbb{H})]\otimes[\mathbb{H}+\mathbb{M}(\mathbb{H})]\right]:J\mathbf{C}^{-1} && \rightarrow \text{eq. (C.116)}\\
&= U_0^*(\mathbf{F},\mathbb{B}(\mathbb{H})) - \mu_0[\mathbb{H}\otimes\mathbb{H}]:J\mathbf{C}^{-1} - \mu_0[\mathbb{H}\otimes\mathbb{M}(\mathbb{H})]:J\mathbf{C}^{-1}\\
&\quad + \frac{\mu_0}{2}[\mathbb{H}\otimes\mathbb{H}]:J\mathbf{C}^{-1} + \frac{\mu_0}{2}[\mathbb{H}\otimes\mathbb{M}(\mathbb{H})]:J\mathbf{C}^{-1}\\
&\quad + \frac{\mu_0}{2}[\mathbb{M}(\mathbb{H})\otimes\mathbb{H}]:J\mathbf{C}^{-1} + \frac{\mu_0}{2}[\mathbb{M}(\mathbb{H})\otimes\mathbb{M}(\mathbb{H})]:J\mathbf{C}^{-1}\\
&= U_0^*(\mathbf{F},\mathbb{B}(\mathbb{H})) - \mu_0[\mathbb{H}\otimes\mathbb{H}]:J\mathbf{C}^{-1}\\
&\quad + \frac{\mu_0}{2}[\mathbb{H}\otimes\mathbb{H}]:J\mathbf{C}^{-1} + \frac{\mu_0}{2}[\mathbb{M}(\mathbb{H})\otimes\mathbb{M}(\mathbb{H})]:J\mathbf{C}^{-1}\\
&= \underbrace{U_0^*(\mathbf{F},\mathbb{B}(\mathbb{H})) + \frac{\mu_0}{2}[\mathbb{M}(\mathbb{H})\otimes\mathbb{M}(\mathbb{H})]:J\mathbf{C}^{-1}}_{U_0(\mathbf{F},\mathbb{H})}\\
&\quad \underbrace{-\frac{\mu_0}{2}[\mathbb{H}\otimes\mathbb{H}]:J\mathbf{C}^{-1}}_{M_0(\mathbf{F},\mathbb{H})} && \text{(C.115)}
\end{aligned}$$

wherein, due to symmetry of the tensors, we exploited the identity

$$[\mathbb{H}\otimes\mathbb{M}]:J\mathbf{C}^{-1} = \frac{1}{2}[\mathbb{H}\otimes\mathbb{M}+\mathbb{M}\otimes\mathbb{H}]:J\mathbf{C}^{-1} \quad . \tag{C.116}$$

This split is chosen based on the observation that there is no magnetisation in the free space.

Free field energy

As an alternate to the above, from equation (5.165) and with the simplification that $\mathbb{M} = \mathbf{0}$ in the free space then

$$M_0(\mathbf{F},\mathbb{H}) = \left[M_0^*(\mathbf{F},\mathbb{B}(\mathbb{H})) - \mathbb{H}\cdot\mathbb{B}(\mathbb{H})\right]\Big|_{\mathbb{M}=\mathbf{0}} \qquad \leftarrow \text{eq. (7.2)}\rightarrow \text{eq. (5.165)}$$

$$
\begin{aligned}
&= \frac{1}{2\mu_0}\left[\mathbb{B}(\mathbb{H})\otimes\mathbb{B}(\mathbb{H})\right]\big|_{\mathbb{M}=\mathbf{0}} : J^{-1}\mathbf{C} - \mathbb{H}\cdot\mathbb{B}(\mathbb{H})\big|_{\mathbb{M}=\mathbf{0}} && \rightarrow \text{eq. (5.74)}\\
&= \frac{1}{2\mu_0}\left[\left[J\mu_0\mathbf{C}^{-1}\cdot\mathbb{H}\right]\otimes\left[J\mu_0\mathbf{C}^{-1}\cdot\mathbb{H}\right]\right] : J^{-1}\mathbf{C} - \mathbb{H}\cdot\left[J\mu_0\mathbf{C}^{-1}\cdot\mathbb{H}\right]\\
&= \frac{\mu_0 J}{2}\left[\left[\mathbf{C}^{-1}\cdot\mathbb{H}\right]\otimes\left[\mathbf{C}^{-1}\cdot\mathbb{H}\right]\right] : \mathbf{C} - J\mu_0\mathbb{H}\cdot\left[\mathbf{C}^{-1}\cdot\mathbb{H}\right]\\
&= \frac{\mu_0 J}{2}\left[\mathbf{C}^{-1}\cdot\mathbb{H}\right]\cdot\mathbf{C}\cdot\left[\mathbf{C}^{-1}\cdot\mathbb{H}\right] - J\mu_0\mathbb{H}\cdot\left[\mathbf{C}^{-1}\cdot\mathbb{H}\right]\\
&= \frac{\mu_0 J}{2}\left[\mathbf{C}^{T}\cdot\mathbf{C}^{-1}\cdot\mathbb{H}\right]\cdot\left[\mathbf{C}^{-1}\cdot\mathbb{H}\right] - J\mu_0\mathbb{H}\cdot\left[\mathbf{C}^{-1}\cdot\mathbb{H}\right]\\
&= \frac{\mu_0 J}{2}\mathbb{H}\cdot\left[\mathbf{C}^{-1}\cdot\mathbb{H}\right] - J\mu_0\mathbb{H}\cdot\left[\mathbf{C}^{-1}\cdot\mathbb{H}\right]\\
&= -\frac{\mu_0}{2}\left[\mathbb{H}\otimes\mathbb{H}\right] : J\mathbf{C}^{-1} \quad , && \text{(C.117)}
\end{aligned}
$$

which aligns with the result given in equation (5.178) and was identified as a component of equation (C.115).

Two-point magnetisation stress tensor

Using the additive split stated in equation (5.214a), we can assign to the contributions of the mechanical and magnetisation stresses

$$
\begin{aligned}
\mathbf{P}^{\mathrm{mech}} + \mathbf{P}^{\mathrm{mag}} &:= \frac{\partial U_0(\mathbf{F},\mathbb{H})}{\partial\mathbf{F}} && \leftarrow \text{eq. (5.214b}_1) \rightarrow \text{eq. (C.115)}\\
&= \frac{\partial}{\partial\mathbf{F}}\left[U_0^{*}(\mathbf{F},\mathbb{B}(\mathbb{H})) + \frac{\mu_0}{2}\left[\mathbb{M}(\mathbb{H})\otimes\mathbb{M}(\mathbb{H})\right] : J\mathbf{C}^{-1}\right]\\
&= \underbrace{\left.\frac{\partial U_0^{*}(\mathbf{F},\mathbb{B}(\mathbb{H}))}{\partial\mathbf{F}}\right|_{\mathbb{B}}}_{\mathbf{P}^{\mathrm{mech}}+\mathbf{P}^{*\mathrm{mag}}} + \underbrace{\left.\frac{\partial U_0^{*}(\mathbf{F},\mathbb{B}(\mathbb{H}))}{\partial\mathbb{B}}\right|_{\mathbf{F}}}_{-\mathbb{M}}\cdot\frac{\partial\mathbb{B}(\mathbb{H})}{\partial\mathbf{F}}\\
&\quad + \frac{\partial}{\partial\mathbf{F}}\left[\frac{\mu_0}{2}\left[\mathbb{M}(\mathbb{H})\otimes\mathbb{M}(\mathbb{H})\right] : J\mathbf{C}^{-1}\right] && \rightarrow \text{eqs. (C.113,C.114,5.74)}\\
&= \mathbf{P}^{\mathrm{mech}} + \mathbf{P}^{*\mathrm{mag}} - \mathbb{M}\cdot\frac{\partial}{\partial\mathbf{F}}\left[\mu_0 J\mathbf{C}^{-1}\cdot\left[\mathbb{H}+\mathbb{M}\right]\right]\\
&\quad + \frac{\partial}{\partial\mathbf{F}}\left[\frac{\mu_0}{2}\left[\mathbb{M}\otimes\mathbb{M}\right] : J\mathbf{C}^{-1}\right]\\
&= \mathbf{P}^{\mathrm{mech}} + \mathbf{P}^{*\mathrm{mag}} - \mu_0\left[\mathbb{M}\otimes\mathbb{H}\right] : \frac{\partial J\mathbf{C}^{-1}}{\partial\mathbf{F}}\\
&\quad - \frac{\mu_0}{2}\left[\mathbb{M}\otimes\mathbb{M}\right] : \frac{\partial J\mathbf{C}^{-1}}{\partial\mathbf{F}}\\
&= \mathbf{P}^{\mathrm{mech}} + \underbrace{\mathbf{P}^{*\mathrm{mag}} - \mu_0\left[\left[\mathbb{M}\otimes\mathbb{H}\right] + \frac{1}{2}\left[\mathbb{M}\otimes\mathbb{M}\right]\right] : \frac{\partial J\mathbf{C}^{-1}}{\partial\mathbf{F}}}_{\mathbf{P}^{\mathrm{mag}}} \quad . && \text{(C.118)}
\end{aligned}
$$

With the assistance of the identity

$$
\mathbf{F}\cdot\mathbf{C}^{-T} = \mathbf{F}\cdot\mathbf{C}^{-1} = \mathbf{F}\cdot\left[\mathbf{F}^{T}\cdot\mathbf{F}\right]^{-1} = \mathbf{F}\cdot\left[\mathbf{F}^{-1}\cdot\mathbf{F}^{-T}\right] = \mathbf{F}^{-T} \quad , \qquad \text{(C.119)}
$$

the third term in equation (C.118) becomes, in index notation,

$$
\begin{aligned}
&-\mu_0\left[[\mathbb{M}_A\mathbb{H}_B]+\frac{1}{2}[\mathbb{M}_A\mathbb{M}_B]\right]\frac{\partial JC^{-1}_{AB}}{\partial F_{iJ}} && \leftarrow \text{eq. (C.118)}\rightarrow \text{eq. (A.45)}\\
&=-\mu_0\left[[\mathbb{M}_A\mathbb{H}_B]+\frac{1}{2}[\mathbb{M}_A\mathbb{M}_B]\right]J\left[C^{-1}_{AB}F^{-1}_{Ji}-C^{-1}_{AJ}F^{-1}_{Bi}-C^{-1}_{BJ}F^{-1}_{Ai}\right] && \rightarrow \text{eq. (5.5)}\\
&=-\mu_0 J\left[\mathbb{M}_A\left[\mathbb{H}_B+\frac{1}{2}\mathbb{M}_B\right]\right]\left[F^{-1}_{Ak}F^{-1}_{Bk}F^{-1}_{Ji}-F^{-1}_{Ak}F^{-1}_{Jk}F^{-1}_{Bi}-F^{-1}_{Bk}F^{-1}_{Jk}F^{-1}_{Ai}\right]\\
&=-\mu_0 J\left[\mathbb{M}_A\left[\mathbb{H}_B+\frac{1}{2}\mathbb{M}_B\right]F^{-1}_{Ak}F^{-1}_{Bk}F^{-1}_{Ji}-\mathbb{M}_A\left[\mathbb{H}_B+\frac{1}{2}\mathbb{M}_B\right]F^{-1}_{Ak}F^{-1}_{Jk}F^{-1}_{Bi}\right.\\
&\quad\left.-\mathbb{M}_A\left[\mathbb{H}_B+\frac{1}{2}\mathbb{M}_B\right]F^{-1}_{Bk}F^{-1}_{Jk}F^{-1}_{Ai}\right]\\
&=-\mu_0 J\left[F^{-1}_{Ak}\mathbb{M}_AF^{-1}_{Bk}\left[\mathbb{H}_B+\frac{1}{2}\mathbb{M}_B\right]F^{-1}_{Ji}-F^{-1}_{Ak}\mathbb{M}_AF^{-1}_{Bi}\left[\mathbb{H}_B+\frac{1}{2}\mathbb{M}_B\right]F^{-1}_{Jk}\right.\\
&\quad\left.-F^{-1}_{Ai}\mathbb{M}_AF^{-1}_{Bk}\left[\mathbb{H}_B+\frac{1}{2}\mathbb{M}_B\right]F^{-1}_{Jk}\right]\quad . && \text{(C.120)}
\end{aligned}
$$

Using this intermediate calculation, we finally arrive at the result that

$$
\begin{aligned}
P^{\text{mag}}_{iJ}&=\mathbb{M}_A\mathbb{B}_AF^{-1}_{Ji}-F^{-1}_{Ai}\mathbb{M}_A\mathbb{B}_J\\
&\quad-\mu_0 J\left[F^{-1}_{Ak}\mathbb{M}_AF^{-1}_{Bk}\left[\mathbb{H}_B+\frac{1}{2}\mathbb{M}_B\right]F^{-1}_{Ji}-F^{-1}_{Ak}\mathbb{M}_AF^{-1}_{Bi}\left[\mathbb{H}_B+\frac{1}{2}\mathbb{M}_B\right]F^{-1}_{Jk}\right.\\
&\quad\left.-F^{-1}_{Ai}\mathbb{M}_AF^{-1}_{Bk}\left[\mathbb{H}_B+\frac{1}{2}\mathbb{M}_B\right]F^{-1}_{Jk}\right] \qquad \leftarrow \text{eqs. (C.118,5.91b,C.120)}\rightarrow \text{eqs. (5.74,5.5)}\\
&=\mu_0 J\left[F^{-1}_{Ak}\mathbb{M}_AF^{-1}_{Bk}\cdot[\mathbb{H}_B+\mathbb{M}_B]F^{-1}_{Ji}-F^{-1}_{Ai}\mathbb{M}_AF^{-1}_{Bk}\cdot[\mathbb{H}_B+\mathbb{M}_B]F^{-1}_{Jk}\right]\\
&\quad-\mu_0 J\left[F^{-1}_{Ak}\mathbb{M}_AF^{-1}_{Bk}\left[\mathbb{H}_B+\frac{1}{2}\mathbb{M}_B\right]F^{-1}_{Ji}-F^{-1}_{Ai}\mathbb{M}_AF^{-1}_{Bk}\left[\mathbb{H}_B+\frac{1}{2}\mathbb{M}_B\right]F^{-1}_{Jk}\right.\\
&\quad\left.-F^{-1}_{Ak}\mathbb{M}_AF^{-1}_{Bi}\left[\mathbb{H}_B+\frac{1}{2}\mathbb{M}_B\right]F^{-1}_{Jk}\right]\\
&=\mu_0 J\left[\frac{1}{2}F^{-1}_{Ak}\mathbb{M}_AF^{-1}_{Bk}\mathbb{M}_BF^{-1}_{Ji}-\frac{1}{2}F^{-1}_{Ai}\mathbb{M}_AF^{-1}_{Bk}\mathbb{M}_BF^{-1}_{Jk}\right.\\
&\quad\left.+F^{-1}_{Ak}\mathbb{M}_AF^{-1}_{Bi}\left[\mathbb{H}_B+\frac{1}{2}\mathbb{M}_B\right]F^{-1}_{Jk}\right]\\
&=\mu_0 J\left[\frac{1}{2}\mathbb{M}_A\mathbb{M}_BC^{-1}_{AB}F^{-1}_{Ji}+F^{-1}_{Bi}\mathbb{H}_B\mathbb{M}_AF^{-1}_{Ak}F^{-1}_{Jk}\right.\\
&\quad\left.-\frac{1}{2}F^{-1}_{Bk}\mathbb{M}_BF^{-1}_{Ai}\mathbb{M}_AF^{-1}_{Jk}+\frac{1}{2}F^{-1}_{Ak}\mathbb{M}_AF^{-1}_{Bi}\mathbb{M}_BF^{-1}_{Jk}\right]\\
&=\mu_0 J\left[\frac{1}{2}\mathbb{M}_A\mathbb{M}_BC^{-1}_{AB}F^{-1}_{Ji}+F^{-1}_{Bi}\mathbb{H}_BC^{-1}_{AJ}\mathbb{M}_A\right]\\
\mathbf{P}^{\text{mag}}&=\mu_0 J\left[\frac{1}{2}\left[[\mathbb{M}\otimes\mathbb{M}]:\mathbf{C}^{-1}\right]\mathbf{F}^{-T}+\left[\mathbf{F}^{-T}\cdot\mathbb{H}\right]\otimes\left[\mathbf{C}^{-1}\cdot\mathbb{M}\right]\right] && \text{(C.121)}
\end{aligned}
$$

which matches that stated in equation (5.214d).

Spatial magnetisation stress tensor

$$\begin{aligned}\boldsymbol{\sigma}^{\mathrm{mag}} &= \mathbf{P}^{\mathrm{mag}} \cdot \operatorname{cof}\left(\mathbf{F}^{-1}\right) = J^{-1}\mathbf{P}^{\mathrm{mag}} \cdot \mathbf{F}^T && \leftarrow \text{eqs. (5.99,A.5)} \rightarrow \text{eq. (5.214d)}\\ &= J^{-1}\mu_0 J\left[\frac{1}{2}\left[[\mathbb{M}\otimes\mathbb{M}] : \mathbf{C}^{-1}\right]\mathbf{F}^{-T} + \left[\mathbf{F}^{-1}\cdot\mathbb{H}\right]\otimes\left[\mathbf{C}^{-1}\cdot\mathbb{M}\right]\right]\cdot\mathbf{F}^T\\ &= \mu_0\left[\frac{1}{2}\left[[\mathbb{M}\otimes\mathbb{M}] : \mathbf{C}^{-1}\right]\mathbf{F}^{-T}\cdot\mathbf{F}^T + \left[\mathbf{F}^{-1}\cdot\mathbb{H}\right]\otimes\left[\mathbb{M}\cdot\mathbf{C}^{-1}\cdot\mathbf{F}^T\right]\right] && \rightarrow \text{eq. (5.5)}\\ &= \mu_0\left[\frac{1}{2}\left[\left[\mathbf{F}^{-T}\cdot\mathbb{M}\right]\cdot\left[\mathbf{F}^{-T}\cdot\mathbb{M}\right]\right]\mathbf{i} + \left[\mathbf{F}^{-1}\cdot\mathbb{H}\right]\otimes\left[\mathbf{F}^{-T}\cdot\mathbb{M}\right]\right] && \rightarrow \text{eq. (5.102)}\\ &= \mu_0\left[\frac{1}{2}\left[\mathbb{m}\cdot\mathbb{m}\right]\mathbf{i} + \mathbb{h}\otimes\mathbb{m}\right] && \text{(C.122)}\end{aligned}$$

Two-point Maxwell stress tensor

$$\begin{aligned}P^{\max}_{iJ} &= \frac{\partial M_0(\mathbf{F},\mathbb{H})}{\partial F_{iJ}} && \leftarrow \text{eq. (5.214b}_2\text{)} \rightarrow \text{eq. (5.178)}\\ &= \frac{\partial}{\partial F_{iJ}}\left[-\frac{\mu_0}{2}\mathbb{H}_A\mathbb{H}_B J C^{-1}_{AB}\right] && \rightarrow \text{eq. (A.45)}\\ &= -\frac{\mu_0}{2}\mathbb{H}_A\mathbb{H}_B\frac{\partial J C^{-1}_{AB}}{\partial F_{iJ}}\\ &= -\frac{1}{2}\mu_0 J\mathbb{H}_A\mathbb{H}_B\left[C^{-1}_{AB}F^{-1}_{Ji} - C^{-1}_{AJ}F^{-1}_{Bi} - C^{-1}_{BJ}F^{-1}_{Ai}\right]\\ &= -\frac{1}{2}\mu_0 J\left[\mathbb{H}_A\mathbb{H}_B C^{-1}_{AB}F^{-1}_{Ji} - F^{-1}_{Bi}\mathbb{H}_B C^{-1}_{AJ}\mathbb{H}_A - F^{-1}_{Ai}\mathbb{H}_A\mathbb{H}_B C^{-1}_{BJ}\right]\\ &= -\frac{1}{2}\mu_0 J\left[\mathbb{H}_A\mathbb{H}_B C^{-1}_{AB}F^{-1}_{Ji} - 2F^{-1}_{Ai}\mathbb{H}_A C^{-1}_{BJ}\mathbb{H}_B\right]\\ &= -\mu_0 J\left[\frac{1}{2}\mathbb{H}_A\mathbb{H}_B C^{-1}_{AB}F^{-1}_{Ji} - F^{-1}_{Ai}\mathbb{H}_A C^{-1}_{BJ}\mathbb{H}_B\right]\\ \mathbf{P}^{\max} &= \frac{\partial M_0(\mathbf{F},\mathbb{H})}{\partial \mathbf{F}}\\ &= -\mu_0 J\left[\frac{1}{2}\left[[\mathbb{H}\otimes\mathbb{H}] : \mathbf{C}^{-1}\right]\mathbf{F}^{-T} - \left[\mathbf{F}^{-T}\cdot\mathbb{H}\right]\otimes\left[\mathbf{C}^{-1}\cdot\mathbb{H}\right]\right] && \text{(C.123)}\end{aligned}$$

Spatial Maxwell stress tensor

$$\begin{aligned}\boldsymbol{\sigma}^{\max} &= \mathbf{P}^{\max} \cdot \operatorname{cof}\left(\mathbf{F}^{-1}\right) = J^{-1}\mathbf{P}^{\max} \cdot \mathbf{F}^T && \leftarrow \text{eqs. (5.99,A.5)} \rightarrow \text{eq. (5.214e)}\\ &= -J^{-1}\mu_0 J\left[\frac{1}{2}\left[[\mathbb{H}\otimes\mathbb{H}] : \mathbf{C}^{-1}\right]\mathbf{F}^{-T} - \left[\mathbf{F}^{-T}\cdot\mathbb{H}\right]\otimes\left[\mathbf{C}^{-1}\cdot\mathbb{H}\right]\right]\cdot\mathbf{F}^T\\ &= -\mu_0\left[\frac{1}{2}\left[[\mathbb{H}\otimes\mathbb{H}] : \mathbf{C}^{-1}\right]\mathbf{F}^{-T}\cdot\mathbf{F}^T - \left[\mathbf{F}^{-T}\cdot\mathbb{H}\right]\otimes\left[\mathbb{H}\cdot\mathbf{C}^{-1}\cdot\mathbf{F}^T\right]\right]\\ &= -\mu_0\left[\frac{1}{2}\left[\mathbf{F}^{-T}\cdot\mathbb{H}\right]\cdot\left[\mathbf{F}^{-T}\cdot\mathbb{H}\right]\mathbf{i} - \left[\mathbf{F}^{-T}\cdot\mathbb{H}\right]\otimes\left[\mathbf{F}^{-T}\cdot\mathbb{H}\right]\right] && \rightarrow \text{eq. (5.102)}\\ &= -\mu_0\left[\frac{1}{2}\left[\mathbb{h}\cdot\mathbb{h}\right]\mathbf{i} - \mathbb{h}\otimes\mathbb{h}\right] && \text{(C.124)}\end{aligned}$$

Two-point ponderomotive stress tensor

Using the identities (expressed in index notation)

$$\mathbb{M}_A \mathbb{M}_B - \mathbb{H}_A \mathbb{H}_B = [\mathbb{M}_A + \mathbb{H}_A]\,[\mathbb{M}_B + \mathbb{H}_B] - [\mathbb{M}_A + \mathbb{H}_A]\,\mathbb{H}_B - \mathbb{H}_A\,[\mathbb{M}_B + \mathbb{H}_B] \tag{C.125}$$

$$\begin{aligned}
[\mathbb{M}_A \mathbb{M}_B - \mathbb{H}_A \mathbb{H}_B]\, C^{-1}_{AB} &= [[\mathbb{M}_A + \mathbb{H}_A]\,[\mathbb{M}_B + \mathbb{H}_B] \\
&\quad - [\mathbb{M}_A + \mathbb{H}_A]\,\mathbb{H}_B - \mathbb{H}_A\,[\mathbb{M}_B + \mathbb{H}_B]]\, C^{-1}_{AB} \qquad \rightarrow \text{eqs. (C.125,C.116)} \\
&= [[\mathbb{M}_A + \mathbb{H}_A]\,[\mathbb{M}_B + \mathbb{H}_B] - 2\mathbb{H}_A\,[\mathbb{M}_B + \mathbb{H}_B]]\, C^{-1}_{AB} \qquad \text{(C.126)}
\end{aligned}$$

then

$$\begin{aligned}
\mathbf{P}^{\mathrm{pon}} &= \mathbf{P}^{\mathrm{mag}} + \mathbf{P}^{\mathrm{max}} \qquad \leftarrow \text{eq. (5.214a)} \rightarrow \text{eqs. (5.214d,5.214e)} \\
&= \mu_0 J \left[\frac{1}{2} \left[[\mathbb{M} \otimes \mathbb{M}] : \mathbf{C}^{-1} \right] \mathbf{F}^{-T} + \left[\mathbf{F}^{-T} \cdot \mathbb{H} \right] \otimes \left[\mathbf{C}^{-1} \cdot \mathbb{M} \right] \right] \\
&\quad - \mu_0 J \left[\frac{1}{2} \left[[\mathbb{H} \otimes \mathbb{H}] : \mathbf{C}^{-1} \right] \mathbf{F}^{-T} - \left[\mathbf{F}^{-T} \cdot \mathbb{H} \right] \otimes \left[\mathbf{C}^{-1} \cdot \mathbb{H} \right] \right] \\
&= \mu_0 J \frac{1}{2} \left[\left[[\mathbb{M} \otimes \mathbb{M}] : \mathbf{C}^{-1} \right] \mathbf{F}^{-T} - \left[[\mathbb{H} \otimes \mathbb{H}] : \mathbf{C}^{-1} \right] \mathbf{F}^{-T} \right] \\
&\quad + \mu_0 J \left[\left[\mathbf{F}^{-T} \cdot \mathbb{H} \right] \otimes \left[\mathbf{C}^{-1} \cdot \mathbb{M} \right] + \left[\mathbf{F}^{-T} \cdot \mathbb{H} \right] \otimes \left[\mathbf{C}^{-1} \cdot \mathbb{H} \right] \right] \\
&= \mu_0 J \frac{1}{2} \left[\left[[[\mathbb{M} \otimes \mathbb{M}] - [\mathbb{H} \otimes \mathbb{H}]] : \mathbf{C}^{-1} \right] \mathbf{F}^{-T} \right] \\
&\quad + \left[\mathbf{F}^{-T} \cdot \mathbb{H} \right] \otimes \left[\mu_0 J \mathbf{C}^{-1} \cdot [\mathbb{H} + \mathbb{M}] \right] \qquad \rightarrow \text{eq. (C.126)} \\
&= \mu_0 J \frac{1}{2} \left[\left[[[\mathbb{M} + \mathbb{H}] \otimes [\mathbb{M} + \mathbb{H}] - 2\mathbb{H} \otimes [\mathbb{M} + \mathbb{H}]] : \mathbf{C}^{-1} \right] \mathbf{F}^{-T} \right] + \left[\mathbf{F}^{-T} \cdot \mathbb{H} \right] \otimes \mathbb{B} \\
&= \frac{1}{2} \left[\left[[[\mathbb{M} + \mathbb{H}] - 2\mathbb{H}] \cdot \left[\mu_0 J \mathbf{C}^{-1} \cdot [\mathbb{M} + \mathbb{H}] \right] \right] \mathbf{F}^{-T} \right] + \left[\mathbf{F}^{-T} \cdot \mathbb{H} \right] \otimes \mathbb{B} \qquad \rightarrow \text{eq. (5.74)} \\
&= \frac{1}{2} \left[[[[\mathbb{M} + \mathbb{H}] - 2\mathbb{H}] \cdot \mathbb{B}]\, \mathbf{F}^{-T} \right] + \left[\mathbf{F}^{-T} \cdot \mathbb{H} \right] \otimes \mathbb{B} \\
&= \frac{1}{2} \left[[[\mathbb{M} + \mathbb{H}] \cdot \mathbb{B}]\, \mathbf{F}^{-T} \right] - [\mathbb{H} \cdot \mathbb{B}]\, \mathbf{F}^{-T} + \left[\mathbf{F}^{-T} \cdot \mathbb{H} \right] \otimes \mathbb{B} \\
&= \frac{1}{2} \left[\left[\frac{J^{-1}}{\mu_0} \mathbf{C} \cdot \left[\mu_0 J \mathbf{C}^{-1} \cdot [\mathbb{M} + \mathbb{H}] \right] \cdot \mathbb{B} \right] \mathbf{F}^{-T} \right] - [\mathbb{H} \cdot \mathbb{B}]\, \mathbf{F}^{-T} + \left[\mathbf{F}^{-T} \cdot \mathbb{H} \right] \otimes \mathbb{B} \\
&\qquad \rightarrow \text{eq. (5.74)} \\
&= \left[\frac{J^{-1}}{2\mu_0} \mathbf{C} : [\mathbb{B} \otimes \mathbb{B}] \right] \mathbf{F}^{-T} - [\mathbb{H} \cdot \mathbb{B}]\, \mathbf{F}^{-T} + \left[\mathbf{F}^{-T} \cdot \mathbb{H} \right] \otimes \mathbb{B} \qquad \text{(C.127)} \\
&\equiv \mathbf{P}^{*\mathrm{pon}} \quad . \qquad \leftarrow \text{eq. (5.91a)}
\end{aligned}$$

Spatial ponderomotive stress tensor

Using identity (expressed in index notation)

$$\mathbb{m}_a \mathbb{m}_a - \mathbb{h}_a \mathbb{h}_a = [\mathbb{m}_a + \mathbb{h}_a]\,[\mathbb{m}_a + \mathbb{h}_a] - 2\,[\mathbb{m}_a + \mathbb{h}_a]\,\mathbb{h}_a \tag{C.128}$$

then

$$\begin{aligned}
\boldsymbol{\sigma}^{\text{pon}} &= \boldsymbol{\sigma}^{\text{mag}} + \boldsymbol{\sigma}^{\text{max}} && \leftarrow \text{eq. (5.215a)} \rightarrow \text{eqs. (5.215c,5.215d)}\\
&= \mu_0 \left[\frac{1}{2}\,[\mathbb{m}\cdot\mathbb{m}]\,\mathbf{i} + \mathbb{h}\otimes\mathbb{m}\right] - \mu_0 \left[\frac{1}{2}\,[\mathbb{h}\cdot\mathbb{h}]\,\mathbf{i} - \mathbb{h}\otimes\mathbb{h}\right]\\
&= \frac{\mu_0}{2}\,[[\mathbb{m}\cdot\mathbb{m}]\,\mathbf{i} - [\mathbb{h}\cdot\mathbb{h}]\,\mathbf{i}] + \mu_0\,[\mathbb{h}\otimes\mathbb{h} + \mathbb{h}\otimes\mathbb{m}] && \rightarrow \text{eq. (C.128)}\\
&= \frac{\mu_0}{2}\,[[\mathbb{m}+\mathbb{h}]\cdot[\mathbb{m}+\mathbb{h}] - 2\,[\mathbb{m}+\mathbb{h}]\cdot\mathbb{h}]\,\mathbf{i} + \mathbb{h}\otimes\mathbb{b}\\
&= \frac{1}{2\mu_0}\,[\mu_0\,[\mathbb{m}+\mathbb{h}]\cdot\mu_0\,[\mathbb{m}+\mathbb{h}]]\,\mathbf{i} - [\mu_0\,[\mathbb{m}+\mathbb{h}]\cdot\mathbb{h}]\,\mathbf{i} + \mathbb{h}\otimes\mathbb{b} && \rightarrow \text{eq. (5.118)}\\
&= \frac{1}{2\mu_0}\,[\mathbb{b}\cdot\mathbb{b}]\,\mathbf{i} - [\mathbb{h}\cdot\mathbb{b}]\,\mathbf{i} + \mathbb{h}\otimes\mathbb{b} && \text{(C.129)}\\
&\equiv \boldsymbol{\sigma}^{*\text{pon}} \quad . && \leftarrow \text{eq. (5.131a)}
\end{aligned}$$

C.7 Thermodynamics

C.7.1 Work performed by the Lorentz forces

In the following sections, we determine the contributions to the electromagnetic power derived separately from the magnetic and electric Lorentz forces, which accumulate into the resultant total ponderomotive power.

Total ponderomotive power: Magnetic contribution

$$\begin{aligned}
&\mathscr{P}^{\text{pon (mag)}}_{\text{ext}}(t) && \rightarrow \text{eq. (C.44)}\\
&= \int_{\mathscr{B}_t} \mathbf{v}\cdot\left[\mathbb{j}^{\text{f}}\times\mathbb{b} - \nabla\cdot[[\mathbb{m}\cdot\mathbb{b}]\,\mathbf{i} - \mathbb{m}\otimes\mathbb{b}] + \mathbb{m}\cdot[\nabla\mathbb{b}]\right]\,\text{dv} + \int_{\partial\mathscr{B}_t} \mathbf{v}\cdot[\widehat{\mathbb{j}}\times\{\!\{\mathbb{b}\}\!\}]\,\text{da}\\
&= \int_{\mathscr{B}_t} \left[\mathbf{v}\cdot\left[\mathbb{j}^{\text{f}}\times\mathbb{b} + \mathbb{m}\cdot[\nabla\mathbb{b}]\right] - \mathbf{v}\cdot[\nabla\cdot[[\mathbb{m}\cdot\mathbb{b}]\,\mathbf{i} - \mathbb{m}\otimes\mathbb{b}]]\right]\,\text{dv} + \int_{\partial\mathscr{B}_t} \mathbf{v}\cdot[\widehat{\mathbb{j}}\times\{\!\{\mathbb{b}\}\!\}]\,\text{da}\\
& && \rightarrow \text{eq. (A.35)}\\
&= \int_{\mathscr{B}_t} \left[\mathbf{v}\cdot\left[\mathbb{j}^{\text{f}}\times\mathbb{b} + \mathbb{m}\cdot[\nabla\mathbb{b}]\right] - \nabla\cdot[\mathbf{v}\cdot[[\mathbb{m}\cdot\mathbb{b}]\,\mathbf{i} - \mathbb{m}\otimes\mathbb{b}]] + \nabla\mathbf{v} : [[\mathbb{m}\cdot\mathbb{b}]\,\mathbf{i} - \mathbb{m}\otimes\mathbb{b}]\right]\,\text{dv}\\
&+ \int_{\partial\mathscr{B}_t} \mathbf{v}\cdot[\widehat{\mathbb{j}}\times\{\!\{\mathbb{b}\}\!\}]\,\text{da} && \rightarrow \text{eqs. (B.16,5.131b)}\\
&= \int_{\mathscr{B}_t} \left[\mathbf{v}\cdot\left[\mathbb{j}^{\text{f}}\times\mathbb{b} + \mathbb{m}\cdot[\nabla\mathbb{b}]\right] + \nabla\mathbf{v} : \boldsymbol{\sigma}^{*\text{mag}}\right]\,\text{dv}\\
&+ \int_{\partial\mathscr{B}_t} \mathbf{v}\cdot\left[\widehat{\mathbb{j}}\times\{\!\{\mathbb{b}\}\!\} - [[\mathbb{m}\cdot\mathbb{b}]\,\mathbf{i} - \mathbb{m}\otimes\mathbb{b}]\cdot\mathbf{n}\right]\,\text{da} && \rightarrow \text{eq. (C.46)}
\end{aligned}$$

$$= \int_{\mathscr{B}_t} \left[\mathbf{v} \cdot \mathbf{b}_t^{\text{pon(mag)}} + \boldsymbol{\sigma}^{*\text{mag}} : [\nabla \mathbf{v}]\right] \text{dv} + \int_{\partial \mathscr{B}_t} \mathbf{v} \cdot \mathbf{t}_t^{\text{pon(mag)}} \text{da} \tag{C.130}$$

$$\equiv \int_{\mathscr{B}_t} \left[\mathbf{v} \cdot J \mathbf{b}_0^{\text{pon(mag)}} + \left[\mathbf{P}^{*\text{mag}} \cdot \text{cof}\left(\mathbf{F}^{-1}\right)\right] : \left[\dot{\mathbf{F}} \cdot \mathbf{F}^{-1}\right]\right] \text{dv} + \int_{\partial \mathscr{B}_t} \mathbf{v} \cdot \mathbf{t}_t^{\text{pon(mag)}} \text{da}$$

$\leftarrow$ eqs. (5.127,A.49)$\rightarrow$ eqs. (5.11,5.99,5.10)

$$= \int_{\mathscr{B}_0} \left[\mathbf{v} \cdot \mathbf{b}_0^{\text{pon(mag)}} + \mathbf{P}^{*\text{mag}} : \dot{\mathbf{F}}\right] \text{dV} + \int_{\partial \mathscr{B}_0} \mathbf{v} \cdot \mathbf{t}_0^{\text{pon(mag)}} \text{dA} \tag{C.131}$$

Total ponderomotive power: Electric contribution

Using similar argumentation as for the magnetic case, the contribution to the ponderomotive power from the electric fields is

$$\mathscr{P}_{\text{ext}}^{\text{pon (elec)}}(t) = \int_{\mathscr{B}_t} \mathbf{v} \cdot [\varrho_t \mathbb{e}] \text{ dv} + \int_{\partial \mathscr{B}_t} \mathbf{v} \cdot [\widehat{\varrho}_t \{\!\{\mathbb{e}\}\!\}] \text{ da} \quad \rightarrow \text{eq. (C.39)}$$

$$= \int_{\mathscr{B}_t} \left[\mathbf{v} \cdot \mathbf{b}_t^{\text{pon(elec)}} + \boldsymbol{\sigma}^{\text{pol}} : [\nabla \mathbf{v}]\right] \text{dv} + \int_{\partial \mathscr{B}_t} \mathbf{v} \cdot \mathbf{t}_t^{\text{pon(elec)}} \text{da} \tag{C.132}$$

$$= \int_{\mathscr{B}_0} \left[\mathbf{v} \cdot \mathbf{b}_0^{\text{pon(elec)}} + \mathbf{P}^{*\text{pol}} : \dot{\mathbf{F}}\right] \text{dV} + \int_{\partial \mathscr{B}_0} \mathbf{v} \cdot \mathbf{t}_0^{\text{pon(elec)}} \text{dA} \quad . \tag{C.133}$$

For the sake of brevity, we have excluded the derivations and expressions for the electric ponderomotive body force and traction, and polarisation stress. We refer the reader to [414, 122, 541, 73, 504, 532, 351], among others, for further details regarding these quantities and more that are relevant to problems in electromechanics.

Total ponderomotive power

$$\mathscr{P}_{\text{ext}}^{\text{pon}}(t) = \mathscr{P}_{\text{ext}}^{\text{pon (mag)}}(t) + \mathscr{P}_{\text{ext}}^{\text{pon (elec)}}(t) \quad \rightarrow \text{eq. (C.39)}$$

$$= \int_{\mathscr{B}_t} \mathbf{v} \cdot [\varrho_t \mathbb{e} + \mathbb{j} \times \mathbb{b}] \text{ dv} + \int_{\partial \mathscr{B}_t} \mathbf{v} \cdot [\widehat{\varrho}_t \{\!\{\mathbb{e}\}\!\} + \widehat{\mathbb{j}} \times \{\!\{\mathbb{b}\}\!\}] \text{ da} \quad \rightarrow \text{eqs. (C.130,C.132)}$$

$$= \int_{\mathscr{B}_t} \left[\mathbf{v} \cdot \left[\mathbf{b}_t^{\text{pon(mag)}} + \mathbf{b}_t^{\text{pon(elec)}}\right] + \left[\boldsymbol{\sigma}^{*\text{mag}} + \boldsymbol{\sigma}^{\text{pol}}\right] : [\nabla \mathbf{v}]\right] \text{dv}$$

$$+ \int_{\partial \mathscr{B}_t} \mathbf{v} \cdot \left[\mathbf{t}_t^{\text{pon(mag)}} + \mathbf{t}_t^{\text{pon(elec)}}\right] \text{da}$$

$$= \int_{\mathscr{B}_t} \left[\mathbf{v} \cdot \mathbf{b}_t^{\text{pon}} + \boldsymbol{\sigma}^{*\text{(mag+pol)}} : [\nabla \mathbf{v}]\right] \text{dv} + \int_{\partial \mathscr{B}_t} \mathbf{v} \cdot \mathbf{t}_t^{\text{pon}} \text{da} \tag{C.134}$$

$$= \int_{\mathscr{B}_0} \left[\mathbf{v} \cdot \mathbf{b}_0^{\text{pon}} + \mathbf{P}^{*\text{(mag+pol)}} : \dot{\mathbf{F}}\right] \text{dV} + \int_{\partial \mathscr{B}_0} \mathbf{v} \cdot \mathbf{t}_0^{\text{pon}} \text{dA} \tag{C.135}$$

C.7.2 Boundary contributions to external power

In the following paragraphs, we derive the equivalent volumetric expressions for the boundary terms in the external power contribution to the first law of thermodynamics.

Mechanical contribution

$$\begin{aligned}
\int_{\partial\mathscr{B}_0} \mathbf{v}\cdot\mathbf{t}_0^{\text{mech,ext}}\,\text{dA} &= -\int_{\partial\mathscr{B}_0} [\![\mathbf{P}^{\text{tot}}]\!]\cdot\mathbf{N}^+\,\text{dA} && \leftarrow \text{eq. (5.98)}\rightarrow \text{eq. (C.70)}\\
&= \int_{\partial\mathscr{B}_0} \mathbf{v}\cdot\left[\mathbf{P}^{\text{mech}}\cdot\mathbf{N}-\mathbf{t}_0^{\text{pon}}\right]\,\text{dA} && \leftarrow \text{eqs. (5.194,5.97)}\rightarrow \text{eq. (B.16)}\\
&= \int_{\mathscr{B}_0} \nabla_0\cdot\left[\mathbf{v}\cdot\mathbf{P}^{\text{mech}}\right]\,\text{dV} - \int_{\partial\mathscr{B}_0} \mathbf{v}\cdot\mathbf{t}_0^{\text{pon}}\,\text{dA} && \rightarrow \text{eq. (A.35)}\\
&= \int_{\mathscr{B}_0} \left[\mathbf{v}\cdot\left[\nabla_0\cdot\mathbf{P}^{\text{mech}}\right]+\mathbf{P}^{\text{mech}}:\dot{\mathbf{F}}\right]\,\text{dV} - \int_{\partial\mathscr{B}_0} \mathbf{v}\cdot\mathbf{t}_0^{\text{pon}}\,\text{dA} && \text{(C.136)}
\end{aligned}$$

Thermal contribution
Directly from equation (5.195) using equation (B.16),

$$\int_{\partial\mathscr{B}_0} [-\mathbf{Q}\cdot\mathbf{N}]\,\text{dA} = -\int_{\mathscr{B}_0} \nabla_0\cdot\mathbf{Q}\,\text{dV} \quad . \qquad \text{(C.137)}$$

Electromechanical contribution

$$\begin{aligned}
&\int_{\partial\mathscr{B}_0} \left[\mathbf{v}\cdot\mathbf{t}_0^{\text{pon}}+[\mathbb{E}\times\mathbb{M}]\cdot\mathbf{N}\right]\,\text{dA} && \leftarrow \text{eqs. (5.196)}\rightarrow \text{eqs. (B.16)}\\
&= \int_{\partial\mathscr{B}_0} \mathbf{v}\cdot\mathbf{t}_0^{\text{pon}}\,\text{dA} + \int_{\mathscr{B}_0} \nabla_0\cdot[\mathbb{E}\times\mathbb{M}]\,\text{dV} && \rightarrow \text{eq. (C.139)}\\
&= \int_{\partial\mathscr{B}_0} \mathbf{v}\cdot\mathbf{t}_0^{\text{pon}}\,\text{dA} + \int_{\mathscr{B}_0} \left[\mathbb{E}\cdot\dot{\mathbb{P}}-\mathbb{M}\cdot\dot{\mathbb{B}}-\mathbb{E}\cdot\mathbb{J}^{\text{b}}\right]\,\text{dV} && \text{(C.138)}
\end{aligned}$$

Note that to simplify the last term in the last equation, we exploited the identity

$$\begin{aligned}
\nabla_0\cdot[\mathbb{E}\times\mathbb{M}] &= \mathbb{M}\cdot[\nabla_0\times\mathbb{E}]-\mathbb{E}\cdot[\nabla_0\times\mathbb{M}] && \leftarrow \text{eq. (A.31)}\rightarrow \text{eq. (5.66}_1)\\
&= -\mathbb{M}\cdot\dot{\mathbb{B}}-\mathbb{E}\cdot[\nabla_0\times\mathbb{M}] && \rightarrow \text{tbl. (5.5)}\\
&= -\mathbb{M}\cdot\dot{\mathbb{B}}+\mathbb{E}\cdot\left[\dot{\mathbb{P}}-\mathbb{J}^{\text{b}}\right] \quad . && \text{(C.139)}
\end{aligned}$$

C.7.3 Combined contribution of external mechanical and electromagnetic powers

The total power derived from external mechanical and electromagnetic sources is

$$\begin{aligned}
\mathscr{P}_{\text{ext}}(t) &= \mathscr{P}^{\text{mech}}_{\text{ext}}(t) + \mathscr{P}^{\text{EM}}_{\text{ext}}(t) \\
&= \int_{\mathscr{B}_0} \left[\mathbf{v}\cdot\left[\nabla_0\cdot\mathbf{P}^{\text{mech}}\right] + \mathbf{P}^{\text{mech}} : \dot{\mathbf{F}} + \mathbf{v}\cdot\mathbf{b}_0^{\text{mech}}\right] \mathrm{dV} - \int_{\partial\mathscr{B}_0} \mathbf{v}\cdot\mathbf{t}_0^{\text{pon}}\,\mathrm{dA} && \leftarrow \text{eqs. (C.136,5.194)} \\
&+ \int_{\partial\mathscr{B}_0} \mathbf{v}\cdot\mathbf{t}_0^{\text{pon}}\,\mathrm{dA} + \int_{\mathscr{B}_0} \left[\mathbb{E}\cdot\dot{\mathbb{P}} - \mathbb{M}\cdot\dot{\mathbb{B}} - \mathbb{E}\cdot\mathbb{J}^{\text{b}}\right] \mathrm{dV} && \leftarrow \text{eq. (C.138)} \\
&+ \int_{\mathscr{B}_0} \left[\mathbf{v}\cdot\mathbf{b}_0^{\text{pon}} + \mathbf{P}^{*(\text{mag+pol})} : \dot{\mathbf{F}} + \mathbb{J}^{\text{b}}\cdot\mathbb{E}\right] \mathrm{dV} && \leftarrow \text{eq. (5.196)} \\
&= \int_{\mathscr{B}_0} \left[\mathbf{v}\cdot\left[\nabla_0\cdot\mathbf{P}^{\text{mech}} + \mathbf{b}_0^{\text{pon}} + \mathbf{b}_0^{\text{mech}}\right] + \left[\mathbf{P}^{\text{mech}} + \mathbf{P}^{*(\text{mag+pol})}\right] : \dot{\mathbf{F}}\right] \mathrm{dV} \\
&+ \int_{\mathscr{B}_0} \left[\mathbb{E}\cdot\dot{\mathbb{P}} - \mathbb{M}\cdot\dot{\mathbb{B}}\right] \mathrm{dV} && \rightarrow \text{eq. (5.87)} \\
&= \int_{\mathscr{B}_0} \left[\rho_0\mathbf{v}\cdot\mathbf{a} + \left[\mathbf{P}^{\text{mech}} + \mathbf{P}^{*(\text{mag+pol})}\right] : \dot{\mathbf{F}} - \mathbb{M}\cdot\dot{\mathbb{B}} + \mathbb{E}\cdot\dot{\mathbb{P}}\right] \mathrm{dV} \quad . && \text{(C.140)}
\end{aligned}$$

We may also identify the time-rate of the stored energy function for the magnetic field as

$$\begin{aligned}
\mathrm{D}_t M_0{}^{*(\text{mag})}(\mathbf{F},\mathbb{B}) &= \frac{\partial M_0{}^*}{\partial \mathbf{F}} : \dot{\mathbf{F}} + \frac{\partial M_0{}^*}{\partial \mathbb{B}}\cdot\dot{\mathbb{B}} && \rightarrow \text{eqs. (5.214b}_2\text{,5.165)} \\
&= \mathbf{P}^{*\max(\text{mag})} : \dot{\mathbf{F}} + \frac{\partial}{\partial\mathbb{B}}\left[\frac{1}{2\mu_0}\left[[\mathbb{B}\otimes\mathbb{B}] : J^{-1}\mathbf{C}\right]\right]\cdot\dot{\mathbb{B}} \\
&= \mathbf{P}^{*\max(\text{mag})} : \dot{\mathbf{F}} + \left[\frac{1}{\mu_0}J^{-1}\mathbf{C}\cdot\mathbb{B}\right]\cdot\dot{\mathbb{B}} && \rightarrow \text{eq. (5.74)} \\
&= \mathbf{P}^{*\max(\text{mag})} : \dot{\mathbf{F}} + [\mathbb{H} + \mathbb{M}]\cdot\dot{\mathbb{B}} \\
\implies \quad -\mathbb{M}\cdot\dot{\mathbb{B}} &= \mathbf{P}^{*\max(\text{mag})} : \dot{\mathbf{F}} + \mathbb{H}\cdot\dot{\mathbb{B}} - \mathrm{D}_t M_0{}^{*(\text{mag})} \quad . && \text{(C.141)}
\end{aligned}$$

The electric energy stored in the free field is quantified by [122, 124, 541, 532, 504]

$$M_0{}^{(\text{elec})}(\mathbf{F},\mathbb{E}) := -\frac{1}{2}\varepsilon_0\,[\mathbb{E}\otimes\mathbb{E}] : J\mathbf{C}^{-1} \qquad \text{(C.142)}$$

that, upon re-parameterisation using a Legendre transformation, becomes

$$\begin{aligned}
M_0{}^{*(\text{elec})}(\mathbf{F},\mathbb{D}_\varepsilon) &= M_0{}^{(\text{elec})}(\mathbf{F},\mathbb{E}(\mathbb{D}_\varepsilon)) + \mathbb{D}_\varepsilon\cdot\mathbb{E}(\mathbb{D}_\varepsilon) && \rightarrow \text{eq. (C.142)} \\
&= -\frac{1}{2}\varepsilon_0\,[\mathbb{E}(\mathbb{D}_\varepsilon)\otimes\mathbb{E}(\mathbb{D}_\varepsilon)] : J\mathbf{C}^{-1} + \mathbb{D}_\varepsilon\cdot\mathbb{E}(\mathbb{D}_\varepsilon) && \rightarrow \text{tbl. (5.5)}
\end{aligned}$$

$$= -\frac{1}{2}\varepsilon_0 \left[\left[\frac{1}{\varepsilon_0}J^{-1}\mathbf{C}\cdot\mathbb{D}_\varepsilon\right]\otimes\left[\frac{1}{\varepsilon_0}J^{-1}\mathbf{C}\cdot\mathbb{D}_\varepsilon\right]\right] : J\mathbf{C}^{-1} + \mathbb{D}_\varepsilon\cdot\left[\frac{1}{\varepsilon_0}J^{-1}\mathbf{C}\cdot\mathbb{D}_\varepsilon\right]$$
$$= -\frac{1}{2\varepsilon_0}\left[\mathbb{D}_\varepsilon\otimes\mathbb{D}_\varepsilon\right] : J^{-1}\mathbf{C} + \frac{1}{\varepsilon_0}\left[\mathbb{D}_\varepsilon\otimes\mathbb{D}_\varepsilon\right] : J^{-1}\mathbf{C}$$
$$= \frac{1}{2\varepsilon_0}\left[\mathbb{D}_\varepsilon\otimes\mathbb{D}_\varepsilon\right] : J^{-1}\mathbf{C} \quad . \qquad \text{(C.143)}$$

From this, we may identify the time-rate of the transformed stored energy function for the electric field as

$$\mathrm{D}_t {M_0}^{*(\text{elec})}(\mathbf{F},\mathbb{D}_\varepsilon) = \frac{\partial {M_0}^*}{\partial\mathbf{F}} : \dot{\mathbf{F}} + \frac{\partial {M_0}^*}{\partial\mathbb{D}_\varepsilon}\cdot\dot{\mathbb{D}}_\varepsilon$$
$$= \mathbf{P}^{*\max(\text{elec})} : \dot{\mathbf{F}} + \left[\frac{1}{\varepsilon_0}J^{-1}\mathbf{C}\cdot\mathbb{D}_\varepsilon\right]\cdot\dot{\mathbb{D}}_\varepsilon \qquad \rightarrow \text{tbl. (5.5)}$$
$$= \mathbf{P}^{*\max(\text{elec})} : \dot{\mathbf{F}} + \mathbb{E}\cdot\dot{\mathbb{D}}_\varepsilon \qquad \rightarrow \text{tbl. (5.5)}$$
$$= \mathbf{P}^{*\max(\text{elec})} : \dot{\mathbf{F}} + \mathbb{E}\cdot\mathrm{D}_t[\mathbb{D}-\mathbb{P}]$$
$$\Longrightarrow \quad \mathbb{E}\cdot\dot{\mathbb{P}} = \mathbf{P}^{*\max(\text{elec})} : \dot{\mathbf{F}} + \mathbb{E}\cdot\dot{\mathbb{D}} - \mathrm{D}_t {M_0}^{*(\text{elec})} \qquad \text{(C.144)}$$

where we define the electric component of the Maxwell stress $\mathbf{P}^{*\max(\text{elec})} =: \frac{\partial {M_0}^{*(\text{elec})}}{\partial\mathbf{F}}$ in accordance to its magnetic counterpart introduced in equation $(5.214\text{b})_2$.

Combining the above, we obtain the final result that

$$\mathscr{P}_{\text{ext}}(t) = \int_{\mathscr{B}_0} \Big[\rho_0\mathbf{v}\cdot\mathbf{a} + \left[\mathbf{P}^{\text{mech}} + \mathbf{P}^{*(\text{mag+pol})}\right] : \dot{\mathbf{F}} + \mathbf{P}^{*\max(\text{mag})} : \dot{\mathbf{F}} + \mathbb{H}\cdot\dot{\mathbb{B}} - \mathrm{D}_t {M_0}^{*(\text{mag})}$$
$$+ \mathbf{P}^{*\max(\text{elec})} : \dot{\mathbf{F}} + \mathbb{E}\cdot\dot{\mathbb{D}} - \mathrm{D}_t {M_0}^{*(\text{elec})}\Big]\,\mathrm{dV}$$
$$= \int_{\mathscr{B}_0} \Big[\rho_0\mathbf{v}\cdot\mathbf{a} + \left[\mathbf{P}^{\text{mech}} + \mathbf{P}^{*(\text{mag+pol})} + \left[\mathbf{P}^{*\max(\text{mag})} + \mathbf{P}^{*\max(\text{elec})}\right]\right] : \dot{\mathbf{F}}$$
$$+ \mathbb{H}\cdot\dot{\mathbb{B}} + \mathbb{E}\cdot\dot{\mathbb{D}} - \mathrm{D}_t\left[{M_0}^{*(\text{mag})} + {M_0}^{*(\text{elec})}\right]\Big]\,\mathrm{dV} \qquad \rightarrow \text{eq. (5.90)}$$
$$= \int_{\mathscr{B}_0} \left[\rho_0\mathbf{v}\cdot\mathbf{a} + \mathbf{P}^{\text{tot}} : \dot{\mathbf{F}} + \mathbb{H}\cdot\dot{\mathbb{B}} + \mathbb{E}\cdot\dot{\mathbb{D}} - \mathrm{D}_t {M_0}^*\right]\mathrm{dV} \qquad \text{(C.145)}$$

where the total energy stored in magnetic and electric fields is ${M_0}^* = {M_0}^{*(\text{mag})} + {M_0}^{*(\text{elec})}$, and the total Maxwell stress is $\mathbf{P}^{*\max} = \mathbf{P}^{*\max(\text{mag})} + \mathbf{P}^{*\max(\text{elec})}$.

C.8 Homogenisation

C.8.1 Relationship between macroscopic and microscopic field quantities

Deformation gradient tensor

Assuming the existence of a motion potential field, then

$$\overline{\overline{\mathbf{F}}}\left(\overline{\overline{\mathbf{X}}}\right) = \frac{1}{\mathrm{V}_0}\int_{\mathscr{B}_0} \mathbf{F}\,\mathrm{dV} \qquad \leftarrow \text{eq. } (9.3_1) \rightarrow \text{eq. (5.3)}$$

$$= \frac{1}{V_0} \int_{\mathscr{B}_0} [\nabla_0 \boldsymbol{\varphi}] \; dV \qquad \rightarrow \text{eq. (B.18)}$$

$$= \frac{1}{V_0} \int_{\partial \mathscr{B}_0} \boldsymbol{\varphi} \otimes \mathbf{N} \, dA \qquad \rightarrow \text{eq. } (9.3_3)$$

$$= \lceil \boldsymbol{\varphi} \otimes \mathbf{N} \rfloor_0 \quad . \tag{C.146}$$

Piola stress tensor

Assuming the existence of a motion potential field, then

$$\overline{\overline{\mathbf{P}}}^T \left(\overline{\overline{\mathbf{X}}} \right) = \left\langle \mathbf{P}^T \right\rangle_0 + \left\langle \mathbf{X} \otimes \underbrace{[\nabla_0 \cdot \mathbf{P}]}_{=\mathbf{0}} \right\rangle_0 \qquad \leftarrow \text{eqs. } (9.3_1, 5.87)$$

$$= \left\langle \underbrace{[\nabla_0 \mathbf{X}]}_{=\mathbf{I}} \cdot \mathbf{P}^T + \mathbf{X} \otimes [\nabla_0 \cdot \mathbf{P}] \right\rangle_0 \qquad \rightarrow \text{eq. (A.37)}$$

$$= \langle \nabla_0 \cdot [\mathbf{X} \otimes \mathbf{P}] \rangle_0 \qquad \rightarrow \text{eq. } (B.16_2)$$

$$= \lceil [\mathbf{X} \otimes \mathbf{P}] \cdot \mathbf{N} \rfloor_0 = \lceil \mathbf{X} \otimes [\mathbf{P} \cdot \mathbf{N}] \rfloor_0$$

$$\Rightarrow \quad \overline{\overline{\mathbf{P}}} \left(\overline{\overline{\mathbf{X}}} \right) = \lceil [\mathbf{P} \cdot \mathbf{N}] \otimes \mathbf{X} \rfloor_0 \quad . \tag{C.147}$$

Note that we could deduce the dual formulations for equations (C.146) and (C.147), where instead of assuming the potential field for the motion, a potential field from which the stress is derived is instead considered. This extension is presented in part in [82], and in such a case the expression for $\overline{\overline{\mathbf{P}}}$ is determined through application of the dual potential, while that of $\overline{\overline{\mathbf{F}}}$ is determined through application of the compatibility condition stated in equation (5.94).

Magnetic induction vector

Assuming the existence of a magnetic vector potential, then

$$\overline{\overline{\mathbb{B}}} \left(\overline{\overline{\mathbf{X}}} \right) = \frac{1}{V_0} \int_{\mathscr{B}_0} \mathbb{B} \; dV \qquad \leftarrow \text{eq. } (9.3_1) \rightarrow \text{eq. (5.141)}$$

$$= \frac{1}{V_0} \int_{\mathscr{B}_0} \nabla_0 \times \mathbb{A} \; dV \qquad \rightarrow \text{eq. (B.40)}$$

$$= \frac{1}{V_0} \int_{\partial \mathscr{B}_0} \mathbf{N} \times \mathbb{A} \; dA \qquad \rightarrow \text{eqs. } (9.3_3, A.1)$$

$$= \lceil \mathbf{N} \times \mathbb{A} \rfloor_0 = - \lceil \mathbb{A} \times \mathbf{N} \rfloor_0 \quad . \tag{C.148}$$

When the magnetic induction is not the primary variable, then

$$\overline{\overline{\mathbb{B}}} \left(\overline{\overline{\mathbf{X}}} \right) = \langle \mathbb{B} \rangle_0 + \left\langle \mathbf{X} \underbrace{[\nabla_0 \cdot \mathbb{B}]}_{=0} \right\rangle_0 \qquad \leftarrow \text{eqs. } (9.3_1, 5.67_2)$$

$$= \left\langle \underbrace{[\nabla_0 \mathbf{X}]}_{=\mathbf{I}} \cdot \mathbb{B} + \mathbf{X}\,[\nabla_0 \cdot \mathbb{B}] \right\rangle_0 \qquad \rightarrow \text{eq. (A.33)}$$

$$= \langle \nabla_0 \cdot [\mathbf{X} \otimes \mathbb{B}] \rangle_0 \qquad \rightarrow \text{eq. (B.16}_2\text{)}$$

$$= \lceil [\mathbf{X} \otimes \mathbb{B}] \cdot \mathbf{N} \rfloor_0 = \lceil [\mathbb{B} \cdot \mathbf{N}]\,\mathbf{X} \rfloor_0 \quad . \qquad \text{(C.149)}$$

Magnetic field vector

Assuming the existence of a magnetic scalar potential, then

$$\overline{\overline{\mathbb{H}}}\left(\overline{\overline{\mathbf{X}}}\right) = \frac{1}{V_0} \int\limits_{\mathscr{B}_0} \mathbb{H}\,\mathrm{dV} \qquad \leftarrow \text{eq. (9.3}_1\text{)} \rightarrow \text{eq. (5.146)}$$

$$= \frac{1}{V_0} \int\limits_{\mathscr{B}_0} [-\nabla_0 \Phi]\,\mathrm{dV} \qquad \rightarrow \text{eq. (B.17)}$$

$$= -\frac{1}{V_0} \int\limits_{\partial\mathscr{B}_0} \Phi \mathbf{N}\,\mathrm{dA} \qquad \rightarrow \text{eq. (9.3}_3\text{)}$$

$$= -\lceil \Phi \mathbf{N} \rfloor_0 \quad . \qquad \text{(C.150)}$$

When the magnetic field is not the primary variable, then

$$\overline{\overline{\mathbb{H}}}\left(\overline{\overline{\mathbf{X}}}\right) = \langle \mathbb{H} \rangle_0 + \left\langle \mathbf{X} \times \underbrace{[\nabla_0 \times \mathbb{H}]}_{=\mathbf{0}} \right\rangle_0 \qquad \leftarrow \text{eqs. (9.3}_1\text{,5.84)} \rightarrow \text{eq. (A.40)}$$

$$= \left\langle \left[\underbrace{[\nabla_0 \mathbf{X}]}_{=\mathbf{I}} \cdot \mathbb{H} + \mathbf{X} \cdot \nabla_0 \mathbb{H} \right] - \nabla_0 \mathbb{H} \cdot \mathbf{X} \right\rangle_0 \qquad \rightarrow \text{eq. (A.27)}$$

$$= \left\langle \nabla_0\,[\mathbb{H} \cdot \mathbf{X}] - \left[\nabla_0 \mathbb{H} \cdot \mathbf{X} + \mathbb{H} \cdot \underbrace{[\nabla_0 \cdot \mathbf{X}]}_{=0} \right] \right\rangle_0 \qquad \rightarrow \text{eq. (A.33)}$$

$$= \langle \nabla_0 \cdot [[\mathbb{H} \cdot \mathbf{X}]\,\mathbf{I}] - \nabla_0 \cdot [\mathbb{H} \otimes \mathbf{X}] \rangle_0 \qquad \rightarrow \text{eq. (B.16}_2\text{)}$$

$$= \lceil [\mathbb{H} \cdot \mathbf{X}]\,\mathbf{N} - [\mathbb{H} \otimes \mathbf{X}] \cdot \mathbf{N} \rfloor_0 \qquad \rightarrow \text{eq. (A.15)}$$

$$= \lceil [\mathbb{H} \times \mathbf{N}] \times \mathbf{X} \rfloor_0 \quad . \qquad \text{(C.151)}$$

C.8.2 The Hill–Mandel condition

Quasi-static mechanical problem

From equation (9.11) and assuming a fluctuating micro-scale displacement field as given in equation (9.18), the micro-scale mechanical virtual power is

$$\langle \mathbf{P} : \delta \mathbf{F} \rangle_0 = \left\langle \mathbf{P} : \delta\left[\overline{\overline{\mathbf{F}}} + \widetilde{\mathbf{F}}\right] \right\rangle_0 \qquad \leftarrow \text{eq. (9.19)}$$

$$= \left\langle \mathbf{P} : \delta\overline{\overline{\mathbf{F}}} \right\rangle_0 + \left\langle \mathbf{P} : \delta\widetilde{\mathbf{F}} \right\rangle_0 \qquad \rightarrow \text{eqs. (5.170}_1\text{,5.87)}$$

$$= \langle \mathbf{P} \rangle_0 : \delta\overline{\overline{\mathbf{F}}} + \left\langle \mathbf{P} : [\nabla_0 \delta\widetilde{\boldsymbol{\varphi}}] + \delta\widetilde{\boldsymbol{\varphi}} \cdot \underbrace{[\nabla_0 \cdot \mathbf{P}]}_{=\mathbf{0}} \right\rangle_0 \qquad \rightarrow \text{eqs. (9.5,A.35)}$$

$$= \overline{\overline{\mathbf{P}}} : \delta\overline{\overline{\mathbf{F}}} + \langle \nabla_0 \cdot [\delta\widetilde{\boldsymbol{\varphi}} \cdot \mathbf{P}] \rangle_0 \qquad \rightarrow \text{eq. (B.16)}$$

$$= \overline{\overline{\mathbf{P}}} : \delta\overline{\overline{\mathbf{F}}} + \lceil \delta\widetilde{\boldsymbol{\varphi}} \cdot [\mathbf{P} \cdot \mathbf{N}] \rfloor_0 \quad . \qquad \text{(C.152)}$$

Assuming that the traction condition given in equation (9.23) holds, the micro-scale mechanical virtual power contributing to equation (9.11) can then be rearranged as

$$\langle \mathbf{P} : \delta\mathbf{F} \rangle_0 = \left\langle \mathbf{P} : \delta\mathbf{F} + \delta\boldsymbol{\varphi} \cdot \underbrace{[\nabla_0 \cdot \mathbf{P}]}_{=\mathbf{0}} \right\rangle_0 \qquad \leftarrow \text{eq. (5.87)} \rightarrow \text{eq. } (5.170_1)$$

$$= \langle \mathbf{P} : [\nabla_0 \delta\boldsymbol{\varphi}] + \delta\boldsymbol{\varphi} \cdot [\nabla_0 \cdot \mathbf{P}] \rangle_0 \qquad \rightarrow \text{eq. (A.35)}$$

$$= \langle \nabla_0 \cdot [\delta\boldsymbol{\varphi} \cdot \mathbf{P}] \rangle_0 \qquad \rightarrow \text{eq. (B.16)}$$

$$= \lceil \delta\boldsymbol{\varphi} \cdot [\mathbf{P} \cdot \mathbf{N}] \rfloor_0 \qquad \rightarrow \text{eq. (9.23)}$$

$$= \left\lceil \delta\boldsymbol{\varphi} \cdot \left[\overline{\overline{\mathbf{P}}} \cdot \mathbf{N} \right] \right\rfloor_0$$

$$= \overline{\overline{\mathbf{P}}} : \lceil \delta\boldsymbol{\varphi} \otimes \mathbf{N} \rfloor_0 \qquad \rightarrow \text{eq. (9.4)}$$

$$= \overline{\overline{\mathbf{P}}} : \delta\overline{\overline{\mathbf{F}}} \quad . \qquad \text{(C.153)}$$

Quasi-static magnetic problem: MSP formulation

From equation (9.11) and assuming a fluctuating micro-scale magnetic scalar potential field as given in equation (9.32), the micro-scale magnetic virtual power is

$$-\langle \mathbb{B} \cdot \delta\mathbb{H} \rangle_0 = -\left\langle \mathbb{B} \cdot \delta\left[\overline{\overline{\mathbb{H}}} + \widetilde{\mathbb{H}} \right] \right\rangle_0 \qquad \leftarrow \text{eq. (9.33)}$$

$$= -\left\langle \mathbb{B} \cdot \delta\overline{\overline{\mathbb{H}}} \right\rangle_0 - \langle \mathbb{B} \cdot \delta\widetilde{\mathbb{H}} \rangle_0 \qquad \rightarrow \text{eqs. } (5.184, 5.67_2)$$

$$= -\langle \mathbb{B} \rangle_0 \cdot \delta\overline{\overline{\mathbb{H}}} + \left\langle \mathbb{B} \cdot [\nabla_0 \delta\widetilde{\Phi}] + \delta\widetilde{\Phi} \underbrace{[\nabla_0 \cdot \mathbb{B}]}_{=0} \right\rangle_0 \qquad \rightarrow \text{eqs. } (9.6_2, \text{A.28})$$

$$= -\overline{\overline{\mathbb{B}}} \cdot \delta\overline{\overline{\mathbb{H}}} + \langle \nabla_0 \cdot [\delta\widetilde{\Phi}\mathbb{B}] \rangle_0 \qquad \rightarrow \text{eq. (B.16)}$$

$$= -\overline{\overline{\mathbb{B}}} \cdot \delta\overline{\overline{\mathbb{H}}} + \lceil \delta\widetilde{\Phi} \, [\mathbb{B} \cdot \mathbf{N}] \rfloor_0 \quad . \qquad \text{(C.154)}$$

Assuming that the magnetic induction condition given in equation (9.37) holds, the micro-scale magnetic virtual power contributing to equation (9.11) can then be rearranged as

$$-\langle \mathbb{B} \cdot \delta\mathbb{H} \rangle_0 = \left\langle -\mathbb{B} \cdot \delta\mathbb{H} + \delta\Phi \underbrace{[\nabla_0 \cdot \mathbb{B}]}_{=0} \right\rangle_0 \qquad \leftarrow \text{eq. } (5.67_2) \rightarrow \text{eq. (5.184)}$$

$$= \langle \mathbb{B} \cdot [\nabla_0 \delta\Phi] + \delta\Phi \, [\nabla_0 \cdot \mathbb{B}] \rangle_0 \qquad \rightarrow \text{eq. (A.28)}$$

$$
\begin{aligned}
&= \langle \nabla_0 \cdot [\delta\Phi\mathbb{B}]\rangle_0 && \rightarrow \text{eq. (B.16)}\\
&= \lceil \delta\Phi\, [\mathbb{B}\cdot\mathbf{N}]\rfloor_0 && \rightarrow \text{eq. (9.37)}\\
&= \left\lceil \delta\Phi \left[\overline{\overline{\mathbb{B}}}\cdot\mathbf{N}\right]\right\rfloor_0 = \overline{\overline{\mathbb{B}}}\cdot\lceil\delta\Phi\mathbf{N}\rfloor_0 && \rightarrow \text{eq. } (9.8_1)\\
&= -\overline{\overline{\mathbb{B}}}\cdot\delta\overline{\overline{\mathbb{H}}} \quad . && \text{(C.155)}
\end{aligned}
$$

Quasi-static magnetic problem: MVP formulation

From equation (9.11) and assuming a fluctuating micro-scale magnetic vector potential field as given in equation (9.25), the micro-scale magnetic virtual power is

$$
\begin{aligned}
\langle \mathbb{H}\cdot\delta\mathbb{B}\rangle_0 &= \left\langle \mathbb{H}\cdot\delta\left[\overline{\overline{\mathbb{B}}}+\widetilde{\mathbb{B}}\right]\right\rangle_0 && \leftarrow \text{eq. (9.26)}\\
&= \left\langle \mathbb{H}\cdot\delta\overline{\overline{\mathbb{B}}}\right\rangle_0 + \langle \mathbb{H}\cdot\delta\widetilde{\mathbb{B}}\rangle_0 && \rightarrow \text{eqs. (5.170,5.84)}\\
&= \langle\mathbb{H}\rangle_0\cdot\delta\overline{\overline{\mathbb{B}}} + \left\langle \mathbb{H}\cdot[\nabla_0\times\delta\widetilde{\mathbb{A}}] - \delta\widetilde{\mathbb{A}}\cdot\underbrace{[\nabla_0\times\mathbb{H}]}_{=\mathbf{0}}\right\rangle_0 && \rightarrow \text{eq. (A.31)}\\
&= \overline{\overline{\mathbb{H}}}\cdot\delta\overline{\overline{\mathbb{B}}} + \langle \mathbb{H}\cdot[\nabla_0\times\delta\widetilde{\mathbb{A}}] - \mathbb{H}\cdot[\nabla_0\times\delta\widetilde{\mathbb{A}}] + \nabla_0\cdot[\delta\widetilde{\mathbb{A}}\times\mathbb{H}]\rangle_0\\
&= \overline{\overline{\mathbb{H}}}\cdot\delta\overline{\overline{\mathbb{B}}} + \langle\nabla_0\cdot[\delta\widetilde{\mathbb{A}}\times\mathbb{H}]\rangle_0 && \rightarrow \text{eq. (B.16)}\\
&= \overline{\overline{\mathbb{H}}}\cdot\delta\overline{\overline{\mathbb{B}}} + \lceil\mathbf{N}\cdot[\delta\widetilde{\mathbb{A}}\times\mathbb{H}]\rfloor_0 && \rightarrow \text{eq. (A.12)}\\
&= \overline{\overline{\mathbb{H}}}\cdot\delta\overline{\overline{\mathbb{B}}} + \lceil\delta\widetilde{\mathbb{A}}\cdot[\mathbb{H}\times\mathbf{N}]\rfloor_0 \quad . && \text{(C.156)}
\end{aligned}
$$

Assuming that the magnetic field condition given in equation (9.30) holds, the micro-scale magnetic virtual power contributing to equation (9.11) can then be rearranged as

$$
\begin{aligned}
\langle \mathbb{H}\cdot\delta\mathbb{B}\rangle_0 &= \left\langle \mathbb{H}\cdot\delta\mathbb{B} - \delta\mathbb{A}\cdot\underbrace{[\nabla_0\times\mathbb{H}]}_{=\mathbf{0}}\right\rangle_0 && \leftarrow \text{eq. (5.84)}\rightarrow \text{eq. (5.170)}\\
&= \langle \mathbb{H}\cdot[\nabla_0\times\delta\mathbb{A}] - \delta\mathbb{A}\cdot[\nabla_0\times\mathbb{H}]\rangle_0 && \rightarrow \text{eq. (A.31)}\\
&= \langle\nabla_0\cdot[\delta\mathbb{A}\times\mathbb{H}]\rangle_0 && \rightarrow \text{eq. (B.16)}\\
&= \lceil\mathbf{N}\cdot[\delta\mathbb{A}\times\mathbb{H}]\rfloor_0 && \rightarrow \text{eq. (A.12)}\\
&= \lceil\delta\mathbb{A}\cdot[\mathbb{H}\times\mathbf{N}]\rfloor_0 && \rightarrow \text{eq. (9.30)}\\
&= \left\lceil\delta\mathbb{A}\cdot\left[\overline{\overline{\mathbb{H}}}\times\mathbf{N}\right]\right\rfloor_0 && \rightarrow \text{eq. (A.12)}\\
&= \left\lceil\overline{\overline{\mathbb{H}}}\cdot[\mathbf{N}\times\delta\mathbb{A}]\right\rfloor_0 = \overline{\overline{\mathbb{H}}}\cdot\lceil[\mathbf{N}\times\delta\mathbb{A}]\rfloor_0 && \rightarrow \text{eqs. } (\text{A.1}, 9.7_1)\\
&= \overline{\overline{\mathbb{H}}}\cdot\delta\overline{\overline{\mathbb{B}}} \quad . && \text{(C.157)}
\end{aligned}
$$

C.8.3 Algorithmically consistent tangent moduli

Following the approach presented by Keip and Rambausek [255], the starting point for deriving the algorithmically consistent tangent moduli for the macro-scale problem is equation (9.53). Applying the definitions in equations (9.3) and (9.5), $(9.6)_2$, the macro-scale moduli can be expanded as

$$\overline{\overline{\mathcal{A}}} = \frac{\partial \langle \mathbf{P} \rangle_0}{\partial \overline{\overline{\mathbf{F}}}} = \left\langle \frac{\partial \mathbf{P}}{\partial \overline{\overline{\mathbf{F}}}} \right\rangle_0 \quad , \tag{C.158a}$$

$$\overline{\overline{\mathfrak{L}}} = -\frac{\partial \langle \mathbf{P} \rangle_0}{\partial \overline{\overline{\mathbb{H}}}} = \left\langle -\frac{\partial \mathbf{P}}{\partial \overline{\overline{\mathbb{H}}}} \right\rangle_0 \equiv \left\langle \frac{\partial \mathbb{B}}{\partial \overline{\overline{\mathbf{F}}}} \right\rangle_0^T \quad , \tag{C.158b}$$

$$\overline{\overline{\mathbf{D}}} = \frac{\partial \langle \mathbb{B} \rangle_0}{\partial \overline{\overline{\mathbb{H}}}} = \left\langle \frac{\partial \mathbb{B}}{\partial \overline{\overline{\mathbb{H}}}} \right\rangle_0 \quad . \tag{C.158c}$$

Considering the definitions of the micro-scale fields given in equations (9.19) and (9.33), application of the chain rule to the integrands of the above renders

$$\frac{\partial \mathbf{P}}{\partial \overline{\overline{\mathbf{F}}}} = \frac{\partial \mathbf{P}}{\partial \mathbf{F}} : \frac{\partial \mathbf{F}}{\partial \overline{\overline{\mathbf{F}}}} = \mathcal{A} : \frac{\partial [\overline{\overline{\mathbf{F}}} + \widetilde{\mathbf{F}}]}{\partial \overline{\overline{\mathbf{F}}}} = \mathcal{A} + \mathcal{A} : \frac{\partial \widetilde{\mathbf{F}}}{\partial \overline{\overline{\mathbf{F}}}} \quad , \tag{C.159a}$$

$$-\frac{\partial \mathbf{P}}{\partial \overline{\overline{\mathbb{H}}}} = -\frac{\partial \mathbf{P}}{\partial \mathbb{H}} \cdot \frac{\partial \mathbb{H}}{\partial \overline{\overline{\mathbb{H}}}} = \mathfrak{L} \cdot \frac{\partial [\overline{\overline{\mathbb{H}}} + \widetilde{\mathbb{H}}]}{\partial \overline{\overline{\mathbb{H}}}} = \mathfrak{L} + \mathfrak{L} \cdot \frac{\partial \widetilde{\mathbb{H}}}{\partial \overline{\overline{\mathbb{H}}}} \quad , \tag{C.159b}$$

$$\frac{\partial \mathbb{B}}{\partial \overline{\overline{\mathbb{H}}}} = \frac{\partial \mathbb{B}}{\partial \mathbb{H}} \cdot \frac{\partial \mathbb{H}}{\partial \overline{\overline{\mathbb{H}}}} = \mathbf{D} \cdot \frac{\partial [\overline{\overline{\mathbb{H}}} + \widetilde{\mathbb{H}}]}{\partial \overline{\overline{\mathbb{H}}}} = \mathbf{D} + \mathbf{D} \cdot \frac{\partial \widetilde{\mathbb{H}}}{\partial \overline{\overline{\mathbb{H}}}} \quad , \tag{C.159c}$$

thereby linking the macro-scale tangents to the consistent micro-scale moduli. The above can now be compactly summarised as

$$\begin{aligned} \begin{bmatrix} \Delta \overline{\overline{\mathbf{P}}} \\ -\Delta \overline{\overline{\mathbb{B}}} \end{bmatrix} &= \begin{bmatrix} \overline{\overline{\mathcal{A}}} & -\overline{\overline{\mathfrak{L}}} \\ -\overline{\overline{\mathfrak{L}}}^T & -\overline{\overline{\mathbf{D}}} \end{bmatrix} \cdot \begin{bmatrix} \Delta \overline{\overline{\mathbf{F}}} \\ \Delta \overline{\overline{\mathbb{H}}} \end{bmatrix} \\ &= \left\langle \begin{bmatrix} \mathcal{A} & -\mathfrak{L} \\ -\mathfrak{L}^T & -\mathbf{D} \end{bmatrix} + \begin{bmatrix} \mathcal{A} : \frac{\partial \widetilde{\mathbf{F}}}{\partial \overline{\overline{\mathbf{F}}}} & -\mathfrak{L} \cdot \frac{\partial \widetilde{\mathbb{H}}}{\partial \overline{\overline{\mathbb{H}}}} \\ -\mathfrak{L}^T : \frac{\partial \widetilde{\mathbf{F}}}{\partial \overline{\overline{\mathbf{F}}}} & -\mathbf{D} \cdot \frac{\partial \widetilde{\mathbb{H}}}{\partial \overline{\overline{\mathbb{H}}}} \end{bmatrix} \right\rangle_0 \cdot \begin{bmatrix} \Delta \overline{\overline{\mathbf{F}}} \\ \Delta \overline{\overline{\mathbb{H}}} \end{bmatrix} \quad . \end{aligned} \tag{C.160}$$

Note that the contraction between the respective elements of the matrix and vector elements is a mixture of single and double contractions (for the vector and tensor components, respectively).

What remains is to determine the sensitivities of the micro-scale fluctuation fields to their corresponding macro-scale fields, namely $\frac{\partial \widetilde{\mathbf{F}}}{\partial \overline{\overline{\mathbf{F}}}}$ and $\frac{\partial \widetilde{\mathbb{H}}}{\partial \overline{\overline{\mathbb{H}}}}$; this can be achieved through the application of FEM. For the micro-scale problem, the linear expansion of the residual is given by equation (5.187), wherein the residual is defined by equations (5.183) and (5.185) and the linearisation by equation (5.189). Importantly, it should observed that under the condition of equilibrium the residual linearisation renders

$$0 \doteq \Delta \delta \Pi^{\text{int}} = \Delta \delta_{\boldsymbol{\varphi}} \Pi^{\text{int}} + \Delta \delta_{\Phi} \Pi^{\text{int}} \quad . \tag{C.161}$$

When considering equations (9.19) and (9.33) and assuming that the loading is a dead load, then the components of the linearisation can be reformulated as

$$\Delta\delta_{\boldsymbol{\varphi}}\Pi^{\text{int}} = \int\limits_{\mathscr{D}_0} \left[\delta\mathbf{F} : \mathscr{A} : \left[\Delta\overline{\overline{\mathbf{F}}} + \Delta\widetilde{\mathbf{F}}\right] - \delta\mathbf{F} : \mathfrak{L} \cdot \left[\Delta\overline{\overline{\mathbb{H}}} + \Delta\widetilde{\mathbb{H}}\right]\right] \, \mathrm{dV} \quad , \tag{C.162a}$$

$$\Delta\delta_{\Phi}\Pi^{\text{int}} = \int\limits_{\mathscr{D}_0} \left[-\delta\mathbb{H} \cdot \mathfrak{L}^T : \left[\Delta\overline{\overline{\mathbf{F}}} + \Delta\widetilde{\mathbf{F}}\right] - \delta\mathbb{H} \cdot \mathbf{D} \cdot \left[\Delta\overline{\overline{\mathbb{H}}} + \Delta\widetilde{\mathbb{H}}\right]\right] \, \mathrm{dV} \quad . \tag{C.162b}$$

Next it is necessary to apply a FE ansatz for the field variations given by equations (6.7) and (6.23), and define the analogous discretisation for the fluctuation of the primary fields. The gradients of these fluctuation fields can be expressed by the equivalent of equations (6.6) and (6.22) in conjunction with equation $(5.175)_1$ and (5.191), that is,

$$\begin{aligned} \Delta\widetilde{\mathbf{F}} &= \Delta\widetilde{\mathbf{F}}(\mathbf{X}) = \nabla_0 \Delta\widetilde{\boldsymbol{\varphi}}(\mathbf{X}) \approx \sum_{I}^{n_{\Psi_{\boldsymbol{\varphi}}}} \Delta\widetilde{\boldsymbol{\varphi}}^I \nabla_0 \boldsymbol{\Psi}^I_{\boldsymbol{\varphi}}(\mathbf{X}) \quad , \\ \Delta\widetilde{\mathbb{H}} &= \Delta\widetilde{\mathbb{H}}(\mathbf{X}) = -\nabla_0 \Delta\widetilde{\Phi}(\mathbf{X}) \approx -\sum_{I}^{n_{\Psi_{\Phi}}} \Delta\widetilde{\Phi}^I \nabla_0 \Psi^I_{\Phi}(\mathbf{X}) \quad . \end{aligned} \tag{C.163}$$

With this, the discrete form of the linearisation is

$$\begin{aligned} \Delta\delta_{\boldsymbol{\varphi}}\Pi^{\text{int}} \approx{}& \sum_{I}^{n_{\Psi_{\boldsymbol{\varphi}}}} \delta\widetilde{\boldsymbol{\varphi}}^I \int\limits_{\mathscr{D}_0} \nabla_0 \boldsymbol{\Psi}^I_{\boldsymbol{\varphi}} : \mathscr{A} : \left[\Delta\overline{\overline{\mathbf{F}}} + \sum_{J}^{n_{\Psi_{\boldsymbol{\varphi}}}} \Delta\widetilde{\boldsymbol{\varphi}}^J \nabla_0 \boldsymbol{\Psi}^J_{\boldsymbol{\varphi}}\right] \mathrm{dV} \\ &+ \sum_{I}^{n_{\Psi_{\boldsymbol{\varphi}}}} \delta\widetilde{\boldsymbol{\varphi}}^I \int\limits_{\mathscr{D}_0} -\nabla_0 \boldsymbol{\Psi}^I_{\boldsymbol{\varphi}} : \mathfrak{L} \cdot \left[\Delta\overline{\overline{\mathbb{H}}} - \sum_{J}^{n_{\Psi_{\Phi}}} \Delta\widetilde{\Phi}^J \nabla_0 \Psi^J_{\Phi}\right] \mathrm{dV} \quad , \end{aligned} \tag{C.164a}$$

$$\begin{aligned} \Delta\delta_{\Phi}\Pi^{\text{int}} \approx{}& -\sum_{I}^{n_{\Psi_{\Phi}}} \delta\widetilde{\Phi}^I \int\limits_{\mathscr{D}_0} -\nabla_0 \Psi^I_{\Phi} \cdot \mathfrak{L}^T : \left[\Delta\overline{\overline{\mathbf{F}}} + \sum_{J}^{n_{\Psi_{\boldsymbol{\varphi}}}} \Delta\widetilde{\boldsymbol{\varphi}}^J \nabla_0 \boldsymbol{\Psi}^J_{\boldsymbol{\varphi}}\right] \mathrm{dV} \\ &- \sum_{I}^{n_{\Psi_{\Phi}}} \delta\widetilde{\Phi}^I \int\limits_{\mathscr{D}_0} -\nabla_0 \Psi^I_{\Phi} \cdot \mathbf{D} \cdot \left[\Delta\overline{\overline{\mathbb{H}}} - \sum_{J}^{n_{\Psi_{\Phi}}} \Delta\widetilde{\Phi}^J \nabla_0 \Psi^J_{\Phi}\right] \mathrm{dV} \quad . \end{aligned} \tag{C.164b}$$

After summing these discrete linearisations and with some rearrangement of terms, a sparse linear system results. Noting that $\Delta\overline{\overline{\mathbf{F}}}$ and $\Delta\overline{\overline{\mathbb{H}}}$ are spatially constant fields, this can be expressed in matrix-vector form as

$$0 \doteq \begin{bmatrix} \delta\boldsymbol{\varphi} \\ \delta\Phi \end{bmatrix} \cdot \left[\underbrace{\begin{bmatrix} \boldsymbol{L}_{\boldsymbol{\varphi}\boldsymbol{\varphi}} & \boldsymbol{L}_{\boldsymbol{\varphi}\Phi} \\ \boldsymbol{L}_{\Phi\boldsymbol{\varphi}} & \boldsymbol{L}_{\Phi\Phi} \end{bmatrix}}_{\boldsymbol{L}} \cdot \underbrace{\begin{bmatrix} \Delta\overline{\overline{\mathbf{F}}} \\ \Delta\overline{\overline{\mathbb{H}}} \end{bmatrix}}_{\Delta\overline{\overline{\boldsymbol{s}}}} + \underbrace{\begin{bmatrix} \boldsymbol{K}_{\boldsymbol{\varphi}\boldsymbol{\varphi}} & \boldsymbol{K}_{\boldsymbol{\varphi}\Phi} \\ \boldsymbol{K}_{\Phi\boldsymbol{\varphi}} & \boldsymbol{K}_{\Phi\Phi} \end{bmatrix}}_{\boldsymbol{K}} \cdot \underbrace{\begin{bmatrix} \Delta\widetilde{\boldsymbol{d}}_{\boldsymbol{\varphi}} \\ \Delta\widetilde{\boldsymbol{d}}_{\Phi} \end{bmatrix}}_{\Delta\widetilde{\boldsymbol{d}}} \right] \tag{C.165}$$

where the elements of the matrices $\boldsymbol{L}$ and $\boldsymbol{K}$ are respectively stated in equations (9.60) and (9.61). To accommodate the mixed rank nature of the fields that form the macro-scale field vector $\overline{\overline{\mathbf{s}}}$, Schröder and Keip [477] (as well as [257, 255]) recommend "flattening" these fields using an index ordering strategy in the spirit of either Voigt or Kelvin [393] notation. In this case, the matrix $\boldsymbol{L}$ is of size $n_{\Psi} \times [n_{\mathbf{F}} + n_{\mathbb{H}}]$, where $n_{\mathbf{F}}$ and $n_{\mathbb{H}}$ respectively represent the number of independent components in these tensor- and vector-valued fields, and the vector $\Delta\overline{\overline{\mathbf{s}}}$ has size $[n_{\mathbf{F}} + n_{\mathbb{H}}] \times 1$. By invoking the arbitrariness of the FE test functions and rearranging equation (C.165) as

$$\Delta\widetilde{\boldsymbol{d}} = -\boldsymbol{K}^{-1} \cdot \boldsymbol{L} \cdot \Delta\overline{\overline{\mathbf{s}}} \tag{C.166}$$

which expresses the global approximation of the increments of the fluctuation fields at the support points. Applying the FE approximation of the incremental macro-scale gradient fields

$$\begin{aligned} \Delta\overline{\overline{\mathbf{F}}}\left(\overline{\overline{\mathbf{X}}}\right) &= \nabla_0 \Delta\overline{\overline{\boldsymbol{\varphi}}}\left(\overline{\overline{\mathbf{X}}}\right) \approx \sum_I^{n_{\Psi_{\boldsymbol{\varphi}}}} \Delta\overline{\overline{\varphi}}^I \left(\overline{\overline{\mathbf{X}}}\right) \nabla_0 \boldsymbol{\Psi}^I_{\boldsymbol{\varphi}}(\mathbf{X}) \quad , \\ \Delta\overline{\overline{\mathbb{H}}}\left(\overline{\overline{\mathbf{X}}}\right) &= -\nabla_0 \Delta\overline{\overline{\Phi}}\left(\overline{\overline{\mathbf{X}}}\right) \approx -\sum_I^{n_{\Psi_{\Phi}}} \Delta\overline{\overline{\Phi}}^I \left(\overline{\overline{\mathbf{X}}}\right) \nabla_0 \Psi^I_{\Phi}(\mathbf{X}) \quad , \end{aligned} \tag{C.167}$$

the second term in equation (C.160) can be rewritten as

$$\begin{aligned} \left\langle \begin{bmatrix} \mathscr{A} : \dfrac{\partial \overline{\mathbf{F}}}{\partial \overline{\overline{\mathbf{F}}}} & -\mathfrak{L} \cdot \dfrac{\partial \overline{\mathbb{H}}}{\partial \overline{\overline{\mathbb{H}}}} \\ -\mathfrak{L}^T : \dfrac{\partial \overline{\mathbf{F}}}{\partial \overline{\overline{\mathbf{F}}}} & -\mathbf{D} \cdot \dfrac{\partial \overline{\mathbb{H}}}{\partial \overline{\overline{\mathbb{H}}}} \end{bmatrix} \right\rangle_0 &= \frac{1}{\mathrm{V}_0} \int_{\mathscr{D}_0} \begin{bmatrix} \mathscr{A} : \nabla_0 \boldsymbol{\Psi}^I_{\boldsymbol{\varphi}} & \mathfrak{L} \cdot \nabla_0 \Psi^I_{\Phi} \\ -\mathfrak{L}^T : \nabla_0 \boldsymbol{\Psi}^I_{\boldsymbol{\varphi}} & \mathbf{D} \cdot \nabla_0 \Psi^I_{\Phi} \end{bmatrix} \mathrm{dV} \cdot \begin{bmatrix} \dfrac{\partial \Delta\widetilde{\boldsymbol{d}}_{\boldsymbol{\varphi}}}{\partial \overline{\overline{\mathbf{F}}}} & \dfrac{\partial \Delta\widetilde{\boldsymbol{d}}_{\boldsymbol{\varphi}}}{\partial \overline{\overline{\mathbb{H}}}} \\ \dfrac{\partial \Delta\widetilde{\boldsymbol{d}}_{\Phi}}{\partial \overline{\overline{\mathbf{F}}}} & \dfrac{\partial \Delta\widetilde{\boldsymbol{d}}_{\Phi}}{\partial \overline{\overline{\mathbb{H}}}} \end{bmatrix} \\ &= \frac{1}{\mathrm{V}_0} \begin{bmatrix} \boldsymbol{L}^T_{\boldsymbol{\varphi}\boldsymbol{\varphi}} & \boldsymbol{L}^T_{\Phi\boldsymbol{\varphi}} \\ \boldsymbol{L}^T_{\boldsymbol{\varphi}\Phi} & \boldsymbol{L}^T_{\Phi\Phi} \end{bmatrix} \cdot \begin{bmatrix} \dfrac{\partial \Delta\widetilde{\boldsymbol{d}}_{\boldsymbol{\varphi}}}{\partial \overline{\overline{\mathbf{F}}}} & \dfrac{\partial \Delta\widetilde{\boldsymbol{d}}_{\boldsymbol{\varphi}}}{\partial \overline{\overline{\mathbb{H}}}} \\ \dfrac{\partial \Delta\widetilde{\boldsymbol{d}}_{\Phi}}{\partial \overline{\overline{\mathbf{F}}}} & \dfrac{\partial \Delta\widetilde{\boldsymbol{d}}_{\Phi}}{\partial \overline{\overline{\mathbb{H}}}} \end{bmatrix} \\ &\equiv \frac{1}{\mathrm{V}_0} \boldsymbol{L}^T \cdot \frac{\partial \Delta\widetilde{\boldsymbol{d}}}{\partial \overline{\overline{\mathbf{s}}}} \quad . \end{aligned} \tag{C.168}$$

Finally, exploiting equation (C.166) to compute the partial derivative in the above (a matrix of the same dimensions as $\boldsymbol{L}$), namely through the relation

$$\frac{\partial \Delta\widetilde{\boldsymbol{d}}}{\partial \overline{\overline{\mathbf{s}}}} = -\boldsymbol{K}^{-1} \cdot \boldsymbol{L} \quad , \tag{C.169}$$

the resultant expression for the macro-scale tangent moduli may be resolved. This set of linear systems can be solved without explicitly computing $\boldsymbol{K}^{-1}$ by considering each column $\boldsymbol{l}_i$ of matrix $\boldsymbol{L}$ individually, that is,

$$\left[\frac{\partial \Delta\widetilde{\boldsymbol{d}}}{\partial \overline{\overline{s}}_1}, \frac{\partial \Delta\widetilde{\boldsymbol{d}}}{\partial \overline{\overline{s}}_2}, \ldots, \frac{\partial \Delta\widetilde{\boldsymbol{d}}}{\partial \overline{\overline{s}}_i} \right] = -\boldsymbol{K}^{-1} \cdot [\boldsymbol{l}_1, \boldsymbol{l}_2, \ldots, \boldsymbol{l}_i] \quad \forall\, i \in \{1, \ldots, n_{\mathbf{F}} + n_{\mathbb{H}}\} \quad . \tag{C.170}$$

Inserting the above into equation (C.160), the collection of algorithmically correct decoupled and coupled macro-scale material tangents are

$$\begin{bmatrix} \overline{\overline{\mathscr{A}}} & -\overline{\overline{\mathfrak{L}}} \\ -\overline{\overline{\mathfrak{L}}}^T & -\overline{\overline{\mathbf{D}}} \end{bmatrix} = \left\langle \begin{bmatrix} \mathscr{A} & -\mathfrak{L} \\ -\mathfrak{L}^T & -\mathbf{D} \end{bmatrix} \right\rangle_0 - \frac{1}{V_0} \boldsymbol{L}^T \cdot \boldsymbol{K}^{-1} \cdot \boldsymbol{L} \quad . \tag{C.171}$$

The macro-scale tangents for other valid parameterisations of the energy increment would follow similarly.

C.9 Constitutive modelling

C.9.1 Free energy function derivatives

For a free energy function expressed in terms of invariants, the computation of the kinetic quantities and their linearisations can be compactly expressed through application of the chain rule. Then, for example, the first derivatives of such a function (for an arbitrary number of invariants $\alpha, \beta \in [1, n_I]$) are [504]

$$\frac{\partial \Psi(\mathbf{C}, \mathbb{H})}{\partial \mathbf{C}} = \sum_\alpha \frac{\partial \Psi(\mathbf{C}, \mathbb{H})}{\partial I_\alpha} \frac{\partial I_\alpha}{\partial \mathbf{C}} \quad , \quad \frac{\partial \Psi(\mathbf{C}, \mathbb{H})}{\partial \mathbb{H}} = \sum_\alpha \frac{\partial \Psi(\mathbf{C}, \mathbb{H})}{\partial I_\alpha} \frac{\partial I_\alpha}{\partial \mathbb{H}} \quad , \tag{C.172}$$

and the second derivatives can be succinctly written as

$$\frac{\partial^2 \Psi(\mathbf{C}, \mathbb{H})}{\partial \mathbf{C} \otimes \partial \mathbf{C}} = \sum_\alpha \left[\sum_\beta \frac{\partial^2 \Psi(\mathbf{C}, \mathbb{H})}{\partial I_\beta \partial I_\alpha} \frac{\partial I_\alpha}{\partial \mathbf{C}} \otimes \frac{\partial I_\beta}{\partial \mathbf{C}} + \frac{\partial \Psi(\mathbf{C}, \mathbb{H})}{\partial I_\alpha} \frac{\partial^2 I_\alpha}{\partial \mathbf{C} \otimes \partial \mathbf{C}} \right] \quad , \tag{C.173a}$$

$$\frac{\partial^2 \Psi(\mathbf{C}, \mathbb{H})}{\partial \mathbf{C} \otimes \partial \mathbb{H}} = \sum_\alpha \left[\sum_\beta \frac{\partial^2 \Psi(\mathbf{C}, \mathbb{H})}{\partial I_\beta \partial I_\alpha} \frac{\partial I_\alpha}{\partial \mathbf{C}} \otimes \frac{\partial I_\beta}{\partial \mathbb{H}} + \frac{\partial \Psi(\mathbf{C}, \mathbb{H})}{\partial I_\alpha} \frac{\partial^2 I_\alpha}{\partial \mathbf{C} \otimes \partial \mathbb{H}} \right] \quad , \tag{C.173b}$$

$$\frac{\partial^2 \Psi(\mathbf{C}, \mathbb{H})}{\partial \mathbb{H} \otimes \partial \mathbf{C}} = \sum_\alpha \left[\sum_\beta \frac{\partial^2 \Psi(\mathbf{C}, \mathbb{H})}{\partial I_\beta \partial I_\alpha} \frac{\partial I_\alpha}{\partial \mathbb{H}} \otimes \frac{\partial I_\beta}{\partial \mathbf{C}} + \frac{\partial \Psi(\mathbf{C}, \mathbb{H})}{\partial I_\alpha} \frac{\partial^2 I_\alpha}{\partial \mathbb{H} \otimes \partial \mathbf{C}} \right] \quad , \tag{C.173c}$$

$$\frac{\partial^2 \Psi(\mathbf{C}, \mathbb{H})}{\partial \mathbb{H} \otimes \partial \mathbb{H}} = \sum_\alpha \left[\sum_\beta \frac{\partial^2 \Psi(\mathbf{C}, \mathbb{H})}{\partial I_\beta \partial I_\alpha} \frac{\partial I_\alpha}{\partial \mathbb{H}} \otimes \frac{\partial I_\beta}{\partial \mathbb{H}} + \frac{\partial \Psi(\mathbf{C}, \mathbb{H})}{\partial I_\alpha} \frac{\partial^2 I_\alpha}{\partial \mathbb{H} \otimes \partial \mathbb{H}} \right] \quad . \tag{C.173d}$$

C.9.2 Invariants

Cayley–Hamilton theorem [80]

$$\begin{aligned} \mathbf{0} &= \mathbf{C}^3 - I_1 \mathbf{C}^2 + I_2 \mathbf{C} - I_3 \mathbf{I} \\ \Longrightarrow \quad \mathbf{C}^{-1} &= \frac{1}{I_3} \left[\mathbf{C}^2 - I_1 \mathbf{C} + I_2 \mathbf{I} \right] \end{aligned} \tag{C.174}$$

Reducibility of magnetoelastic pseudo-invariant

$$
\begin{aligned}
I_7 &= [\mathbb{H} \otimes \mathbb{H}] : \mathbf{C}^{-1} && \leftarrow \text{eq. (7.25)} \rightarrow \text{eq. (7.24)} \\
&= [\mathbb{H} \otimes \mathbb{H}] : \left[\frac{1}{I_3}\left[\mathbf{C}^2 - I_1\mathbf{C} + I_2\mathbf{I}\right]\right] && \rightarrow \text{eq. (7.23b)} \\
&= \frac{1}{I_3}\left[I_6 - I_1 I_5 + I_2 I_4\right] && \text{(C.175)}
\end{aligned}
$$

C.9.3 Invariant derivatives [504]

Listed below are the first and second derivatives of the isotropic coupled invariants, as listed in equations (7.23a), (7.23b) and (7.24), with respect to their arguments. Note that the derivatives for the third invariant $I_3(\mathbf{C}) \equiv \widetilde{I_3}(J)$ can be computed from equations (A.47a) and (A.47b).

First derivatives

From equations (7.23a), (7.23b) and (7.24), the first derivatives of the isotropic coupled invariants are

$$\frac{\partial I_1}{\partial C_{AB}} = \delta_{AB} \quad , \quad \frac{\partial I_1}{\partial \mathbb{H}_A} = 0 \quad , \tag{C.176a}$$

$$\frac{\partial I_2}{\partial C_{AB}} = I_1 \delta_{AB} - C_{AB} \quad , \quad \frac{\partial I_2}{\partial \mathbb{H}_A} = 0 \quad , \tag{C.176b}$$

$$\frac{\partial I_3}{\partial C_{AB}} = I_3 C_{AB}^{-1} \quad , \quad \frac{\partial I_3}{\partial \mathbb{H}_A} = 0 \quad , \tag{C.176c}$$

$$\frac{\partial I_4}{\partial C_{AB}} = 0 \quad , \quad \frac{\partial I_4}{\partial \mathbb{H}_A} = 2\mathbb{H}_A \quad , \tag{C.176d}$$

$$\frac{\partial I_5}{\partial C_{AB}} = \mathbb{H}_A \mathbb{H}_B \quad , \quad \frac{\partial I_5}{\partial \mathbb{H}_A} = 2\mathbb{Z}_A \quad , \tag{C.176e}$$

$$\frac{\partial I_6}{\partial C_{AB}} = \mathbb{Z}_A \mathbb{H}_B + \mathbb{H}_A \mathbb{Z}_B \quad , \quad \frac{\partial I_6}{\partial \mathbb{H}_A} = 2 C_{AE} \mathbb{Z}_E \quad , \tag{C.176f}$$

$$\frac{\partial I_7}{\partial C_{AB}} = -\mathbb{Y}_A \mathbb{Y}_B \quad , \quad \frac{\partial I_7}{\partial \mathbb{H}_A} = 2\mathbb{Y}_A \quad , \tag{C.176g}$$

where, for brevity, we have defined the magnetoelastic vectorial quantities

$$\mathbb{Y} := \mathbf{C}^{-1} \cdot \mathbb{H} \quad , \quad \mathbb{Z} := \mathbf{C} \cdot \mathbb{H} \quad .$$

Second derivatives

From equations (C.176a) to (C.176g), the second derivatives of the isotropic coupled invariants are

$$\frac{\partial^2 I_1}{\partial C_{CD} \partial C_{AB}} = 0 \quad , \quad \frac{\partial^2 I_1}{\partial \mathbb{H}_B \partial \mathbb{H}_A} = 0 \quad , \quad \frac{\partial^2 I_1}{\partial \mathbb{H}_C \partial C_{AB}} = 0 \quad , \tag{C.177a}$$

$$\frac{\partial^2 I_2}{\partial C_{CD}\partial C_{AB}} = \delta_{AB}\delta_{CD} - \mathscr{I}_{ABCD} \quad , \quad \frac{\partial^2 I_2}{\partial \mathbb{H}_B \partial \mathbb{H}_A} = 0 \quad , \quad \frac{\partial^2 I_2}{\partial \mathbb{H}_C \partial C_{AB}} = 0 \quad , \tag{C.177b}$$

$$\frac{\partial^2 I_3}{\partial C_{CD}\partial C_{AB}} = I_3\left[C^{-1}_{AB}C^{-1}_{CD} - \frac{1}{2}\left[C^{-1}_{AC}C^{-1}_{BD} + C^{-1}_{AD}C^{-1}_{BC}\right]\right] \quad , \quad \frac{\partial^2 I_3}{\partial \mathbb{H}_B \partial \mathbb{H}_A} = 0 \quad , \quad \frac{\partial^2 I_3}{\partial \mathbb{H}_C \partial C_{AB}} = 0 \quad , \tag{C.177c}$$

$$\frac{\partial^2 I_4}{\partial C_{CD}\partial C_{AB}} = 0 \quad , \quad \frac{\partial^2 I_4}{\partial \mathbb{H}_B \partial \mathbb{H}_A} = 2\delta_{AB} \quad , \quad \frac{\partial^2 I_4}{\partial \mathbb{H}_C \partial C_{AB}} = 0 \quad , \tag{C.177d}$$

$$\frac{\partial^2 I_5}{\partial C_{CD}\partial C_{AB}} = 0 \quad , \quad \frac{\partial^2 I_5}{\partial \mathbb{H}_B \partial \mathbb{H}_A} = 2C_{AB} \quad , \quad \frac{\partial^2 I_5}{\partial \mathbb{H}_C \partial C_{AB}} = \delta_{AC}\mathbb{H}_B + \mathbb{H}_A\delta_{BC} \quad , \tag{C.177e}$$

$$\frac{\partial^2 I_6}{\partial C_{CD}\partial C_{AB}} = \frac{1}{2}\left[\mathbb{H}_A\mathbb{H}_D\delta_{BC} + \mathbb{H}_A\mathbb{H}_C\delta_{BD} + \delta_{AC}\mathbb{H}_B\mathbb{H}_D + \delta_{AD}\mathbb{H}_B\mathbb{H}_C\right] \quad ,$$

$$\frac{\partial^2 I_6}{\partial \mathbb{H}_B \partial \mathbb{H}_A} = 2C_{AE}C_{EB} \quad , \quad \frac{\partial^2 I_6}{\partial \mathbb{H}_C \partial C_{AB}} = \delta_{AC}\mathbb{Z}_B + \mathbb{Z}_A\delta_{BC} + \mathbb{H}_A C_{BC} + C_{AC}\mathbb{H}_B \quad , \tag{C.177f}$$

$$\frac{\partial^2 I_7}{\partial C_{CD}\partial C_{AB}} = \frac{1}{2}\left[\mathbb{Y}_A\mathbb{Y}_C C^{-1}_{BD} + \mathbb{Y}_A\mathbb{Y}_D C^{-1}_{BC} + C^{-1}_{AD}\mathbb{Y}_B\mathbb{Y}_C + C^{-1}_{AC}\mathbb{Y}_B\mathbb{Y}_D\right] \quad ,$$

$$\frac{\partial^2 I_7}{\partial \mathbb{H}_B \partial \mathbb{H}_A} = 2C^{-1}_{AB} \quad , \quad \frac{\partial^2 I_7}{\partial \mathbb{H}_C \partial C_{AB}} = -C^{-1}_{AC}\mathbb{Y}_B - \mathbb{Y}_A C^{-1}_{BC} \quad . \tag{C.177g}$$

C.9.4 Free space stored energy function derivatives

Note that what follows is for the parameterisation in terms of the magnetic field, namely the function defined in equation (5.178). For convenience (with specific application to the quasi-incompressible coupled formulation detailed in Section 11.3), we will compute some of the derivatives in a partially decoupled manner. Therefore, using the chain rule we can express the first derivatives of the stored energy function as

$$\frac{\partial M_0}{\partial \mathbf{C}} = \left.\frac{\partial M_0}{\partial \mathbf{C}}\right|_J + \left.\frac{\partial M_0}{\partial J}\right|_{\mathbf{C}}\frac{\partial J}{\partial \mathbf{C}} \quad , \quad \frac{\partial M_0}{\partial \mathbb{H}} = -\frac{\mu_0}{2}J\frac{\partial I_7}{\partial \mathbb{H}} \tag{C.178}$$

and the second derivatives by

$$\frac{\partial^2 M_0}{\partial \mathbf{C}\otimes\partial \mathbf{C}} = \left.\frac{\partial^2 M_0}{\partial \mathbf{C}\otimes\partial \mathbf{C}}\right|_J + \left.\frac{\partial^2 M_0}{\partial J\otimes\partial J}\right|_{\mathbf{C}}\frac{\partial J}{\partial \mathbf{C}}\otimes\frac{\partial J}{\partial \mathbf{C}} + \left.\frac{\partial M_0}{\partial J}\right|_{\mathbf{C}}\frac{\partial^2 J}{\partial \mathbf{C}\otimes\partial \mathbf{C}} \quad , \tag{C.179a}$$

$$\frac{\partial^2 M_0}{\partial \mathbf{C}\otimes\partial \mathbb{H}} = \left.\frac{\partial^2 M_0}{\partial \mathbf{C}\otimes\partial \mathbb{H}}\right|_J + \frac{\partial J}{\partial \mathbf{C}}\otimes\left.\frac{\partial^2 M_0}{\partial \mathbb{H}\partial J}\right|_{\mathbf{C}} \quad , \tag{C.179b}$$

$$\frac{\partial^2 M_0}{\partial \mathbb{H}\otimes\partial \mathbb{H}} = -\frac{\mu_0}{2}J\frac{\partial^2 I_7}{\partial \mathbb{H}\otimes\partial \mathbb{H}} \quad . \tag{C.179c}$$

Conveniently, the derivatives of the volumetric Jacobian are already provided in equations (A.47a) and (A.47b), and those of I_7 in the previous appendix. It is thus only necessary to define only a few more partial derivatives to fully define the linearisation of the energy function.

First partial derivatives

From equation (5.178), the first partial derivatives of the free space stored energy are

$$\left.\frac{\partial M_0}{\partial C_{AB}}\right|_J = -\frac{\mu_0}{2} J \frac{\partial I_7}{\partial C_{AB}} \quad , \quad \left.\frac{\partial M_0}{\partial J}\right|_{\mathbf{C}} = -\frac{\mu_0}{2} I_7 \quad . \tag{C.180}$$

Second partial derivatives

From equation (5.178), the first partial derivatives of the free space stored energy are

$$\left.\frac{\partial^2 M_0}{\partial C_{CD} \partial C_{AB}}\right|_J = -\frac{\mu_0}{2} J \frac{\partial^2 I_7}{\partial C_{CD} \partial C_{AB}} \quad , \quad \frac{\partial^2 M_0}{\partial J \partial J} = 0 \quad , \tag{C.181}$$

$$\left.\frac{\partial^2 M_0}{\partial \mathbb{H}_C \partial C_{AB}}\right|_J = -\frac{\mu_0}{2} J \frac{\partial^2 I_7}{\partial \mathbb{H}_C \partial C_{AB}} \quad , \quad \left.\frac{\partial^2 M_0}{\partial \mathbb{H}_A \partial J}\right|_{\mathbf{C}} = -\frac{\mu_0}{2} \frac{\partial I_7}{\partial \mathbb{H}_A} \quad . \tag{C.182}$$

C.9.5 Volumetric / deviatoric split of free energy function

Deviatoric projection tensor

For the dimension independent problem, where $\mathrm{d} \in \{2, 3\}$,

$$\begin{aligned}
\frac{\mathrm{d}\overline{\mathbf{C}}}{\mathrm{d}\mathbf{C}} &= \frac{\mathrm{d}}{\mathrm{d}\mathbf{C}} \left[J^{-\frac{2}{\mathrm{d}}} \mathbf{C} \right] && \rightarrow \text{eq. (7.15)} \\
&= \mathbf{C} \otimes \frac{\mathrm{d}\left[J^{-\frac{2}{\mathrm{d}}}\right]}{\mathrm{d}\mathbf{C}} + J^{-\frac{2}{\mathrm{d}}} \frac{\mathrm{d}\mathbf{C}}{\mathrm{d}\mathbf{C}} && \rightarrow \text{eq. (A.48)} \\
&= -\frac{1}{\mathrm{d}} J^{-\frac{2}{\mathrm{d}}} \mathbf{C} \otimes \mathbf{C}^{-1} + J^{-\frac{2}{\mathrm{d}}} \mathscr{I} \\
&= J^{-\frac{2}{\mathrm{d}}} \left[-\frac{1}{\mathrm{d}} \mathbf{C} \otimes \mathbf{C}^{-1} + \mathscr{I} \right] =: \widehat{\mathscr{P}}
\end{aligned} \tag{C.183}$$

which is the definition given in equation (7.20). Note that for an arbitrary second-order symmetric tensor $\mathbf{T}$,

$$\begin{aligned}
\mathbf{T} : \widehat{\mathscr{P}} &= \mathbf{T} : J^{-\frac{2}{\mathrm{d}}} \left[\mathscr{I} - \frac{1}{\mathrm{d}} \mathbf{C} \otimes \mathbf{C}^{-1} \right] \\
&= J^{-\frac{2}{\mathrm{d}}} \left[\mathbf{T} : \mathscr{I} - \frac{1}{\mathrm{d}} \mathbf{T} : \left[\mathbf{C} \otimes \mathbf{C}^{-1} \right] \right] \\
&\equiv J^{-\frac{2}{\mathrm{d}}} \left[\mathbf{T} - \frac{1}{\mathrm{d}} \left[\mathbf{T} : \mathbf{C} \right] \mathbf{C}^{-1} \right] \quad .
\end{aligned} \tag{C.184}$$

Component of the deviatoric part of material elasticity tensor

From the last term in equation (7.17), denoting $\mathbf{T} := \frac{\partial \overline{U}_0(\overline{\mathbf{C}}, \mathbb{H})}{\partial \overline{\mathbf{C}}}$, we compute the contraction of a symmetric second-order tensor with the derivative of the referential projection tensor (with respect to the right Cauchy–Green deformation tensor) as

$$
\begin{aligned}
T_{CD}\frac{\mathrm{d}\widehat{\mathscr{P}}_{CDAB}}{\mathrm{d}C_{EF}} &= T_{CD}\frac{\mathrm{d}}{\mathrm{d}C_{EF}}\left[J^{-\frac{2}{\mathrm{d}}}\left[\mathscr{I}_{CDAB} - \frac{1}{\mathrm{d}}C_{CD}C^{-1}_{AB}\right]\right] \qquad \leftarrow \text{eq. (7.20)}\\
&= T_{CD}\left[\frac{\mathrm{d}\left[J^{-\frac{2}{\mathrm{d}}}\right]}{\mathrm{d}C_{EF}}\left[\mathscr{I}_{CDAB} - \frac{1}{\mathrm{d}}C_{CD}C^{-1}_{AB}\right] + J^{-\frac{2}{\mathrm{d}}}\frac{\mathrm{d}}{\mathrm{d}C_{EF}}\left[\mathscr{I}_{CDAB} - \frac{1}{\mathrm{d}}C_{CD}C^{-1}_{AB}\right]\right]\\
&\qquad \rightarrow \text{eq. (A.48)}\\
&= T_{CD}\left[-\frac{1}{\mathrm{d}}J^{-\frac{2}{\mathrm{d}}}C^{-1}_{EF}\left[\mathscr{I}_{CDAB} - \frac{1}{\mathrm{d}}C_{CD}C^{-1}_{AB}\right] - \frac{1}{\mathrm{d}}J^{-\frac{2}{\mathrm{d}}}\left[\frac{\mathrm{d}C_{CD}}{\mathrm{d}C_{EF}}C^{-1}_{AB} + C_{CD}\frac{\mathrm{d}C^{-1}_{AB}}{\mathrm{d}C_{EF}}\right]\right]\\
&= -\frac{1}{\mathrm{d}}J^{-\frac{2}{\mathrm{d}}}T_{CD}\left[C^{-1}_{EF}\left[\mathscr{I}_{CDAB} - \frac{1}{\mathrm{d}}C_{CD}C^{-1}_{AB}\right] + \left[\mathscr{I}_{CDEF}C^{-1}_{AB} + C_{CD}\frac{\mathrm{d}C^{-1}_{AB}}{\mathrm{d}C_{EF}}\right]\right]\\
&= -\frac{1}{\mathrm{d}}J^{-\frac{2}{\mathrm{d}}}\left[C^{-1}_{EF}\left[T_{CD}\mathscr{I}_{CDAB} - \frac{1}{\mathrm{d}}T_{CD}C_{CD}C^{-1}_{AB}\right]\right.\\
&\quad \left. + \left[T_{CD}\mathscr{I}_{CDEF}C^{-1}_{AB} + T_{CD}C_{CD}\frac{\mathrm{d}C^{-1}_{AB}}{\mathrm{d}C_{EF}}\right]\right]\\
&= -\frac{1}{\mathrm{d}}J^{-\frac{2}{\mathrm{d}}}\left[T_{AB}C^{-1}_{EF} - \frac{1}{\mathrm{d}}\left[T_{CD}C_{CD}\right]C^{-1}_{AB}C^{-1}_{EF} + \left[C^{-1}_{AB}T_{EF} + \left[T_{CD}C_{CD}\right]\frac{\mathrm{d}C^{-1}_{AB}}{\mathrm{d}C_{EF}}\right]\right]\\
&= \frac{1}{\mathrm{d}}J^{-\frac{2}{\mathrm{d}}}\left[-\left[T_{AB}C^{-1}_{EF} + C^{-1}_{AB}T_{EF}\right] + \left[T_{CD}C_{CD}\right]\left[\frac{1}{\mathrm{d}}C^{-1}_{AB}C^{-1}_{EF} - \frac{\mathrm{d}C^{-1}_{AB}}{\mathrm{d}C_{EF}}\right]\right]\\
\mathbf{T} : \frac{\mathrm{d}\widehat{\mathscr{P}}}{\mathrm{d}\mathbf{C}} &= \frac{1}{\mathrm{d}}J^{-\frac{2}{d}}\left[\left[\mathbf{T} : \mathbf{C}\right]\left[\frac{1}{\mathrm{d}}\mathbf{C}^{-1}\otimes\mathbf{C}^{-1} - \frac{\mathrm{d}\mathbf{C}^{-1}}{\mathrm{d}\mathbf{C}}\right] - \left[\mathbf{T}\otimes\mathbf{C}^{-1} + \mathbf{C}^{-1}\otimes\mathbf{T}\right]\right] \quad . \qquad (C.185)
\end{aligned}
$$

Note that, using a rearrangement of equation (C.184), further manipulation of this equation leads to the same result given in [559, equation (3.253)]. We retain this form simply because there is then no need to compute the isochoric equivalent of **T**.

C.10 Time integrators for rate-dependent materials described by internal variables

The following approach outlines a simple, but general Euler-method time integration routine, assuming a constant time step size, for arbitrary rank internal variables. It can therefore be applied to the evolution law governing either elastic or magnetic internal variables, as were introduced in Chapter 7. Alternative approaches that offer better accuracy and stability criteria, as well as the ability to better solve stiff systems, include the time-adaptive Runge–Kutta methods [126], predictor-corrector methods or Richardson extrapolation [436]. In a numerical context, robust frameworks for solving non-linear systems of ordinary differential equations may be employed [200].

Introducing a fractional time stepping procedure, at time t we approximate the first-order ordinary differential equation by

$$\partial_t \mathbf{Q}^t = \mathbf{f}\left(\mathbf{A}, \mathbf{Q}^t\right) := \theta \mathbf{f}\left(\mathbf{A}, \mathbf{Q}^t\right) + [1-\theta]\,\mathbf{f}\left(\mathbf{A}, \mathbf{Q}^{t-1}\right) \tag{C.186}$$

where the choice of θ defines the time stepping method. For the purpose of generality, $\mathbf{A} = \{\mathbf{A}_1, \mathbf{A}_2, \ldots\}$ represents the set of global kinematic variables. Note that values of the (local) internal variables $\mathbf{Q}^t$ remain unknown while $\mathbf{Q}^{t-1}$ are final equilibrium values known from the previous time steps. Valid (that is to say, with some proof of stability and convergence) common choices for θ are:

$\theta = 0$: Explicit method (conditional stability, $\mathcal{O}(\Delta t)$)
$\theta = \frac{1}{2}$: Crank–Nicholson (unconditional stability, $\mathcal{O}(\Delta t^2)$)
$\theta = 1$: Implicit method (unconditional stability, $\mathcal{O}(\Delta t)$)

Using a finite difference approximation for the time derivative (the left-hand side of equation (C.186)), the rate of evolution of the internal variable can be written as

$$\begin{aligned} \partial_t \mathbf{Q} &\approx \mathbf{g}_1\left(\mathbf{Q}^t, \Delta t\right) + \mathbf{g}_2\left(\mathbf{Q}^{t-1}, \mathbf{Q}^{t-2}, \ldots, \Delta t\right) \\ &= g_1^*(\Delta t)\, \mathbf{Q}^t + \mathbf{g}_2\left(\mathbf{Q}^{t-1}, \mathbf{Q}^{t-2}, \ldots, \Delta t\right) \quad ; \end{aligned} \tag{C.187}$$

some choices for g_1^* and $\mathbf{g}_2$ are stated in the concluding summary provided in Table C.2. Combining the two sets of equations renders the implicit, non-linear relationship

$$g_1^*(\Delta t)\, \mathbf{Q}^t + \mathbf{g}_2\left(\mathbf{Q}^{t-1}, \mathbf{Q}^{t-2}, \ldots, \Delta t\right) = \theta \mathbf{f}\left(\mathbf{A}, \mathbf{Q}^t\right) + [1-\theta]\,\mathbf{f}\left(\mathbf{A}, \mathbf{Q}^{t-1}\right) \quad . \tag{C.188}$$

This non-linear problem can be solved using an iterative method, such as the Newton–Raphson method. Linearisation of a residual equation $\mathbf{R}^{t,i}(\mathbf{A}, \mathbf{Q})$, a function of the external kinematic variable $\mathbf{A}$ and the internal variable $\mathbf{Q}$, at time-step t and Newton iterate i leads to

$$\mathbf{0} = \mathbf{R}^{t,i}(\mathbf{A}, \mathbf{Q}) + \left.\frac{\partial \mathbf{R}^{t,i}}{\partial \mathbf{A}}\right|_{\mathbf{Q}} \cdot \Delta \mathbf{A}^{t,i} + \left.\frac{\partial \mathbf{R}^{t,i}}{\partial \mathbf{Q}}\right|_{\mathbf{A}} \cdot \Delta \mathbf{Q}^{t,i} \quad .$$

Noting that the kinematic variable $\mathbf{A}$ is non-local and remains fixed during the solution procedure, this results in the simplification

$$\mathbf{0} = \mathbf{R}^{t,i}(\mathbf{A}, \mathbf{Q}) + \left.\frac{\partial \mathbf{R}^{t,i}}{\partial \mathbf{Q}}\right|_{\mathbf{A}} \cdot \Delta \mathbf{Q}^{t,i} \quad , \tag{C.189}$$

which in turn leads to the solution of the internal variable update being stated as

$$\Delta \mathbf{Q}^{t,i} = \left[\left.\frac{\partial \mathbf{R}^{t,i}}{\partial \mathbf{Q}}\right|_{\mathbf{A}}\right]^{-1} \cdot \left[-\mathbf{R}^{t,i}(\mathbf{A}, \mathbf{Q})\right] \tag{C.190}$$

with the incremental update

$$\mathbf{Q}^{t,i+1} = \mathbf{Q}^{t,i} + \Delta\mathbf{Q}^{t,i} \quad . \tag{C.191}$$

In this instance, from equation (C.188) we define the residual equation as

$$\mathbf{R}^{t,i}(\mathbf{A},\mathbf{Q}) := g_1^*(\Delta t)\,\mathbf{Q}^t + \mathbf{g}_2\left(\mathbf{Q}^{t-1},\mathbf{Q}^{t-2},\ldots,\Delta t\right) - \left[\theta\mathbf{f}\left(\mathbf{A},\mathbf{Q}^t\right) + [1-\theta]\,\mathbf{f}\left(\mathbf{A},\mathbf{Q}^{t-1}\right)\right] \tag{C.192}$$

which has the linearisation around the current value of $\mathbf{Q}$

$$\left.\frac{\partial\mathbf{R}^{t,i}}{\partial\mathbf{Q}^t}\right|_{\mathbf{A}} = g_1^*(\Delta t)\,\frac{\partial\mathbf{Q}^t}{\partial\mathbf{Q}^t} - \theta\frac{\partial\mathbf{f}(\mathbf{A},\mathbf{Q}^t)}{\partial\mathbf{Q}^t} = g_1^*(\Delta t)\,\mathscr{I} - \theta\frac{\partial\mathbf{f}(\mathbf{A},\mathbf{Q}^t)}{\partial\mathbf{Q}^t} \quad . \tag{C.193}$$

For the purpose of the calculation of material tangents, linearisation of the internal variable around the global kinematic variable(s) is also necessary. This occurs once the equilibrium position for $\mathbf{Q}$ is found using equation (C.189). To achieve this, we return to equation (C.188) and linearise it. Noting that the previous history of the internal variable $\mathbf{Q}^{t-1}$ is not influenced by the current kinematic state, this results in

$$\begin{aligned}
\frac{\mathrm{d}}{\mathrm{d}\mathbf{A}}\left[g_1^*(\Delta t)\,\mathbf{Q}^t + \mathbf{g}_2\left(\mathbf{Q}^{t-1},\mathbf{Q}^{t-2},\ldots,\Delta t\right)\right]\Big|_{\mathbf{Q}} &= \frac{\mathrm{d}}{\mathrm{d}\mathbf{A}}\left[\theta\mathbf{f}\left(\mathbf{A},\mathbf{Q}^t\right) + [1-\theta]\,\mathbf{f}\left(\mathbf{A},\mathbf{Q}^{t-1}\right)\right]\Big|_{\mathbf{Q}} \\
\implies \quad g_1^*(\Delta t)\,\frac{\mathrm{d}\mathbf{Q}^t}{\mathrm{d}\mathbf{A}} &= \theta\frac{\mathrm{d}\mathbf{f}(\mathbf{A},\mathbf{Q}^t)}{\mathrm{d}\mathbf{A}} + [1-\theta]\,\frac{\mathrm{d}\mathbf{f}(\mathbf{A},\mathbf{Q}^{t-1})}{\mathrm{d}\mathbf{A}} \\
&= \theta\left[\left.\frac{\partial\mathbf{f}(\mathbf{A},\mathbf{Q}^t)}{\partial\mathbf{A}}\right|_{\mathbf{Q}^t} + \left.\frac{\partial\mathbf{f}(\mathbf{A},\mathbf{Q}^t)}{\partial\mathbf{Q}^t}\right|_{\mathbf{A}}\cdot\frac{\mathrm{d}\mathbf{Q}^t}{\mathrm{d}\mathbf{A}}\right] + [1-\theta]\,\frac{\partial\mathbf{f}(\mathbf{A},\mathbf{Q}^{t-1})}{\partial\mathbf{A}}
\end{aligned} \tag{C.194}$$

which resolves to

$$\begin{aligned}
\frac{\mathrm{d}\mathbf{Q}^t}{\mathrm{d}\mathbf{A}} &= \left[g_1^*(\Delta t)\,\mathscr{I} - \theta\frac{\partial\mathbf{f}(\mathbf{A},\mathbf{Q}^t)}{\partial\mathbf{Q}^t}\right]^{-1}\left[\theta\frac{\partial\mathbf{f}(\mathbf{A},\mathbf{Q}^t)}{\partial\mathbf{A}} + [1-\theta]\,\frac{\partial\mathbf{f}(\mathbf{A},\mathbf{Q}^{t-1})}{\partial\mathbf{A}}\right] \\
&\equiv \left[g_1^*(\Delta t)\,\mathscr{I} - \theta\left.\frac{\partial\mathbf{f}(\mathbf{A},\mathbf{Q})}{\partial\mathbf{Q}}\right|_{\mathbf{Q}=\mathbf{Q}^t}\right]^{-1}\left[\theta\left.\frac{\partial\mathbf{f}(\mathbf{A},\mathbf{Q})}{\partial\mathbf{A}}\right|_{\mathbf{Q}=\mathbf{Q}^t} + [1-\theta]\left.\frac{\partial\mathbf{f}(\mathbf{A},\mathbf{Q})}{\partial\mathbf{A}}\right|_{\mathbf{Q}=\mathbf{Q}^{t-1}}\right] \quad .
\end{aligned} \tag{C.195}$$

For example, with $\mathbf{A} = \{\mathbf{C},\mathbb{H}\}$ and $\mathbf{Q} = \{\mathbf{C}_{\mathrm{v}},\mathbb{H}_{\mathrm{v}}\}$ this provides the definition of $\frac{\partial\mathbf{C}_{\mathrm{v}}}{\partial\mathbf{C}}$, $\frac{\partial\mathbf{C}_{\mathrm{v}}}{\partial\mathbb{H}}$, $\frac{\partial\mathbb{H}_{\mathrm{v}}}{\partial\mathbf{C}}$, and $\frac{\partial\mathbb{H}_{\mathrm{v}}}{\partial\mathbb{H}}$ as is required in equations (7.46) to (7.49). Consideration of the finite difference formula generalisation algorithm described by Fornberg [149] would render a equivalent expression for an arbitrary time stepping scheme.

Summary of calculations for incremental update of internal variable

In summary, Table C.1 outlines the list of computations that must be performed in order to perform a single incremental update of an internal variable that is governed by a non-linear evolution law, as well as its linearisation that contributes towards the overall material tangent. A non-exhaustive list of coefficients for the finite difference approximation to the rate of evolution of the internal variable (assuming a constant time step size) is provided in Table C.2.

Table C.1: Summary of terms required for the non-linear solution algorithm for internal variables.

Term defined by	Value	Equations requiring this term
Finite difference approximation	$g_1^*(\Delta t)$	C.192,C.193,C.195
	$\mathbf{g}_2\left(\mathbf{Q}^{t-1}, \mathbf{Q}^{t-2}, \ldots, \Delta t\right)$	C.192
Evolution law	$\mathbf{f}(\mathbf{A}, \mathbf{Q}^t)$	C.192
	$\mathbf{f}\left(\mathbf{A}, \mathbf{Q}^{t-1}\right)$	C.192
	$\frac{\partial \mathbf{f}(\mathbf{A}, \mathbf{Q}^t)}{\partial \mathbf{Q}^t}$	C.193, C.195
	$\frac{\partial \mathbf{f}(\mathbf{A}, \mathbf{Q}^t)}{\partial \mathbf{A}}$	C.195
	$\frac{\partial \mathbf{f}\left(\mathbf{A}, \mathbf{Q}^{t-1}\right)}{\partial \mathbf{A}}$	C.195

Table C.2: Summary of coefficients for the finite difference approximation for $\partial_t \mathbf{Q}$ assuming a constant time step size.

Order	$g_1^*(\Delta t)$	$\mathbf{g}_2\left(\mathbf{Q}^{t-1}, \mathbf{Q}^{t-2}, \ldots, \Delta t\right)$
1st	$\frac{1}{\Delta t}$	$\frac{1}{\Delta t}\left[-\mathbf{Q}^{t-1}\right]$
2nd	$\frac{3}{2\Delta t}$	$\frac{1}{2\Delta t}\left[-4\mathbf{Q}^{t-1} + \mathbf{Q}^{t-2}\right]$
3rd	$\frac{11}{6\Delta t}$	$\frac{1}{6\Delta t}\left[-18\mathbf{Q}^{t-1} + 9\mathbf{Q}^{t-2} - 2\mathbf{Q}^{t-3}\right]$

Bibliography

[1] C. Abert, L. Exl, F. Bruckner, A. Drews, and D. Suess. magnum.fe: A micromagnetic finite-element simulation code based on FEniCS. *J. Magn. Magn. Mater.*, 345:29–35, November 2013. DOI: 10.1016/j.jmmm.2013.05.051.

[2] C. W. Abert. *Discrete Mathematical Concepts in Micromagnetic Computations*. PhD thesis, Universität Hamburg, Von-Melle-Park 3, 20146 Hamburg, 2013. URL http://ediss.sub.uni-hamburg.de/volltexte/2013/6458. Last visited: September 2018.

[3] J. Aboudi. Micromechanical analysis of fully coupled electro-magneto-thermo-elastic multiphase composites. *Smart Mater. Struct.*, 10(5):867–877, August 2001. DOI: 10.1088/0964-1726/10/5/303.

[4] M. A. Abshinova, N. E. Kazantseva, P. Sáha, I. Sapurina, J. Kovářová, and J. Stejskal. The enhancement of the oxidation resistance of carbonyl iron by polyaniline coating and consequent changes in electromagnetic properties. *Polym. Degrad. Stab.*, 93(10):1826–1831, October 2008. DOI: 10.1016/j.polymdegradstab.2008.07.008.

[5] R. Albanese and G. Rubinacci. Integral formulation for 3D eddy-current computation using edge elements. *IEE Proc. A Phys. Sci., Meas. Instrum. Manag. Educ., Reviews*, 135(7):457–462, 1988. DOI: 10.1049/ip-a-1.1988.0072.

[6] M. Alnaes, J. Blechta, J. Hake, A. Johansson, B. Kehlet, A. Logg, C. Richardson, J. Ring, M. E. Rognes, and G. N. Wells. The FEniCS project version 1.5. *Arch. Numer. Softw.*, 3(100):9–23, December 2015. DOI: 10.11588/ans.2015.100.20553.

[7] S. Aloui and M. Klüppel. Magneto-rheological response of elastomer composites with hybrid-magnetic fillers. *Smart Mater. Struct.*, 24(2):025016, December 2014. DOI: 10.1088/0964-1726/24/2/025016.

[8] J. Alsayednoor, P. Harrison, and Z. Guo. Large strain compressive response of 2-D periodic representative volume element for random foam microstructures. *Mech. Mater.*, 66:7–20, November 2013. DOI: 10.1016/j.mechmat.2013.06.006.

[9] T. Alshuth, M. Ramspeck, R. H. Schuster, B. Halbedel, and F. Zschunke. Magnetorheologische Elastomere: Einfluss der Partikelausrichtung auf die Schaltbarkeit. *Kautsch. Gummi Kunstst.*, 60(9):448–455, September 2007.

[10] G. Alzetta, D. Arndt, W. Bangerth, V. Boddu, B. Brands, D. Davydov, R. Gassmoeller, T. Heister, L. Heltai, K. Kormann, M. Kronbichler, M. Maier, J.-P. Pelteret, B. Turcksin, and D. Wells. The `deal.II` library, version 9.0. *J. Numer. Math.*, 2018. DOI: 10.1515/jnma-2018-0054.

[11] I. V. Andrianov, V. V. Danishevs'kyy, and D. Weichert. Simple estimation on effective transport properties of a random composite material with cylindrical fibres. *Z. Angew. Math. Phys.*, 59(5):889–903, September 2008. DOI: 10.1007/s00033-007-6146-3.

[12] A. Anthoine. Derivation of the in-plane elastic characteristics of masonry through homogenization theory. *Int. J. Solids Struct.*, 32(2):137–163, January 1995. DOI: 10.1016/0020-7683(94)00140-r.

[13] A. Antoulas, D. Sorensen, K. A. Gallivan, P. Van Dooren, A. Grama, C. Hoffmann, and A. Sameh. Model reduction of large-scale dynamical systems. In M. Bubak, G. D. van Albada, P. M. A. Sloot, and J. Dongarra, editors, *Computational Science – ICCS 2004*, pages 740–747. Springer Berlin Heidelberg, 2004. DOI: 10.1007/978-3-540-24688-6_96.

[14] F. Armero and I. Romero. On the formulation of high-frequency dissipative time-stepping algorithms for nonlinear dynamics. Part I: low-order methods for two model problems and nonlinear elastodynamics. *Comput. Methods Appl. Mech. Eng.*, 190(20-21):2603–2649, February 2001. DOI: 10.1016/s0045-7825(00)00256-5.

https://doi.org/10.1515/9783110418576-016

[15] D. Arndt, W. Bangerth, T. C. Clevenger, D. Davydov, M. Fehling, D. Garcia-Sanchez, G. Harper, T. Heister, L. Heltai, M. Kronbichler, R. M. Kynch, M. Maier, J-P. Pelteret, B. Turcksin, and D. Wells. The deal.II library, version 9.1. *J. Numer. Math.*, 2019. DOI: 10.1515/jnma-2019-0064.

[16] D. N. Arnold and G. Awanou. The serendipity family of finite elements. *Found. Comput. Math.*, 11(3):337–344, June 2011. DOI: 10.1007/s10208-011-9087-3.

[17] E. M. Arruda and M. C. Boyce. A three-dimensional constitutive model for the large stretch behavior of rubber elastic materials. *J. Mech. Phys. Solids*, 41(2):389–412, February 1993. DOI: 10.1016/0022-5096(93)90013-6.

[18] A. Ask, A. Menzel, and M. Ristinmaa. Phenomenological modeling of viscous electrostrictive polymers. *Int. J. Non-Linear Mech.*, 47(2):156–165, March 2012. DOI: 10.1016/j.ijnonlinmec.2011.03.020.

[19] A. Ask, A. Menzel, and M. Ristinmaa. Electrostriction in electro-viscoelastic polymers. *Mech. Mater.*, 50:9–21, July 2012. DOI: 10.1016/j.mechmat.2012.01.009.

[20] A. Ask, R. Denzer, A. Menzel, and M. Ristinmaa. Inverse-motion-based form finding for quasi-incompressible finite electroelasticity. *Int. J. Numer. Methods Eng.*, 94(6):554–572, February 2013. DOI: 10.1002/nme.4462.

[21] S. N. Atluri. On constitutive relations at finite strain: Hypo-elasticity and elasto-plasticity with isotropic or kinematic hardening. *Comput. Methods Appl. Mech. Eng.*, 43(2):137–171, April 1984. DOI: 10.1016/0045-7825(84)90002-1.

[22] I. Babuška and B. Q. Guo. Regularity of the solution of elliptic problems with piecewise analytic data. Part I. Boundary value problems for linear elliptic equation of second order. *SIAM J. Math. Anal.*, 19(1):172–203, January 1988. DOI: 10.1137/0519014.

[23] I. Babuška and M. Suri. The *h-p* version of the finite element method with quasiuniform meshes. *ESAIM: Math. Model. Numer. Anal.*, 21(2):199–238, 1987. DOI: 10.1051/m2an/1987210201991.

[24] I. Babuška and M. Suri. The *p-* and *h-p* versions of the finite element method, an overview. *Comput. Methods Appl. Mech. Eng.*, 80(1-3):5–26, June 1990. DOI: 10.1016/0045-7825(90)90011-a.

[25] I. Babuška, R. Tempone, and G. E. Zouraris. Galerkin finite element approximations of stochastic elliptic partial differential equations. *SIAM J. Numer. Anal.*, 42(2):800–825, January 2004. DOI: 10.1137/S0036142902418680.

[26] I. Babuška, R. Tempone, and G. E. Zouraris. Solving elliptic boundary value problems with uncertain coefficients by the finite element method: the stochastic formulation. *Comput. Methods Appl. Mech. Eng.*, 194(12–16):1251–1294, April 2005. DOI: 10.1016/j.cma.2004.02.026.

[27] D. Balzani, A. Gandhi, A. Klawonn, M. Lanser, O. Rheinbach, and J. Schröder. *Software for Exascale Computing - SPPEXA 2013–2015*, chapter One-way and fully-coupled FE^2 methods for heterogeneous elasticity and plasticity problems: parallel scalability and an application to thermo-elastoplasticity of dual-phase steels, pages 91–112. Springer Nature, 2016. ISBN 978-3-319-40528-5. DOI: 10.1007/978-3-319-40528-5_5.

[28] W. Bangerth and O. Kayser-Herold. Data structures and requirements for *hp* finite element software. *ACM Trans. Math. Softw.*, 36(1):4/1–4/31, March 2009. DOI: 10.1145/1486525.1486529.

[29] W. Bangerth, R. Hartmann, and G. Kanschat. deal.II – A general purpose object oriented finite element library. *ACM Trans. Math. Softw.*, 33(4):24/1–24/27, August 2007. DOI: 10.1145/1268776.1268779.

[30] W. Bangerth, C. Burstedde, T. Heister, and M. Kronbichler. Algorithms and data structures for massively parallel generic adaptive finite element codes. *ACM Trans. Math. Softw.*, 38(2):14:1–14:28, December 2012. DOI: 10.1145/2049673.2049678.

[31] S. Bargmann, B. Klusemann, J. Markmann, J. E. Schnabel, K. Schneider, C. Soyarslan, and J. Wilmers. Generation of 3D representative volume elements for heterogeneous materials: A review. *Prog. Mater. Sci.*, 96:322–384, July 2018. DOI: 10.1016/j.pmatsci.2018.02.003.

[32] M. I. Barham, D. A. White, and D. J. Steigmann. Finite element modeling of the deformation of magnetoelastic film. *J. Comput. Phys.*, 229(18):6193–6207, September 2010. DOI: 10.1016/j.jcp.2010.04.007.

[33] A. Bartels, T. Bartel, M. Canadija, and J. Mosler. On the thermomechanical coupling in dissipative materials: A variational approach for generalized standard materials. *J. Mech. Phys. Solids*, 82:218–234, September 2015. DOI: 10.1016/j.jmps.2015.04.011.

[34] R. A. Bartlett, D. M. Gay, and E. T. Phipps. Automatic differentiation of C++ codes for large-scale scientific computing. In V. N. Alexandrov, G. D. van Albada, P. M. A. Sloot, and J. Dongarra, editors, *Computational Science – ICCS 2006*, pages 525–532. Springer, Springer Berlin Heidelberg, 2006. DOI: 10.1007/11758549_73.

[35] M. L. Barton and Z. J. Cendes. New vector finite elements for three-dimensional magnetic field computation. *J. Appl. Phys.*, 61(8):3919–3921, April 1987. DOI: 10.1063/1.338584.

[36] K.-J. Bathe and M. M. I. Baig. On a composite implicit time integration procedure for nonlinear dynamics. *Comput. Struct.*, 83(31–32):2513–2524, December 2005. DOI: 10.1016/j.compstruc.2005.08.001.

[37] C. Bauer, A. Frink, and R. Kreckel. Introduction to the GiNaC framework for symbolic computation within the C++ programming language. *J. Symb. Comput.*, 33(1):1–12, January 2002. DOI: 10.1006/jsco.2001.0494.

[38] A. Bayat and F. Gordaninejad. A new computational method for homogenized tangent moduli of a soft magnetoelastic composite. *Smart Mater. Struct.*, 24(7):075010, June 2015. DOI: 10.1088/0964-1726/24/7/075010.

[39] Y. Bazilevs, V. M. Calo, T. J. R. Hughes, and Y. Zhang. Isogeometric fluid-structure interaction: Theory, algorithms, and computations. *Comput. Mech.*, 43(1):3–37, December 2008. DOI: 10.1007/s00466-008-0315-x.

[40] T. Belytschko, J. S.-J. Ong, W. K. Liu, and J. M. Kennedy. Hourglass control in linear and nonlinear problems. *Comput. Methods Appl. Mech. Eng.*, 43(3):251–276, May 1984. DOI: 10.1016/0045-7825(84)90067-7.

[41] Y. Benveniste. A new approach to the application of Mori-Tanaka's theory in composite materials. *Mech. Mater.*, 6(2):147–157, June 1987. DOI: 10.1016/0167-6636(87)90005-6.

[42] M. Benzi, G. H. Golub, and J. Liesen. Numerical solution of saddle point problems. *Acta Numer.*, 14:1–137, May 2005. DOI: 10.1017/S0962492904000212.

[43] J. S. Bergström and M. C. Boyce. Constitutive modeling of the large strain time-dependent behavior of elastomers. *J. Mech. Phys. Solids*, 46(5):931–954, May 1998. DOI: 10.1016/S0022-5096(97)00075-6.

[44] J. S. Bergström and M. C. Boyce. Mechanical behavior of particle filled elastomers. *Rubber Chem. Technol.*, 72(4):633–656, September 1999. DOI: 10.5254/1.3538823.

[45] P. Betsch and P. Steinmann. Conservation properties of a time FE method–part II: Time-stepping schemes for non-linear elastodynamics. *Int. J. Numer. Methods Eng.*, 50(8):1931–1955, March 2001. DOI: 10.1002/nme.103.

[46] R. B. Bird, R. C. Armstrong, and O. Hassager. *Dynamics of Polymeric Liquids, volume 1: Fluid Mechanics*. John Wiley and Sons Inc., New York, New York, USA, 1987. ISBN 978-0471802457.

[47] M. Blagojevic. *Gravitation and Gauge Symmetries*. Series in High Energy Physics, Cosmology and Gravitation. CRC Press, 2001. ISBN 9780750307673.

[48] A. Boczkowska and S. F. Awietjan. Smart composites of urethane elastomers with carbonyl iron. *J. Mater. Sci.*, 44(15):4104–4111, August 2009. DOI: 10.1007/s10853-009-3592-7.

[49] A. Boczkowska, S. F. Awietjan, and R. Wroblewski. Microstructure–property relationships of urethane magnetorheological elastomers. *Smart Mater. Struct.*, 16(5):1924–1930, September 2007. DOI: 10.1088/0964-1726/16/5/049.

[50] D. Boffi, L. Gastaldi, L. Heltai, and C. S. Peskin. On the hyper-elastic formulation of the immersed boundary method. *Comput. Methods Appl. Mech. Eng.*, 197(25–28):2210–2231, April 2008. DOI: 10.1016/j.cma.2007.09.015.

[51] J. Bonet and R. Wood. *Nonlinear Continuum Mechanics for Finite Element Analysis*. Cambridge University Press, New York, USA, 1997. ISBN 0-521-57272-X. DOI: 10.1017/cbo9780511755446.

[52] J. Bonet, A. J. Gil, and R. Ortigosa. A computational framework for polyconvex large strain elasticity. *Comput. Methods Appl. Mech. Eng.*, 283:1061–1094, January 2015. DOI: 10.1016/j.cma.2014.10.002.

[53] L. Borcea and O. Bruno. On the magneto-elastic properties of elastomer-ferromagnet composites. *J. Mech. Phys. Solids*, 49(12):2877–2919, December 2001. DOI: 10.1016/S0022-5096(01)00108-9.

[54] D. Y. Borin and G. V. Stepanov. Elastomer with magneto- and electrorheological properties. *J. Intell. Mater. Syst. Struct.*, 26(14):1893–1898, April 2015. DOI: 10.1177/1045389x15581521.

[55] H. Böse, R. Rabindranath, and J. Ehrlich. Soft magnetorheological elastomers as new actuators for valves. *J. Intell. Mater. Syst. Struct.*, 23(9):989–994, January 2012. DOI: 10.1177/1045389X11433498.

[56] A. Bossavit. *Computational Electromagnetism. Variational Formulations, Complementarity, Edge Elements*, volume 2 of *Electromagnetism Series*. Academic Press, San Diego, 1st edition, 1998. ISBN 978-0121187101.

[57] M. C. Boyce and E. M. Arruda. Constitutive models of rubber elasticity: A review. *Rubber Chem. Technol.*, 73(3):504–523, July 2000. DOI: 10.5254/1.3547602.

[58] H. Brauer and E. Kriegel. Kornbewegung bei der Sedimentation. *Chem. Ing. Tech.*, 38(3):321–330, March 1966. DOI: 10.1002/cite.330380326.

[59] H. Brauer and H. Thiele. Bewegung von Partikelschwärmen. *Chem. Ing. Tech.*, 45(13):909–912, July 1973. DOI: 10.1002/cite.330451317.

[60] I. A. Brigadnov and A. Dorfmann. Mathematical modeling of magneto-sensitive elastomers. *Int. J. Solids Struct.*, 40(18):4659–4674, September 2003. DOI: 10.1016/s0020-7683(03)00265-8.

[61] U. Brink and E. Stein. On some mixed finite element methods for incompressible and nearly incompressible finite elasticity. *Comput. Mech.*, 19(1):105–119, November 1996. DOI: 10.1007/BF02824849.

[62] S. Brisard and L. Dormieux. FFT-based methods for the mechanics of composites: A general variational framework. *Comput. Mater. Sci.*, 49(3):663–671, September 2010. DOI: 10.1016/j.commatsci.2010.06.009.

[63] W. E. Brown. Random number generation in C++11. Technical Report WG21 N3551, Standard C++ Foundation, 2013.

[64] W. F. Brown. *Magnetoelastic Interactions*. Springer Berlin Heidelberg, 1966. ISBN 978-3-642-87398-0. DOI: 10.1007/978-3-642-87396-6.

[65] W. F. Brown. *Micromagnetics*. Krieger Pub Co., New York, USA, 1978. ISBN 978-0882756653.

[66] F. Bruckner, C. Vogler, B. Bergmair, T. Huber, M. Fuger, D. Suess, M. Feischl, T. Fuehrer, M. Page, and D. Praetorius. Combining micromagnetism and magnetostatic Maxwell equations for multiscale magnetic simulations. *J. Magn. Magn. Mater.*, 343:163–168, October 2013. DOI: 10.1016/j.jmmm.2013.04.085.

[67] A. Büschel, S. Klinkel, and W. Wagner. Dielectric elastomers–numerical modeling of nonlinear visco-electroelasticity. *Int. J. Numer. Methods Eng.*, 93(8):834–856, February 2013. DOI: 10.1002/nme.4409.

[68] R. Bustamante. Transversely isotropic nonlinear magneto-active elastomers. *Acta Mech.*, 210(3–4):183–214, March 2010. DOI: 10.1007/s00707-009-0193-0.

[69] R. Bustamante and R. W. Ogden. Nonlinear magnetoelastostatics: Energy functionals and their second variations. *Math. Mech. Solids*, 18(7):760–772, September 2013. DOI: 10.1177/1081286512448347.

[70] R. Bustamante and K. R. Rajagopal. On a new class of electro-elastic bodies. II. Boundary value problems. *Proc. R. Soc. A, Math. Phys. Eng. Sci.*, 469(2155):1–20, July 2013. DOI: 10.1098/rspa.2013.0106.

[71] R. Bustamante, A. Dorfmann, and R. W. Ogden. Universal relations in isotropic nonlinear magnetoelasticity. *Q. J. Mech. Appl. Math.*, 59(3):435–450, August 2006. DOI: 10.1093/qjmam/hbl010.

[72] R. Bustamante, A. Dorfmann, and R. W. Ogden. A nonlinear magnetoelastic tube under extension and inflation in an axial magnetic field: numerical solution. *J. Eng. Math.*, 59(1):139–153, September 2007. DOI: 10.1007/s10665-006-9088-4.

[73] R. Bustamante, A. Dorfmann, and R. W. Ogden. Nonlinear electroelastostatics: a variational framework. *Z. Angew. Math. Phys.*, 60(1):154–177, January 2009. DOI: 10.1007/s00033-007-7145-0.

[74] R. Bustamante, A. Dorfmann, and R. W. Ogden. Numerical solution of finite geometry boundary-value problems in nonlinear magnetoelasticity. *Int. J. Solids Struct.*, 48(6):874–883, March 2011. DOI: 10.1016/j.ijsolstr.2010.11.021.

[75] M. Canadija and J. Mosler. On the thermomechanical coupling in finite strain plasticity theory with non-linear kinematic hardening by means of incremental energy minimization. *Int. J. Solids Struct.*, 48(7–8):1120–1129, April 2011. DOI: 10.1016/j.ijsolstr.2010.12.018.

[76] J. D. Carlson and M. R. Jolly. MR fluid, foam and elastomer devices. *Mechatronics*, 10(4–5):555–569, June 2000. DOI: 10.1016/S0957-4158(99)00064-1.

[77] C. J. Carpenter. Comparison of alternative formulations of 3-dimensional magnetic-field and eddy-current problems at power frequencies. In *Proceedings of the Institution of Electrical Engineers*, volume 124, pages 1026–1034, 1977. DOI: 10.1049/piee.1977.0211.

[78] C. Carstensen, F. Ebobisse, A. T. McBride, B. Daya Reddy, and P. Steinmann. *Mathematical Modelling in Solid Mechanics*, chapter Flow relations and yield functions for dissipative strain-gradient plasticity, pages 31–42. Springer Singapore, 2017. ISBN 978-981-10-3764-1. DOI: 10.1007/978-981-10-3764-1_3.

[79] O. Čertík, I. Fernando, S. Garg, and T. Rathnayake et al. SymEngine: A fast symbolic manipulation library, 2013–. URL https://github.com/symengine/symengine. Online; Last visited: September 2018.

[80] P. Chadwick. *Continuum Mechanics: Concise Theory and Problems*. Dover Publications Inc., Mineola, New York, USA, 2nd edition, 1998. ISBN 9780486401805.

[81] N. Charalambakis. Homogenization techniques and micromechanics. A survey and perspectives. *Appl. Mech. Rev.*, 63(3):030803, July 2010. DOI: 10.1115/1.4001911.

[82] G. Chatzigeorgiou, A. Javili, and P. Steinmann. Unified magnetomechanical homogenization framework with application to magnetorheological elastomers. *Math. Mech. Solids*, 19(2):193–211, March 2014. DOI: 10.1177/1081286512458109.

[83] G. Chatzigeorgiou, A. Javili, and P. Steinmann. Surface magnetoelasticity theory. *Arch. Appl. Mech.*, 85(9–10):1265–1288, September 2015. DOI: 10.1007/s00419-014-0938-7.

[84] L. Chen, X. L. Gong, and W. H. Li. Microstructures and viscoelastic properties of anisotropic magnetorheological elastomers. *Smart Mater. Struct.*, 16(6):2645–2650, November 2007. DOI: 10.1088/0964-1726/16/6/069.

[85] L. Chen, Z. Chang, and T. Qin. Elastic wave propagation in simple-sheared hyperelastic materials with different constitutive models. *Int. J. Solids Struct.*, 126–127:1–7, November 2017. DOI: 10.1016/j.ijsolstr.2017.07.027.

[86] R. M. Christensen. A critical evaluation for a class of micro-mechanics models. *J. Mech. Phys. Solids*, 38(3):379–404, January 1990. DOI: 10.1016/0022-5096(90)90005-o.

[87] R. M. Christensen and K. H. Lo. Solutions for effective shear properties in three phase sphere and cylinder models. *J. Mech. Phys. Solids*, 27(4):315–330, August 1979. DOI: 10.1016/0022-5096(79)90032-2.

[88] P. Ciarletta and M. Destrade. Torsion instability of soft solid cylinders. *IMA J. Appl. Math.*, 79(5):804–819, October 2014. DOI: 10.1093/imamat/hxt052.

[89] F. Clément, L. Bokobza, and L. Monnerie. Investigation of the Payne Effect and its temperature dependence on silica-filled polydimethylsiloxane networks. Part I: Experimental results. *Rubber Chem. Technol.*, 78(2):211–231, May 2005. DOI: 10.5254/1.3547879.

[90] E. W. C. Coenen, V. G. Kouznetsova, E. Bosco, and M. G. D. Geers. A multi-scale approach to bridge microscale damage and macroscale failure: A nested computational homogenization-localization framework. *Int. J. Fract.*, 178(1–2):157–178, November 2012. DOI: 10.1007/s10704-012-9765-4.

[91] B. D. Coleman and M. E. Gurtin. Thermodynamics with internal state variables. *J. Chem. Phys.*, 47(2):597–613, 1967. DOI: 10.1063/1.1711937.

[92] B. D. Coleman and W. Noll. The thermodynamics of elastic materials with heat conduction and viscosity. *Arch. Ration. Mech. Anal.*, 13(1):167–178, December 1963. DOI: 10.1007/BF01262690.

[93] M. Colussi, F. Berto, K. Mori, and F. Narita. Fracture behavior of cracked giant magnetostrictive materials in three-point bending under magnetic fields: Strain energy density criterion. *Adv. Eng. Mater.*, 18(12):2063–2069, December 2016. DOI: 10.1002/adem.201500533.

[94] E. Comellas, F. J. Bellomo, and S. Oller. A generalized finite-strain damage model for quasi-incompressible hyperelasticity using hybrid formulation. *Int. J. Numer. Methods Eng.*, 105(10):781–800, March 2015. DOI: 10.1002/nme.5118.

[95] E. Coquelle, G. Bossis, D. Szabo, and F. Giulieri. Micromechanical analysis of an elastomer filled with particles organized in chain-like structure. *J. Mater. Sci.*, 41(18):5941–5953, September 2006. DOI: 10.1007/s10853-006-0329-8.

[96] R. Cottereau. A stochastic-deterministic coupling method for multiscale problems. Application to numerical homogenization of random materials. *Proc. IUTAM*, 6:35–43, 2013. DOI: 10.1016/j.piutam.2013.01.004.

[97] C. A. Coulomb. *Premier mémoire sur l'electricité et le magnétisme*. Académie Royale des sciences, 1785.

[98] C. A. Coulomb. *Second mémoire sur l'electricité et le magnétisme*. Académie Royale des sciences, 1785.

[99] J. L. Coulomb. Finite elements three dimensional magnetic field computation. *IEEE Trans. Magn.*, 17(6):3241–3246, November 1981. DOI: 10.1109/TMAG.1981.1061587.

[100] T. Crestaux, O. Le Maître, and J.-M. Martinez. Polynomial chaos expansion for sensitivity analysis. *Reliab. Eng. Syst. Saf.*, 94(7):1161–1172, July 2009. DOI: 10.1016/j.ress.2008.10.008.

[101] A. D. Crowell. Motion of the Earth as viewed from the moon and the Y-suspended pendulum. *Am. J. Phys.*, 49(5):452–454, May 1981. DOI: 10.1119/1.12477.

[102] E. Cueto, F. Chinesta, and A. Huerta. *Separated Representations and PGD-Based Model Reduction: Fundamentals and Applications*, chapter Model order reduction based on proper orthogonal decomposition, pages 1–26. Springer Vienna, 2014. ISBN 978-3-7091-1794-1. DOI: 10.1007/978-3-7091-1794-1_1.

[103] B. D. Cullity and C. D. Graham. *Introduction to Magnetic Materials*. Wiley-IEEE Press, 2008. ISBN 978-0-471-47741-9. DOI: 10.1002/9780470386323.

[104] N. Daher and G. A. Maugin. The method of virtual power in continuum mechanics application to media presenting singular surfaces and interfaces. *Acta Mech.*, 60(3–4):217–240, July 1986. DOI: 10.1007/bf01176354.

[105] K. Danas, S. V. Kankanala, and N. Triantafyllidis. Experiments and modeling of iron-particle-filled magnetorheological elastomers. *J. Mech. Phys. Solids*, 60(1):120–138, January 2012. DOI: 10.1016/j.jmps.2011.09.006.

[106] M. d'Aquino, C. Serpico, G. Miano, and C. Forestiere. A novel formulation for the numerical computation of magnetization modes in complex micromagnetic systems. *J. Comput. Phys.*, 228(17):6130–6149, September 2009. DOI: 10.1016/j.jcp.2009.05.026.

[107] T. A. Davis. Algorithm 832: UMFPACK V4.3—an unsymmetric-pattern multifrontal method. *ACM Trans. Math. Softw.*, 30(2):196–199, June 2004. DOI: 10.1145/992200.992206.

[108] A. M. de Villiers, A. T. McBride, B. D. Reddy, T. Franz, and B. S. Spottiswoode. A validated patient-specific FSI model for vascular access in haemodialysis. *Biomech. Model. Mechanobiol.*, 17(2):479–497, April 2018. DOI: 10.1007/s10237-017-0973-8.

[109] M. K. Deb. *Solution of stochastic partial differential equations (SPDEs) using Galerkin method: theory and applications*. PhD thesis, University of Texas at Austin, 2000.

[110] M. K. Deb, I. M. Babuška, and J. T. Oden. Solution of stochastic partial differential equations using Galerkin finite element techniques. *Comput. Methods Appl. Mech. Eng.*, 190(48):6359–6372, September 2001. DOI: 10.1016/s0045-7825(01)00237-7.

[111] L. Delannay, I. Doghri, and O. Pierard. Prediction of tension–compression cycles in multiphase steel using a modified incremental mean-field model. *Int. J. Solids Struct.*, 44(22–23):7291–7306, November 2007. DOI: 10.1016/j.ijsolstr.2007.04.013.

[112] J. Dellinger, D. Vasicek, and C. Sondergeld. Kelvin notation for stabilizing elastic-constant inversion. *Rev. Inst. Fr. Pét.*, 53(5):709–719, September 1998. DOI: 10.2516/ogst:1998063.

[113] F. Dell'Isola and A. Romano. On the derivation of thermomechanical balance equations for continuous systems with a nonmaterial interface. *Int. J. Eng. Sci.*, 25(11–12):1459–1468, 1987. DOI: 10.1016/0020-7225(87)90023-1.

[114] H. X. Deng and X. L. Gong. Adaptive tuned vibration absorber based on magnetorheological elastomer. *J. Intell. Mater. Syst. Struct.*, 18(12):1205–1210, December 2007. DOI: 10.1177/1045389x07083128.

[115] M. Destrade, B. Mac Donald, J. G. Murphy, and G. Saccomandi. At least three invariants are necessary to model the mechanical response of incompressible, transversely isotropic materials. *Comput. Mech.*, 52(4):959–969, October 2013. DOI: 10.1007/s00466-013-0857-4.

[116] L. S. Dimas, T. Giesa, and M. J. Buehler. Coupled continuum and discrete analysis of random heterogeneous materials: Elasticity and fracture. *J. Mech. Phys. Solids*, 63:481–490, February 2014. DOI: 10.1016/j.jmps.2013.07.006.

[117] I. Doghri and A. Ouaar. Homogenization of two-phase elasto-plastic composite materials and structures: Study of tangent operators, cyclic plasticity and numerical algorithms. *Int. J. Solids Struct.*, 40(7):1681–1712, April 2003. DOI: 10.1016/s0020-7683(03)00013-1.

[118] M. Doi and S. F. Edwards. *The Theory of Polymer Dynamics*, volume 73. Oxford University Press, 1988. ISBN 978-0198520337.

[119] A. Dorfmann and R. W. Ogden. Magnetoelastic modelling of elastomers. *Eur. J. Mech. A, Solids*, 22(4):497–507, July 2003. DOI: 10.1016/S0997-7538(03)00067-6.

[120] A. Dorfmann and R. W. Ogden. Nonlinear magnetoelastic deformations. *Q. J. Mech. Appl. Math.*, 57(4):599–622, November 2004. DOI: 10.1093/qjmam/57.4.599.

[121] A. Dorfmann and R. W. Ogden. Nonlinear magnetoelastic deformations of elastomers. *Acta Mech.*, 167(1–2):13–28, January 2004. DOI: 10.1007/s00707-003-0061-2.

[122] A. Dorfmann and R. W. Ogden. Nonlinear electroelasticity. *Acta Mech.*, 174(3–4):167–183, March 2005. DOI: 10.1007/s00707-004-0202-2.

[123] A. Dorfmann and R. W. Ogden. Some problems in nonlinear magnetoelasticity. *Z. Angew. Math. Phys.*, 56(4):718–745, July 2005. DOI: 10.1007/s00033-004-4066-z.

[124] A. Dorfmann and R. W. Ogden. Nonlinear electroelastic deformations. *J. Elast.*, 82(2):99–127, February 2006. DOI: 10.1007/s10659-005-9028-y.

[125] L. Dorfmann and R. W. Ogden. *Nonlinear Theory of Electroelastic and Magnetoelastic Interactions*. Springer Science & Business Media, 2014. ISBN 978-1-4614-9595-6. DOI: 10.1007/978-1-4614-9596-3.

[126] J. R. Dormand and P. J. Prince. A family of embedded Runge-Kutta formulae. *J. Comput. Appl. Math.*, 6(1):19–26, March 1980. DOI: 10.1016/0771-050x(80)90013-3.

[127] D. Drummer. *Verarbeitung und Eigenschaften Kunststoffgebundener Dauermagnete*. Habilitation, University of Erlangen-Nuremberg, 2004.

[128] M. L. Dunn and M. Taya. Micromechanics predictions of the effective electroelastic moduli of piezoelectric composites. *Int. J. Solids Struct.*, 30(2):161–175, 1993. DOI: 10.1016/0020-7683(93)90058-f.

[129] M. L. Dunn and M. Taya. An analysis of piezoelectric composite materials containing ellipsoidal inhomogeneities. *Proc. R. Soc. A, Math. Phys. Eng. Sci.*, 443(1918):265–287, November 1993. DOI: 10.1098/rspa.1993.0145.

[130] F. Durst. *Fluid Mechanics: An Introduction to the Theory of Fluid Flows*. Springer Science & Business Media, 2008. ISBN 978-3-540-71342-5. DOI: 10.1007/978-3-540-71343-2.

[131] A. Einstein. *Special Theory of Relativity*, chapter 5: On the electrodynamics of moving bodies, pages 187–218. Pergamon Press Ltd., 1970. DOI: 10.1016/b978-0-08-006995-1.50014-4.

[132] K. Eriksson, D. Estep, P. Hansbo, and C. Johnson. *Computational Differential Equations, volume 1*. Cambridge University Press, 1996. ISBN 978-0521567381.

[133] A. C. Eringen and G. A. Maugin. *Electrodynamics of Continua I: Foundations and Solid Media*. Springer Berlin Heidelberg, 1990. ISBN 978-1-4612-7923-5. DOI: 10.1007/978-1-4612-3226-1.

[134] J. D. Eshelby. The determination of the elastic field of an ellipsoidal inclusion, and related problems. *Proc. R. Soc. A, Math. Phys. Eng. Sci.*, 241(1226):376–396, August 1957. DOI: 10.1098/rspa.1957.0133.

[135] J. D. Eshelby. The elastic field outside an ellipsoidal inclusion. *Proc. R. Soc. A, Math. Phys. Eng. Sci.*, 252(1271):561–569, October 1959. DOI: 10.1098/rspa.1959.0173.

[136] A. Esmaeili, P. Steinmann, and A. Javili. Surface plasticity: theory and computation. *Comput. Mech.*, 62(4):617–634, October 2018. DOI: 10.1007/s00466-017-1517-x.

[137] G. Ethiraj. *Computational modeling of ferromagnetics and magnetorheological elastomers*. PhD thesis, Universität Stuttgart, 2014. URL https://elib.uni-stuttgart.de/handle/11682/626.

[138] G. Ethiraj, A. Sridhar, and C. Miehe. Variational modeling and homogenization in dissipative magneto-mechanics. *GAMM-Mitt.*, 38(1):75–101, April 2015. DOI: 10.1002/gamm.201510004.

[139] R. H. Ewoldt. Defining nonlinear rheological material functions for oscillatory shear. *J. Rheol.*, 57(1):177–195, January 2013. DOI: 10.1122/1.4764498.

[140] R. H. Ewoldt, A. E. Hosoi, and G. H. McKinley. New measures for characterizing nonlinear viscoelasticity in large amplitude oscillatory shear. *J. Rheol.*, 52(6):1427–1458, November 2008. DOI: 10.1122/1.2970095.

[141] R. H. Ewoldt, A. E. Hosoi, and G. H. McKinley. Nonlinear viscoelastic biomaterials: meaningful characterization and engineering inspiration. *Integr. Comp. Biol.*, 49(1):40–50, July 2009. DOI: 10.1093/icb/icp010.

[142] R. H. Ewoldt, P. Winter, J. Maxey, and G. H. McKinley. Large amplitude oscillatory shear of pseudoplastic and elastoviscoplastic materials. *Rheol. Acta*, 49(2):191–212, February 2010. DOI: 10.1007/s00397-009-0403-7.

[143] F. F. Fang, H. J. Choi, and Y. Seo. Sequential coating of magnetic carbonyliron particles with polystyrene and multiwalled carbon nanotubes and its effect on their magnetorheology. *ACS Appl. Mater. Interfaces*, 2(1):54–60, January 2010. DOI: 10.1021/am900577w.

[144] S. Federico and W. Herzog. On the anisotropy and inhomogeneity of permeability in articular cartilage. *Biomech. Model. Mechanobiol.*, 7(5):367–378, October 2008. DOI: 10.1007/s10237-007-0091-0.

[145] C. Feuersänger. *Manual for Package PGFPLOTS*, 2011. URL http://mirrors.ctan.org/graphics/pgf/contrib/pgfplots/doc/pgfplots.pdf.

[146] J. A. Fike and J. J. Alonso. The development of hyper-dual numbers for exact second-derivative calculations. In *49th AIAA Aerospace Sciences Meeting including the New Horizons Forum and Aerospace Exposition*. American Institute of Aeronautics and Astronautics, January 2011. DOI: 10.2514/6.2011-886.

[147] G. Filipcsei, I. Csetneki, A. Szilágyi, and M. Zrínyi. *Oligomers – Polymer Composites – Molecular Imprinting*, volume 206 of *Advances in Polymer Science*, chapter Magnetic field-responsive smart polymer composites, pages 137–189. Springer Berlin Heidelberg, 2007. ISBN 978-3-540-46829-5. DOI: 10.1007/12_2006_104.

[148] W. P. Fletcher and A. N. Gent. Nonlinearity in the dynamic properties of vulcanized rubber compounds. *Rubber Chem. Technol.*, 27(1):209–222, March 1954. DOI: 10.5254/1.3543472.

[149] B. Fornberg. Generation of finite difference formulas on arbitrarily spaced grids. *Math. Comput.*, 51(184):699–706, October 1988. DOI: 10.1090/s0025-5718-1988-0935077-0.

[150] B. Fraeijs de Veubeke. Diffusion des inconnues hyperstatiques dans les voilures á longerons couplés. *Bull. Serv. Techn. Aeronaut.*, 24:1–56, 1951. URL http://hdl.handle.net/2268/37343. pp. 56.

[151] T.-P. Fries and T. Belytschko. The extended/generalized finite element method: An overview of the method and its applications. *Int. J. Numer. Methods Eng.*, 84(3):253–304, October 2010. DOI: 10.1002/nme.2914.

[152] R. Fuhrer, C. M. Schumacher, M. Zeltner, and W. J. Stark. Soft iron/silicon composite tubes for magnetic peristaltic pumping: Frequency-dependent pressure and volume flow. *Adv. Funct. Mater.*, 23(31):3845–3849, March 2013. DOI: 10.1002/adfm.201203572.

[153] M. Fujita, T. Tokumasu, T. Yamada, T. Hirose, Y. Tanaka, N. Kumagai, and S. Uchida. 3-dimensional electromagnetic analysis and design of an eddy-current rail brake system. *IEEE Trans. Magn.*, 34(5):3548–3551, September 1998. DOI: 10.1109/20.717837.

[154] C. Gabriel and H. M. Laun. Combined slit and plate–plate magnetorheometry of a magnetorheological fluid (MRF) and parameterization using the Casson model. *Rheol. Acta*, 48(7):755–768, August 2009. DOI: 10.1007/s00397-009-0369-5.

[155] E. Galipeau. *Non-linear homogenization of magnetorheological elastomers at finite strain*. PhD thesis, University of Pennsylvania, 2012. URL http://repository.upenn.edu/edissertations/634.

[156] E. Galipeau and P. Ponte Castañeda. The effect of particle shape and distribution on the macroscopic behavior of magnetoelastic composites. *Int. J. Solids Struct.*, 49(1):1–17, January 2012. DOI: 10.1016/j.ijsolstr.2011.08.014.

[157] E. Galipeau and P. Ponte Castañeda. A finite-strain constitutive model for magnetorheological elastomers: Magnetic torques and fiber rotations. *J. Mech. Phys. Solids*, 61(4):1065–1090, April 2013. DOI: 10.1016/j.jmps.2012.11.007.

[158] E. Galipeau, S. Rudykh, and P. Ponte Castañeda. Magnetoactive elastomers with periodic and random microstructures. *Int. J. Solids Struct.*, 51(18):3012–3024, September 2014. DOI: 10.1016/j.ijsolstr.2014.04.013.

[159] T. Gamblin, M. Culpo, G. Becker, M. Legendre, G. Lee, E. Fischer, and B. Hegner. *Managing HPC Software Complexity with Spack*. Colorado Convention Center, Denver, USA, November 2017. URL http://spack.io/slides/Spack-SC17-Tutorial.pdf. Tutorial presented at Supercomputing 2017. Online; Last visited: September 2018.

[160] M. Gărăjeu, J. C. Michel, and P. Suquet. A micromechanical approach of damage in viscoplastic materials by evolution in size, shape and distribution of voids. *Comput. Methods Appl. Mech. Eng.*, 183(3–4):223–246, March 2000. DOI: 10.1016/s0045-7825(99)00220-0.

[161] T. C. Gasser, R. W. Ogden, and G. A. Holzapfel. Hyperelastic modelling of arterial layers with distributed collagen fibre orientations. *J. R. Soc. Interface*, 3(6):15–35, February 2006. DOI: 10.1098/rsif.2005.0073.

[162] D. M. Gay. *Mixed Integer Nonlinear Programming*, chapter Using expression graphs in optimization algorithms, pages 247–262. Springer New York, 2012. ISBN 978-1-4614-1927-3. DOI: 10.1007/978-1-4614-1927-3_8.

[163] M. W. Gee, C. M. Siefert, J. J. Hu, R. S. Tuminaro, and M. G. Sala. ML 5.0 smoothed aggregation user's guide. Technical Report SAND2006-2649, Sandia National Laboratories, 2006.

[164] M. G. D. Geers, V. G. Kouznetsova, and W. A. M. Brekelmans. Multi-scale computational homogenization: Trends and challenges. *J. Comput. Appl. Math.*, 234(7):2175–2182, August 2010. DOI: 10.1016/j.cam.2009.08.077.

[165] M. G. D. Geers, V. G. Kouznetsova, and W. A. M. Brekelmans. *Multiscale Modelling of Plasticity and Fracture by Means of Dislocation Mechanics*, chapter Computational homogenization, pages 327–394. Springer Vienna, 2010. ISBN 978-3-7091-0282-4. DOI: 10.1007/978-3-7091-0283-1_7.

[166] R. G. Ghanem and P. D. Spanos. *Stochastic Finite Elements: A Spectral Approach*. Springer New York, 1991. ISBN 978-1-4612-7795-8. DOI: 10.1007/978-1-4612-3094-6.

[167] A. J. Giacomin and J. M. Dealy. *Techniques in Rheological Measurement*, chapter 4: Large-amplitude oscillatory shear, pages 99–121. Chapman and Hall London & New York, 1993. ISBN 978-94-010-4937-5. DOI: 10.1007/978-94-011-2114-9_4.

[168] J. W. Gibbs. Thermodynamics. In *The Scientific Papers*. Longmans, Green, and Co., 1906.

[169] A. J. Gil and R. Ortigosa. A new framework for large strain electromechanics based on convex multi-variable strain energies: Variational formulation and material characterisation. *Comput. Methods Appl. Mech. Eng.*, 302:293–328, April 2016. DOI: 10.1016/j.cma.2015.11.036.

[170] T. L. Gilbert. A phenomenological theory of damping in ferromagnetic materials. *IEEE Trans. Magn.*, 40(6):3443–3449, November 2004. DOI: 10.1109/tmag.2004.836740.

[171] K. T. Gillen. Effect of cross-links which occur during continuous chemical stress-relaxation. *Macromolecules*, 21(2):442–446, February 1988. DOI: 10.1021/ma00180a027.

[172] J. M. Ginder. *Encyclopedia of Applied Physics*, volume 16, chapter Rheology controlled by magnetic fields, pages 487–503. Wiley-VCH Verlag GmbH & Co KGaA, New York, New York, USA, 1996. DOI: 10.1002/3527600434.eap402.

[173] J. M. Ginder, M. E. Nichols, L. D. Elie, and J. L. Tardiff. Magnetorheological elastomers: properties and applications. In *SPIE Proceedings, Smart Structures and Materials*, volume 3675, pages 131–138, 1999. DOI: 10.1117/12.352787.

[174] J. M. Ginder, M. E. Nichols, L. D. Elie, and S. M. Clark. Controllable-stiffness components based on magnetorheological elastomers. In *SPIE Proceedings, Smart Structures and Materials*, volume 3985. SPIE, June 2000. DOI: 10.1117/12.388844.

[175] J. M. Ginder, W. F. Schlotter, and M. E. Nichols. Magnetorheological elastomers in tunable vibration absorbers. In *SPIE Proceedings, Smart Structures and Materials*, volume 4331, July 2001. DOI: 10.1117/12.432694.

[176] F. S. Göküzüm and M.-A. Keip. An algorithmically consistent macroscopic tangent operator for FFT-based computational homogenization. *Int. J. Numer. Methods Eng.*, 113(4):581–600, January 2018. DOI: 10.1002/nme.5627.

[177] O. Gonzalez. Exact energy and momentum conserving algorithms for general models in nonlinear elasticity. *Comput. Methods Appl. Mech. Eng.*, 190(13–14):1763–1783, December 2000. DOI: 10.1016/s0045-7825(00)00189-4.

[178] C. Goodyear. Improvements in India-Rubber fabrics, 1844. URL https://www.google.com/patents/US3633.

[179] A. Griewank, D. Juedes, and J. Utke. Algorithm 755: ADOL-C: a package for the automatic differentiation of algorithms written in C/C++. *ACM Trans. Math. Softw.*, 22(2):131–167, June 1996. DOI: 10.1145/229473.229474.

[180] D. J. Griffiths. *Introduction to Electrodynamics*. Prentice Hall, Upper Saddle River, New Jersey, USA, 3rd edition, 1999. ISBN 0-13-805326-X.

[181] E. A. Guggenheim. The thermodynamics of interfaces in systems of several components. *Trans. Faraday Soc.*, 35:397–412, 1940. DOI: 10.1039/tf9403500397.

[182] B. Guo and I. Babuška. The *h-p* version of the finite element method. *Comput. Mech.*, 1(1):21–41, March 1986. DOI: 10.1007/bf00298636.

[183] F. Guo, C. b. Du, and R. p. Li. Viscoelastic parameter model of magnetorheological elastomers based on Abel Dashpot. *Adv. Mech. Eng.*, 6(0):629386–629386, February 2015. DOI: 10.1155/2014/629386.

[184] A. Gupta and D. J. Steigmann. Balance laws. In J. Merodio and G. Saccomandi, editors, *Continuum Mechanics: Encyclopedia of Life Support Systems, volume 1*. EOLSS Publishers, 2011. ISBN 978-1-84826-822-7.

[185] M. E. Gurtin and A. I. Murdoch. A continuum theory of elastic material surfaces. *Arch. Ration. Mech. Anal.*, 57(4):291–323, December 1975. DOI: 10.1007/bf00261375.

[186] E. Guth. Theory of filler reinforcement. *Rubber Chem. Technol.*, 18(3):596–604, September 1945. DOI: 10.5254/1.3546754.

[187] R. Haj-Ali and J. Aboudi. Nonlinear micromechanical formulation of the high fidelity generalized method of cells. *Int. J. Solids Struct.*, 46(13):2577–2592, June 2009. DOI: 10.1016/j.ijsolstr.2009.02.004.

[188] R. Haj-Ali and J. Aboudi. Formulation of the high-fidelity generalized method of cells with arbitrary cell geometry for refined micromechanics and damage in composites. *Int. J. Solids Struct.*, 47(25–26):3447–3461, December 2010. DOI: 10.1016/j.ijsolstr.2010.08.022.

[189] Y. Han, W. Hong, and L. E. Faidley. Field-stiffening effect of magneto-rheological elastomers. *Int. J. Solids Struct.*, 50(14–15):2281–2288, 2013. DOI: 10.1016/j.ijsolstr.2013.03.030.

[190] Z. Hashin and S. Shtrikman. A variational approach to the theory of the elastic behaviour of multiphase materials. *J. Mech. Phys. Solids*, 11(2):127–140, March 1963. DOI: 10.1016/0022-5096(63)90060-7.

[191] C. Heinrich, M. Aldridge, A. S. Wineman, J. Kieffer, A. M. Waas, and K. W. Shahwan. The role of curing stresses in subsequent response, damage and failure of textile polymer composites. *J. Mech. Phys. Solids*, 61(5):1241–1264, May 2013. DOI: 10.1016/j.jmps.2012.12.005.

[192] M. A. Heroux. AztecOO user guide. Technical Report SAND2004-3796, Sandia National Laboratories, 2007. URL https://trilinos.org/oldsite/packages/aztecoo/AztecOOUserGuide.pdf.

[193] M. A. Heroux, R. A. Bartlett, V. E. Howle, R. J. Hoekstra, J. J. Hu, T. G. Kolda, R. B. Lehoucq, K. R. Long, R. P. Pawlowski, E. T. Phipps, A. G. Salinger, H. K. Thornquist, R. S. Tuminaro, J. M. Willenbring, A. Williams, and K. S. Stanley. An overview of the Trilinos project. *ACM Trans. Math. Softw.*, 31(3):397–423, September 2005. DOI: 10.1145/1089014.1089021.

[194] R. Hertel. Computational micromagnetism of magnetization processes in nickel nanowires. *J. Magn. Magn. Mater.*, 249(1–2):251–256, August 2002. DOI: 10.1016/s0304-8853(02)00539-5.

[195] C. Hesch, A. J. Gil, A. Arranz Carreño, J. Bonet, and P. Betsch. A mortar approach for fluid-structure interaction problems: Immersed strategies for deformable and rigid bodies. *Comput. Methods Appl. Mech. Eng.*, 278:853–882, August 2014. DOI: 10.1016/j.cma.2014.06.004.

[196] R. Hill. Elastic properties of reinforced solids: Some theoretical principles. *J. Mech. Phys. Solids*, 11(5):357–372, September 1963. DOI: 10.1016/0022-5096(63)90036-x.

[197] R. Hill. Continuum micro-mechanics of elastoplastic polycrystals. *J. Mech. Phys. Solids*, 13(2):89–101, April 1965. DOI: 10.1016/0022-5096(65)90023-2.

[198] R. Hill. The essential structure of constitutive laws for metal composites and polycrystals. *J. Mech. Phys. Solids*, 15(2):79–95, March 1967. DOI: 10.1016/0022-5096(67)90018-x.

[199] R. Hill. On constitutive macro-variables for heterogeneous solids at finite strain. *Proc. R. Soc. A, Math. Phys. Eng. Sci.*, 326(1565):131–147, January 1972. DOI: 10.1098/rspa.1972.0001.

[200] A. C. Hindmarsh, P. N. Brown, K. E. Grant, S. L. Lee, R. Serban, D. E. Shumaker, and C. S. Woodward. SUNDIALS: Suite of nonlinear and differential/algebraic equation solvers. *ACM Trans. Math. Softw.*, 31(3):363–396, September 2005. DOI: 10.1145/1089014.1089020.

[201] C. B. Hirschberger, N. Sukumar, and P. Steinmann. Computational homogenization of material layers with micromorphic mesostructure. *Philos. Mag.*, 88(30–32):3603–3631, October 2008. DOI: 10.1080/14786430802502567.

[202] N. Hoang, N. Zhang, W. H. Li, and H. Du. Development of a torsional dynamic absorber using a magnetorheological elastomer for vibration reduction of a powertrain test rig. *J. Intell. Mater. Syst. Struct.*, 24(16):2036–2044, May 2013. DOI: 10.1177/1045389x13489361.

[203] M. Hojjati, A. Johnston, S. V. Hoa, and J. Denault. Viscoelastic behavior of Cytec FM73 adhesive during cure. *J. Appl. Polym. Sci.*, 91(4):2548–2557, February 2004. DOI: 10.1002/app.13416.

[204] G. A. Holzapfel. *Nonlinear Solid Mechanics: A Continuum Approach for Engineering*. John Wiley & Sons Ltd., West Sussex, England, 2007. ISBN 0-471-82304-X.

[205] G. A. Holzapfel and T. C. Gasser. A viscoelastic model for fiber-reinforced composites at finite strains: Continuum basis, computational aspects and applications. *Comput. Methods Appl. Mech. Eng.*, 190(34):4379–4403, May 2001. DOI: 10.1016/s0045-7825(00)00323-6.

[206] G. A. Holzapfel and R. W. Ogden. Constitutive modelling of arteries. *Proc. R. Soc. A, Math. Phys. Eng. Sci.*, 466(2118):1551–1597, June 2010. DOI: 10.1098/rspa.2010.0058.

[207] G. A. Holzapfel and J. C. Simo. A new viscoelastic constitutive model for continuous media at finite thermomechanical changes. *Int. J. Solids Struct.*, 33(20–22):3019–3034, August 1996. DOI: 10.1016/0020-7683(95)00263-4.

[208] M. Hossain and P. Steinmann. *Advances in Applied Mechanics*, volume 48, chapter 3: Continuum physics of materials with time-dependent properties, pages 141–259. Elsevier BV, 2015. ISBN 978-0-12-802128-6. DOI: 10.1016/bs.aams.2015.10.003.

[209] M. Hossain, G. Possart, and P. Steinmann. A small-strain model to simulate the curing of thermosets. *Comput. Mech.*, 43(6):769–779, May 2009. DOI: 10.1007/s00466-008-0344-5.

[210] M. Hossain, G. Possart, and P. Steinmann. A finite strain framework for the simulation of polymer curing. Part I: elasticity. *Comput. Mech.*, 44(5):621–630, October 2009. DOI: 10.1007/s00466-009-0397-0.

[211] M. Hossain, G. Possart, and P. Steinmann. A finite strain framework for the simulation of polymer curing. Part II. Viscoelasticity and shrinkage. *Comput. Mech.*, 46(3):363–375, August 2010. DOI: 10.1007/s00466-010-0479-z.

[212] M. Hossain, D. K. Vu, and P. Steinmann. Experimental study and numerical modelling of VHB 4910 polymer. *Comput. Mater. Sci.*, 59:65–74, 2012. DOI: 10.1016/j.commatsci.2012.02.027.

[213] M. Hossain, G. Chatzigeorgiou, F. Meraghni, and P. Steinmann. A multi-scale approach to model the curing process in magneto-sensitive polymeric materials. *Int. J. Solids Struct.*, 69–70:34–44, September 2015. DOI: 10.1016/j.ijsolstr.2015.06.011.

[214] M. Hossain, P. Saxena, and P. Steinmann. Modelling the mechanical aspects of the curing process of magneto-sensitive elastomeric materials. *Int. J. Solids Struct.*, 58:257–269, April 2015. DOI: 10.1016/j.ijsolstr.2015.01.010.

[215] M. Hossain, P. Saxena, and P. Steinmann. Modelling the curing process in magneto-sensitive polymers: Rate-dependence and shrinkage. *Int. J. Non-Linear Mech.*, 74:108–121, September 2015. DOI: 10.1016/j.ijnonlinmec.2015.04.008.

[216] M. Hossain, D. K. Vu, and P. Steinmann. A comprehensive characterization of the electro-mechanically coupled properties of VHB 4910 polymer. *Arch. Appl. Mech.*, 85(4):523–537, April 2015d. DOI: 10.1007/s00419-014-0928-9.

[217] S.-Y. Hsu and C.-L. Chang. Mesh deformation based on fully stressed design: the method and 2-D examples. *Int. J. Numer. Methods Eng.*, 72(5):606–629, October 2007. DOI: 10.1002/nme.2027.

[218] H. Hu, L. Onyebueke, and A. Abatan. Characterizing and modeling mechanical properties of nanocomposites-review and evaluation. *J. Miner. Mater. Charact. Eng.*, 9(4):275–319, April 2010. DOI: 10.4236/jmmce.2010.94022.

[219] W. Hu, G. Z. Lum, M. Mastrangeli, and M. Sitti. Small-scale soft-bodied robot with multimodal locomotion. *Nature*, 554(7690):81–85, February 2018. DOI: 10.1038/nature25443.

[220] J. H. Huang and W.-S. Kuo. The analysis of piezoelectric/piezomagnetic composite materials containing ellipsoidal inclusions. *J. Appl. Phys.*, 81(3):1378–1386, February 1997. DOI: 10.1063/1.363874.

[221] N. Huber and C. Tsakmakis. Finite deformation viscoelasticity laws. *Mech. Mater.*, 32(1):1–18, January 2000. DOI: 10.1016/S0167-6636(99)00045-9.

[222] T. J. Hughes. *The Finite Element Method: Linear Static and Dynamic Finite Element Analysis*. Dover Publications Inc., New York, USA, 2000. ISBN 978-0486411811.

[223] J. D. Humphrey and F. C. P. Yin. On constitutive relations and finite deformations of passive cardiac tissue: I. A pseudostrain-energy function. *J. Biomech. Eng.*, 109(4):298–304, November 1987. DOI: 10.1115/1.3138684.

[224] K. Hyun, M. Wilhelm, C. O. Klein, K. S. Cho, J. G. Nam, K. H. Ahn, S. J. Lee, R. H. Ewoldt, and G. H. McKinley. A review of nonlinear oscillatory shear tests: Analysis and application of large amplitude oscillatory shear (LAOS). *Prog. Polym. Sci.*, 36(12):1697–1753, December 2011. DOI: 10.1016/j.progpolymsci.2011.02.002.

[225] A. Hück, J. Utke, and C. Bischof. Source transformation of C++ codes for compatibility with operator overloading. *Proc. Comput. Sci.*, 80:1485–1496, 2016. DOI: 10.1016/j.procs.2016.05.470.

[226] J. H. Irving and J. G. Kirkwood. The statistical mechanical theory of transport processes. IV. The equations of hydrodynamics. *J. Chem. Phys.*, 18(6):817–829, June 1950. DOI: 10.1063/1.1747782.

[227] M. Itskov and N. Aksel. A class of orthotropic and transversely isotropic hyperelastic constitutive models based on a polyconvex strain energy function. *Int. J. Solids Struct.*, 41(14):3833–3848, July 2004. DOI: 10.1016/j.ijsolstr.2004.02.027.

[228] M. Itskov and V. N. Khiêm. A polyconvex anisotropic free energy function for electro- and magneto-rheological elastomers. *Math. Mech. Solids*, 21(9):1126–1137, October 2016. DOI: 10.1177/1081286514555140.

[229] M. Jabareen. Finite element modeling of the electromechanical coupling in electroactive polymers. In *2nd ECCOMAS Young Investigators Conference (YIC 2013)*, pages 1–4, Bordeaux, France, September 2013. ECCOMAS. URL https://hal.archives-ouvertes.fr/hal-00855859.

[230] J. D. Jackson. *Classical Electrodynamics*. John Wiley & Sons, Inc., 3rd edition, 1999. ISBN 978-0471309321.

[231] A. Janutka. Ordering in rolled-up single-walled ferromagnetic nanomembranes. *J. Magn. Magn. Mater.*, 419:282–291, December 2016. DOI: 10.1016/j.jmmm.2016.06.030.

[232] H. Jasak and Ž. Tuković. Automatic mesh motion for the unstructured finite volume method. *Trans. FAMENA*, 30(2):1–20, 2007.

[233] A. Javili. *Thermomechanics of solids accounting for surfaces and interfaces*. PhD thesis, Friedrich-Alexander Universität Erlangen-Nürnberg, 2012.

[234] A. Javili, G. Chatzigeorgiou, and P. Steinmann. Computational homogenization in magneto-mechanics. *Int. J. Solids Struct.*, 50(25–26):4197–4216, December 2013. DOI: 10.1016/j.ijsolstr.2013.08.024.

[235] A. Javili, A. McBride, J. Mergheim, P. Steinmann, and U. Schmidt. Micro-to-macro transitions for continua with surface structure at the microscale. *Int. J. Solids Struct.*, 50(16–17):2561–2572, August 2013. DOI: 10.1016/j.ijsolstr.2013.03.022.

[236] A. Javili, G. Chatzigeorgiou, A. T. McBride, P. Steinmann, and C. Linder. Computational homogenization of nano-materials accounting for size effects via surface elasticity. *GAMM-Mitt.*, 38(2):285–312, September 2015. DOI: 10.1002/gamm.201510016.

[237] W.-q. Jiang, J.-j. Yao, X.-l. Gong, and L. Chen. Enhancement in magnetorheological effect of magnetorheological elastomers by surface modification of iron particles. *Chin. J. Chem. Phys.*, 21(1):87–92, February 2008. DOI: 10.1088/1674-0068/21/01/87-92.

[238] M. Johlitz. On the representation of ageing phenomena. *J. Adhes.*, 88(7):620–648, 2012. DOI: 10.1080/00218464.2012.682905.

[239] A. A. Johnson and T. E. Tezduyar. Mesh update strategies in parallel finite element computations of flow problems with moving boundaries and interfaces. *Comput. Methods Appl. Mech. Eng.*, 119(1–2):73–94, November 1994. DOI: 10.1016/0045-7825(94)00077-8.

[240] M. R. Jolly, J. D. Carlson, and B. C. Muñoz. A model of the behaviour of magnetorheological materials. *Smart Mater. Struct.*, 5(5):607–614, October 1996. DOI: 10.1088/0964-1726/5/5/009.

[241] M. R. Jolly, J. D. Carlson, B. C. Muñoz, and T. A. Bullions. The magnetoviscoelastic response of elastomer composites consisting of ferrous particles embedded in a polymer matrix. *J. Intell. Mater. Syst. Struct.*, 7(6):613–622, November 1996. DOI: 10.1177/1045389x9600700601.

[242] E. Jones, T. Oliphant, P. Peterson, et al. SciPy: Open source scientific tools for Python, 2001–. URL http://www.scipy.org/. Online; Last visited: September 2018.

[243] M. Kailasam and P. Ponte Castañeda. A general constitutive theory for linear and nonlinear particulate media with microstructure evolution. *J. Mech. Phys. Solids*, 46(3):427–465, March 1998. DOI: 10.1016/s0022-5096(97)00095-1.

[244] K. A. Kalina, P. Metsch, and M. Kästner. Microscale modeling and simulation of magnetorheological elastomers at finite strains: A study on the influence of mechanical preloads. *Int. J. Solids Struct.*, 102–103:286–296, December 2016. DOI: 10.1016/j.ijsolstr.2016.10.019.

[245] M. Kaliske and H. Rothert. Formulation and implementation of three-dimensional viscoelasticity at small and finite strains. *Comput. Mech.*, 19(3):228–239, February 1997. DOI: 10.1007/s004660050171.

[246] M. Kallio, T. Lindroos, S. Aalto, E. Järvinen, T. Kärnä, and T. Meinander. Dynamic compression testing of a tunable spring element consisting of a magnetorheological elastomer. *Smart Mater. Struct.*, 16(2):506–514, March 2007. DOI: 10.1088/0964-1726/16/2/032.

[247] M. Kamiński. *The Stochastic Perturbation Method for Computational Mechanics*. John Wiley & Sons, Ltd, 2013. ISBN 9780470770825. DOI: 10.1002/9781118481844.

[248] S. V. Kankanala and N. Triantafyllidis. On finitely strained magnetorheological elastomers. *J. Mech. Phys. Solids*, 52(12):2869–2908, December 2004. DOI: 10.1016/j.jmps.2004.04.007.

[249] S. V. Kankanala and N. Triantafyllidis. Magnetoelastic buckling of a rectangular block in plane strain. *J. Mech. Phys. Solids*, 56(4):1147–1169, April 2008. DOI: 10.1016/j.jmps.2007.10.008.

[250] P. Kanouté, D. P. Boso, J. L. Chaboche, and B. A. Schrefler. Multiscale methods for composites: A review. *Arch. Comput. Methods Eng.*, 16(1):31–75, March 2009. DOI: 10.1007/s11831-008-9028-8.

[251] C. W. Karl, J. McIntyre, T. Alshuth, and M. Klüppel. Magneto-rheological elastomers with switchable mechanical properties. *Kautsch. Gummi Kunstst.*, 66(1):46–53, February 2013.
[252] G. Karniadakis and S. Sherwin. *Spectral/hp Element Methods for Computational Fluid Dynamics*. Oxford University Press, 2nd edition, June 2005. ISBN 9780198528692. DOI: 10.1093/acprof:oso/9780198528692.001.0001.
[253] G. Karypis and V. Kumar. *Metis: A software Package for Partitioning Unstructured Graphs, Partitioning Meshes, and Computing Fill-Reducing Orderings of Sparse Matrices*. Department of Computer Science and Engineering, University of Minnesota, 1998.
[254] S. Kashima, F. Miyasaka, and K. Hirata. Novel soft actuator using magnetorheological elastomer. *IEEE Trans. Magn.*, 48(4):1649–1652, April 2012. DOI: 10.1109/tmag.2011.2173669.
[255] M.-A. Keip and M. Rambausek. A multiscale approach to the computational characterization of magnetorheological elastomers. *Int. J. Numer. Methods Eng.*, 107(4):338–360, July 2016. DOI: 10.1002/nme.5178.
[256] M.-A. Keip and M. Rambausek. Computational and analytical investigations of shape effects in the experimental characterization of magnetorheological elastomers. *Int. J. Solids Struct.*, 121:1–20, August 2017. DOI: 10.1016/j.ijsolstr.2017.04.012.
[257] M.-A. Keip, P. Steinmann, and J. Schröder. Two-scale computational homogenization of electro-elasticity at finite strains. *Comput. Methods Appl. Mech. Eng.*, 278:62–79, August 2014. DOI: 10.1016/j.cma.2014.04.020.
[258] R. B. Kellogg. On the poisson equation with intersecting interfaces. *Appl. Anal.*, 4(2):101–129, 1974. DOI: 10.1080/00036817408839086.
[259] D. W. Kelly, J. P. de S. R. Gago, O. C. Zienkiewicz, and I. Babuška. *a posteriori* error analysis and adaptive processes in the finite element method: Part I—error analysis. *Int. J. Numer. Methods Eng.*, 19(11):1593–1619, November 1983. DOI: 10.1002/nme.1620191103.
[260] E. Khanmirza, A. Jamalpoor, and A. Kiani. Nano-scale mass sensor based on the vibration analysis of a magneto-electro-elastic nanoplate resting on a visco-Pasternak substrate. *Eur. Phys. J. Plus*, 132(10):422, October 2017. DOI: 10.1140/epjp/i2017-11736-9.
[261] M. S. Kiasat. *Curing shrinkage and residual stresses in viscoelastic thermosetting resins and composites*. PhD thesis, Delft University of Technology, 2000. URL https://repository.tudelft.nl/islandora/object/uuidOnline;Lastvisited:September2018.
[262] P. Kinsler, A. Favaro, and M. W. McCall. Four Poynting theorems. *Eur. J. Phys.*, 30(5):983–993, July 2009. DOI: 10.1088/0143-0807/30/5/007.
[263] E. Kirkinis and R. W. Ogden. On extension and torsion of a compressible elastic circular cylinder. *Math. Mech. Solids*, 7(4):373–392, August 2002. DOI: 10.1177/108128028476.
[264] M. Kleiber and T. D. Hien. *The Stochastic Finite Element Method: Basic Perturbation Technique and Computer Implementation*. Wiley, 1st edition, 1992. ISBN 978-0471936268.
[265] S. Klinge, A. Bartels, and P. Steinmann. Modeling of curing processes based on a multi-field potential. Single- and multiscale aspects. *Int. J. Solids Struct.*, 49(17):2320–2333, September 2012. DOI: 10.1016/j.ijsolstr.2012.04.034.
[266] S. Klinge, A. Bartels, and P. Steinmann. The multiscale approach to the curing of polymers incorporating viscous and shrinkage effects. *Int. J. Solids Struct.*, 49(26):3883–3900, December 2012. DOI: 10.1016/j.ijsolstr.2012.08.016.
[267] S. Klinkel, C. Sansour, and W. Wagner. An anisotropic fibre-matrix material model at finite elastic-plastic strains. *Comput. Mech.*, 35(6):409–417, May 2005. DOI: 10.1007/s00466-004-0629-2.
[268] M. Klüppel. Evaluation of viscoelastic master curves of filled elastomers and applications to fracture mechanics. *J. Phys. Condens. Matter*, 21(3):035104, dec 2008. DOI: 10.1088/0953-8984/21/3/035104.

[269] B. Klusemann, H. J. Böhm, and B. Svendsen. Homogenization methods for multi-phase elastic composites with non-elliptical reinforcements: Comparisons and benchmarks. *Eur. J. Mech. A, Solids*, 34:21–37, July 2012. DOI: 10.1016/j.euromechsol.2011.12.002.

[270] B. Klusemann, T. Yalçinkaya, M. G. D. Geers, and B. Svendsen. Application of non-convex rate dependent gradient plasticity to the modeling and simulation of inelastic microstructure development and inhomogeneous material behavior. *Comput. Mater. Sci.*, 80:51–60, December 2013. DOI: 10.1016/j.commatsci.2013.04.016.

[271] P. Knupp, L. Freitag-Diachin, and B. Tidwell. *MESQUITE Mesh Quality Improvement Toolkit User's Guide*, 2013. URL https://software.sandia.gov/mesquite/doc-2.99/users-guide.pdf. Online; Last visited: September 2018.

[272] G. Kofod and P. Sommer-Larsen. Silicone dielectric elastomer actuators: Finite-elasticity model of actuation. *Sens. Actuators A, Phys.*, 122(2):273–283, August 2005. DOI: 10.1016/j.sna.2005.05.001.

[273] N. Koprowski-Theiss, M. Johlitz, and S. Diebels. Characterizing the time dependence of filled EPDM. *Rubber Chem. Technol.*, 84(2):147–165, June 2011. DOI: 10.5254/1.3570527.

[274] J. Korelc and P. Wriggers. *Automation of Finite Element Methods*. Springer Nature, 2016. ISBN 978-3-319-39003-1. DOI: 10.1007/978-3-319-39005-5.

[275] V. Kouznetsova, M. G. D. Geers, and W. A. M. Brekelmans. Multi-scale constitutive modelling of heterogeneous materials with a gradient-enhanced computational homogenization scheme. *Int. J. Numer. Methods Eng.*, 54(8):1235–1260, July 2002. DOI: 10.1002/nme.541.

[276] V. G. Kouznetsova, M. G. D. Geers, and W. A. M. Brekelmans. Multi-scale second-order computational homogenization of multi-phase materials: a nested finite element solution strategy. *Comput. Methods Appl. Mech. Eng.*, 193(48–51):5525–5550, December 2004. DOI: 10.1016/j.cma.2003.12.073.

[277] A. Kovetz. *Electromagnetic Theory*. Oxford University Press, Oxford and New York, 2000. ISBN 0-19-850603-1.

[278] G. Kraus. Mechanical losses in carbon-black-filled rubbers. *J. Appl. Polym. Sci., Appl. Polym. Symp.*, 39:75–92, 1984.

[279] M. Kronbichler and K. Kormann. A generic interface for parallel cell-based finite element operator application. *Comput. Fluids*, 63:135–147, June 2012. DOI: 10.1016/j.compfluid.2012.04.012.

[280] H. Kronmüller and M. Fähnle. *Micromagnetism and the Microstructure of Ferromagnetic Solids*. Cambridge University Press, 2003. ISBN 9780521331357.

[281] S. Kuznetsov. *Homogenization methods for problems with multiphysics, temporal and spatial coupling*. PhD thesis, Columbia University, 2012. URL https://academiccommons.columbia.edu/doi/10.7916/D85Q536N.

[282] Sandia National Laboratories. *Cubit 13.2 User Documentation*. Sandia Corporation, 2012. URL https://cubit.sandia.gov/public/13.2/help_manual/WebHelp/cubithelp.htm. Online; Last visited: September 2018.

[283] D. C. Lagoudas, A. C. Gavazzi, and H. Nigam. Elastoplastic behavior of metal matrix composites based on incremental plasticity and the Mori-Tanaka averaging scheme. *Comput. Mech.*, 8(3):193–203, 1991. DOI: 10.1007/bf00372689.

[284] E. R. Laithwaite. Linear electric machines—a personal view. *Proc. IEEE*, 63(2):250–290, February 1975. DOI: 10.1109/proc.1975.9734.

[285] L. D. Landau and E. Lifshitz. On the theory of the dispersion of magnetic permeability in ferromagnetic bodies. *Phys. Z. Sowjetunion*, 8:153–169, 1935.

[286] C. Lang, A. Doostan, and K. Maute. Extended stochastic FEM for diffusion problems with uncertain material interfaces. *Comput. Mech.*, 51(6):1031–1049, June 2013. DOI: 10.1007/s00466-012-0785-8.

[287] C. Lang, A. Sharma, A. Doostan, and K. Maute. Heaviside enriched extended stochastic FEM for problems with uncertain material interfaces. *Comput. Mech.*, 56(5):753–767, November 2015. DOI: 10.1007/s00466-015-1199-1.

[288] J. Läuger and H. Stettin. Differences between stress and strain control in the non-linear behavior of complex fluids. *Rheol. Acta*, 49(9):909–930, September 2010. DOI: 10.1007/s00397-010-0450-0.

[289] J. Läuger, K. Wollny, and S. Huck. Direct strain oscillation: a new oscillatory method enabling measurements at very small shear stresses and strains. *Rheol. Acta*, 41(4):356–361, January 2002. DOI: 10.1007/s00397-002-0231-5.

[290] J. Läuger, K. Wollny, H. Stettin, and S. Huck. A new device for the full rheological characterization of magneto-rheological fluids. *Int. J. Mod. Phys. B*, 19(07n09):1353–1359, April 2005. DOI: 10.1142/S0217979205030293.

[291] H. M. Laun and C. Gabriel. Measurement modes of the response time of a magneto-rheological fluid (MRF) for changing magnetic flux density. *Rheol. Acta*, 46(5):665–676, May 2007. DOI: 10.1007/s00397-006-0155-6.

[292] H. M. Laun, C. Kormann, and N. Willenbacher. Rheometry on magnetorheological (MR) fluids. *Rheol. Acta*, 35(5):417–432, September 1996. DOI: 10.1007/BF00368993.

[293] H. M. Laun, G. Schmidt, C. Gabriel, and C. Kieburg. Reliable plate–plate MRF magnetorheometry based on validated radial magnetic flux density profile simulations. *Rheol. Acta*, 47(9):1049–1059, December 2008. DOI: 10.1007/s00397-008-0305-0.

[294] H. M. Laun, C. Gabriel, and C. Kieburg. Magnetorheological fluid in oscillatory shear and parameterization with regard to MR device properties. *J. Intell. Mater. Syst. Struct.*, 21(15):1479–1489, October 2010. DOI: 10.1177/1045389X09351759.

[295] H. M. Laun, C. Gabriel, and C. Kieburg. Wall material and roughness effects on transmittable shear stresses of magnetorheological fluids in plate–plate magnetorheometry. *Rheol. Acta*, 50(2):141–157, February 2011. DOI: 10.1007/s00397-011-0531-8.

[296] J. L. Leblanc. *Filled Polymers: Science and Industrial Applications*. CRC Press, 2009. ISBN 9781439800423. DOI: 10.1201/9781439800430.

[297] J. L. Leblanc and C. de la Chapelle. Updating a torsional dynamic rheometer for Fourier transform rheometry on rubber materials. *Rubber Chem. Technol.*, 76(2):287–298, May 2003. DOI: 10.5254/1.3547743.

[298] W. Leclerc, P. Karamian-Surville, and A. Vivet. An efficient stochastic and double-scale model to evaluate the effective elastic properties of 2D overlapping random fibre composites. *Comput. Mater. Sci.*, 69:481–493, March 2013. DOI: 10.1016/j.commatsci.2012.10.036.

[299] J. Lee, J. G. Boyd, and D. C. Lagoudas. Effective properties of three-phase electro-magneto-elastic composites. *Int. J. Eng. Sci.*, 43(10):790–825, June 2005. DOI: 10.1016/j.ijengsci.2005.01.004.

[300] R. B. Lehoucq and A. Von Lilienfeld-Toal. Translation of Walter Noll's " derivation of the fundamental equations of continuum thermodynamics from statistical mechanics" . *J. Elast.*, 100(1–2):5–24, June 2010. DOI: 10.1007/s10659-010-9246-9.

[301] E. Lenz. Ueber die Bestimmung der Richtung der durch elektrodynamische Vertheilung erregten galvanischen Ströme. *Ann. Phys. Chem.*, 107(31):483–494, 1834.

[302] B. T. Lester, Y. Chemisky, and D. C. Lagoudas. Transformation characteristics of shape memory alloy composites. *Smart Mater. Struct.*, 20(9):094003, September 2011. DOI: 10.1088/0964-1726/20/9/094002.

[303] J. Y. Li. Magnetoelectroelastic multi-inclusion and inhomogeneity problems and their applications in composite materials. *Int. J. Eng. Sci.*, 38(18):1993–2011, December 2000. DOI: 10.1016/s0020-7225(00)00014-8.

[304] J. Y. Li and M. L. Dunn. Anisotropic coupled-field inclusion and inhomogeneity problems. *Philos. Mag. A*, 77(5):1341–1350, 1998. DOI: 10.1080/01418619808214256.

[305] J. Y. Li and M. L. Dunn. Micromechanics of magnetoelectroelastic composite materials: average fields and effective behavior. *J. Intell. Mater. Syst. Struct.*, 9(6):404–416, June 1998. DOI: 10.1177/1045389x9800900602.

[306] S. Li and P. A. Mataga. Dynamic crack propagation in piezoelectric materials—Part I. Electrode solution. *J. Mech. Phys. Solids*, 44(11):1799–1830, November 1996. DOI: 10.1016/0022-5096(96)00055-5.

[307] X. Li, S.-Q. Wang, and X. Wang. Nonlinearity in large amplitude oscillatory shear (LAOS) of different viscoelastic materials. *J. Rheol.*, 53(5):1255–1274, September 2009. DOI: 10.1122/1.3193713.

[308] Y. Li, J. Li, T. Tian, and W. Li. A highly adjustable magnetorheological elastomer base isolator for applications of real-time adaptive control. *Smart Mater. Struct.*, 22(9):095020, August 2013. DOI: 10.1088/0964-1726/22/9/095020.

[309] Y. Li, J. Li, W. Li, and H. Du. A state-of-the-art review on magnetorheological elastomer devices. *Smart Mater. Struct.*, 23(12):123001, December 2014. DOI: 10.1088/0964-1726/23/12/123001.

[310] C. Liebl, M. Johlitz, B. Yagimli, and A. Lion. Three-dimensional chemo-thermomechanically coupled simulation of curing adhesives including viscoplasticity and chemical shrinkage. *Comput. Mech.*, 49(5):603–615, May 2012. DOI: 10.1007/s00466-011-0663-9.

[311] C. Liebl, M. Johlitz, B. Yagimli, and A. Lion. Simulation of curing-induced viscoplastic deformation: a new approach considering chemo-thermomechanical coupling. *Arch. Appl. Mech.*, 82(8):1133–1144, August 2012. DOI: 10.1007/s00419-012-0639-z.

[312] C. Linder, M. Tkachuk, and C. Miehe. A micromechanically motivated diffusion-based transient network model and its incorporation into finite rubber viscoelasticity. *J. Mech. Phys. Solids*, 59(10):2134–2156, October 2011. DOI: 10.1016/j.jmps.2011.05.005.

[313] A. Lion. A physically based method to represent the thermo-mechanical behaviour of elastomers. *Acta Mech.*, 123(1–4):1–25, March 1997. DOI: 10.1007/bf01178397.

[314] A. Lion and P. Höfer. On the phenomenological representation of curing phenomena in continuum mechanics. *Arch. Mech.*, 59(1):59–89, 2007.

[315] A. Lion and M. Johlitz. On the representation of chemical ageing of rubber in continuum mechanics. *Int. J. Solids Struct.*, 49(10):1227–1240, May 2012. DOI: 10.1016/j.ijsolstr.2012.01.014.

[316] A. Lion, B. Yagimli, G. Baroud, and U. Goerke. Constitutive modelling of PMMA-based bone cement: a functional model of viscoelasticity and its approximation for time domain investigations. *Arch. Mech.*, 60(3):221–242, 2008.

[317] G. R. Liu, W. Zeng, and H. Nguyen-Xuan. Generalized stochastic cell-based smoothed finite element method (GS_CS-FEM) for solid mechanics. *Finite Elem. Anal. Des.*, 63(Supplement C):51–61, January 2013. DOI: 10.1016/j.finel.2012.08.007.

[318] L. P. Liu, R. D. James, and P. H. Leo. Magnetostrictive composites in the dilute limit. *J. Mech. Phys. Solids*, 54(5):951–974, May 2006. DOI: 10.1016/j.jmps.2005.11.006.

[319] Y. Liu and P. Ponte Castañeda. Second-order theory for the effective behavior and field fluctuations in viscoplastic polycrystals. *J. Mech. Phys. Solids*, 52(2):467–495, February 2004. DOI: 10.1016/s0022-5096(03)00078-4.

[320] A. Logg, K.-A. Mardal, and G. Wells, editors. *Automated Solution of Differential Equations by the Finite Element Method*. Springer Berlin Heidelberg, 2012. ISBN 978-3-642-23098-1. DOI: 10.1007/978-3-642-23099-8.

[321] R. Löhner and C. Yang. Improved ALE mesh velocities for moving bodies. *Commun. Numer. Methods Eng.*, 12(10):599–608, October 1996. DOI: 10.1002/(SICI)1099-0887(199610)12:10<599::AID-CNM1>3.0.CO;2-Q.

[322] M. A. R. Loja, C. M. Mota Soares, and J. I. Barbosa. Optimization of magneto-electro-elastic composite structures using differential evolution. *Compos. Struct.*, 107:276–287, January 2014. DOI: 10.1016/j.compstruct.2013.08.005.

[323] M. Lokander and B. Stenberg. Performance of isotropic magnetorheological rubber materials. *Polym. Test.*, 22(3):245–251, May 2003. DOI: 10.1016/s0142-9418(02)00043-0.

[324] M. Lokander, T. Reitberger, and B. Stenberg. Oxidation of natural rubber-based magnetorheological elastomers. *Polym. Degrad. Stab.*, 86(3):467–471, December 2004. DOI: 10.1016/j.polymdegradstab.2004.05.019.

[325] H. Lorentz. *The Theory of Electrons: and Its Applications to the Phenomena of Light and Radiant Heat.* Dover Books on Physics. Dover Publications, 2nd edition, 1952. ISBN 978-0486495583.

[326] H. A. Lorentz. *Special Theory of Relativity*, chapter 3: Electromagnetic phenomena in a system moving with any velocity less than that of light, pages 119–144. The Commonwealth and International Library: Selected Readings in Physics. Pergamon Press Ltd., 1970. ISBN 978-0-08-006995-1. DOI: 10.1016/B978-0-08-006995-1.50012-0.

[327] J. Lubliner. A model of rubber viscoelasticity. *Mech. Res. Commun.*, 12(2):93–99, March 1985. DOI: 10.1016/0093-6413(85)90075-8.

[328] V. Lucas, J.-C. Golinval, S. Paquay, V.-D. Nguyen, L. Noels, and L. Wu. A stochastic computational multiscale approach; application to MEMS resonators. *Comput. Methods Appl. Mech. Eng.*, 294:141–167, September 2015. DOI: 10.1016/j.cma.2015.05.019.

[329] G. Z. Lum, Z. Ye, X. Dong, H. Marvi, O. Erin, W. Hu, and M. Sitti. Shape-programmable magnetic soft matter. *Proc. Natl. Acad. Sci.*, 113(41):E6007–E6015, September 2016. DOI: 10.1073/pnas.1608193113.

[330] D. Łydżba. *Effective Properties of Composites: Introduction to Micromechanics*. PRINTPAP Łódź, Wrocław, 2011. ISBN 978-83-62098-48-4.

[331] R. Mahnken. Thermodynamic consistent modeling of polymer curing coupled to visco–elasticity at large strains. *Int. J. Solids Struct.*, 50(13): 2003–2021, June 2013. DOI: 10.1016/j.ijsolstr.2013.01.033.

[332] K. K. Mandadapu, A. Sengupta, and P. Papadopoulos. A homogenization method for thermomechanical continua using extensive physical quantities. *Proc. R. Soc. A, Math. Phys. Eng. Sci.*, 468(2142):1696–1715, June 2012. DOI: 10.1098/rspa.2011.0578.

[333] M. Mansuripur, A. R. Zakharian, and E. M. Wright. Electromagnetic-force distribution inside matter. *Phys. Rev. A*, 88(2), August 2013. DOI: 10.1103/physreva.88.023826.

[334] J. E. Martin, R. A. Anderson, D. Read, and G. Gulley. Magnetostriction of field-structured magnetoelastomers. *Phys. Rev. E*, 74(5):051507, November 2006. DOI: 10.1103/physreve.74.051507.

[335] K. Matouš, M. G. D. Geers, V. G. Kouznetsova, and A. Gillman. A review of predictive nonlinear theories for multiscale modeling of heterogeneous materials. *J. Comput. Phys.*, 330:192–220, February 2017. DOI: 10.1016/j.jcp.2016.10.070.

[336] T. Matsuo, N. Mimuro, and M. Shimasaki. A micromagnetic study of domain structure modeling. *J. Magn. Magn. Mater.*, 320(20):e1029–e1033, October 2008. DOI: 10.1016/j.jmmm.2008.04.106.

[337] G. Maugin. *Configurational Forces: Thermomechanics, Physics, Mathematics, and Numerics*. CRC Series: Modern Mechanics and Mathematics. Chapman and Hall/CRC, 2016. ISBN 9781439846124.

[338] G. A. Maugin. A continuum theory of deformable ferrimagnetic bodies. I. Field equations. *J. Math. Phys.*, 17(9):1727–1738, September 1976. DOI: 10.1063/1.523101.

[339] G. A. Maugin. A continuum theory of deformable ferrimagnetic bodies. II. Thermodynamics, constitutive theory. *J. Math. Phys.*, 17(9):1739–1751, September 1976. DOI: 10.1063/1.523102.

[340] G. A. Maugin. *Continuum Mechanics of Electromagnetic Solids*. North-Holland, Amsterdam, The Netherlands, 1988. ISBN 978-0444703996.

[341] G. A. Maugin and A. C. Eringen. On the equations of the electrodynamics of deformable bodies of finite extent. *J. Mech.*, 16(1):101–147, 1977.

[342] J. C. Maxwell. *A Treatise on Electricity and Magnetism*, volume 1 of *Dover books on advanced mathematics*. Dover Publications Inc., 3rd edition, 1954a. ISBN 0486606368. DOI: 10.1017/cbo9780511709333.

[343] J. C. Maxwell. *A Treatise on Electricity and Magnetism*, volume 2 of *Dover books on advanced mathematics*. Dover Publications Inc., 3rd edition, 1954b. ISBN 0486606376. DOI: 10.1017/CBO9780511709340.

[344] J. C. Maxwell. *The Scientific Papers of James Clerk Maxwell, volume 1*. Dover Publications Inc., New York, New York, USA, 1965. ISBN 9780486781662. DOI: 10.1017/cbo9780511698095.

[345] A. McBride, A. Javili, P. Steinmann, and B. D. Reddy. A finite element implementation of surface elasticity at finite strains using the deal.II library. arXiv, 2015. URL http://arxiv.org/abs/1506.01361v1. Online; Last visited: September 2018.

[346] A. T. McBride. *Formulation, analysis and solution algorithms for a model of gradient plasticity within a discontinuous Galerkin framework*. PhD thesis, University of Cape Town, Formulation, analysis and solution algorithms for a model, 2008. URL https://open.uct.ac.za/handle/11427/10924. Online; Last visited: September 2018.

[347] K. McGough, S. Ahmed, M. Frecker, and Z. Ounaies. Finite element analysis and validation of dielectric elastomer actuators used for active origami. *Smart Mater. Struct.*, 23(9):094002, August 2014. DOI: 10.1088/0964-1726/23/9/094002.

[348] D. Meeker. *Finite Element Method Magnetics. Version 4.2. User's Manual*, October 2015. URL http://www.femm.info/Archives/doc/manual42.pdf.

[349] A. P. Meera, S. Said, Y. Grohens, and S. Thomas. Nonlinear viscoelastic behavior of silica-filled natural rubber nanocomposites. *J. Phys. Chem. C*, 113(42):17997–18002, September 2009. DOI: 10.1021/jp9020118.

[350] M. Mehnert, M. Hossain, and P. Steinmann. Towards a thermo-magneto-mechanical coupling framework for magneto-rheological elastomers. *Int. J. Solids Struct.*, 128:117–132, December 2017. DOI: 10.1016/j.ijsolstr.2017.08.022.

[351] M. Mehnert, J.-P. Pelteret, and P. Steinmann. Numerical modelling of nonlinear thermo-electro-elasticity. *Math. Mech. Solids*, 22(11):2196–2213, November 2017. DOI: 10.1177/1081286517729867.

[352] L.-C. Meng and C.-F. Gao. Effects of surface electrostatic force on piezoelectric fracture. *Int. J. Theor. Appl. Mult. Mech.*, 2(1):46–61, 2011. DOI: 10.1504/ijtamm.2011.041173.

[353] J. Mergheim, G. Possart, and P. Steinmann. Modelling and computation of curing and damage of thermosets. *Comput. Mater. Sci.*, 53(1):359–367, February 2012. DOI: 10.1016/j.commatsci.2011.09.013.

[354] R. C. Mesquita and J. P. A. Bastos. An incomplete gauge formulation for 3D nodal finite-element magnetostatics. *IEEE Trans. Magn.*, 28(2):1044–1047, March 1992. DOI: 10.1109/20.123860.

[355] P. Metsch, K. A. Kalina, C. Spieler, and M. Kästner. A numerical study on magnetostrictive phenomena in magnetorheological elastomers. *Comput. Mater. Sci.*, 124:364–374, November 2016. DOI: 10.1016/j.commatsci.2016.08.012.

[356] A. Meurer, C. P. Smith, M. Paprocki, O. Čertík, S. B. Kirpichev, M. Rocklin, A. Kumar, S. Ivanov, J. K. Moore, S. Singh, T. Rathnayake, S. Vig, B. E. Granger, R. P. Muller, F. Bonazzi, H. Gupta, S. Vats, F. Johansson, F. Pedregosa, M. J. Curry, A. R. Terrel, Š. Roučka, A. Saboo, I. Fernando, S. Kulal, R. Cimrman, and A. Scopatz. SymPy: symbolic computing in python. *PeerJ Comput. Sci.*, 3:e103, January 2017. DOI: 10.7717/peerj-cs.103.

[357] J. C. Michel, H. Moulinec, and P. Suquet. Effective properties of composite materials with periodic microstructure: a computational approach. *Comput. Methods Appl. Mech. Eng.*, 172(1–4):109–143, April 1999. DOI: 10.1016/s0045-7825(98)00227-8.

[358] C. Miehe. Aspects of the formulation and finite element implementation of large strain isotropic elasticity. *Int. J. Numer. Methods Eng.*, 37(12):1981–2004, June 1994. DOI: 10.1002/nme.1620371202.

[359] C. Miehe. Numerical computation of algorithmic (consistent) tangent moduli in large-strain computational inelasticity. *Comput. Methods Appl. Mech. Eng.*, 134(3–4):223–240, August 1996. DOI: 10.1016/0045-7825(96)01019-5.

[360] C. Miehe. Strain-driven homogenization of inelastic microstructures and composites based on an incremental variational formulation. *Int. J. Numer. Methods Eng.*, 55(11):1285–1322, December 2002. DOI: 10.1002/nme.515.

[361] C. Miehe. Computational micro-to-macro transitions for discretized micro-structures of heterogeneous materials at finite strains based on the minimization of averaged incremental energy. *Comput. Methods Appl. Mech. Eng.*, 192(5–6):559–591, January 2003. DOI: 10.1016/s0045-7825(02)00564-9.

[362] C. Miehe. A multi-field incremental variational framework for gradient-extended standard dissipative solids. *J. Mech. Phys. Solids*, 59(4):898–923, April 2011. DOI: 10.1016/j.jmps.2010.11.001.

[363] C. Miehe. Variational gradient plasticity at finite strains. Part I: Mixed potentials for the evolution and update problems of gradient-extended dissipative solids. *Comput. Methods Appl. Mech. Eng.*, 268:677–703, January 2014. DOI: 10.1016/j.cma.2013.03.014.

[364] C. Miehe and C. G. Bayreuther. On multiscale FE analyses of heterogeneous structures: from homogenization to multigrid solvers. *Int. J. Numer. Methods Eng.*, 71(10):1135–1180, September 2007. DOI: 10.1002/nme.1972.

[365] C. Miehe and G. Ethiraj. A geometrically consistent incremental variational formulation for phase field models in micromagnetics. *Comput. Methods Appl. Mech. Eng.*, 245–246:331–347, October 2012. DOI: 10.1016/j.cma.2012.03.021.

[366] C. Miehe and S. Göktepe. A micro–macro approach to rubber-like materials. Part II: The micro-sphere model of finite rubber viscoelasticity. *J. Mech. Phys. Solids*, 53(10):2231–2258, October 2005. DOI: 10.1016/j.jmps.2005.04.006.

[367] C. Miehe, J. Schotte, and J. Schröder. Computational micro–macro transitions and overall moduli in the analysis of polycrystals at large strains. *Comput. Mater. Sci.*, 16(1–4):372–382, December 1999. DOI: 10.1016/s0927-0256(99)00080-4.

[368] C. Miehe, J. Schröder, and J. Schotte. Computational homogenization analysis in finite plasticity simulation of texture development in polycrystalline materials. *Comput. Methods Appl. Mech. Eng.*, 171(3–4):387–418, April 1999. DOI: 10.1016/s0045-7825(98)00218-7.

[369] C. Miehe, J. Schröder, and M. Becker. Computational homogenization analysis in finite elasticity: material and structural instabilities on the micro- and macro-scales of periodic composites and their interaction. *Comput. Methods Appl. Mech. Eng.*, 191(44):4971–5005, October 2002. DOI: 10.1016/s0045-7825(02)00391-2.

[370] C. Miehe, B. Kiefer, and D. Rosato. An incremental variational formulation of dissipative magnetostriction at the macroscopic continuum level. *Int. J. Solids Struct.*, 48(13):1846–1866, June 2011. DOI: 10.1016/j.ijsolstr.2011.02.011.

[371] C. Miehe, D. Rosato, and B. Kiefer. Variational principles in dissipative electro-magneto-mechanics: A framework for the macro-modeling of functional materials. *Int. J. Numer. Methods Eng.*, 86(10):1225–1276, June 2011. DOI: 10.1002/nme.3127.

[372] C. Miehe, F. E. Hildebrand, and L. Boger. Mixed variational potentials and inherent symmetries of the Cahn-Hilliard theory of diffusive phase separation. *Proc. R. Soc. A, Math.*

Phys. Eng. Sci., 470(2164):20130641, April 2014. DOI: 10.1098/rspa.2013.0641.

[373] C. Miehe, S. Mauthe, and F. E. Hildebrand. Variational gradient plasticity at finite strains. Part III: Local–global updates and regularization techniques in multiplicative plasticity for single crystals. *Comput. Methods Appl. Mech. Eng.*, 268:735–762, January 2014. DOI: 10.1016/j.cma.2013.08.022.

[374] C. Miehe, S. Mauthe, and H. Ulmer. Formulation and numerical exploitation of mixed variational principles for coupled problems of Cahn-Hilliard-type and standard diffusion in elastic solids. *Int. J. Numer. Methods Eng.*, 99(10):737–762, September 2014. DOI: 10.1002/nme.4700.

[375] C. Miehe, F. Welschinger, and F. Aldakheel. Variational gradient plasticity at finite strains. Part II: Local–global updates and mixed finite elements for additive plasticity in the logarithmic strain space. *Comput. Methods Appl. Mech. Eng.*, 268:704–734, January 2014d. DOI: 10.1016/j.cma.2013.07.015.

[376] C. Miehe, D. Vallicotti, and D. Zäh. Computational structural and material stability analysis in finite electro-elasto-statics of electro-active materials. *Int. J. Numer. Methods Eng.*, 102(10):1605–1637, June 2015. DOI: 10.1002/nme.4855.

[377] C. Miehe, D. Vallicotti, and S. Teichtmeister. Homogenization and multiscale stability analysis in finite magneto-electro-elasticity. Application to soft matter EE, ME and MEE composites. *Comput. Methods Appl. Mech. Eng.*, 300:294–346, March 2016. DOI: 10.1016/j.cma.2015.10.013.

[378] A. Mielke. Formulation of thermoelastic dissipative material behavior using GENERIC. *Contin. Mech. Thermodyn.*, 23(3):233–256, May 2011. DOI: 10.1007/s00161-010-0179-0.

[379] A. Mielke. Dissipative quantum mechanics using GENERIC. In A. Johann, H.-P. Kruse, F. Rupp, and S. Schmitz, editors, *Recent Trends in Dynamical Systems*, pages 555–585, Basel, 2013. Springer Basel. ISBN 978-3-0348-0451-6. DOI: 10.1007/978-3-0348-0451-6_21.

[380] K. Miled, K. Sab, and R. Le Roy. Effective elastic properties of porous materials: Homogenization schemes vs experimental data. *Mech. Res. Commun.*, 38(2):131–135, March 2011. DOI: 10.1016/j.mechrescom.2011.01.009.

[381] H. Minkowski. Die Grundgleichungen für die elektromagnetischen Vorgänge in bewegten Körpern. *Math. Ann.*, 68(4):472–525, December 1910. DOI: 10.1007/bf01455871.

[382] W. F. Mitchell. A collection of 2D elliptic problems for testing adaptive algorithms. Technical Report NISTIR 7668, National Institute of Standards and Technology, 2010.

[383] W. F. Mitchell and M. A. McClain. A comparison of *hp*-adaptive strategies for elliptic partial differential equations. *ACM Trans. Math. Softw.*, 41(1):1–39, October 2014. DOI: 10.1145/2629459.

[384] G. P. Moeckel. Thermodynamics of an interface. *Arch. Ration. Mech. Anal.*, 57(3):255–280, September 1975. DOI: 10.1007/bf00280158.

[385] N. Moës, M. Cloirec, P. Cartraud, and J.-F. Remacle. A computational approach to handle complex microstructure geometries. *Comput. Methods Appl. Mech. Eng.*, 192(28–30):3163–3177, July 2003. DOI: 10.1016/s0045-7825(03)00346-3.

[386] K. Mori, Y. Shindo, and F. Narita. Dynamic electromagneto-mechanical behavior of clamped-free giant magnetostrictive/piezoelectric laminates under AC electric fields. *Smart Mater. Struct.*, 21(11):115003, September 2012. DOI: 10.1088/0964-1726/21/11/115003.

[387] T. Mori and K. Tanaka. Average stress in matrix and average elastic energy of materials with misfitting inclusions. *Acta Metall.*, 21(5):571–574, May 1973. DOI: 10.1016/0001-6160(73)90064-3.

[388] H. Moulinec and P. Suquet. A FFT-based numerical method for computing the mechanical properties of composites from images of their microstructures. In *IUTAM Symposium on Microstructure-Property Interactions in Composite Materials*,

pages 235–246, Springer Netherlands. Dordrecht, 1995. ISBN 978-94-011-0059-5. DOI: 10.1007/978-94-011-0059-5_20.

[389] H. Moulinec and P. Suquet. A numerical method for computing the overall response of nonlinear composites with complex microstructure. *Comput. Methods Appl. Mech. Eng.*, 157(1–2):69–94, April 1998. DOI: 10.1016/s0045-7825(97)00218-1.

[390] R. Müller, B. X. Xu, D. Gross, M. Lyschik, D. Schrade, and S. Klinkel. Deformable dielectrics – optimization of heterogeneities. *Int. J. Eng. Sci.*, 48(7–8):647–657, July 2010. DOI: 10.1016/j.ijengsci.2010.03.001.

[391] L. Mullins. Softening of rubber by deformation. *Rubber Chem. Technol.*, 42(1):339–362, March 1969. DOI: 10.5254/1.3539210.

[392] A. I. Murdoch. A thermodynamical theory of elastic material interfaces. *Q. J. Mech. Appl. Math.*, 29(3):245–275, August 1976. DOI: 10.1093/qjmam/29.3.245.

[393] T. Nagel, U.-J. Görke, K. M. Moerman, and O. Kolditz. On advantages of the Kelvin mapping in finite element implementations of deformation processes. *Environ. Earth Sci.*, 75(11):937, June 2016. DOI: 10.1007/s12665-016-5429-4.

[394] J. C. Nagtegaal, D. M. Parks, and J. R. Rice. On numerically accurate finite element solutions in the fully plastic range. *Comput. Methods Appl. Mech. Eng.*, 4(2):153–177, September 1974. DOI: 10.1016/0045-7825(74)90032-2.

[395] F. Narita, K. Shikanai, Y. Shindo, and K. Mori. Three-point bending fracture behavior of cracked giant magnetostrictive materials under magnetic fields. *J. Test. Eval.*, 44(4):1454–1460, July 2016. DOI: 10.1520/jte20140439.

[396] J.-C. Nédélec. Mixed finite elements in $\mathbb{R}^3$. *Numer. Math.*, 35(3):315–341, September 1980. DOI: 10.1007/BF01396415.

[397] J.-C. Nédélec. A new family of mixed finite elements in $\mathbb{R}^3$. *Numer. Math.*, 50(1):57–81, January 1986. DOI: 10.1007/BF01389668.

[398] B. Nedjar. An anisotropic viscoelastic fibre–matrix model at finite strains: Continuum formulation and computational aspects. *Comput. Methods Appl. Mech. Eng.*, 196(9–12):1745–1756, February 2007. DOI: 10.1016/j.cma.2006.09.009.

[399] S. Nemat-Nasser. Averaging theorems in finite deformation plasticity. *Mech. Mater.*, 31(8):493–523, August 1999. DOI: 10.1016/s0167-6636(98)00073-8.

[400] R. Niekamp, D. Markovic, A. Ibrahimbegovic, H. G. Matthies, and R. L. Taylor. Multi-scale modelling of heterogeneous structures with inelastic constitutive behavior: Part II – software coupling implementation aspects. *Eng. Comput.*, 26(1/2):6–28, 2009. DOI: 10.1108/02644400910924780.

[401] W. Noll. Die herleitung der grundgleichungen der thermomechanik der kontinua aus der statistischen mechanik. *Indiana Univ. Math. J.*, 4(5):627–646, 1955. DOI: 10.1512/iumj.1955.4.54022.

[402] M. Norouzi, S. M. S. Alehashem, H. Vatandoost, Y. Q. Ni, and M. M. Shahmardan. A new approach for modeling of magnetorheological elastomers. *J. Intell. Mater. Syst. Struct.*, 27(8):1121–1135, nov 2015. DOI: 10.1177/1045389x15615966.

[403] A. Nouy and A. Clément. eXtended Stochastic Finite Element Method for the numerical simulation of heterogeneous materials with random material interfaces. *Int. J. Numer. Methods Eng.*, 83(10):1312–1344, September 2010. DOI: 10.1002/nme.2865.

[404] A. Nouy, A. Clement, F. Schoefs, and N. Moës. An extended stochastic finite element method for solving stochastic partial differential equations on random domains. *Comput. Methods Appl. Mech. Eng.*, 197(51–52):4663–4682, October 2008. DOI: 10.1016/j.cma.2008.06.010.

[405] R. W. Ogden. *Non-Linear Elastic Deformations*. Dover Publications, 1997. ISBN 978-0486696485.

[406] R. W. Ogden. *Mechanics and Electrodynamics of Magneto- and Electro-elastic Materials*,

volume 527 of *CISM Courses and Lectures*, chapter Magnetostatics: from basic principles to nonlinear interactions in deformable media, pages 107–152. Springer Vienna, Vienna, 1st edition, 2011. ISBN 978-3-7091-0700-3. DOI: 10.1007/978-3-7091-0701-0_3.

[407] A. T. Onisan, A. N. Bogdanov, and U. K. Rößler. Domain models for ferromagnetic shape-memory materials. *Acta Mater.*, 58(13):4378–4386, August 2010. DOI: 10.1016/j.actamat.2010.04.032.

[408] OPEN CASCADE S. A. S. Open CASCADE Technology, 1999. URL https://dev.opencascade.org/doc/overview/html/index.html. Online; Last visited: September 2018.

[409] R. Ortigosa and A. J. Gil. A new framework for large strain electromechanics based on convex multi-variable strain energies: Finite Element discretisation and computational implementation. *Comput. Methods Appl. Mech. Eng.*, 302:329–360, April 2016. DOI: 10.1016/j.cma.2015.12.007.

[410] M. Ortiz and L. Stainier. The variational formulation of viscoplastic constitutive updates. *Comput. Methods Appl. Mech. Eng.*, 171(3–4):419–444, April 1999. DOI: 10.1016/s0045-7825(98)00219-9.

[411] M. Otténio, M. Destrade, and R. W. Ogden. Incremental magnetoelastic deformations, with application to surface instability. *J. Elast.*, 90(1):19–42, January 2008. DOI: 10.1007/s10659-007-9120-6.

[412] I. Özdemir, W. A. M. Brekelmans, and M. G. D. Geers. FE^2 computational homogenization for the thermo-mechanical analysis of heterogeneous solids. *Comput. Methods Appl. Mech. Eng.*, 198(3–4):602–613, December 2008. DOI: 10.1016/j.cma.2008.09.008.

[413] N. Pakam and A. Arockiarajan. Study on effective properties of 1-3-2 type magneto-electro-elastic composites. *Sens. Actuators A, Phys.*, 209:87–99, March 2014. DOI: 10.1016/j.sna.2014.01.009.

[414] Y. H. Pao. Electromagnetic forces in deformable continua. In S. Nemat-Nasser, editor, *Mechanics Today*, volume 4 of *Pergamon Mechanics Today Series*, chapter IV, pages 209–305. Elsevier, New York, 1978. ISBN 978-0-08-021792-5. DOI: 10.1016/b978-0-08-021792-5.50012-4.

[415] A. R. Payne. The dynamic properties of carbon black loaded natural rubber vulcanizates. Part II. *J. Appl. Polym. Sci.*, 6(21):368–372, May 1962. DOI: 10.1002/app.1962.070062115.

[416] J.-P. V. Pelteret and B. D. Reddy. Computational model of soft tissues in the human upper airway. *Int. J. Numer. Methods Biomed. Eng.*, 28(1):111–132, January 2012. DOI: 10.1002/cnm.1487.

[417] J.-P. V. Pelteret, D. Davydov, A. McBride, D. K. Vu, and P. Steinmann. Computational electro-elasticity and magneto-elasticity for quasi-incompressible media immersed in free space. *Int. J. Numer. Methods Eng.*, 108(11):1307–1342, December 2016. DOI: 10.1002/nme.5254.

[418] J.-P. V. Pelteret, B. Walter, and P. Steinmann. Application of metaheuristic algorithms to the identification of nonlinear magneto-viscoelastic constitutive parameters. *J. Magn. Magn. Mater.*, 464:116–131, October 2018. DOI: 10.1016/j.jmmm.2018.02.094.

[419] C. S. Peskin. The immersed boundary method. *Acta Numer.*, 11:479–517, January 2002. DOI: 10.1017/s0962492902000077.

[420] H. E. Pettermann, A. F. Plankensteiner, H. J. Böhm, and F. G. Rammerstorfer. A thermo-elasto-plastic constitutive law for inhomogeneous materials based on an incremental Mori–Tanaka approach. *Comput. Struct.*, 71(2):197–214, April 1999. DOI: 10.1016/s0045-7949(98)00208-9.

[421] H. E. Pettermann, C. O. Huber, M. H. Luxner, S. Nogales, and H. J. Böhm. An incremental Mori-Tanaka homogenization scheme for finite strain thermoelastoplasticity of MMCs. *Materials*, 3(1):434–451, January 2010. DOI: 10.3390/ma3010434.

[422] C. Pheatt. Intel® threading building blocks. *J. Comput. Sci. Coll.*, 23(4):298, April 2008. Last visited: September 2018.

[423] E. Phipps and R. Pawlowski. Efficient expression templates for operator overloading-based automatic differentiation. In S. Forth, P. Hovland, E. Phipps, J. Utke, and A. Walther, editors, *Recent Advances in Algorithmic Differentiation*, volume 73 of *Lecture Notes in Computational Science and Engineering*, pages 309–319. Springer Berlin Heidelberg, Berlin, Heidelberg, 2012. ISBN 978-3-642-30023-3. DOI: 10.1007/978-3-642-30023-3_28.

[424] M.-J. Pindera, H. Khatam, A. S. Drago, and Y. Bansal. Micromechanics of spatially uniform heterogeneous media: A critical review and emerging approaches. *Composites, Part B, Eng.*, 40(5):349–378, July 2009. DOI: 10.1016/j.compositesb.2009.03.007.

[425] D. Pivovarov and P. Steinmann. Modified SFEM for computational homogenization of heterogeneous materials with microstructural geometric uncertainties. *Comput. Mech.*, 57(1):123–147, January 2016. DOI: 10.1007/s00466-015-1224-4.

[426] D. Pivovarov and P. Steinmann. On stochastic FEM based computational homogenization of magneto-active heterogeneous materials with random microstructure. *Comput. Mech.*, 58(6):981–1002, December 2016. DOI: 10.1007/s00466-016-1329-4.

[427] M. Planck. Zur Dynamik bewegter Systeme. *Ann. Phys.*, 331(6):1–34, 1908. DOI: 10.1002/andp.19083310602.

[428] P. Ponte Castañeda and E. Galipeau. Homogenization-based constitutive models for magnetorheological elastomers at finite strain. *J. Mech. Phys. Solids*, 59(2):194–215, February 2011. DOI: 10.1016/j.jmps.2010.11.004.

[429] P. Ponte Castañeda. Exact second-order estimates for the effective mechanical properties of nonlinear composite materials. *J. Mech. Phys. Solids*, 44(6):827–862, June 1996. DOI: 10.1016/0022-5096(96)00015-4.

[430] P. Ponte Castañeda. Second-order homogenization estimates for nonlinear composites incorporating field fluctuations: I — theory. *J. Mech. Phys. Solids*, 50(4):737–757, April 2002a. DOI: 10.1016/s0022-5096(01)00099-0.

[431] P. Ponte Castañeda. Second-order homogenization estimates for nonlinear composites incorporating field fluctuations: II — applications. *J. Mech. Phys. Solids*, 50(4):759–782, April 2002b. DOI: 10.1016/s0022-5096(01)00098-9.

[432] P. Ponte Castañeda and E. Tiberio. A second-order homogenization method in finite elasticity and applications to black-filled elastomers. *J. Mech. Phys. Solids*, 48(6–7):1389–1411, June 2000. DOI: 10.1016/s0022-5096(99)00087-3.

[433] P. Ponte Castañeda and M. Zaidman. The finite deformation of nonlinear composite materials — i. Instantaneous constitutive relations. *Int. J. Solids Struct.*, 33(9):1271–1286, April 1996. DOI: 10.1016/0020-7683(95)00099-2.

[434] R. Poya, A. J. Gil, and P. D. Ledger. A computational framework for the analysis of linear piezoelectric beams using *hp*-FEM. *Comput. Struct.*, 152:155–172, May 2015. DOI: 10.1016/j.compstruc.2015.01.012.

[435] J. H. Poynting. On the transfer of energy in the electromagnetic field. *Philos. Trans. R. Soc. Lond.*, 175:343–361, 1884. DOI: 10.1098/rstl.1884.0016.

[436] W. H. Press, S. A. Teukolsky, W. T. Vetterling, and B. P. Flannery. *Numerical Recipes: The Art of Scientific Computing*. Cambridge University Press, 3rd edition, 2007. ISBN 978-0-521-88068-8.

[437] A. Prohl. *Computational Micromagnetism*. Vieweg+Teubner Verlag, 2001. ISBN 978-3-519-00358-8. DOI: 10.1007/978-3-663-09498-2.

[438] F. Qiu and B. J. Nelson. Magnetic helical micro- and nanorobots: Toward their biomedical applications. *Engineering*, 1(1):021–026, March 2015. DOI: 10.15302/j-eng-2015005.

[439] J. Qu and M. Cherkaoui. *Fundamentals of Micromechanics of Solids*. John Wiley & Sons, Inc., Hoboken, New Jersey, USA, 2006. ISBN 978-0-471-46451-8.

[440] R. Rabindranath and H. Böse. On the mobility of iron particles embedded in elastomeric silicone matrix. *J. Phys. Conf. Ser.*, 412:012034. IOP Publishing, 2013. DOI: 10.1088/1742-6596/412/1/012034.

[441] K. R. Rajagopal and A. R. Srinivasa. A Gibbs-potential-based formulation for obtaining the response functions for a class of viscoelastic materials. *Proc. R. Soc. Lond. A: Math. Phys. Eng. Sci.*, 467(2125):39–58, January 2011.

[442] M. Rambausek and M.-A. Keip. Micro- and macrostructural magneto-electric coupling in soft composites. In *Proceedings of the 7th GACM Colloquium on Computational Mechanics for Young Scientists from Academia and Industry*, pages 600–604, 2017.

[443] E. Rank and O. C. Zienkiewicz. A simple error estimator in the finite element method. *Commun. Appl. Numer. Methods*, 3(3):243–249, May 1987. DOI: 10.1002/cnm.1630030311.

[444] B. D. Reddy. *Introductory Functional Analysis: With Applications to Boundary Value Problems and Finite Elements*, volume 27 of *Texts in Applied Mathematics*. Springer-Verlag New York, 1998. ISBN 978-0-387-98307-3. DOI: 10.1007/978-1-4612-0575-3.

[445] B. D. Reddy. Some theoretical and computational aspects of single-crystal strain-gradient plasticity. *Z. Angew. Math. Mech.*, 93(12):844–867, July 2013. DOI: 10.1002/zamm.201200101.

[446] N. H. Reddy and P. Saxena. Limit points in the free inflation of a magnetoelastic toroidal membrane. *Int. J. Non-Linear Mech.*, 95:248–263, October 2017. DOI: 10.1016/j.ijnonlinmec.2017.06.017.

[447] S. Reese and S. Govindjee. Theoretical and numerical aspects in the thermo-viscoelastic material behaviour of rubber-like polymers. *Mech. Time-Depend. Mater.*, 1(4):357–396, December 1997.

[448] S. Reese and S. Govindjee. A theory of finite viscoelasticity and numerical aspects. *Int. J. Solids Struct.*, 35(26–27):3455–3482, September 1998. DOI: 10.1016/s0020-7683(97)00217-5.

[449] S. Reese, P. Wriggers, and B. D. Reddy. A new locking-free brick element technique for large deformation problems in elasticity. *Comput. Struct.*, 75(3):291–304, April 2000. DOI: 10.1016/S0045-7949(99)00137-6.

[450] X. Ren, H. Hou, and Y. Xu. Study on electro-magneto-elastic fiber composites with periodically distributed reinforced phases under antiplane deformation mode and its applications. *Open Mater. Sci. J.*, 5:248–255, 2011. DOI: 10.2174/1874088x01105010248.

[451] A. Reuss. Berechnung der fließgrenze von mischkristallen auf grund der plastizitätsbedingung für einkristalle. *Z. Angew. Math. Mech.*, 9(1):49–58, 1929. DOI: 10.1002/zamm.19290090104.

[452] J. F. Richardson and W. N. Zaki. The sedimentation of a suspension of uniform spheres under conditions of viscous flow. *Chem. Eng. Sci.*, 3(2):65–73, April 1954. DOI: 10.1016/0009-2509(54)85015-9.

[453] S. Ricker, J. Mergheim, P. Steinmann, and R. Müller. A comparison of different approaches in the multi-scale computation of configurational forces. *Int. J. Fract.*, 166(1–2):203–214, November 2010. DOI: 10.1007/s10704-010-9525-2.

[454] Z. Rigbi and L. Jilken. The response of an elastomer filled with soft ferrite to mechanical and magnetic influences. *J. Magn. Magn. Mater.*, 37(3):267–276, July 1983. DOI: 10.1016/0304-8853(83)90055-0.

[455] Z. Rigbi and J. E. Mark. Effects of a magnetic field applied during the curing of a polymer loaded with magnetic filler. *J. Polym. Sci., Polym. Phys. Ed.*, 23(6):1267–1269, June 1985. DOI: 10.1002/pol.1985.180230618.

[456] R. Rodríguez-Ramos, B. E. Pobedria, P. Padilla, J. Bravo-Castillero, R. Guinovart-Díaz, and G. A. Maugin. Variational principles for nonlinear piezoelectric materials. *Arch. Appl. Mech.*, 74(3–4):191–200, December 2004. DOI: 10.1007/BF02637195.

[457] S. Rudykh and K. Bertoldi. Stability of anisotropic magnetorheological elastomers in finite deformations: A micromechanical approach. *J. Mech. Phys. Solids*, 61(4):949–967, April 2013. DOI: 10.1016/j.jmps.2012.12.008.

[458] S. Rudykh, K. Bhattacharya, and G. deBotton. Multiscale instabilities in soft heterogeneous dielectric elastomers. *Proc. R. Soc. A, Math. Phys. Eng. Sci.*, 470(2162):20130618, February 2014. DOI: 10.1098/rspa.2013.0618.

[459] J. L. Rui and B. Bell. CppADCodeGen: CppAD 2017 (version v2.2.0). Zenodo, 2016. URL https://github.com/joaoleal/CppADCodeGen/tree/v2.2.0. Online; Last visited: September 2018.

[460] E. Ruiz and F. Trochu. Thermomechanical properties during cure of glass-polyester RTM composites: Elastic and viscoelastic modeling. *J. Compos. Mater.*, 39(10):881–916, May 2005. DOI: 10.1177/0021998305048732.

[461] Y. Saad. *Iterative Methods for Sparse Linear Systems*. Society for Industrial and Applied Mathematics, Philadelphia, Pennsylvania, USA, 2nd edition, 2003. ISBN 978-0898715347. DOI: 10.1137/1.9780898718003.

[462] Y. Saad. Schur complement preconditioners for distributed general sparse linear systems. In O. B. Widlund and Keyes D. E., editors, *Domain Decomposition Methods in Science and Engineering XVI*, volume 55 of *Lecture Notes in Computational Science and Engineering*, pages 127–138. Springer Berlin Heidelberg, 2007. ISBN 978-3-540-34468-1. DOI: 10.1007/978-3-540-34469-8_11.

[463] S. Saeb, P. Steinmann, and A. Javili. Aspects of computational homogenization at finite deformations: A unifying review from Reuss' to Voigt's Bound. *Appl. Mech. Rev.*, 68(5):050801–050834, September 2016. DOI: 10.1115/1.4034024.

[464] S. Sakata and F. Ashida. Hierarchical stochastic homogenization analysis of a particle reinforced composite material considering non-uniform distribution of microscopic random quantities. *Comput. Mech.*, 48(5):529–540, November 2011. DOI: 10.1007/s00466-011-0604-7.

[465] S. Sakata, F. Ashida, and T. Kojima. Stochastic homogenization analysis on elastic properties of fiber reinforced composites using the equivalent inclusion method and perturbation method. *Int. J. Solids Struct.*, 45(25–26):6553–6565, December 2008. DOI: 10.1016/j.ijsolstr.2008.08.017.

[466] E. Salas and R. Bustamante. Numerical solution of some boundary value problems in nonlinear magneto-elasticity. *J. Intell. Mater. Syst. Struct.*, 26(2):156–171, 2015. DOI: 10.1177/1045389X14522533.

[467] C. Sansour. On the physical assumptions underlying the volumetric-isochoric split and the case of anisotropy. *Eur. J. Mech. A, Solids*, 27(1):28–39, January 2008. DOI: 10.1016/j.euromechsol.2007.04.001.

[468] P. Saxena. Finite deformations and incremental axisymmetric motions of a magnetoelastic tube. *Math. Mech. Solids*, 23(6):950–983, June 2018. DOI: 10.1177/1081286517697502.

[469] P. Saxena, M. Hossain, and P. Steinmann. A theory of finite deformation magneto-viscoelasticity. *Int. J. Solids Struct.*, 50(24):3886–3897, November 2013. DOI: 10.1016/j.ijsolstr.2013.07.024.

[470] P. Saxena, M. Hossain, and P. Steinmann. Nonlinear magneto-viscoelasticity of transversally isotropic magneto-active polymers. *Proc. R. Soc. A, Math. Phys. Eng. Sci.*, 470(2166):20140082, June 2014. DOI: 10.1098/rspa.2014.0082.

[471] P. Saxena, D. K. Vu, and P. Steinmann. On rate-dependent dissipation effects in electro-elasticity. *Int. J. Non-Linear Mech.*, 62:1–11, June 2014. DOI: 10.1016/j.ijnonlinmec.2014.02.002.

[472] P. Saxena, J.-P. V. Pelteret, and P. Steinmann. Modelling of iron-filled magneto-active polymers with a dispersed chain-like microstructure. *Eur. J. Mech. A, Solids*, 50:132–151, March 2015. DOI: 10.1016/j.euromechsol.2014.10.005.

[473] A. Schneebeli, An *H*-(curl; Ω)-conforming FEM: Nédélec's elements of the first type, University of Basel, 2003. URL https://www.dealii.org/reports/nedelec/nedelec.pdf.

[474] K. Schneider, B. Klusemann, and S. Bargmann. Automatic three-dimensional geometry and mesh generation of periodic representative volume elements for matrix-inclusion composites. *Adv. Eng. Softw.*, 99:177–188, September 2016. DOI: 10.1016/j.advengsoft.2016.06.001.

[475] K. Schneider, B. Klusemann, and S. Bargmann. Fully periodic RVEs for technological relevant composites: Not worth the effort! *J. Mech. Mater. Struct.*, 12(4):471–484, 2017. DOI: 10.2140/jomms.2017.12.471.

[476] J. Schröder. A numerical two-scale homogenization scheme: the FE^2-method. In *Plasticity and Beyond*, volume 550 of *CISM International Centre for Mechanical Sciences*, pages 1–64. Springer Vienna, 2014. ISBN 978-3-7091-1624-1. DOI: 10.1007/978-3-7091-1625-8_1.

[477] J. Schröder and M.-A. Keip. Two-scale homogenization of electromechanically coupled boundary value problems. *Comput. Mech.*, 50(2):229–244, August 2012. DOI: 10.1007/s00466-012-0715-9.

[478] J. Schröder and P. Neff. Invariant formulation of hyperelastic transverse isotropy based on polyconvex free energy functions. *Int. J. Solids Struct.*, 40(2):401–445, January 2003. DOI: 10.1016/S0020-7683(02)00458-4.

[479] J. Schröder, P. Wriggers, and D. Balzani. A new mixed finite element based on different approximations of the minors of deformation tensors. *Comput. Methods Appl. Mech. Eng.*, 200(49–52):3583–3600, December 2011. DOI: 10.1016/j.cma.2011.08.009.

[480] J. Schröder, M. Labusch, and M.-A. Keip. Algorithmic two-scale transition for magneto-electro-mechanically coupled problems: FE2-scheme: Localization and homogenization. *Comput. Methods Appl. Mech. Eng.*, 302:253–280, April 2016. DOI: 10.1016/j.cma.2015.10.005.

[481] G. Schubert and P. Harrison. Equi-biaxial tension tests on magneto-rheological elastomers. *Smart Mater. Struct.*, 25(1):015015, January 2016. DOI: 10.1088/0964-1726/25/1/015015.

[482] A. Schwarz, J. Schröder, and G. Starke. A modified least-squares mixed finite element with improved momentum balance. *Int. J. Numer. Methods Eng.*, 81(3):286–306, January 2010. DOI: 10.1002/nme.2692.

[483] M. D. Semon and J. R. Taylor. Thoughts on the magnetic vector potential. *Am. J. Phys.*, 64(11):1361–1369, November 1996. DOI: 10.1119/1.18400.

[484] Y. M. Shabana and M. Ristinmaa. Micromechanical modeling of smart composites considering debonding of reinforcements. *Int. J. Solids Struct.*, 48(22–23):3209–3216, November 2011. DOI: 10.1016/j.ijsolstr.2011.07.011.

[485] M. S. Shephard. Linear multipoint constraints applied via transformation as part of a direct stiffness assembly process. *Int. J. Numer. Methods Eng.*, 20(11):2107–2112, November 1984. DOI: 10.1002/nme.1620201112.

[486] Y. Shindo, K. Mori, and F. Narita. Electromagneto-mechanical fields of giant magnetostrictive /piezoelectric laminates. *Acta Mech.*, 212(3–4):253–261, July 2010. DOI: 10.1007/s00707-009-0259-z.

[487] M. H. Siboni and P. Ponte Castañeda. A magnetically anisotropic, ellipsoidal inclusion subjected to a non-aligned magnetic field in an elastic medium. *C. R., Méc.*, 340(4–5):205–218, April 2012. DOI: 10.1016/j.crme.2012.02.003. Recent Advances in Micromechanics of Materials.

[488] D. Silvester and A. Wathen. Fast iterative solution of stabilised stokes systems part II: Using general block preconditioners. *SIAM J. Numer. Anal.*, 31(5):1352–1367, October 1994. DOI: 10.1137/0731070.

[489] J. C. Simo. On a fully three-dimensional finite-strain viscoelastic damage model: Formulation and computational aspects. *Comput. Methods Appl. Mech. Eng.*, 60(2):153–173, February 1987. DOI: 10.1016/0045-7825(87)90107-1.

[490] J. C. Simo and F. Armero. Geometrically non-linear enhanced strain mixed methods and the method of incompatible modes. *Int. J. Numer. Methods Eng.*, 33(7):1413–1449, May 1992. DOI: 10.1002/nme.1620330705.

[491] J. C. Simo and T. J. R. Hughes. *Computational Inelasticity*, volume 7 of *Interdisciplinary Applied Mathematics*. Springer Science & Business Media, 2000. ISBN 978-0-387-97520-7. DOI: 10.1007/b98904.

[492] J. C. Simo and J. E. Marsden. On the rotated stress tensor and the material version of the Doyle-Ericksen formula. *Arch. Ration. Mech. Anal.*, 86(3):213–231, September 1984. DOI: 10.1007/BF00281556.

[493] J. C. Simo and N. Tarnow. The discrete energy-momentum method. Conserving algorithms for nonlinear elastodynamics. *Z. Angew. Math. Phys.*, 43(5):757–792, September 1992. DOI: 10.1007/bf00913408.

[494] J. C. Simo and R. L. Taylor. Quasi-incompressible finite elasticity in principal stretches. Continuum basis and numerical algorithms. *Comput. Methods Appl. Mech. Eng.*, 85(3):273–310, February 1991. DOI: 10.1016/0045-7825(91)90100-K.

[495] J. C. Simo, R. L. Taylor, and K. S. Pister. Variational and projection methods for the volume constraint in finite deformation elasto-plasticity. *Comput. Methods Appl. Mech. Eng.*, 51(1–3):177–208, September 1985. DOI: 10.1016/0045-7825(85)90033-7.

[496] S. Skatulla, C. Sansour, and A. Arockiarajan. A multiplicative approach for nonlinear electro-elasticity. *Comput. Methods Appl. Mech. Eng.*, 245–246:243–255, October 2012. DOI: 10.1016/j.cma.2012.07.002.

[497] A. J. M. Spencer. *Continuum Physics, volume 1 – Mathematics*, chapter Theory of invariants, pages 239–353. Academic Press, New York, New York, USA, 1971. ISBN 9780122408038. DOI: 10.1016/B978-0-12-240801-4.50008-X.

[498] C. Spieler, M. Kästner, J. Goldmann, J. Brummund, and V. Ulbricht. XFEM modeling and homogenization of magnetoactive composites. *Acta Mech.*, 224(11):2453–2469, November 2013. DOI: 10.1007/s00707-013-0948-5.

[499] A. Sridhar, M.-A. Keip, and C. Miehe. Homogenization in micro-magneto-mechanics. *Comput. Mech.*, 58(1):151–169, July 2016. DOI: 10.1007/s00466-016-1286-y.

[500] L. Stainier and M. Ortiz. Study and validation of a variational theory of thermo-mechanical coupling in finite visco-plasticity. *Int. J. Solids Struct.*, 47(5):705–715, March 2010. DOI: 10.1016/j.ijsolstr.2009.11.012.

[501] G. Stefanou. Simulation of heterogeneous two-phase media using random fields and level sets. *Front. Struct. Civ. Eng.*, 9(2):114–120, June 2015. DOI: 10.1007/s11709-014-0267-5.

[502] D. J. Steigmann. Equilibrium theory for magnetic elastomers and magnetoelastic membranes. *Int. J. Non-Linear Mech.*, 39(7):1193–1216, September 2004. DOI: 10.1016/j.ijnonlinmec.2003.08.002.

[503] D. J. Steigmann and R. W. Ogden. Elastic surface—substrate interactions. *Proc. R. Soc., Math. Phys. Eng. Sci.*, 455(1982):437–474, February 1999. DOI: 10.1098/rspa.1999.0320.

[504] P. Steinmann. Computational nonlinear electro-elasticity – getting started. In R. W. Ogden and D. J. Steigmann, editors, *Mechanics and Electrodynamics of Magneto- and Electro-elastic Materials*, volume 527 of *CISM Courses and Lectures*, pages 181–230. Springer-Verlag Wien, 1st edition, 2011. ISBN 978-3-7091-0700-3. DOI: 10.1007/978-3-7091-0701-0.

[505] P. Steinmann, C. Miehe, and E. Stein. Comparison of different finite deformation inelastic damage models within multiplicative elastoplasticity for ductile materials. *Comput. Mech.*, 13(6):458–474, March 1994. DOI: 10.1007/bf00374241.

[506] G. V. Stepanov, S. S. Abramchuk, D. A. Grishin, L. V. Nikitin, E. Y. Kramarenko, and A. R. Khokhlov. Effect of a homogeneous magnetic field on the viscoelastic behavior of magnetic elastomers. *Polymer*, 48(2):488–495, January 2007. DOI: 10.1016/j.polymer.2006.11.044.

[507] C. J. Stimpson, C. D. Ernst, P. Knupp, P. P. Pébay, and D. Thompson. The verdict geometric quality library. techreport SAND2007-1751, Sandia National Laboratories, 2007.

[508] K. Suzuki, Y. Miyano, and T. Kunio. Change of viscoelastic properties of epoxy resin in the curing process. *J. Appl. Polym. Sci.*, 21(12):3367–3379, December 1977. DOI: 10.1002/app.1977.070211215.

[509] H. Szambolics, L. D. Buda-Prejbeanu, J. C. Toussaint, and O. Fruchart. A constrained finite element formulation for the Landau–Lifshitz–Gilbert equations. *Comput. Mater. Sci.*, 44(2):253–258, December 2008. DOI: 10.1016/j.commatsci.2008.03.019.

[510] J. Takács. A phenomenological mathematical model of hysteresis. *Compel*, 20(4):1002–1015, 2001. DOI: 10.1108/eum0000000005771.

[511] P. Tan and L. Tong. Modeling for the electro-magneto-thermo-elastic properties of piezoelectric-magnetic fiber reinforced composites. *Composites, Part A, Appl. Sci. Manuf.*, 33(5):631–645, May 2002. DOI: 10.1016/s1359-835x(02)00015-5.

[512] M. Tanaka, M. Fujikawa, D. Balzani, and J. Schröder. Robust numerical calculation of tangent moduli at finite strains based on complex-step derivative approximation and its application to localization analysis. *Comput. Methods Appl. Mech. Eng.*, 269:454–470, February 2014. DOI: 10.1016/j.cma.2013.11.005.

[513] M. Taya. Micromechanics modeling of smart composites. *Composites, Part A, Appl. Sci. Manuf.*, 30(4):531–536, April 1999. DOI: 10.1016/s1359-835x(98)00146-8.

[514] İ. Temizer and P. Wriggers. An adaptive method for homogenization in orthotropic nonlinear elasticity. *Comput. Methods Appl. Mech. Eng.*, 196(35–36):3409–3423, July 2007. DOI: 10.1016/j.cma.2007.03.017.

[515] İ. Temizer and P. Wriggers. On the computation of the macroscopic tangent for multiscale volumetric homogenization problems. *Comput. Methods Appl. Mech. Eng.*, 198(3–4):495–510, December 2008. DOI: 10.1016/j.cma.2008.08.018.

[516] İ. Temizer and P. Wriggers. Homogenization in finite thermoelasticity. *J. Mech. Phys. Solids*, 59(2):344–372, February 2011. DOI: 10.1016/j.jmps.2010.10.004.

[517] İ. Temizer and P Wriggers. An adaptive multiscale resolution strategy for the finite deformation analysis of microheterogeneous structures. *Comput. Methods Appl. Mech. Eng.*, 200(37–40):2639–2661, September 2011. DOI: 10.1016/j.cma.2010.06.013.

[518] K. Terada and N. Kikuchi. A class of general algorithms for multi-scale analyses of heterogeneous media. *Comput. Methods Appl. Mech. Eng.*, 190(40–41):5427–5464, July 2001. DOI: 10.1016/s0045-7825(01)00179-7.

[519] K. Terada, I. Saiki, K. Matsui, and Y. Yamakawa. Two-scale kinematics and linearization for simultaneous two-scale analysis of periodic heterogeneous solids at finite strain. *Comput. Methods Appl. Mech. Eng.*, 192(31–32):3531–3563, August 2003. DOI: 10.1016/s0045-7825(03)00365-7.

[520] H. F. Tiersten. Coupled magnetomechanical equations for magnetically saturated insulators. *J. Math. Phys.*, 5(9):1298–1318, September 1964. DOI: 10.1063/1.1704239.

[521] N. Triantafyllidis and K. Danas. Magnetorheological elastomers. In *MECAMAT*, France, January 2012. URL https://hal-polytechnique.archives-ouvertes.fr/hal-00760758. Online; Last visited: September 2018.

[522] C. Truesdell and W. Noll. *Encyclopedia of Physics, volume 3*, chapter 1: The non-linear field theories of mechanics, pages 1–602. Springer-Verlag Berlin Heidelberg, 1965. ISBN 978-3-642-46017-3. DOI: 10.1007/978-3-642-46015-9.

[523] C. Truesdell and R. Toupin. *Encyclopedia of Physics: Principles of Thermodynamics and Statics, volume 1*, chapter 2: The classical field theories, pages 226–794. Springer-Verlag Berlin Heidelberg, 1960. ISBN 978-3-540-02547-4. DOI: 10.1007/978-3-642-45943-6.

[524] I. D. Vagner, B. I. Lembrikov, and P. Wyder. *Electrodynamics of Magnetoactive Media*. Solid-State Sciences. Springer-Verlag Berlin Heidelberg, 2004. ISBN 978-3-642-07825-5. DOI: 10.1007/978-3-662-06941-7.

[525] J. D. van der Waals. The thermodynamic theory of capillarity under the hypothesis of a continuous variation of density. *J. Stat. Phys.*, 20(2):200–244, February 1979. DOI: 10.1007/bf01011514.

[526] N. P. van Dijk. Formulation and implementation of stress-driven and/or strain-driven computational homogenization for finite strain. *Int. J. Numer. Methods Eng.*, 107(12):1009–1028, September 2016. DOI: 10.1002/nme.5198.

[527] J. van Welij. Calculation of Eddy currents in terms of H on hexahedra. *IEEE Trans. Magn.*, 21(6):2239–2241, November 1985. DOI: 10.1109/TMAG.1985.1064199.

[528] Z. Varga, G. Filipcsei, and M. Zrínyi. Smart composites with controlled anisotropy. *Polymer*, 46(18):7779–7787, August 2005. DOI: 10.1016/j.polymer.2005.03.102.

[529] Z. Varga, G. Filipcsei, and M. Zrínyi. Magnetic field sensitive functional elastomers with tuneable elastic modulus. *Polymer*, 47(1):227–233, January 2006. DOI: 10.1016/j.polymer.2005.10.139.

[530] F. Vogel. *On the Modeling and Computation of Electro- and Magneto-Active Polymers*. PhD thesis, Friedrich-Alexander Universität Erlangen-Nürnberg, 2015.

[531] F. Vogel, R. Bustamante, and P. Steinmann. On some mixed variational principles in magneto-elastostatics. *Int. J. Non-Linear Mech.*, 51:157–169, May 2013. DOI: 10.1016/j.ijnonlinmec.2012.12.005.

[532] F. Vogel, S. Göktepe, P. Steinmann, and E. Kuhl. Modeling and simulation of viscous electro-active polymers. *Eur. J. Mech. A, Solids*, 48:112–128, November 2014. DOI: 10.1016/j.euromechsol.2014.02.001.

[533] F. Vogel, J.-P. V. Pelteret, S. Kaessmair, and P. Steinmann. Magnetic force and torque on particles subject to a magnetic field. *Eur. J. Mech. A, Solids*, 48:23–37, November 2014. DOI: 10.1016/j.euromechsol.2014.03.007.

[534] W. Voigt. Ueber die Beziehung zwischen den beiden Elasticitätsconstanten isotroper Körper. *Ann. Phys.*, 274(12):573–587, 1889. DOI: 10.1002/andp.18892741206.

[535] K. Y. Volokh. On electromechanical coupling in elastomers. *J. Appl. Mech.*, 79(4):044507–1–044507–5, July 2012. DOI: 10.1115/1.4006057.

[536] D. K. Vu. *A study on nonlinear electro-elastostatics: Theory and numerical simulation*. Habilitation, Friedrich-Alexander Universität Erlangen-Nürnberg, 2015.

[537] D. K. Vu and P. Steinmann. Nonlinear electro- and magneto-elastostatics: Material and spatial settings. *Int. J. Solids Struct.*, 44(24):7891–7905, December 2007. DOI: 10.1016/j.ijsolstr.2007.05.017.

[538] D. K. Vu and P. Steinmann. Material and spatial motion problems in nonlinear electro- and magneto-elastostatics. *Math. Mech. Solids*, 15(2):239–257, March 2010. DOI: 10.1177/1081286508098583.

[539] D. K. Vu and P. Steinmann. A 2-D coupled BEM–FEM simulation of electro-elastostatics at large strains. *Comput. Methods Appl. Mech. Eng.*, 199(17–20):1124–1133, March 2010. DOI: 10.1016/j.cma.2009.12.001.

[540] D. K. Vu and P. Steinmann. On 3-D coupled BEM–FEM simulation of nonlinear electro-elastostatics. *Comput. Methods Appl. Mech. Eng.*, 201–204:82–90, January 2012. DOI: 10.1016/j.cma.2011.08.024.

[541] D. K. Vu, P. Steinmann, and G. Possart. Numerical modelling of non-linear electroelasticity. *Int. J. Numer. Methods Eng.*, 70(6):685–704, May 2007. DOI: 10.1002/nme.1902.

[542] B. L. Walter. *Scherrheologische Untersuchungen zum Einfluss magnetischer Felder auf das dynamisch-mechanische Werkstoffverhalten von ferromagnetisch gefüllten Elastomeren*. PhD thesis, Friedrich-Alexander Universität Erlangen-Nürnberg, 2019. To be submitted.

[543] B. L. Walter, J.-P. Pelteret, J. Kaschta, D. W. Schubert, and P. Steinmann. On the wall slip phenomenon of elastomers in oscillatory shear measurements using parallel-plate rotational rheometry – I. Detecting wall slip. *Polym. Test.*, 61:430–440, August 2017. DOI: 10.1016/j.polymertesting.2017.05.035.

[544] B. L. Walter, J.-P. Pelteret, J. Kaschta, D. W. Schubert, and P. Steinmann. On the wall slip phenomenon of elastomers in oscillatory shear measurements using parallel-plate rotational rheometry – II. Influence of experimental conditions. *Polym. Test.*, 61:455–463, August 2017. DOI: 10.1016/j.polymertesting.2017.05.036.

[545] B. L. Walter, J.-P. Pelteret, J. Kaschta, D. W. Schubert, and P. Steinmann. Preparation of magnetorheological elastomers and their slip-free characterization by means of parallel-plate rotational rheometry. *Smart Mater. Struct.*, 26(8):085004, August 2017. DOI: 10.1088/1361-665X/aa6b63.

[546] H. Wang, M. Lei, and S. Cai. Viscoelastic deformation of a dielectric elastomer membrane subject to electromechanical loads. *J. Appl. Phys.*, 113(21):213508–1–213508–7, June 2013. DOI: 10.1063/1.4807911.

[547] P. J. Wang and S. J. Chiueh. Analysis of Eddy-current brakes for high speed railway. *IEEE Trans. Magn.*, 34(4):1237–1239, July 1998. DOI: 10.1109/20.706507.

[548] R. L. Warley, D. L. Feke, and I. Manas-Zloczower. Transient effects in dynamic modulus measurement of silicone elastomers 1. Zero mean strain measurements. *J. Appl. Polym. Sci.*, 98(3):1001–1009, November 2005. DOI: 10.1002/app.22209.

[549] R. L. Warley, D. L. Feke, and I. Manas-Zloczower. Transient effects in dynamic modulus measurement of silicone rubber, part 2: Effect of mean strains and strain history. *J. Appl. Polym. Sci.*, 104(4):2197–2204, May 2007. DOI: 10.1002/app.25136.

[550] B. Wei, X. Gong, W. Jiang, L. Qin, and Y. Fan. Study on the properties of magnetorheological gel based on polyurethane. *J. Appl. Polym. Sci.*, 118(5):2765–2771, December 2010. DOI: 10.1002/app.32688.

[551] M. Wilhelm. Fourier-transform rheology. *Macromol. Mater. Eng.*, 287(2):83–105, February 2002. DOI: 10.1002/1439-2054(20020201)287:2<83::aid-mame83>3.3.co;2-2.

[552] M. Wilhelm, D. Maring, and H.-W. Spiess. Fourier-transform rheology. *Rheol. Acta*, 37(4):399–405, August 1998. DOI: 10.1007/s003970050126.

[553] M. Wilhelm, P. Reinheimer, and M. Ortseifer. High sensitivity Fourier-transform rheology. *Rheol. Acta*, 38(4):349–356, October 1999. DOI: 10.1007/s003970050185.

[554] M. Wissler and E. Mazza. Mechanical behavior of an acrylic elastomer used in dielectric elastomer actuators. *Sens. Actuators A, Phys.*, 134(2):494–504, March 2007. DOI: 10.1016/j.sna.2006.05.024.

[555] M. T. Wissler. *Modeling dielectric elastomer actuators*. PhD thesis, Swiss Federal Institute of Technology in Zurich, 2007.

[556] K. Wollny, J. Läuger, and S. Huck. Magneto sweep - A method for characterizing the viscoelastic properties of magneto-rheological fluids. *Appl. Rheol.*, 12:25–31, 2002. DOI: 10.3933/ApplRheol-12-25.

[557] C. Wrana. *Polymerphysik: Eine physikalische Beschreibung von Elastomeren und ihren anwendungsrelevanten Eigenschaften*. Springer Spektrum, 1st edition, 2014. ISBN 978-3-642-45075-4. DOI: 10.1007/978-3-642-45076-1.

[558] P. Wriggers. *Computational Contact Mechanics*. Springer Berlin Heidelberg, 2nd edition, 2006. ISBN 978-3-540-32608-3. DOI: 10.1007/978-3-540-32609-0.

[559] P. Wriggers. *Nonlinear Finite Element Methods*. Springer Berlin Heidelberg, 2008. ISBN 978-3-540-71000-4. DOI: 10.1007/978-3-540-71001-1.

[560] K. x. Wei, G. Meng, W. m. Zhang, and S. s. Zhu. Experimental investigation on vibration characteristics of sandwich beams with magnetorheological elastomers cores. *J. Cent. South Univ. Technol.*, 15(S1):239–242, September 2008. DOI: 10.1007/s11771-008-0354-7.

[561] T. Xu, J. Yu, X. Yan, H. Choi, and L. Zhang. Magnetic actuation based motion control for microrobots: An overview. *Micromachines*, 6(9):1346–1364, September 2015. DOI: 10.3390/mi6091346.

[562] B. Yagimli and A. Lion. Experimental investigations and material modelling of curing processes under small deformations. *Z. Angew. Math. Mech.*, 91(5):342–359, May 2011. DOI: 10.1002/zamm.201000096.

[563] E. Yang, M. Frecker, and E. Mockensturm. Finite element and experimental analyses of non-axisymmetric dielectric elastomer actuators. In *SPIE Proceedings, Smart Structures and Materials*, volume 6168, San Diego, California, USA, 2006. DOI: 10.1117/12.658740.

[564] J. S. Yang and R. C. Batra. Mixed variational principles in non-linear electroelasticity. *Int. J. Non-Linear Mech.*, 30(5):719–725, September 1995. DOI: 10.1016/0020-7462(95)00027-L.

[565] Q. Yang, L. Stainier, and M. Ortiz. A variational formulation of the coupled thermo-mechanical boundary-value problem for general dissipative solids. *J. Mech. Phys. Solids*, 54(2):401–424, February 2006. DOI: 10.1016/j.jmps.2005.08.010.

[566] H. M. Yin and L. Z. Sun. Magnetoelastic modelling of composites containing randomly dispersed ferromagnetic particles. *Philos. Mag.*, 86(28):4367–4395, February 2006. DOI: 10.1080/14786430600724421.

[567] H. M. Yin, L. Z. Sun, and J. S. Chen. Micromechanics-based hyperelastic constitutive modeling of magnetostrictive particle-filled elastomers. *Mech. Mater.*, 34(8):505–516, August 2002. DOI: 10.1016/s0167-6636(02)00178-3.

[568] H. M. Yin, L. Z. Sun, and J. S. Chen. Magneto-elastic modeling of composites containing chain-structured magnetostrictive particles. *J. Mech. Phys. Solids*, 54(5):975–1003, May 2006. DOI: 10.1016/j.jmps.2005.11.007.

[569] K. W. Yung, P. B. Landecker, and D. D. Villani. An analytic solution for the force between two magnetic dipoles. *Magn. Electr. Sep.*, 9(1):39–52, 1998. DOI: 10.1155/1998/79537.

[570] R. Zabihyan, J. Mergheim, A. Javili, and P. Steinmann. Aspects of computational homogenization in magneto-mechanics: Boundary conditions, RVE size and microstructure composition. *Int. J. Solids Struct.*, 130–131:105–121, January 2018. DOI: 10.1016/j.ijsolstr.2017.10.009.

[571] S. Zaglmayr. *High order finite element methods for electromagnetic field computation*. PhD thesis, Johannes Kepler Universität Linz, 2006. URL https://www.applied.math.tugraz.at/zaglmayr/pub/szthesis.pdf.

[572] M. Zaidman and P. Ponte Castañeda. The finite deformation of nonlinear composite materials —II. Evolution of the microstructure. *Int. J. Solids Struct.*, 33(9):1287–1303, April 1996. DOI: 10.1016/0020-7683(95)00100-x.

[573] Z. K. Zhang and A. K. Soh. Micromechanics predictions of the effective moduli of magnetoelectroelastic composite materials. *Eur. J. Mech. A, Solids*, 24(6):1054–1067, November 2005. DOI: 10.1016/j.euromechsol.2005.07.005.

[574] Q.-S. Zheng. Theory of representations for tensor functions—a unified invariant approach to constitutive equations. *Appl. Mech. Rev.*, 47(11):545–587, November 1994. DOI: 10.1115/1.3111066.

[575] G. Y. Zhou and Q. Wang. Design of a smart piezoelectric actuator based on a magnetorheological elastomer. *Smart Mater. Struct.*, 14(4):504–510, May 2005. DOI: 10.1088/0964-1726/14/4/007.

[576] G. Y. Zhou, K. C. Lin, and Q. Wang. Finite element studies on field-dependent rigidities of sandwich beams with magnetorheological elastomer cores. *Smart Mater. Struct.*, 15(3):787–791, April 2006. DOI: 10.1088/0964-1726/15/3/014.

[577] O. C. Zienkiewicz and R. L. Taylor. *The Finite Element Method, volume 1: The basis*. Butterworth-Heinemann, 5th edition, 2000. ISBN 0-7506-5049-4.

[578] O. C. Zienkiewicz and R. L. Taylor. *The Finite Element Method, volume 2: Solid mechanics*. Butterworth-Heinemann, 5th edition, 2000. ISBN 0-7506-5055-9.

[579] O. C. Zienkiewicz and J. Z. Zhu. A simple error estimator and adaptive procedure for practical engineerng analysis. *Int. J. Numer. Methods Eng.*, 24(2):337–357, feb 1987. DOI: 10.1002/nme.1620240206.

[580] T. I. Zohdi and P. Wriggers. *An Introduction to Computational Micromechanics*. Lecture Notes in Applied and Computational Mechanics. Springer-Verlag Berlin Heidelberg, 2008. ISBN 978-3-540-77482-2. DOI: 10.1007/978-3-540-32360-0.

[581] S. Zwecker, S. Klinkel, and R. Müller. Nonlinear finite element simulation of thin dielectric elastomer structures. In N. Balle, F. Marheineke, F. Kienle, K. Schmidt, and N. Tschauder, editors, *Proceedings of 1st Young Researcher Symposium by Center for Mathematical and Computational Modelling*, pages 33–38, Kaiserslautern, Germany, February 2011. URL http://ceur-ws.org/Vol-750/YRS2011_proceedings.pdf. Online; Last visited: September 2018.

Image reproduction

Figures 8.7 to 8.11 reprinted from *International Journal of Non-Linear Mechanics*, vol. 74, Hossain, M. and Saxena, P. and Steinmann, P., Modelling the curing process in magneto-sensitive polymers: Ratedependence and shrinkage, 108–121 (figures 4, 7b, 8b, 9b, 15, 16), copyright (2015), with permission from Elsevier.

Figures 9.5, 9.7 and 9.8 reprinted from *International Journal of Solids and Structures*, vol. 69–70, Hossain, M. and Chatzigeorgiou, G. and Meraghni, F. and Steinmann, P., A multi-scale approach to model the curing process in magneto-sensitive polymeric materials, 34–44 (figures 2a, 2c, 10, 11), copyright (2015), with permission from Elsevier.

Figures 11.5, 11.13 and 11.14 reprinted from *International Journal for Numerical Methods in Engineering*, vol. 108, Pelteret, J.-P. V. and Davydov, D. and McBride, A. and Vu, D. K. and Steinmann, P., Computational electro-elasticity and magneto-elasticity for quasi-incompressible media immersed in free space, 1307–1342 (figures 5, 15b, 20a), copyright (2016), with permission from JohnWiley & Sons, Ltd..

Figures 9.12, 9.14 and 9.15 reprinted from *Computational Mechanics*, vol. 57, D. Pivovarov and P. Steinmann, Modified SFEM for computational homogenization of heterogeneous materials with microstructural geometric uncertainties, 123–147 (figures 14c, 29e, 43, 45), copyright (2016), with permission from Springer Nature.

Figures 9.16 to 9.18 reprinted from *Computational Mechanics*, vol. 58, D. Pivovarov and P. Steinmann, On Stochastic FEM based computational homogenization of magneto-active heterogeneous materials with random microstructure, 981–1002 (figures 17, 18, 21, 24b, 25, 26, 27), copyright (2016), with permission from Springer Nature.

Figures 7.1 and 7.2 reprinted from *International Journal of Solids and Structures*, vol. 50, Saxena, P. and Hossain, M. and Steinmann, P., A theory of finite deformation magneto-viscoelasticity, 3886–3897 (figures 2, 10), copyright (2013), with permission from Elsevier.

Figures 11.7 and 11.8 reprinted from *European Journal of Mechanics – A/Solids*, vol. 50, Saxena, P. and Pelteret, J.-P. V. and Steinmann, P., Modelling of iron-filled magneto-active polymers with a dispersed chain-like microstructure, 132–151 (figures 18a, 18b, 18c, 19a, 19b), copyright (2015), with permission from Elsevier.

Figures 10.5 and 10.7 reprinted from *European Journal of Mechanics – A/Solids*, vol. 48, Vogel, F. and Pelteret, J.-P. V. and Kaessmair, S. and Steinmann, P., Magnetic force and torque on particles subject to a magnetic field, 23–37 (figures 13, 14, 16), copyright (2014), with permission from Elsevier.

Figures 3.8 to 3.10 reprinted from *Polymer Testing*, vol. 61, Walter, B. L. and Pelteret, J.-P. and Kaschta, J. and Schubert, D. W. and Steinmann, P., On the wall slip phenomenon of elastomers in oscillatory shear measurements using parallel-plate rota-

https://doi.org/10.1515/9783110418576-017

tional rheometry – I. Detecting wall slip, 430–440 (figures 3a, 7a, 7b, 7c, 10a, II), copyright (2017), with permission from Elsevier.

Figure 3.11 reprinted from *Polymer Testing*, vol. 61, Walter, B. L. and Pelteret, J.-P. and Kaschta, J. and Schubert, D. W. and Steinmann, P., On the wall slip phenomenon of elastomers in oscillatory shear measurements using parallel-plate rotational rheometry – II. Influence of experimental conditions, 455–463 (figures 1, 6), copyright (2017), with permission from Elsevier.

Figure 2.4 reprinted from *Smart Materials and Structures*, vol. 26, Walter, B. L. and Pelteret, J.-P. and Kaschta, J. and Schubert, D. W. and Steinmann, P., Preparation of magnetorheological elastomers and their slip-free characterization by means of parallel-plate rotational rheometry, 085004 (figure 1a), copyright (2017), with permission from IOP Publishing.

Figures 9.3 to 9.5 reprinted from *International Journal of Solids and Structures*, vol. 130–131, Zabihyan, R. and Mergheim, J. and Javili, A. and Steinmann, P., Aspects of computational homogenization in magneto-mechanics: Boundary conditions, RVE size and microstructure composition, 105–121 (figures 4, 10, 14), copyright (2018), with permission from Elsevier.

Index

Bei Fragen zur Produktsicherheit wenden Sie sich bitte an:
If you have any questions regarding product safety,
please contact:

Walter de Gruyter GmbH
Genthiner Straße 13
10785 Berlin
productsafety@degruyterbrill.com